STRUCTURE AND FUNCTION
OF PROPRIOCEPTORS IN THE
INVERTEBRATES

Structure and function of proprioceptors in the invertebrates

EDITED BY

P. J. MILL

Senior Lecturer
Department of Pure and Applied Zoology
University of Leeds

LONDON
CHAPMAN AND HALL

A Halsted Press Book
John Wiley & Sons, New York

First published 1976
by Chapman and Hall Ltd
11 New Fetter Lane, London EC4P 4EE

© 1976 Chapman and Hall Ltd

Typeset by Santype International Ltd.
and printed in Great Britain by
Whitstable Litho Ltd, Whitstable, Kent

ISBN 0 412 12890 X

Distributed in the U.S.A. by Halsted Press,
a Division of John Wiley & Sons, Inc., New York

Library of Congress Cataloging in Publication Data

Main entry under title:
Structure and function of proprioceptors in the invertebrates.

    "A Halsted Press book."
    Includes index.
    1. Proprioceptors. 2. Invertebrates--Physiology.
3. Invertebrates--Anatomy. I. Mill, Peter John.
[DNIM: 1. Proprioception. 2. Invertebrates. QL364. S927]
QL364.S77 1976   592'.01'88   75-40417
ISBN 0-470-14987-6

Contents

Contributors

B.-U. Budelmann   Fachbereich Biologie, Universität Regensburg, Federal Republic of Germany.

B. M. H. Bush   Department of Physiology, University of Bristol, U.K.

F. Clarac   Institut de Neurophysiologie et de Psychophysiologie, C.N.R.S., Marseille, France.

D. A. Dorsett   Department of Marine Biology, Marine Science Laboratories, University College of North Wales, Anglesey, U.K.

H. L. Fields   Department of Neurology, School of Medicine, University of California, San Francisco, California, U.S.A.

L. H. Finlayson   Department of Zoology and Comparative Physiology, University of Birmingham, U.K.

M. S. Laverack   Gatty Marine Laboratory, University of St. Andrews, U.K.

D. L. Macmillan   Department of Zoology, University of Melbourne, Victoria, Australia.

P. J. Mill   Department of Pure and Applied Zoology, University of Leeds, U.K.

M. Moulins   Biologie Générale Physiologie des Invertébrés, Université de Province Marseille, France.

R. N. Price   Department of Zoology, University of Nottingham, U.K.

D. C. Sandeman   Research School of Biological Sciences, Australian National University, Canberra, Australia.

W. Wales   Dunstaffnage Marine Research Laboratory, Scottish Marine Biological Association, Oban, U.K.

M. J. Wells   Department of Zoology, University of Cambridge, U.K.

C. A. G. Wiersma   Division of Biology, California Institute of Technology, California, U.S.A.

B. R. Wright   Department of Zoology, University of Newcastle, U.K.

Acknowledgements

Thanks are due to the following for permission to reproduce various figures, detailed acknowledgements being given at the end of the relevant captions: American Society of Physiologists; Cold Spring Harbour Laboratory; Company of Biologists; Gordon and Breach; John Wiley & Sons; Macmillan Press Ltd.; North Holland Publishing Company; Oliver & Boyd; Pergamon Press; Plenum Press; Royal Entomological Society; Royal Society of London; Springer-Verlag; The Journal of the Marine Biological Association, U.K.; W. B. Saunders & Company; Winter Institute Press.

Lack of space precludes a mention here of individuals who have kindly granted permission for illustrations to be reproduced; they are acknowledged at the end of the relevant legends.

Preface

The term 'proprioceptor' was introduced by Sherrington (1906) since when it has been defined and redefined on innumerable occasions. However, perhaps the most useful and widely accepted definition is that of Lissmann (1950) who described proprioceptors as 'Sense organs capable of registering continuously deformation (changes in length) and stress (tensions, decompressions) in the body, which can arise from the animal's own movements or may be due to its weight or other external mechanical forces'.

It is the intention here to follow Lissmann's definition in its widest sense and, as will be seen from the ensuing chapters, only those mechanoreceptors which are exclusively sensitive to touch or to sound, do not fit within its bounds. However, the border lines are by no means clear; many tactile receptors have a proprioceptive function under certain conditions and the detection of high frequency vibrations is difficult to separate from that of low frequency sound.

Chapter 1 (Laverack) commences with a more detailed elaboration of the nature of proprioception within the framework of Lissmann's definition and then proceeds to consider those exteroreceptors which may be considered as providing the animal with proprioceptive information. As will be seen this involves a considerable variety of structures and provides an introduction to many subsequent chapters.

Chapters 2 to 10 deal with arthropod proprioceptors. Chapter 2 (Fields) is concerned with the crustacean abdominal and thoracic receptors which all are, or are derived from, muscle receptor organs. The only other crustacean muscle receptor organs so far described are located at the bases of the appendages. Since these latter are unusual in that their sensory cell bodies lie within the central nervous system and information is transmitted as decremental non-propagated potentials, rather than action potentials, these are the subject of a separate chapter (Chapter 3—Bush). The thoracic and abdominal receptors of the other arthropodan classes, and which include connective tissue stretch receptors, muscle receptor organs and chordotonal organs, are dealt with in Chapter 4 (Finlayson). Chapter 5 (Wales) describes those receptors which are associated with the mouthparts and gut; information on which is so far only available for crustaceans and insects.

Chapter 6 (Mill) is concerned with joint chordotonal and myochordotonal organs of the crustacean appendages; with the exception of two which are located near the base of the limbs and which have a specialized function concerned with limb autotomy. These are discussed in Chapter 7 (Clarac). The internal and, to some extent, the external limb receptors of the other arthropods are dealt with in Chapter 8 (Wright). As can be seen by comparing the contents of the above chapters, the restriction of chordotonal organs to the appendages and of muscle receptor organs to the body and bases of the appendages, applies only to the crustaceans.

Owing to the variety and complexity of chordotonal organs a separate chapter (Chapter 9—Moulins) is devoted to their ultra-structure. Fairly recently, tension receptors associated with the tendons of arthropod limb muscles have been described and these are dealt with separately in Chapter 10 (Macmillan).

The soft-bodied invertebrates continue to produce problems as far as describing and analysing proprioceptors are concerned and Chapter 11 (Dorsett) is devoted to these problems.

The following two chapters are concerned with the equilibrium of arthropods and molluscs, about which two phyla a considerable body of knowledge has been accumulated. In the first of these (Chapter 12—Sandman) the general principles of position, velocity and linear and angular acceleration are introduced and followed by an account of the detection of acceleration in arthropods. Chapter 13 (Budelmann) follows this with details of the receptor systems responsible for the detection of acceleration in molluscs.

Chapter 14 (Wells) deals with the question of whether proprioceptive information is, or can be, used in learning situations and discusses the problems pertaining to both hard-bodied and soft-bodied invertebrates. Chapter 15 (Mill and Price) discusses the methods and problems of analysing proprioceptive information in a quantitative manner. Then, in Chapter 16 (Wiersma) the fate of the proprioceptive input to the central nervous system is discussed. A brief summary follows. Possibly 'summary' is too optimistic a term. Rather a few broad issues are dealt with: the classification of proprioceptors, proprioceptor function, transduction and the optimal stimulus and the site of spike initiation.

I would like to thank all of the contributors, without whose collaboration this volume would not have been written. Considerable efforts were made to meet the deadline for manuscripts and my task of editing was eased by fast and helpful responses to my various queries and suggestions. In all of the chapters much that is new and previously unpublished has been included and this should certainly enhance the usefulness of this book to the reader and it is hoped that it will meet the needs of undergraduate courses as well as providing a valuable source to the research worker.

My thanks are also due to Mr. D. Ingram of Chapman and Hall for his constant interest in this project and for all the useful advice which he has given me; to those others at Chapman and Hall who have helped in any way; and to my wife for all of her help and understanding throughout the various stages of preparation of this volume.

University of Leeds
August 1975   P.J.M.

1 External proprioceptors

M. S. LAVERACK

1.1 The nature of proprioception

If the animal body is considered as a solid object capable of orientation in space relative to gravity, then it is of adaptive value that the parameters of position, movement, gravity and velocity should be monitored. There are few animals without motile phases in their life histories. Even the most sedentary of adult sponges, protozoa, annelids and crustacea usually possess a mobile larval or juvenile phase. The activities of these immature organisms are adaptive, both towards the demands of the dispersive stage, and also the requirements of the immobile, metamorphosed adult. Numerous observations have indicated responses of larvae to various physical events, e.g. alterations in light intensity, and positive or negative responses to gravity, and have shown that these change at metamorphosis. Nonetheless within this early active phase responses take place (e.g. site selection, Crisp, 1974) that affect all of the adult activity since, once taken, the decisions are irreversible. It is therefore imperative for survival that correct information be available.

For the majority of free living adult animals it is also of great significance that correct sensory information is presented to the central nervous system to enable modifications to be made to continuing motor output, or to initiate or trigger new patterns of behaviour, reflexive or directed. Sensory input may be derived from many physico-chemical sources, classified in a variety of ways (Laverack, 1968): among these are the mechanical events that continually impinge upon the animal.

Correct behaviour, however this may be defined, depends upon appropriate responses to the information presented. A number of externally placed receptors may be mechanical event monitors, but it is not usual to consider them as imparting proprioceptive information except in the widest of all possible definitions, namely that *any* information relative to changes in environmental forces may be pertinent to bodily orientation. Thus water flow across a surface, or wind on the head of a fly may be significant. This may be especially true where detectable differences in mechanical input occur between the two sides of a body (in a bilaterally symmetrical arrangement). As stated in the opening remarks of this chapter, modification of the orientation of the body, especially in response to mechanical forces, occurs in a variety of ways. It is my intention here to propose that proprioception is a function of many external mechanoreceptors where stimulation can be construed as providing information regarding position and movement.

Descriptions of sense organs exist in the literature that suggest that proprioception is a major if not the sole purpose of the organ described. In some instances, anatomical evidence only is available, with no experimental evidence at all to verify the hypothesis. However, whilst functional evidence is crucial, sufficient is now known and understood about proprioception in well established cases, that some conjecture may be tolerated in those situations lacking experimental verification. The demonstration of a reflex resulting from such input may clinch the argument in Sherringtonian terms, but is often a late addition to details of sensory investigations.

The essential ingredient in the phenomenon of proprioception is space, and the bodily adaptations towards monitoring the precise position of either the whole, or parts within that space. As other authors in this volume will argue, this sensory information may be generated by very special organs, e.g. statocysts or halteres, but in the simplest form, special aggregations of sense endings may not be required, only a certain disposition relative to some other structure.

Thus our first task is to say something of the aspects of the environment that may give rise to information classifiable as proprioceptive.

Since life is adapted to existence on the planet earth it is always under the influence of gravity (only escaping such in highly artificial circumstances to which there has as yet been no evolutionary adaptation!). Orientation towards gravity is therefore a prime response of living material, but it may not always dominate the attitude and orientation of the body (e.g. flies may walk upside down, but usually fly in a correct dorso-ventral manner). The body may therefore have a more direct response to a substrate than it does to gravity.

Proprioception in the sense of monitoring position can encompass the action of determining the position of the substrate (e.g. the precise position of the distal end of a leg relative to the effective floor) or even the position relative to a fluid medium such as air or water (as in flying insects or swimming invertebrates). Stress (force/unit area) and strain (change in length/original length) are both likely to affect the surfaces of animals. It is difficult to conceive of stress which is not likely to bring about changes in length of associated regions (even micro-changes) that are better defined as strain. Such forces may be set up by external events, or by internal muscle contractions or even perhaps by blood pressure or gut content (as in haemocoelic animals).

1.1.1   Modes of stimulation

(a) Perhaps the simplest form of proprioceptive sensory input derives from the distortion of parts of the body (as in the case especially of all soft-bodied organisms). Mechanically sensitive sensory endings may be associated with deformable cuticle (e.g. earthworms, Knapp and Mill, 1971), or with underlying dermal tissues (crustacea, Alexandrowicz, 1933; Laverack, unpublished observations). These units could be stimulated by strains established in the body, either by muscle contraction or passive deformation.

(b) Since the body moves relative to the space around it, it follows that structures upon the body surface in contact with that space are also subject to movement. Any interference with these structures will give information regarding the relative movement and position of that particular point. Thus if a seta projects from the body surface and is flexed by contact with another solid object, information is derived about the position of the seta (body part) relative to the solid object.

(c) By the same token as in (b) contact may be made or broken at a substrate interface. It is no coincidence that sensors are located in the extremities of limbs that are continually moving with regard to the solid substrate.

(d) A further special case of (b) is derived from situations in which external sensors are stimulated by other portions of the body. Hairs, strategically placed to be deflected when another limb or joint is moved are ideally situated to provide information regarding the position of those two component parts. But precise indication of which part moves relative to the other requires more input from parallel channels (Fig. 1.1).

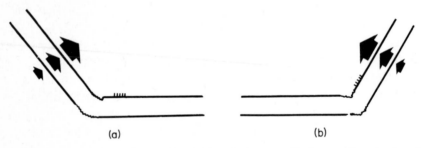

Fig. 1.1   Two (a, b) situations in which a single row of hairs may be stimulated at a joint, but without discriminating which part of the limb is actually moving. The arrows show which portion is moved.

1.2   Monitoring gross environmental movements

Movements occurring in the environment external to the animal may range from earthquakes to zephyr breezes. Movements generated by the animal move the environment relative to the surface, as well as the surface relative to the environment. All may affect the mechanical senses of invertebrates, but not all are proprioceptive stimuli.

1.2.1   Movements relative to substrate

Consider a body pressed against a substrate (e.g. an earthworm in a burrow) or one in contact with a substrate via its legs. Information regarding the absolute disposition of the body is imperative in order that minor adjustments of muscular activity may be made to cope with irregularities in the surface. Projections from the body may be functionally linked with nervous elements that are strain sensitive. Any restrictions placed on the movement of such projections would be monitored by activity in the attendant sensors.

Any rod-like structure contacting a solid substrate may act in two major ways in response to physical events; either (1) as a simple contactor upon which the weight of the body falls, or (2) as a lever over which the body passes in an arc. In case (1) proprioceptors at the end of the strut may respond to impact, stress or pressure; in case (2) proprioceptors may respond to movement at the base of the lever (i.e. the joint, or the body wall insertion). Receptors falling into category (1) include the decapod crustacean dactyl receptors inserted into the distal epicuticular cap which is deformed when stepping takes place (Shelton and Laverack, 1968) and those in (2) include the joint chordotonal organs of the decapod limb (Burke, 1954) (Section 6.2.1), at the chaetal base in polychaetes (Horridge, 1963; Dorsett, 1964) (Section 11.4), and probably at the base of echinoderm spines. In case (1) stimulation occurs when the lever moves relative to the substrate, and in case (2) when the body moves relative to the lever.

1.2.2   Movements relative to medium

All animals live in fluid media (air or water) movements of which occur as currents, eddies, air streams or wind. These fluid movements give rise to turbulence and agitation at surfaces. Numerous organs monitor variations in the flow, ranging from stiff cilia (ctenophores; Horridge, 1966) to hairs (head of locust; Camhi, 1969a) and fans (Crustacea; Laverack, 1963). All are pivoted in such a way as to respond to movement with great sensitivity. Some compound organs, such as halteres (diptera) are composed of aggregations of many simpler organs (campaniform organs) and have an imposed movement (vibration) that confers properties (gyroscopic) over and above those of the component parts.

The face of the locust is covered with hairs. These are arranged in fields and respond to movement of the air across the surface of the locust. Each hair responds maximally to wind from a specific direction (Fig. 1.2), with the optimal direction being determined by

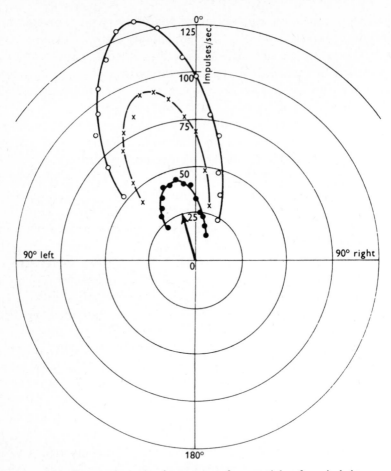

Fig. 1.2   This figure plots the frequency of potentials of a wind detector on the locust head, relative to the direction of the wind. Sensory response depends on direction (about 15° left) and is approximately linearly related to wind speed (From Camhi, 1969).

the angle of curvature of the hair shaft. The total directionality of the response is a function of asymmetry of the hair shaft, the elastic arrangements of the hair socket and the eccentric attachments of the dendrite (see Gaffal and Hansen, 1972). The receptors are slow-adapting. The hairs represent a form of directional proprioceptive information that is valuable in controlling lift during flight, and hence the orientation of the flying animal (Camhi, 1969a, b). Rather similar hairs are found on the cercus of the cockroach (Nicklaus, 1965).

An analogous receptor system exists on the antennae of the spiny lobster *Palinurus vulgaris* (Clarac and Vedel, 1973). This system, consisting of 2 rows of plumose hairs (Fig. 1.3), has been termed the hydrodynamic receptor and is found on the carpopodite of the antenna. Each hair is innervated by three neurons and stimulated by water movement around the limb. Sensitivity is also shown to vibrational stimuli transmitted through the water.

Fig. 1.3   A single plumose hair of the antennal hydrodynamic organ of *Palinurus vulgaris.* (Photograph supplied by Clarac, Vedel, and Moulins; to whom I am grateful.)

Responses to water movements and vibrations in Crustacea were first shown physiologically by Laverack (1962, 1963). Hair fans (Fig. 1.4) and hair pegs, supplied with at least two (usually more) neurons are directionally sensitive to water currents and respond in a synchronous manner to water vibrations up to 100 Hz. These organs are found on the legs, chelae and carapace of lobsters. Mellon (1963) also reported single setae on the carapace of crayfish that show directional mechanical responses from a dual innervation.

Fig. 1.4   Hair fan from the carapace of *Homarus gammarus* (From Shelton and Laverack, 1970).

Orientation towards vibrations in water has been seen to be an attribute of superficial receptors in the water bug *Notonecta* (Wiese, 1972; Murphy and Mendenhall, 1973). These receptors consist of rows of hairs arranged on the last three abdominal segments. Ablation of the hairs decreases the probability of turns toward the operated side for large target (prey) angles. Other receptor populations, probably external hairs, on the pro- and meso-thoracic legs are also implicated in these responses.

1.2.3   Movements relative to other body parts

Length detectors
Any animal that is flexible, mobile, and possesses longitudinal muscle, is capable of shortening; and thus it is of adaptive value that the length of the animal be in some way monitored so that comparative measurements may be made as to the disposition of the body. Where no joints are involved, other means of gaining information relating to length are likely to have been evolved.

Discussion of the effects of length on the discharge of nerve fibres *per se* has been given by Bullock, Cohen and Faulstick (1950), who concluded that stretch and relaxation of the nerve cord in *Lumbricus* did not affect the activity of the component neurons.

More recently, however, it has become evident that although the central neurons of a nervous system may not in themselves be influenced by stretch there are specific receptor neurons that do respond to such stimulation. It must be accepted that length detectors in the nerve cord are not, in themselves, externally placed, and hence are not strictly within the compass of this review. Nonetheless, the change in length is an external event, and the derivation of the CNS from ectodermal layers makes the area likely to possess other ectodermal derivations.

Hughes and Wiersma (1960) reported that stretch of the crayfish nerve cord specifically stimulated units located in the sheath. These observations were extended by Kennedy, Evoy and Fields (1966) and subsequently fully analysed by Grobstein (1973). Grobstein argues that these units are primary mechanoreceptor axons, rather than, say, sensitive interneurons. When the crayfish is fully extended from the fully flexed position the abdominal nerve cord increases in length by about 40 per cent, and tonic receptors are active over the whole range. These slowly adapting receptors are responsive to stretch of the sheath of the connectives between ganglia, whilst phasic receptors respond to movement around the ganglia themselves. These units may be inhibitory on their fellows, and are also responsible for tonic excitation of the extensor motor neurons of the abdomen.

That such units may be more widespread than previously suspected is indicated by the work of Smith and Page (1974). These authors demonstrate that in *Hirudo* the nerve cord sheath contains receptors that initiate activities within the rapid-conduction pathway of the leech. This is effectively a single median 'giant' fibre that conducts both caudally and rostrally (Laverack, 1969) and excites the longitudinal motor neurons of the animal in each segment. Activity in the sheath units leads to immediate shortening of the whole animal. This direct reflex pathway could be effective in compensating for gross extension. Leeches, however, are capable of considerable natural extension (as in looping motion) without immediate rapid shortening brought about by neurons. Thus, this pathway could represent a fail safe mechanism.

It is probable that investigation of a variety of other phyla the members of which exhibit great plasticity of form and shape, (e.g.

Nemertini, Mollusca, Priapuloidea, Echiuroidea, and Echinodermata (Holothuroidea)) would reveal that length monitoring is the norm.

Flexible region receptors
The presence of receptors sensitive to mechanical stimuli could be expected in flexible areas that are deformed by movement. Most groups (e.g. Annelids, Nemertini, Priapuloidea, Echinodermata and, less obviously, Arthropoda, Diplopoda and Chilopoda, have many areas ideally suited for general innervation.

Pabst and Kennedy (1967) described soft cuticle receptors in the abdomen of the crayfish (Fig. 1.5). The cell bodies lie within the segmental roots of the abdominal ganglia with dendrites that project peripherally (see also Wilson and Sherman, 1975), but details of the endings were not described. I believe, however, that the name applied by Pabst and Kennedy may be a misnomer. To qualify as a soft *cuticle* receptor it seems necessary that the receptor be shown to be physically connected to the exoskeleton, or to the overlying cuticle of non-arthropods. The dendrites appear in fact to end in the hypodermis with *no* connection to the cuticle. Pabst and Kennedy (1967) showed that the cuticular layers could be removed, and although the sensitivity of the units decreased they were not destroyed. The cuticle may mechanically amplify the signal without being physically connected to the receptors. Until details of the terminations are known, I propose that these units be henceforward termed hypodermal mechanoreceptors. It may be suspected that such units are ubiquitous. They are the fundamental unit of a component receptor: the crustacean mandibular MRO (Chapter 5). Other situations in which the long dendrites of bipolar cells terminate in unspecified areas are known, e.g. at the base of the antenna, within the anterior thorax and also close to the telson of *Homarus* and *Palinurus* (Laverack, unpublished observations).

Receptors responding to changing positions of sclerites (exoskeletal plates), and hence analogous to limb proprioceptors, have been described by Chapple (1966) in the hermit crab *Pagurus granosimanus*. The first and second segmental roots of the abdominal ganglia carry afferent fibres from receptors located in the hypodermis. The first root innervates the pleopods and ventro-lateral area while the second root innervates the sclerite of the next segment and a region dorsal and posterior to it (Fig. 1.6). The sensitivity of the left side is greater than that of the right. Units which respond to

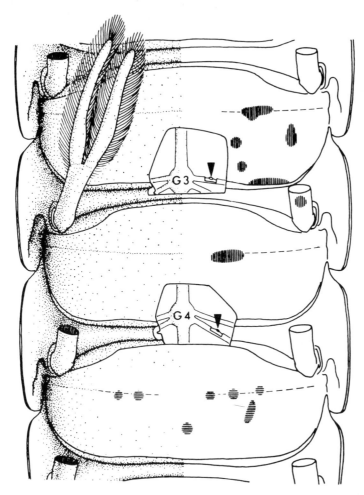

Fig. 1.5   The location of hypodermal receptors in the abdomen of the crayfish. Arrow heads show position of cell bodies; hatched areas the sites of maximal sensitivity G3, G4 are the third and fourth abdominal ganglia respectively. (From Pabst and Kennedy, 1967.)

stretch were found to be localized in the epidermis; but no muscle receptor organs were observed, although such organs have been discovered in larger hermit crabs (Alexandrowicz, 1952; Pilgrim, 1960).

Proprioceptive units in the flexible body wall of soft-bodied animals are probably legion, but few have been shown either anatomically or physiologically. In molluscs, stretch sensitive units in

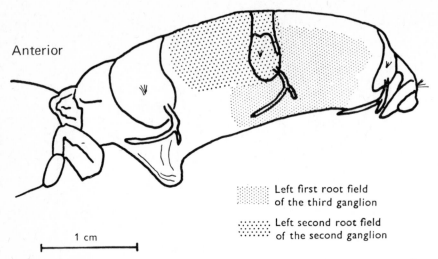

Anterior

:::::::: Left first root field
:::::::: of the third ganglion

.:.:.:.: Left second root field
.:.:.:.: of the second ganglion

1 cm

Fig. 1.6   The mechano-sensory fields of the left first root of the third ganglion and left second root of the second ganglion of the abdomen of a hermit crab (From Chapple, 1966).

the siphon of the whelk *Buccinum undatum* have been shown electrophysiologically (Fig. 1.7) (Laverack and Bailey, 1963), but their anatomical appearance is not known. Thus whilst a truly surface position is not confirmed, it seems likely that it will be found to be similar to the situation in annelids (Chapter 11). Horridge (1963) and Dorsett (1964) both report bipolar cells on the ventral side of the body in polychaetes; the former author claiming that they end in the basement membrane of the epithelium. Dorsett also reports receptors near the groove between segments.

The cuticle of oligochaetes is relatively thick. Lying amongst the fibres of the cuticle are numerous ciliary endings that extend in a number of directions from the surface of receptor cells (Fig. 1.8, Section 11.14). These are well placed to be presumptive flexible region receptors (Knapp and Mill, 1971; Chapter 11).

The localisation of the sensory field of mechanoreceptors has been well shown in arthropods (Hughes and Wiersma, 1960; Chapple, 1966—see above), annelids (Nicholls and Baylor, 1968; Knapp and Mill, 1971) and molluscs (Janse, 1974). All show fairly discrete areas subserved by single peripheral nerve trunks or neurons. These units thus have a kind of specificity which is as informative as those of single projecting sensilla. Janse (1974) has carried out an extensive

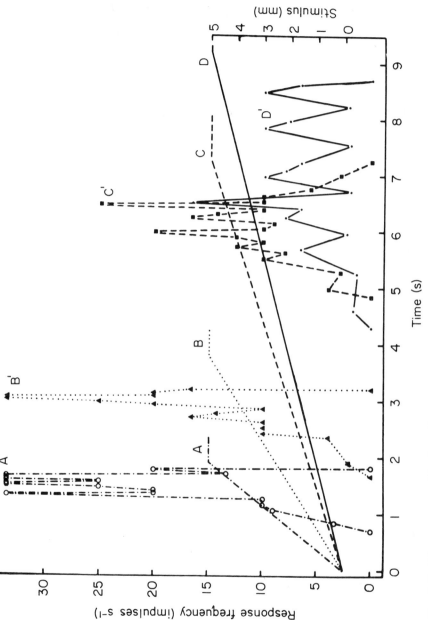

Fig. 1.7   The response of body wall receptors in *Buccinum undatum*. Four velocities (A, B, C, D) of stretch to the same amplitude of deformation (5mm) are shown together with the consequent frequency of response in mechano-receptors (A′, B′, C′, D′) (From Laverack and Bailey, 1963).

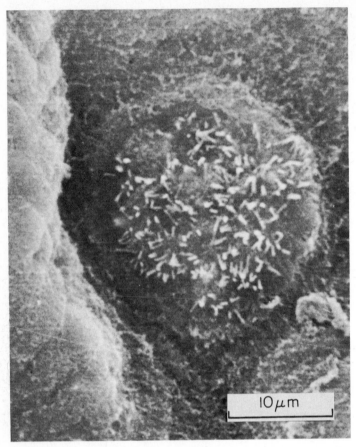

Fig. 1.8   Stereoscan electron micrograph of an epidermal sense organ from *Allolobophora longa* (Annelida: Oligochaeta) showing the raised hillock of epidermal and sensory cells. Note the cilia projecting from the latter. (Scale 10 μm.) See also Fig. 11.14 which shows the structure of the two types of sensory cell present; Fig. 11.12c an epidermal sense cell from *Nereis*.

analysis of the various fields of innervation of nerves serving the body wall and muscles of the pond snail *Lymnaea*. He gives details of the neurophysiological responses of several types of peripheral mechanoreceptors, including tactile and stretch sensitive units. Reconstructions of possible branching patterns are available, but these have not been correlated with known anatomical structures, and what external manifestations there are, remain unkown. One might anticipate that ciliated endings will be found within the cuticular layers. Olivo (1970) working on the razor shell *Ensis* reported finding no highly sensitive vibration receptors on the foot, but plenty of localized tactile receptors which respond to displacement. Again, however, there is no anatomical detail and it is not clear precisely what type of ending is involved.

Distortion of the cuticle is also a property of the claw edge of *Limulus* (Wyse, 1971), deformation leading to stimulation of mechanoreceptors localized in the underlying hypodermis. These receptors are multidendritic, and respond to stimulation over a considerable length of the claw edge, but with a maximum sensitivity over a relatively limited area (Fig. 1.9). Information could be conveyed concerning the force exerted at the joint of the claw.

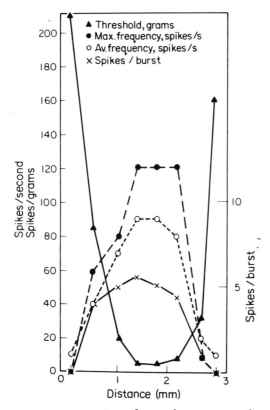

Fig. 1.9   Four response parameters of a mechanoreceptor from the pad of the chela of *Limulus*. The number of spikes and their maximum and average frequency in a burst all have maxima corresponding to the minimum threshold of the unit. Distance along the pad (mm) is correlated with the discharge pattern of the receptor (From Wyse, 1971).

Limulus apparently does not possess joint sensors responsive to opening or the open position, and this information could be obtained from the force receptors. Thus sensors responding to surface events may play a part in overall orientational behaviour.

Stress in rigid structures

It is apparent from the foregoing discussion that various areas of flexible structure may be innervated and capable of signalling deformation. The box-like exoskeleton of arthropods may be very rigid and brittle (e.g. calcified crustacean cuticle; see Hepburn, Joffe, Green and Nelson, 1975), but stress at a point may be referred elsewhere as a strain in the cuticle (see p. 3). These strains may be concentrated at certain places due to the morphology and anatomy of the limb or body. It is at such places that length changes may be maximal and where special organs, adapted to respond to small strain stimuli, are found. In this group are included campaniform organs and slit sensilla. These organs may occur singly, but are often aggregated into complex arrays and fields.

Pressure detectors have been hypothesized to account for the sensitivity of numerous marine animals (Sleigh and McDonald, 1972). Is it possible that increasing and decreasing pressures produce minute compressions of the external hair sensilla of such animals as copepods and chaetognaths? Fleminger (1973) describes the characteristic arrays of sensilla present on the copepod *Eucalanus*. Stress applied as Force (i.e. pressure) per unit area could be a potent stimulus upon such sense organs.

1.3   Primitive requirements

Reference to Fig. 1.1 will readily show that it is of adaptive significance that displacement of a body, or part thereof, should be monitored in respect of two major parameters, namely, position and movement. The absolute spatial relationship between body parts is important for the appropriate function of those parts, e.g. that a leg should be placed correctly during walking, or that the head is carried precisely on the neck. It the position is disturbed, either passively or actively (i.e. by an imposed movement or by muscular activity respectively) then it is important that the newly adopted position be signalled. However, information concerning the final position may not be enough for generating sufficient information and it may be necessary to have further information on how that position was attained. For this it would be advantageous to have details concerning the actual movement taking place.

1.3.1 Position detectors

It has become more and more evident that the organelle known as the cilium, prevalent throughout the animal kingdom, has become modified in many ways to subserve numerous functions. The (presumably) basic motile cilium has a remarkably uniform structure wherever it occurs and in whatever group from Protozoa to Mammalia, but the appearance of cilia in modified form often heralds a sensory function. However, not all sensory cilia have a very specialized structure (Barber, 1974).

Even normally motile cilia are sensitive to bending (i.e. an imposed mechanical force) as shown by Kinosita and Kamada (1939). Thurm (1968) subsequently demonstrated that the abfrontal cilia of *Mytilus* (Mollusca, Lamellibranchiata) show mechanosensitivity and beat after passive bending, and that calculated shearing forces are focused at the basal body.

If the cilium may be taken as at least a simple starting point for sense organ structure we may look for receptors even amongst the protozoa. Sensitivity towards physico-chemical events is well known, but specialized receptors much less so. One such candidate for a primary sense organ concerned with monitoring position has been described for the holotrichous ciliate, *Nassula*, by Tucker (1968, 1971), who proposes that projecting membranelles composed of modified cilia are mechanoreceptors. In the case of *Nassula* they may be concerned with manipulating food (blue-green algae) rather than monitoring position as such, but should contact be made with other solid objects then there is at least the likelihood of positional information. Transmission seems to be a function of structural connectives leading eventually to a contraction of the cytopharynx.

Thus mechanoreception seems indicated in lowly forms. Such projecting cilia, often stiff, rigid and long, are known from several sites, and have been shown to respond to mechanical events—or at least bring about reflexive movement after the stimulation of the end organ.

Stiff cilia on the surface of ctenophores were first described in separate papers in the same year (Chun, 1880; Eimer, 1880; Hertwig,

1880). Horridge (1966) points out that they are sensitive to water displacement since stimulation was followed by movement of mobile 'fingers' of the animal. In *Leucothea* these cilia have specializations of the basal regions that make them remarkable as sensory structures (Fig. 1.10). Hernandez-Nicaise (1974) has confirmed the appearance of these receptors on *Beroe*, *Pleurobrachia*, *Lampetia* and *Callianira*. This author claims that the receptors are monosynaptically linked to neighbouring epidermal glandular cells. An immediate surface response due to movement of water across the receptors may thus lead to a change in behaviour. It seems unlikely that this has anything to do with orientation but it does seem probable that such actions are concomitant with feeding.

There is a wide range of examples of stiff cilia: ctenophores (*Leucothea*; Horridge, 1966); chaetognaths (*Spadella*; Horridge and Boulton, 1967); echinoderms (pedicellaria of *Echinus*; Cobb, 1968) and they may be suspected in other groups such as polychaetous annelids (e.g. *Sabella*; Santer and Laverack, 1971), cnidaria (e.g. hydroids; Josephson, 1961) and nemertines. Indeed, non-motile

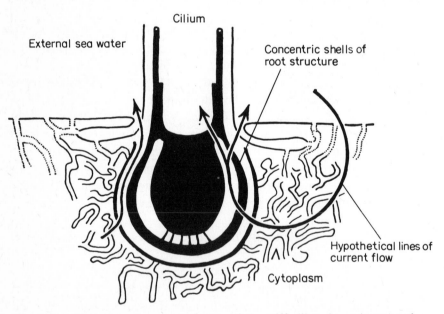

Fig. 1.10   Diagrammatic representation of the stiff ciliary bases in *Leucothea*. Horridge suggests that shearing movement of the concentric shells of the root structure at the base brings about depolarization. Arrows indicate hypothetical lines of current flow. (From Horridge, 1965: Courtesy of the Royal Society)

cilia are found on papillae on the palps of *Harmothoë* (Daly, 1973), and also on other appendages of this polychaete such as the dorsal, tentacular and anal cirri and the antennae. It is likely that these form the basis of the sensitivity to vibrational sources in the water shown by *Harmothoë*. Ockelmann and Vahl (1970) also found vibration a potent stimulus in the polychaete *Glycera*.

Bending of innervated projecting structures is certain to provide information regarding the force involved. In the case of CAP organs (see p. 36) the linear arrangement of the pegs of the individual organs could give a precise input concerning the absolute position of the adjacent articulating membrane. Hair plates also signal the position of the neighbouring bodily parts of insects (Markl, 1966) and they may do so in some crustacea (Clarac, Vedel and Moulins, personal communication).

1.3.2   Movement detectors

The question of position sensitivity is coupled with that of movement. Movement of the surrounding environment (air, water, or even solid substrate) often has sufficient power to bring about stimulation of external receptors regardless of the absolute or relative position of the animal or its parts. In highly-developed proprioceptors such as chordotonal organs, movement receptors may be separated from position receptors (Wiersma, 1959) as a discrete class of units within the system. Movement receptors are sensitive to velocity and direction, position receptors to final position and, to varying extents, to movement (e.g. Hartmann and Boettiger, 1967; Section 6.5).

1.3.3   External versus internal sensors

Evolutionary trends in several groups show a gradual removal of proprioceptors from the surface to a deep or internal placement. This is demonstrable in vertebrates (e.g. the change in position of the acoustico-lateralis system in fish and amphibia) and in some invertebrates. Insufficient detail is available for the majority of invertebrate phyla although one might hypothesise that receptors associated with the bases of polychaete annelid chaetae represent a specialization of primitively placed epidermal cuticular ciliated receptors (Dorsett, 1964) and that these may have been functionally superseded in leeches by receptors with centrally placed cell bodies (Fig. 1.11) (Nicholls and Baylor, 1968). A middle course in which both external and central receptors are known is found in oligochaetes (Mill and Knapp, 1971; Günther, 1970, 1971). Further work, however, may reveal that this is the case in all annelids. It is indeed rather curious that the peripheral nervous system of leeches has not been investigated properly in recent years. Kristan (1974) comments on dorsal and ventral stretch receptors but provides no anatomical evidence.

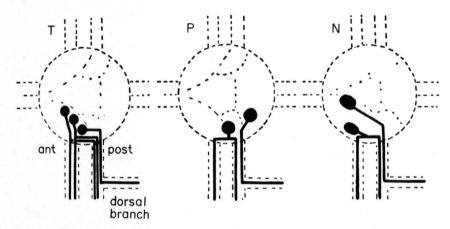

Fig. 1.11   Diagram showing the postions of the T (tactile), P (pressure), and N (nociceptor) cells in each hemi-ganglion of *Hirudo*. A total of 6T, 4P and 4N cells is found in each ganglion located in regular, repeated patterns along the whole ventral nerve cord. It is conceivable that P cells especially are concerned with body wall movements and hence with proprioception (From Nicholls and Baylor, 1968).

A more satisfying example comes from decapod crustacea; CAP organs (Wales, Clarac, Dando and Laverack, 1971; Alexandrowicz, 1972) the organization of which is described on p. 36 of this article. The distribution of this system amongst decapods leads one to comment upon the internal proprioceptor arrangements that parallel these external organs. The position (Fig. 1.12) closely corresponds with the internal position of proprioceptors known conventionally as MCO1, MC2 and CP2 (Section 6.2). The chordotonal organs are well described in decapods belonging to the Astacura, Palinura Anomura and Brachyura. In Astacura and Palinura the layout of the CAP

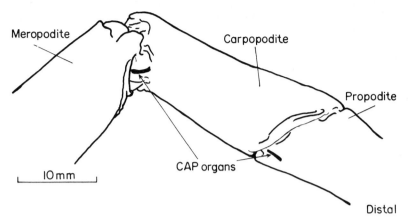

Fig. 1.12   Diagram to show the position of CAP organs on the MC and CP joints of a lobster walking leg (Palinura) (After Alexandrowicz, 1972).

organs is precise and many units are observed; in Anomura the numbers are smaller, the sensilla are small, and scattered, and in some cases totally absent from some joints (see Table 1.1); Brachyurans do not show these CAP organs at all. The natantian *Pasiphaea* on the other hand has bunches of hairs where CAP organs might be expected (Laverack, unpublished).

Alexandrowicz (1972) pointed out that certain differences exist in the organisation of CP2 and MC2 in the various decapod groups (Table 1.2). Some features of anomuran patterns are intermediate but they mostly resemble the Brachyura in their style.

These features argue for the proposition that the chordotonal organs of decapod Crustacea may have originated (in an evolutionary sense) from groups of hairs, very similar to hair plates of insects, of which the individual sensilla have shortened, lost their contact with the surface, and finally been incorporated in a connective tissue strand or sheet (Section 6.2). The remaining vestiges of hairs are evident as scolopidia (Chapter 9). This argument has been advanced previously, but not on these comparative morphological grounds. Schmidt (1973) has recently summarised the evidence available for insects, and demonstrates that fine structure, ontogeny and moulting all support the hypothesis that the cuticular sheath of sensory hairs and campaniform sensilla are homologous to the extracellular cap or tube of scolopidia.

Table 1.1   Distribution of CAP sensilla in various families of decapod crustacean IM, ischio-meropodite; MC, mero-carpopodite; CP, carpo-propodite joint. (Laverack, unpublished).

Decapoda sub-order	Section		Maxillipeds				Chela			Pereiopods												
			2		3					2			3			4			5		
			MC	CP	MC	CP	IM	MC	CP	IM	MC	CP	IM	MC	CP	IM	MC	CP	IM	MC	CP
Natantia		Penaeidae	+	+	+	+	+	+	+	+	+	+	+	+	+	+	+	+	+	+	+
		Caridea	+	+	+	+	+	+	+	+	+	+	+	+	+	+	+	+	+	+	+
Reptantia	Palinura	Eryonidae	+	?	+	+	?	+	+	?	+	+	?	+	+	?	+	+	?	+	+
		Scyllaridae	+	+	+	+	+	+	+	+	+	+	+	+	+	+	+	+	+	+	+
	Astacura	Nephropsidae	+	+	+	+	+	+	+	+	+	+	+	+	+	+	+	+	+	+	+
		Callianassidae	+	+	+	+	+	+	+	+	+	+	+	+	+	+	+	+	+	+	+
	Anomura	*Litbodes*	+	+	+	+	+	+	+	−	+	+	−	+	+	−	+	+	−	+	+
		Galatheidae	−	−	−	−	−	+	+	−	+	+	−	+	+	−	+	+	−	+	+
		Porcellanidae	−	−	−	+	−	−	−	−	−	−	−	−	−	−	−	−	−	−	−
	Brachyura	Portunidae	−	−	−	−	−	−	−	−	−	−	−	−	−	−	−	−	−	−	−
		Corystidae	−	−	−	−	−	−	−	−	−	−	−	−	−	−	−	−	−	−	−

Table 1.2   Comparison of anatomy of chordotonal organs in Astacura, Palinura and Brachyura. Anomuran examples are somewhat intermediate but more closely resemble Brachyura (information from Alexandrowicz, 1972).

Feature	Astacura/Palinura	Brachyura
position	distal to articulation	spanning articulation
large cells	distal	proximal
scolopidia	in canals of CAP	in connective tissue strand
attachment	not to muscle	to tendon of muscle

Multipolar v. bipolar

Two quite distinctive and characteristic neuron types are found associated with surface proprioceptors in invertebrates. These are bipolar and multipolar.

Bipolar cells are often associated with a surface accessory structure, or something that can be derived from such a feature (e.g. hairs, campaniform organs, scolopidia). They normally have relatively short dendrites, although in PSN receptors of the gastric mill in decapods they may be 2 or 3 cms in length (Dando and Laverack, 1969) and 400 μm in chordotonal organs (Hartmann and Boettiger, 1967). Their terminations may be on connective tissue (PSN neurones; chordotonal organs) or at the bases of hair sensilla (hair plates), but are always made at a single point. Details of origins are not always clear, but in insects new bipolar neurons added at a moult derive from epidermal cells (Wigglesworth, 1961), and this class may represent truly superficial elements.

Multipolar cells are less commonly reported, and are often located on or near accessory structure having only a distant connection with the surface. Thus the annelid chaetae, secreted by epidermal cells in pits (Bouligand, 1967), nonetheless project from an apparently deep-lying situation and the receptor neurons are multidendritic (Fig. 11.9). Similarly the receptors of the crustacean MPR system (see Wales, this volume) are multidendritic and insert upon an internally derived connective tissue system. Possibly these neurons represent truly internal receptors which have become secondarily associated with structures influenced by external events.

Thus there can be little doubt that receptors have arisen as truly external structures projecting from the suface of the animal; as truly internal structures which are often associated with internal accessory material; and in some cases have evolved from an external to an internal position. Growth and development studies may show something of the stages of internalization of receptors and the derivations of the sensory cells from either the ectoderm or via the central nervous system.

1.4   Types of sense organs

Numerous types of sense organs have evolved that are stimulated by mechanical events.

1.4.1   Strain gauges

Slit sensilla
Spiders and scorpions are well supplied with specialized modifications of the cuticle that appear as slits in the surface. These organs may be single (slit sensilla) or compound (lyriform organs), and are widespread over the body and limbs. Lyriform organs comprise a number of slit sensilla arranged in parallel. The distribution of such organs was described by McIndoo (1911) though this author mistakenly believed them to be chemoreceptors. Details of histological structure were given by Kaston (1935), but the fine structure and information on innervation has been supplied by Salpeter and Walcott (1960), Walcott (1969) and by Barth (1969, 1970).

The so-called slit is really a shallow trough or groove in the outer cuticular layer surrounded by specially thickened reinforcements of the cuticle. In the spider *Cupiennius* each slit is innervated by 2 dendrites (Fig. 1.13). One of these ends at the inner opening of the cuticular canal beneath the slit, whilst the second passes distally to the inner membrane of the slit. The most distal portion of each dendrite is composed of a modified cilium and microtubules. Salpeter and Walcott (1960) working on *Achaearanes tepidariorum* describe only a single innervation. There is no apparent difference between slit sensilla and lyriform organs except in the number of slits present.

Distribution The most comprehensive analysis of the arrangement of slit sensilla has been made by Barth and Libera (1970) for *Cupiennius.* There is a total of about 3300 slit organs, of which approximately half are single slits, and the remainder form 144 lyriform organs

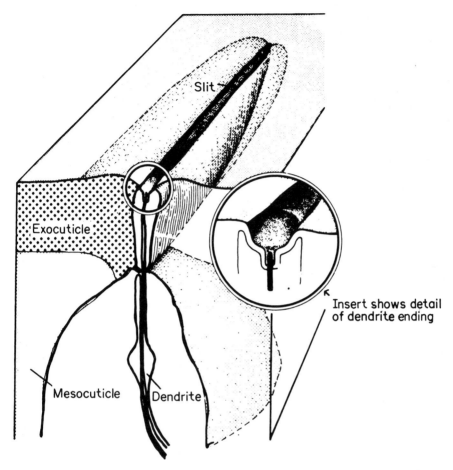

Fig. 1.13   EM diagram of the detailed structure of a slit sensillum from *Cupiennius*. Note the trough-like depression in the cuticle surface that forms the slit and the projection of one dendrite to meet the cuticle, whilst the second ends about 10μm below this level (After Barth, 1971a).

composed of between 2 and 29 parallel slits. All of the complex organs are found on the appendages, the pedipalps and walking legs, usually close to joints (Fig. 1.14). Most slits are arranged parallel to the long axis of the appendage with the exception of those of the organ on the metatarsus which lies dorsally and at right angles to the leg axis. Single slits occur on all regions of the body, but not next to the joints, and on the ventral abdomen there is a marked relationship to the attachment of the muscles.

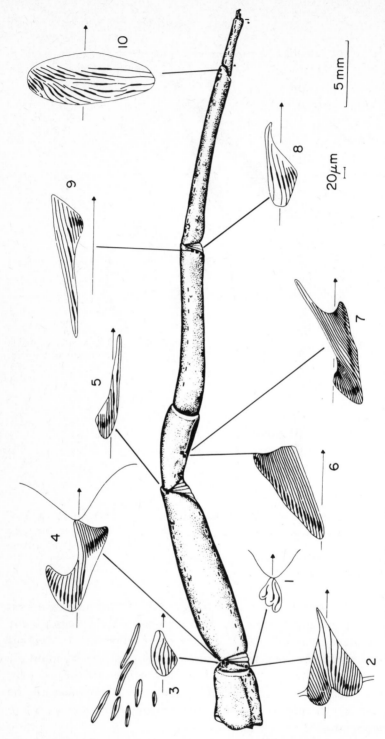

Fig. 1.14   The distribution of slit sensilla on a walking leg of *Cupiennius* seen from the rear aspect. Locations are indicated by black squares, and detailed orientation given by the detailed inserts. (From Barth and Libera, 1970.)

In spiders slit sensilla are not the only surface structure that may give rise to proprioceptive information. Such animals require positional information regarding the disposition of the appendages especially the legs in relation to the web. Foelix and Chu-Wand (1973) found that in these arthropods, as in others, numerous hairs are innervated, typically by 3 bipolar neurons (*Lycosa, Araneus*) and Christian (1971) found in *Tegenaria* that each trichobothrium receives 4 neurons. In Acarines (ticks) the hairs are doubly innervated (Foelix and Axtell, 1971; Chu-Wang and Axtell, 1973).

Mechanism of action Pringle (1955) suggested that in the scorpion *Heterometris* and the amblypygid *Phrynichus* the effective stimulus was likely to be a tension component of shear applied at right angles to the slits, thus giving rise to compression parallel to the slits, and hence a distinct resemblance to the sensitivity of campaniform organs (Section 1.4.1). He further suggested that stretch of the dendrites was the transduction mechanism leading to depolarization. In a series of elegant papers Barth (1967, 1971a, b, 1972) has examined the responses of the slit sensilla apparatus and demonstrated the relationship to structural deformation in araneids. Barth (1971b) showed by the use of models that the slit is deformed most when stress is applied at right angles to it. Stress parallel to the slit is small, and minimized by the various cuticular thickenings around it. The deformation of single slits is greatest in the middle, and the overall tendency to deform increases with length (Fig. 1.15).

These physical events are sufficient to focus the stimulus onto the receptor dendrite. Pringle (1955) believed it to be the opening of the slit that leads to stimulation. Barth (1972) showed conclusively that the stimulus is slit compression. The organ response is slowly adapting and represents the activity of the longer dendrite which ends where membrane curvature is greatest and thus where the moment of bending is greatest. It appears that the slit compression is transmitted as a monaxial compression of the dendrite in a direction perpendicular to the long axis of the slit. Stretch of the dendrite brought about by lifting is not an effective stimulus. However, as we have seen, two dendrites are located at each slit and Barth does not comment upon the adequate stimulus for the second which ends at a different site.

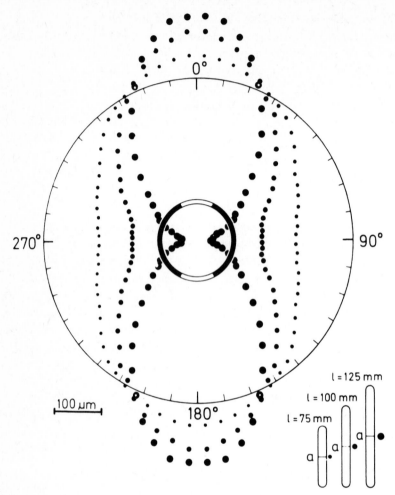

Fig. 1.15   A model reconstruction of forces exerted on a slit sensillam. The maximum focus of strain occurs at right angles (90°) to the slit. Slit lengths are shown in the inset (From Barth, 1971b).

Campaniform sensilla

The structure of campaniform sensilla has been described by Chevalier (1969) and Smith (1969) for those located on the halteres of a fly (*Calliphora*), by Moran, Chapman and Ellis (1971) for those on the cockroach leg (*Blaberus discoidalis*) and by Schmidt (1973) for various others. In these cases the overlying cuticular cap is served by a single dendritic ending. This dendrite is attached to the cap via a complex structure containing numerous microtubules. These are grouped together in a more or less orderly fashion in the ending and in total may number several hundred. The microtubules derive from a ciliary basal body and there is a short connecting cilium of 9 + 0 filaments found between the basal body and microtubules (Fig. 1.16; Chapter 9).

In surface view a campaniform sensillum appears as a small (10μm) diameter, domed cuticular cap. The sensilla are arranged on various limbs of insects in groups with distinctive distributions (Fig. 1.17). Pringle (1938a, b) demonstrated 11 distinct groups of such organs on the legs of *Periplaneta* and Chapman (1965) found a single organ at the base of large tactile spines on the tibia. They also occur in many places where strain could be expected in the cuticle. Those of the haltere (Smith, 1969) provide sensory information about its vibration and hence allow the gyroscopic mechanism of the organ, which is essential to its function as an equilibrium sensor (Section 14.4.2).

Campaniform sensilla from Crustacea were only described as late as 1968 (Shelton and Laverack). For many years the existence of so-called funnel canals had been known, but their function was uncertain. They are particularly well developed at the epicuticular cap of the dactyl of the limb, where the actual composition of the leg ending is of a different nature to the calcified cuticle elsewhere. Electrophysiological recording in the area demonstrated the presence of two closely aligned and parallel units that serve the canal. Shelton and Laverack redescribed these, not as canals (implying a secretory aperture), but as crustacean campaniform organs. Their fine structure is not yet known, but their anatomy is illustrated in Fig. 1.18.

Recent studies with the SEM have revealed other groups of campaniform sensilla in crustaceans that are similar externally to those of insects. A typical group of these is shown in Fig. 1.19a and a single one in Fig. 1.20b. These examples are taken from the outer ramus of the antennule of *Homarus gammarus* and are located at the base of the guard hairs (Laverack, 1964) which flank the aesthetasc chemoreceptor apparatus. In the lobster there are between 2 and 6 found at each individual hair base (in a situation very reminiscent of those described by Chapman (1965) in the cockroach). In *Nephrops*, groups of three are found (Fig. 20c), and in *Crangon* a single such organ appears close to the annular boundary (Fig. 20d), but not in conjunction with an external hair structure. This 'row' of sensilla is found only on the lateral aspect of the antennule (Fig. 1.20).

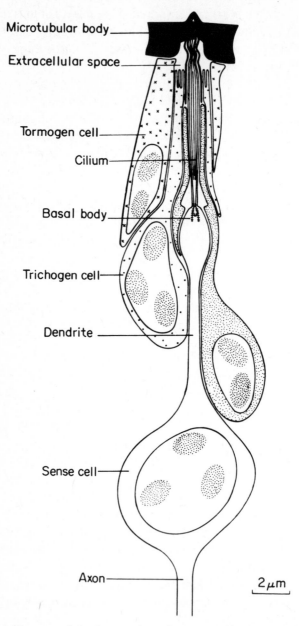

Microtubular body

Extracellular space

Tormogen cell

Cilium

Basal body

Trichogen cell

Dendrite

Sense cell

Axon

2 μm

Fig. 1.16   A diagram of the innervation of a campaniform sensillum from the pedicel of the Ephemeropteran *Cloeon* (From Schmidt, 1974).

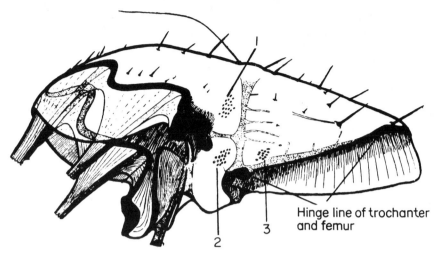

Fig. 1.17 The distribution of campaniform sensilla on the trochanter of the 3rd right leg of *Periplaneta* 1,2,3 = groups of campaniform sensillas. (From Pringle, 1938a.)

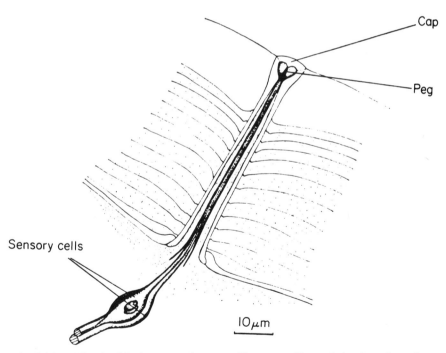

Fig. 1.18 The doubly innervated campaniform sensillum of the dactylopodite of crabs (From Shelton and Laverack, 1968).

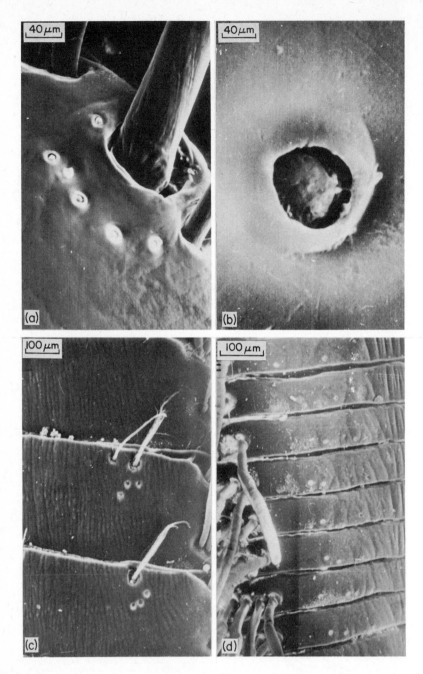

Fig. 1.19 SEM photographs of campaniform sensilla from the antennule of Crustacea. (a, b) From *Homarus gammarus*. (a) Showing the sensilla arranged around the base of a large guard hair. (b) Detail of a single sensillum. (c) From *Nephrops norvegicus* in which campaniform organs are arranged in triplets. (d) From *Crangon vulgaris*, with single and double sensilla asymmetrically arranged upon the antennular flagellum.

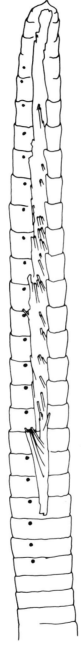

Fig. 1.20   The distribution of campaniform sensilla of the antennule of *Crangon vulgaris* (traced from an SEM photograph) Sensilla marked as black dots. (Laverack, unpublished observations.)

The arrangement of campaniform sensilla on antennae, which may be common to all arthropods, is associated with apparent flexibility of the contiguous cuticle. Masson (1972), working on ants, shows a circumferential ring of sensillae close to the boundary between the second and third segments, and Slifer and Sekhon (1973) state that, in Embioptera (Insecta), the sensilla are arranged at the distal end of nearly every subsegment, where they lie in the region where thick cuticle gives way to the thinner cuticle of the annular membrane. Dumpert (1972) found campaniform sensilla on ant (*Lasius fuliginosus*) antennae.

Hair plates
Hair plates are collections of hair sensilla arranged in characteristic groups (Fig. 1.21) In insects, hair plates (especially those of the head) have been implicated in a number of proprioceptive reflexes. Haskell (1959) found that in locusts the cervical (neck) plates respond to head movement and position, whilst hairs in a fringe on the pronotum edge respond only to head movement. These observations were followed by those of Goodman (1965) who destroyed the

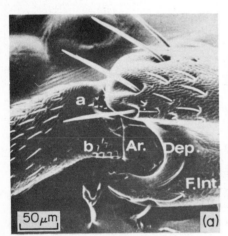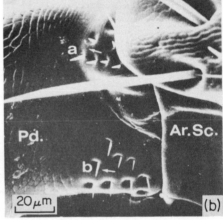

Fig. 1.21   The hair plates at the junction between scapus and pedicel of an antenna of the ant, *Camponotus*. a, b—hair plates; Ar—articulating membrane; Ar. Sc—scapus articulation; Pd—pedicel; Dep—internal depression of scapus F Int, x c. 400. (From Masson, 1972.)

cervical hair plates and discovered that a locust is no longer able to hold its body in line with its head after this treatment. The receptors are essential for maintenance of head position. Shepheard (1973, 1974) commenting on the control of head movements in locusts shows several parallel receptor systems that may affect the control of nystagmus by modifying the central nervous output. Wendler (1966) also removed hair plates from the base of the legs of *Carausius* and found a loss of gravity responses and overstepping during normal locomotion (Chapter 8).

Spencer (1974) suggests that the hair sensillae of the trochanter of the cockroach leg may act as phasic tactile receptors when the animal walks, although this has not been shown unequivocally in free-moving animals. The hair plates are capable of synchronous responses up to 300 Hz.

Rather more complex activities are also at least partly affected by hair plate groups. In this case the stimulus is gravity affecting a large mass (head or thorax) relative to the adjacent part of the body. Honey-bees trained to keep a fixed angle of 60° to gravity are unable to do so if all hair plates at the neck, petiole and coxal joints are non-functional. If only neck, or petiole, or petiole and coxal plates together, are interfered with then a gravity compass action still occurs. The neck plates are the most important single group and gravity responses can occur if just this one area remains, but not in its absence (Markl, 1966).

Well defined hair plates are located at the junction between the head and the scapus of the antenna of ants (*Camponotus vagus*; Masson, 1972) and between the scapus and the pedicel (Fig. 1.2.1) Complex rotational movements are possible at these joints. A ring of campaniform sensilla is found on the proximal side of the junction of the 2nd and 3rd antennal segments. These have a close relationship with the dendrites of Johnston's organ, which inserts at this level. Monitoring of the movements of these various parts probably occurs through these various organs.

Clarac, Vedel and Moulins (personal communication) have located a hair plate organ in a crustacean, *Palinurus vulgaris*. A collection of hairs lies on the internal (medial) side of the antennal base in a position over which the adjacent articulating membrane passes during movement. This may be a group homologous to those of CAP organs on the pereiopods (see Section 1.4.1) and subject to similar forms of stimulation.

Cuticular articulated pegs (CAP organs)

Whilst campaniform sensilla are found in particular orientations, and have a specified distribution relative to other structures, they appear to be present where strain may occur in the cuticle. This may be a directional stimulus but usually occurs in a rather general manner through the surrounding cuticle, being focussed at the site of the sense organ (a similar situation is found as we have seen with lyriform organs).

CAP organs (Wales, Clarac, Dando and Laverack, 1970; Alexandrowicz, 1972; Laverack, unpublished) are located in very precise positions. They are invariably on the distal side of a joint, and very close to the articulation. They occur at the carpo-propodite, mero-carpopodite and, in some limbs, the ischio-meropodite joints (Fig. 1.22a, b). They are typically composed of a line, or fan-shaped group, of sensilla (Fig. 1.23). Amongst natatory forms (*Crangon, Penaeus, Pandalus*) the sensillae are disposed in a very precise line extending away from the articulation (Fig. 1.23b). Amongst primitive large decapods, such as *Panulirus*, the sensillae are found in a duplicate line (Fig. 1.23a), and in the Astacura *Homarus* and *Nephrops*, the appearance is of a broad fan with numerous individual sensilla. Numbers vary with age and are added to at every moult throughout life in *Homarus* (Laverack, unpublished observations).

The nature of the stimulus has not been properly established. Wiersma (1959) reported that these areas respond to mechanical stimulation, and this seems to be the most likely modality, but his evidence as to the precision of the stimulus leaves something to be desired and subsequent attempts to stimulate these organs singly have not been successful (Wales *et al.*, 1970). Mechanical probing is not in any case likely to be the natural stimulus, and it has been proposed that the correct stimulus is the articulating membrane, rolling up over the sensilla when the joint is flexed. This could deflect the rather short pegs distally in a sequential manner, thus giving rise to a linear analogue of the position of the joint.

The merostomatous arthropod *Limulus* possesses a number of spines (6) arranged along the edges of the opisthosoma. Tactile stimulation of these is followed by movement of the tailspine towards the side stimulated, and thence to the dorsal midline. Each spine is innervated by at least 2 and maybe up to 4 units. These are multidendritic receptors (Eagles, 1973). The responses of these sensors show a great range of adaptation rates. In Fig. 1.24 the phasic

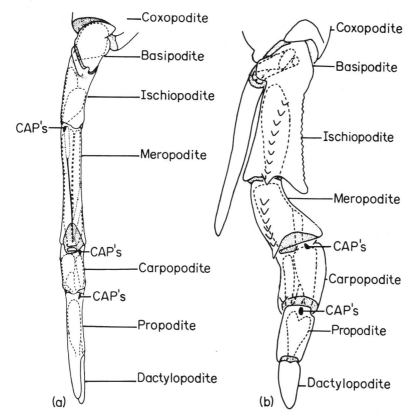

Fig. 1.22 The location of CAP organs on (a) a pereiopod and (b) the third maxilliped. Both limbs are seen from the posterior aspect. Co—Coxopodite; B—Basipodite I—Ischiopodite; M—Meropodite; Ca—Carpodite; P—Propopodite; D—Dactylopodite. (From Wales, Clarac, Dando and Laverack, 1970.)

response indicates the log of the velocity of movement, and the tonic response, the amplitude of spine displacement. In the absence of any specific organs of equilibrium, Eagles suggests that the lateral spines could be of value in determining the orientation of the animal relative to the substratum.

1.4.2 Effects of moult on arthropod receptors

Moulting is a dramatic crisis recurring throughout life in many arthropods, and at these times the external accessory apparatus and muscle apodemes are shed. Remarkably little has so far been done on how the peripheral nervous system compensates for a moult.

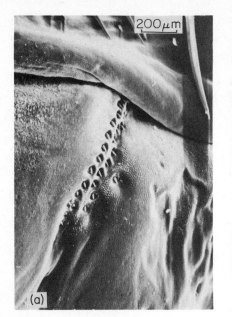

Fig. 1.23   Typical arrangements of CAP organs at joints of decapod crustacea. (a) *Panulirus argus* Carpopropodite joint 2nd maxilliped. (b) *Pandalus borealis* Mero-carpopodite joint leg 3.

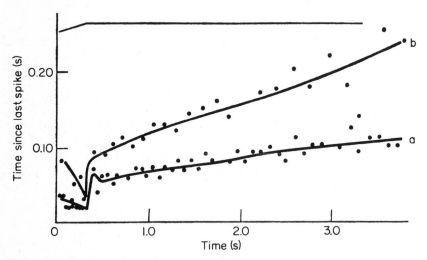

Fig. 1.24   Phasic and tonic response of two mechanoreceptors in the same lateral spine to movement of the spine as represented at the top of the figure. Stimulus amplitude was 15°. a, represents the activity of the more slowly adapting unit; b, that of the more rapidly adapting unit (From Eagles, 1973).

Slit sensilla

The number of slits, both single and those of complex lyriform organs, increases between the fourth and final moult by 1.6x in *Cupiennius* (Barth, 1970).

CAP organs

In the macruran decapods the total number of sensilla rises by 8—10 per cent at each early moult (Laverack, unpublished), from the swimming larval stage I through to the establishment of a benthic juvenile lobster (stage IV) and continuing throughout life (Table 1.3).

Table 1.3   The effect on the CAP organs moulting in *Homarus gammarus*. Note the increase in numbers of sensilla at each moult of the larva, and the much greater number in the 5 cm juvenile and the adult. IM, ischio-meropodite; MC, mero-carpopodite; CP, carpo-propodite joint. (Laverack, unpublished observations).

Larva	Joint	Maxped 2	Maxped 3	Chela	Leg 2	Leg 3	Leg 4	Leg 5
Stage I	MC	5	5	7	7	4	5	—
	CP	4	6	3	3	4	3	—
Stage II	MC	4	8	9	6	4	4	3
	CP	—	4	4	4	4	4	3
Stage IV	IM			3	4	5		
	MC	7	10	11	8	8	6	
	CP	5	7	7	7	5		
Stage V	MC		11	7+	9	5	7	—
	CP			10	8	6	7	7
5 cm	MC	17	12	15	21	18	16	>10
	CP	11		12	7	10	13	14
Adult	IM			29	33	36	44	32
	MC	21	54	40	46	41	38	39
	CP	24	51	90	41	48	80	38

Campaniform sensilla

Moran (1971) showed that at moult the sensory process (micro-tubular region) is lost. At some pre-moult stage this process may become non-functional and a new connection formed at the new cuticle. It may also be possible, however, that a new cilium could grow first and become functional before the original is lost.

Hair sensilla
Hair sensilla similar in type to hair plate sensilla, but located on the telson of crayfish, increase in number throughout life. As is the situation in CAP organs, additions may be made in an identifiable manner (Kennedy, 1974).

Chordotonal organ
The scolopidial cap is not lost at ecdysis (Schmidt, 1974). This makes an interesting contrast with the situation regarding surface placed organs. All hairs are lost, and replaced or added to.

1.4.3   The effective stimulus

Early work on the mechanical sensitivity of neurons revealed that stretch was a potent stimulus leading to depolarisation of an axon. As a consequence almost all mechanoreceptors were postulated to function in this manner.

It is now apparent, however, that although stretch may be a stimulus for some types of receptor it is not by any means a unique source of energy. As more receptor types are fully analysed it becomes apparent that there are several stimulus modalities. At present we may categorize in particular: (1) stretch, (2) muscle tension, (3) relaxation, (4) compression.

Of these we can immediately dispose of (2) and (3) by saying that as far as we know they are demonstrated only by arthropod internal proprioceptors (e.g. muscle tension: abdominal and mandibular muscle receptor organs, apodeme receptors—Chapters 2, 5 and 10); relaxation stimulation appears to be a property of some chordotonal neurons from evidence based on anatomical studies (Hartmann and Boettiger, 1967; Mill and Lowe, 1973), although details of the transduction processes are not available.

Categories (1) and (4), however, are diametrically opposed in action, and have been shown to have relevance to receptors associated with external structure. Stretch as a stimulus mechanism seems indicated in nerve cord sheath detectors, and in flexible cuticle receptors (e.g. Chapple, 1966), because the only evidence available suggests that stretch is the appropriate modality.

Pringle (1938a, b) hypothesised that stretch was also active in stimulating mechanoreceptor neurons at the base of hair sensillae, and campaniform organs. Other workers followed this line of thought until Thurm (1965) analysed the response of hair-plate sensilla in conjunction with the functional morphology of a single sensillum. This work clearly showed that the effective stimulus was *compression* of the dendritic ending rather than stretch (Fig. 1.25). Maximal stimulation occurred when the compression influence amounted to 0.1μm, or 15 per cent of the unstimulated diameter. Threshold was reached when values of only 3nm or 0.5 per cent occurred. The resulting depolarization propagates as an action potential through the sense cell body. The action potential originates very close to the dendritic insertion (Mellon and Kennedy, 1964) and may thus be a direct action of the compression of the microtubules at the ending (this conclusion rests, however, on the assumption that all arthropod hair sensilla mechanoreceptors have similar anatomy to that of insects). A dendritic ending origin for the impulse traffic has also been shown to occur in the sensory cells of joint chordotonal organs close to the scolopidium insertion (Hartmann and Boettiger, 1967). In this case the endings are modified cilia without micro-tubule endings (Whitear, 1962; Mill and Lowe, 1973; Lowe, Mill and Knapp, 1973).

Compression as the potent stimulus has also been demonstrated in spider slit sensilla (Barth, 1972), and Thurm (1965, 1968) has argued that campaniform and chordotonal organs are also affected in the same way. Görner (1965) working on trichobothria (hair sensilla) of spiders, showed that deformation of nerve endings took place by inclination and compression of the nerve over the border of the hair base.

A number of investigations report that the dendritic insertion of mechanosensitive units at the base of hairs in insects occurs asymmetrically and that the internal structure is arranged in parallel arrays of microtubules (see Gaffal and Hansen, 1972, for a summary).

The compression theory outlined above seems clearly indicated in a number of systems, but Rice, Galun and Finlayson (1972) have recently challenged its validity and returned to the proposal that stretch is potent. These authors propose that, at least in some insect hair mechanoreceptors, it is bending translated to receptor terminal stretching that is the stimulatory mechanism. The fine structure of such endings lacks a microtubular array and possesses instead a tubular cytoskeleton over which the membrane stretches. Comparable details are not available for Crustacea and the possibility of

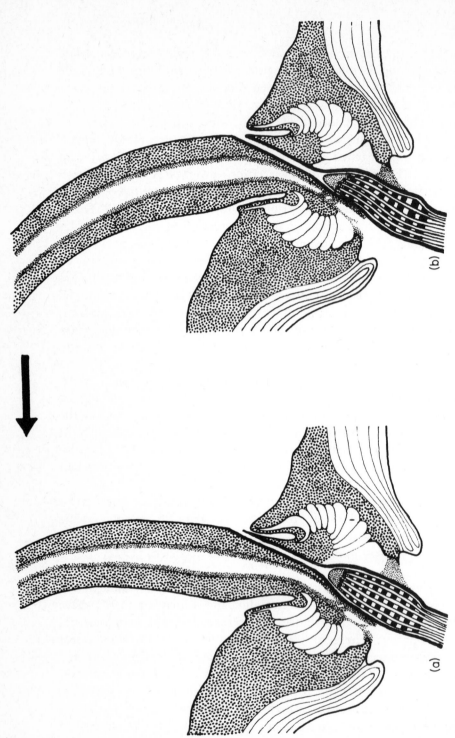

Fig. 1.25 Diagram to show the influence of bending the hair shaft on the microtubular portion of the receptor ending in an insect hair sensillum. Arrow shows direction of movements producing the compressional distortion shown in (b) (cf. Fig. 9.10). (From Gaffal and Hansen, 1972.)

the theory being generally applicable remains to be shown. It does, however, provide an alternative mechanism of stimulation, and it is possible that both situations are found in different examples.

1.5 Innervation

The innervation of receptors is a variable phenomenon. In a relatively small number of cases we can be sure of the total number of mechanically sensitive neurons active at a given site, but in many other cases it is impossible at present to be specific.

1.5.1 Single innervation

The occurrence of but one mechanoreceptive ending has been recorded for campaniform sensilla on the legs, in the halteres, and on the pedicel of insects (Moran, Chapman & Ellis, 1969; Chevalier, 1969; Smith, 1969; Schmidt, 1974). Single sensory units are also reported for hair-plate sensilla (Thurm, 1965), and this may be widespread amongst other forms of hairs on the bodies of insects, whether or not they are part of a proprioceptive arrangement.

Ctenophore stiff cilia (Horridge, 1965), earthworm cuticle cilia (Mill and Knapp, 1971), crustacean hypodermal stretch neurons (Laverack, unpublished observations), and nerve cord sheath receptors (Smith and Page, 1974) may all be considered as singly innervated systems even though in each case multiple parallel neurons may be present in the total population. The dorsal notopodial bristles of *Harmothoë* (Horridge, 1963) are also innervated by one neuron.

1.5.2 Dual innervation

The presence of two dendrites ending at the same sensillum may indicate possible directional sensitivity. For example, the crustacean campaniform sensilla of the leg (Shelton and Laverack, 1968) responds to movement in two directions, with one neuron active for each stimulus direction.

Similar dual directionality is displayed by hair sensilla on the crustacean thorax (Fig. 1.26) (Mellon, 1963; Laverack, 1963). Short hairs on various appendages of Crustacea are heavily supplied with neurons at their base, but little is known of the details of insertion

Fig. 1.26 Record from sensory cells innervating the thoracic receptors of a crayfish. Small spikes appear in response to movement in an anterior direction, large impulses in response to movements in a posterior direction. Lower trace shows a 100 ms rectangular pulse to the mechanical stimulator (From Mellon, 1963).

(Laverack, unpublished observations). It seems possible, however, that double sensory innervation of surface mechanoreceptors (including CAP organs) may be general in Crustacea. This reasoning stems from the observations above coupled with those of Mill and Lowe (1973) on chordotonal organs (Chapter 6). These latter contain numerous receptor cells arranged in parallel. There are both large and small cells (Alexandrowicz, 1972; Hartmann and Boettiger, 1967), and the dendritic endings are encased in scolopidia (Chapter 9) *in pairs*. Each dendrite has different structural properties to its partner (Mill and Lowe, 1973). Since it has been argued here that these chordotonal organs have evolved from hair sensilla (via CAP organs) it may be that each scolopidium represents one hair, and the innervation is consequently double and directionally sensitive. If this is so, CAP organs may be directionally sensitive and give rise to information regarding not only the occlusion of the joint, but also its opening.

Scolopidia with two dendrites inserted in each ending are reported in mayflies. Some scolopidia are innervated by 3 dendrites (Schmidt, 1974), as also are those in Johnston's organ of the termite *Zootermopsis* (Howse, 1968). Dual innervation is also seen in spider slit sensilla (Barth, 1969). One dendrite ends at the level of the cuticle; the other lies deeper. Two neurons are located at the notopodial bristles of large *Harmothoë* (Horridge, 1963) and of *Nereis virens* and *N. diversicolor* and in the neuropodium of *Nereis* (Dorsett, 1964). The numerous bristles of the ventral neuropodium in *Harmothoë* are innervated by several (4 or 5) multipolar sensory neurones (Horridge, 1963).

1.6   Similarity of organization of external sensilla

It is apparent that a requirement for information regarding the position and displacement of the portions of the body has led to the evolution of receptors with common properties.
These are:

(1) The development of either (a) sensilla in the form of projections such as cilia, setae or pegs, or arrangements of neurons associated with accessory projections (e.g. chaetae), or (b) strain detectors such as campaniform or slit sensilla.

(2) The aggregation and concentration of such organs at strategic points on the body surface.

(3) A sensitivity to either (a) movement of another portion of the body, or (b) strains set up due to movement of the body, including muscular contraction.

Two disadvantages of having externally placed detectors for proprioceptive purposes may have placed adaptive significance upon the subsequent development of parallel, internal proprioceptors:

(1) A lack of discrimination between stimulation generated by movement of the body and that generated by external tactile events.
(2) A vulnerability to wear and damage. A superficial placement is bound to expose hairs and pegs to abrasion and other accidents.

1.7   Ultrastructure

As we have mentioned previously (Section 1.5) the neuronal endings of mechanoreceptors may be of several kinds. It is therefore not surprising to discover that the fine structure of such endings also shows a range of types.

We may consider these in an order similar to that adopted in the morphological discussion. First the externally placed receptors.

1.7.1   Cilia

The ciliary ending is the basis of many forms of receptor including a variety of mechanoreceptors. The stiff cilia that project from animal surfaces, and the incorporation of cilia into the innervation of accessory structures such as hairs and bristles, suggests a long evolutionary history of derivation and adaptability for this organelle (see for example, Sleigh, 1974).

Many cilia adapted to a sensory function demonstrate a modified internal and basal structure. Some clues, circumstantial and presumptive, but not at this stage definite, may be deduced from these structures. On the other hand absence or presence of certain portions of the classic cilium do not necessarily give any indication of the mode of function of such cilia.

Protozoan cilia are often dismissed as being all of similar type, but as Tucker (1971) has shown, even small modifications may enable one to postulate different functions for different populations of cilia. Organelles such as the *Euglena* eyespot (Wolken 1961), which are associated with cilia, are fairly well known and may be correlated with a specific sense (in this case photosensitivity), but the clues for mechanoreceptor sensitivity may not be so obvious.

Structural modifications of the shaft of the cilia has been reported many times but it is still not known how the missing or additional parts may assist in the sensory function of such cilia. For example, cilia bearing bladder-like extensions of the cell membrane have been reported (Chapter 11), but most descriptions have concentrated on the occurrence of the internal filaments and their arrangement. Thus the presence of 9 peripheral doublets + 2 central filaments (typical of mobile cilia) is also a feature of some presumptive mechanoreceptors (e.g. Ctenophora, *Leucothea*, Horridge, 1966; Chaetognatha, *Spadella*, Horridge and Boulton, 1967; *Priapulus, Rhynchelmis*, Moritz and Storch, 1971; Nemertini, *Lineus*, Storch and Moritz, 1971) but not of others where 9 + 0 is the pattern (e.g. Crustacea, *Astacus*, statocyst hairs, Schöne and Steinbrecht, 1968; *Carcinus*, chordotonal organs, Mill and Lowe, 1973). Where central filaments are in place, their orientation relative to the direction of stimulus may be important as seems to be the case in motile cilia (Barber, 1974).

Significance has also been attributed to the orientation and organization of the basal regions of cilia. Specializations of the basal body (as in Ctenophora, Horridge, 1966; Hernandez-Nicaise, 1974) have been postulated as significant in membrane current flow patterns resulting from shearing force at the basal body, but no experimental evidence exists on this point. The basal body frequently possesses a side arm or basal foot and in some examples, amongst vertebrates especially, the position has been demonstrated to be correlated with the directionality of the receptor response (e.g. lateral line organs, Flock, 1965). Excitation occurs when movement is towards the basal foot. It may be presumed that directionality is a feature where basal foot structures occur in demonstrated invertebrate mechanoreceptors. It cannot yet be stated that the basal foot *confers* directionality despite the obvious asymmetry its presence affords. Directionality may only be a fortuitous consequence of its presence.

Terminal modifications of ciliary organelles have also been reported and take a number of forms (e.g. as in chordotonal organs; Chapter 8), but those projecting to external regions are perhaps less modified. In cases where the cilium contacts an accessory structure (arthropod hairs) there is frequently a change in internal structure. Thürm (1968), Görner (1965), Schmidt (1973), Moran, Chapman and Ellis (1971) have all commented on the multiplication of numbers of internal filaments in mechanoreceptors.

Modified ciliary endings seem typical of the dendrite terminations of bipolar cells. Amongst arthropods many examples are now known of bipolar mechanosensory cells that end at hair bases or in scolopidia and that take the form of cilia. Annelids, molluscs and echinoderms (Cobb, 1968) all appear to have similar types of endings.

1.7.2   Other endings

Virtually nothing is known of the ultrastructure of the other types of endings discussed in this paper, namely the multipolar cells. By analogy with the findings of Nadol and de Lorenzo (1969) on the abdominal MRO it can be proposed that cilia will not be found, but that various cytoplasmic peculiarities will be present.

1.8   Physiology

The definition we have used throughout this article has followed that of Lissmann (1950) namely that proprioceptors are 'sense organs capable of registering continuously deformations (changes in length) and stresses (tensions, decompressions) in the body. These can arise from the animal's own movements or may be due to its weight or to other external mechanical forces'.

So far we have concerned ourselves only with the morphology, anatomy and funciton of such sense organs. I want briefly now to mention some basic physiological findings. In only a few examples have successful electrophysiological recordings been achieved using external organs. These all fall within the phyla Mollusca, Arthropoda and Annelida. Observations made on these groups may, however, be relevant to other phyla.

Certain basic principles seem well established and will probably be demonstrated in other phyla. These characteristic responses may also be considered as the natural antecedents of the similar, if not identical, activity of internal proprioceptors. If the thesis that many internal receptors may derive from external receptors, is valid, then it would be anticipated that the properties of all mechanoreceptors will be similar. Variety may be expected as a result largely of anatomical rather than physiological attributes.

1.8.1   Response types

Phasic
The phasic properties of units are the attributes of very short latencies and response times and rapid adaptation. These units often show rapid frequency of discharge (see Fig. 1.27) and are common amongst superficially placed mechanoreceptors. These attributes are useful for monitoring dynamic change in conditions, adapting quickly, and then responding to further changes. Small transient influences cause stimulation but long-maintained influences do not produce stimulation beyond the dynamic change. A few examples are: the hairs of crayfish thorax (Mellon, 1963); hairs of scorpion (Sanjeeva-Reddy, 1971); the *T* cells of leeches (Nicholls and Baylor, 1968); accessory setae of crustacean antennules (Laverack, 1964), and body wall neurones of molluscs (Olivo, 1970; Janse, 1974). Such units may be equivalent to movement sensors.

Tonic
Tonic units are characterized by slowly adapting discharges which respond to the static phase of stimulation. Examples are similar to those of phasic units, and include various crustacean hairs, *P* cells of leeches, some molluscan peripheral neurones (Bailey and Laverack, 1966) and the opisthosomal spines of *Limulus* (Eagles, 1973). Final position may be monitored by such receptors.

Phaso-tonic
These are intermediate units with attributes of both phasic and tonic receptors mixed in varying degrees according to site. These responses reflect anatomical and physiological properties of membrane potential, firing thresholds, adaptation rates, visco elastic structure and are better dealt with amongst chordotonal or muscle receptor organs (Nakajima and Onodera, 1969; Chapters 2, 6).

Sensitivity
The sensitivity of mechanoreceptors is a function of all aspects of its structure and physiology. Mechanical advantage conferred by elongate setae, connective tissue strands, cuticle, cupulae, sand grains etc. adds to the total properties of the sensillum not solely those of the receptor cell. Thus Pabst and Kennedy (1967) showed that soft area (hypodermal) receptors have greater sensitivity and a larger receptive field when the overlying cuticle is intact; when it is stripped

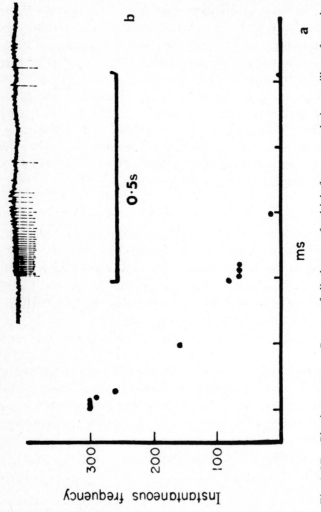

Fig. 1.27   Phasic response. Course of discharge of a high frequency hair sensillum from the antennule of *Panulirus argus*. Frequency is plotted as the reciprocal of the interval between individual impulses. Inset is an electrical record of the discharge (From Laverack, 1964).

away the unit still responds, but not so well, and only over a restricted area. Receptor sensitivity must be related to its site of insertion and ramification and this can be small (as in arthropod hairs) or large (e.g. molluscan body wall, Janse, 1974; and annelid body wall, Prosser, 1935; Nicholls and Baylor, 1968).

Sensitivity is also a question of firing thresholds and the magnitude of stimulation required to reach this threshold. The best invertebrate preparation to demonstrate this is the MRO (Eyzaguirre and Kuffler, 1955) which consists of two neurons of differing levels of excitability. Diagrammatically the situation is shown in Fig. 1.28 (Florey, 1966).

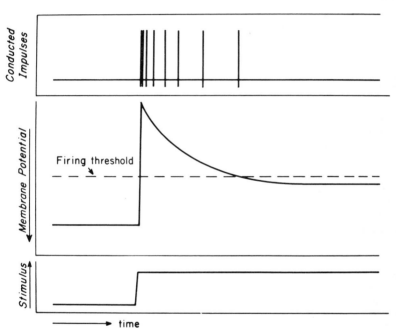

Fig. 1.28   Relation between stimulus, membrane potential at the generator region and impulse frequency. The value of the firing threshold is critical in determining sensitivity (From Florey, 1966).

Directionality

Variablity of response with the direction of stimulus application is notable for many external receptors. Assuming that all receptors in a population have similar structure and physiological properties (although this need not be so) directionality is conferred solely by anatomical restraints. Thus, movements of a seta in one direction may stimulate one neuron, whilst an adjacent unit remains silent, but a diametrically opposed movement may reverse the roles of these units (see Fig. 1.26). These observations indicate that (i) the terminations of the dendrites are symmetrical and opposite, (ii) the effective stimulus for both cells is the same whilst its opposite sign is not (i.e. either compression or stretch occurs but when the stimulus alters direction, and hence switches to the opposite mode, this does *not* stimulate).

Variations on these responses, and multiplication of directions, are found with increasing numbers of sensory units within one organ. So far as is known in the peripheral nervous system no lateral effects of one mechanoreceptor on another occur, but it should be noted that in extended field units in leeches (T, P, N cells) there are complex central inhibitory interactions of cells with similar properties upon one another (Fig. 1.29) and on other segmental mechanoreceptors

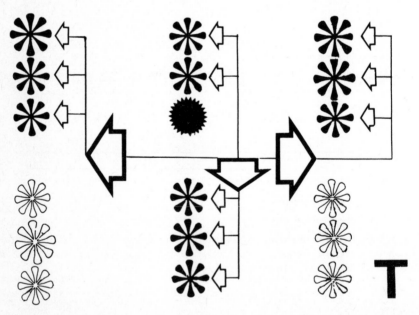

Fig. 1.29   The inhibitory action of 1 stimulated T (tactile) cell of *Hirudo* upon the other (5) T cells in the same ganglion and upon the 3 ipsilateral cells in adjacent ganglia: contralateral cells in are adjacent ganglia are not affected ● stimulated cell; ✳ inhibited cells; ✾ unaffected cells. (Drawn from information in Baylor and Nicholls, 1969.)

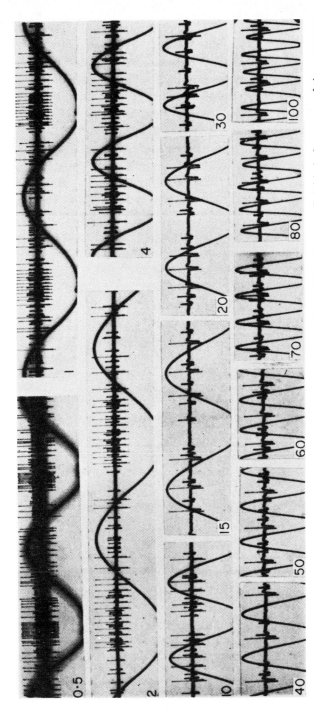

Fig. 1.30   The effects of a water wave at various frequencies, upon the activity of a superficial hair fan organ of the carapace of the lobster *Homarus gammarus.* Synchrony of the burst to the stimulus is notable at all frequencies up to 80 Hz, but becomes erratic at 100 Hz and above. Figure on each trace indicates frequency. (From Laverack, 1963).

(Nicholls and Baylor, 1968). This may also be true for oligochaetes (Günther, 1971).

Vibration

Vibration is a well known stimulus for highly sensitive mechano-receptors and can be shown to influence the behaviour of many animals from many phyla. The sense organs responsible for monitoring vibrational stimuli have to be of high sensitivity.

Regular movements may be treated as transient in nature but more regular movement may come under the heading of vibration (Section 6.8), either heavily damped through the substrate, or as pressure waves (near-field or far-field, van Bergeijk, 1964) through the medium.

Externally placed sensors such as stiff cilia or hair fans respond to vibrational stimuli, and will usually do so in a manner dependent upon the frequency of vibration. Frequency linked discharges have been shown in crustacea (Fig. 1.30) (Laverack, 1963; Taylor, 1967) and annelids (Laverack, 1968). Substrate borne vibrational effects on proprioceptors have been reported for molluscs (Olivo, 1970), crustaceans (Burke, 1954; Horch and Salmon, 1972); insects (Florentine, 1967) and others.

1.9   Summary

The necessity for orientation with regard to the substrate and the surrounding medium is evidently a primitive requirement since it is demonstrated throughout the animal kingdom. Gradual improvement and sophistication of receptor mechanisms, motor responses, and behavioural consequences lead to an evolution of proprioceptive and equilibrium organs that proceed along several lines.

Initially sense organs are superficial in placement. They respond to external influences and are usually tuned to a specific modality. Mechanoreceptors take the form of setae, hairs, stiff cilia and projections from the surface, or alternatively are embedded within the epithelial and dermal layers so that distortion by movement is a potent stimulus mechanism. Projection from the surface and the mechanical linkage with the external medium conveys sensitivity towards movements of the medium and movement of the body relative to the medium. Subsidiary properties of the medium such as propagation of pressure waves, vibration, and currents may all be effective stimuli.

Certain portions of the animal body are more flexible and mobile than others and aggregations of mechanical stress sensitive units occur where these strains are localized and concentrated. Sense *organs* are such localized collections of receptors. They are capable of delivering precise information regarding site of stimulation, direction from which it is delivered, speed of action, and frequency of events. In lowly organisms such as nemertines one might expect rather poorly developed sense organ systems, concerned only with monitoring the distortion of extremely flexible bodies; in annelids the provision of chaetae allows greater concentration of receptors aligned with these lever-like structures, whilst maintaining units within the body wall. Such body wall units are observed also throughout the molluscs (all classes as far as reported). They are probably ubiquitous amongst all groups of Metazoa, more advanced than Porifera.

Superficial projections acting as primary receptors are widespread, but subject to localizations as the process of metamerism and jointing became more complex. Thus in annelids ciliary projections may be widespread but tend to be more numerous at some sites; in molluscs external sensors are virtually undescribed, but in arthropods they are numerous, and localized, especially in advanced forms.

The vulnerability of external organs has, in some cases, led to the withdrawal of superficial units to a more protected internal site, hence the development of certain kinds of internal proprioceptors (chordotonal organs) which are discussed elsewhere in this volume.

Acknowledgements

I am grateful to Drs D. Neil and D. Macmillan for their constructive comments on the manuscript. It is the better for their remarks.

References

Alexandrowicz, J. S. (1933) Innervation des branches de *Squilla mantis*. *Archives de Zoologie Experimentale et générale*, 75, 21–34.

Alexandrowicz, J. S. (1952) Muscle receptor organs in the Paguridae. *Journal of the Marine Biological Association of the United Kingdom*, 31, 277–282.

Alexandrowicz, J. S. (1972) The comparative anatomy of leg proprioceptors in some decapod crustacea. *Journal of the Marine Biological Association of the United Kingdom*, 52, 605–634.

Bailey, D. F. and Laverack, M. S. (1966) Aspects of the neurophysiology of *Buccinum undatum* L. (Gastropoda). *Journal of Experimental Biology*, **44**, 131—148.

Barber, V. C. (1974) Cilia in sense organs. In *Cilia and Flagella*, (ed.) Sleigh, M., Ch. 15, pp. 403—443. Academic Press, New York, and London.

Barth, F. G. (1967) Ein einzelnes Spaltsinnesorgane auf dem Spinnentarsus: seine Erregung in Abhängigkeit von den Parametern des Luftschallreizes. *Zeitschrift für vergleichende Physiologie*, **55**, 407—449.

Barth, F. G. (1969) Die Feinstruktur des Spinneninteguments I. Die cuticula des haufbeins adulter haütungs ferner Tiere (*Cupiennius salei* Keys). *Zeitschrift für Zellforschung*, **97**, 137—159.

Barth, F. G. (1970) Die Feinstruktur des Spinneninteguments. II. Die räumliche Anordnung der Mikrofasern in der lamellierten Cuticula und ihre Beziehung zur gestalt der Porenkanale (*Cupiennius salei* Keys). *Zeitschrift für Zellforschung*, **104**, 87—106.

Barth, F. G. (1971a) Der sensorische Apparat der Spaltsinnesorgane (*Cupiennius salei* Keys, Araneae). *Zeitschrift für Zellforschung*, **112**, 212—246.

Barth, F. G. (1971b) Die Physiologie der Spaltsinnesorgane. I. Modellversuche zur Rolle des cuticularen Spaltes beim Reiztransport. *Journal of Comparative Physiology*, **78**, 415—436.

Barth, F. G. (1972) Die Physiologie der Spaltsinnesorgane. II. Funktionelle Morphologie eines Mechanorezeptors. *Journal of Comparative Physiology*, **81**, 159—186.

Barth, F. G. and Libera, W. (1970) Ein atlas der Spaltsinnesorgane von *Cupiennius salei* Keys. Chelicerata (Araneae). *Zeitschrift für Morphologie der Tiere*, **68**, 343—369.

Baylor, D. A. and Nicholls, J. G. (1969) Chemical and electrical synaptic connections between cutaneous mechanoreceptor neurones in the central nervous system of the leech. *Journal of Physiology*, **203**, 591—609.

Bergeijk, W. A. van (1964) Directional and non-directional hearing in fish. In *Marine Bio-Acoustics*, (ed.) Tavolga, W. N. pp. 281—299, Pergamon Press, Oxford.

Bouligand, Y. (1967) Les soies et les cellules associées chez deux Annèlides Polychètes. *Zeitschrift für Zellforschung*, **79**, 332—363.

Bullock, T. H., Cohen, M. J. and Faulstick, D. (1950) Effect of stretch on conduction in single nerves fibres. *Biological Bulletin, Woods Hole*, **99**, 320.

Burke, W. (1954) An organ for proprioception and vibration sense in *Carcinus maenas*. *Journal of Experimental Biology*, **31**, 127—138.

Camhi, J. (1969a) Locust wind receptors. 1. Transducer mechanics and sensory responses. *Journal of Experimental Biology*, **50**, 335—348.

Camhi, J. (1969b) Locust wind receptors. 2. Interneurons in the cervical connective. *Journal of Experimental Zoology*, **50**, 349—362.

Chapman, K. M. (1965) Campaniform sensilla on the tactile spines of the legs of the cockroach. *Journal of Experimental Biology*, **42**, 191—203.

Chapple, W. D. (1966) Sensory modalities and receptive fields in the abdominal nervous system of the hermit crab, *Pagurus granosimanus* (Stimpson). *Journal of Experimental Biology*, **44**, 209—223.

Chevalier, R. L. (1969) The fine structure of campaniform sensilla on the halteres of *Drosophila melanogaster*. *Journal of Morphology*, **128**, 443–464.

Christian, U. (1971) Zur feinstruktur der Trichobothrium der Winkelspinne *Tegeneria derhami* (Scopoli) (Agenelidae, Araneae). *Cytobiologie*, **4**, 172–185.

Chun, C. (1880) *Die Ctenophoren des Golfes von Neapel, und der augranzenden Meeres-Abschrifte*, Leipzig, Engelmann.

Chu-Wang, I. W. and Axtell, R. C. (1973) Comparative fine structure of the claw sensilla of the soft tick *Argas (Persicargas) arboreus* Kaiser, Hoogstraal and Kohls and a hard tick *Amblyomma americanum*. *Journal of Parasitology*, **59**, 545–555.

Clarac, F. and Vedel, J. P. (1973) Etude électrophysiologique du récepteur hydrodynamique de l'antenne de la Langouste *Palinurus vulgaris*. *Comptes Rendues Academie de Science, Paris*, **276**, 603–606.

Cobb, J. L. S. (1968) The fine structure of the pedicellariae of *Echinus esculentus* (L.). II. The sensory system. *Journal of the Royal Microscopical Society*, **88**, 223–233.

Crisp, D. J. (1974) Factors influencing the settlement of marine invertebrate larvae. In *Chemoreception in Marine Organisms*, (eds.) Grant, P. T. and Mackie, A. M., pp. 177–265, Academic Press, London.

Daly, J. M. (1973) The ability to locate a source of vibrations as a prey-capture mechanism in *Harmothoë imbricata* (Annelida Polychaeta). *Marine Behaviour and Physiology*, **1**, 305–322.

Dando, M. R. and Laverack, M. S. (1969) The anatomy and physiology of the posterior stomach nerve (p.s.n.) in some decapod crustacea. *Proceedings of the Royal Society B*, **171**, 465–482.

Dorsett, D. A. (1964) The sensory and motor innervation of *Nereis*. *Proceedings of the Royal Society B*, **159**, 652–667.

Dumpert, K. (1972) Bau und Verteilung der Sensillen auf der Antennengeisel von *Lasius fuliginosus* (Latr) (Hymenoptera, Formicidae). *Zeitschrift für Morphologie der Tiere*, **73**, 95–119.

Eagles, D. A. (1973) Lateral spine mechanoreceptors in *Limulus polyphemus*. *Comparative Biochemistry and Physiology*, **44A**, 557–575.

Eimer, T. von (1880) Ueber tastapparate bei *Eucharis multicornis*. *Archiv für Mikroskopische Anatomie*, **17**, 342–345.

Eyzaguirre, C. and Kuffler, S. W. (1955) Processes of excitation in the dendrites and in the soma of single isolated sensory nerve cells of the lobster and crayfish. *Journal of General Physiology*, **39**, 87–119.

Fleminger, A. (1973) Pattern, number, variability and taxonomic significance of integumental organs (sensilla and glandular pores) in the genus *Eucalanus* (Copepoda, Calanoida), *Fishery Bulletin*, **71**, 965–1010.

Flock, Å. (1965) Transducing mechanisms in the lateral line canal organ receptors. *Cold Spring Harbor Symposia on Quantitative Biology*, **30**, 133–145.

Florentine, G. J. (1967) An abdominal receptor of the American cockroach, *Periplaneta americana* (L.), and its response to airborne sound. *Journal of Insect Physiology,* 13, 215—218.

Florey, E. (1966) *An Introduction to General and Comparative Animal Physiology.* 713 pp, W. B. Saunders, London.

Foelix, R. F. and Chu-Wang, I. W. (1973) The morphology of spider sensilla. 1. Mechanoreceptors. *Tissue and Cell,* 5, 451—460.

Foelix, R. F. and Axtell, R. C. (1971) Fine structure of tarsal sensilla in the tick *Amblyomma americanum* (L.) *Zeitschrift für Zellforschung,* 114, 22—37.

Gaffal, K. P. and Hansen, K. (1972) Mechanorezeptive strukturen der antennalen haarsensillen der Baumwollanze *Dysdercus intermedius* Dist. *Zeitschrift für Zellforschung,* 132, 79—94.

Goodman, L. J. (1965) The role of certain optomotor reactions in regulating stability in the rolling plane during flight in the desert locust, *Schistocerca gregaria. Journal of Experimental Biology,* 42, 385—407.

Görner, P. (1965) A proposed transducing mechanism for a multiply-innervated mechanoreceptor (Trichobothrium) in spiders. *Cold Spring Harbor Symposia on Quantitative Biology,* 30, 69—73.

Grobstein, P. (1973) Extension sensitivity in the crayfish abdomen. 1. Neurons monitoring nerve cord length. *Journal of Comparative Physiology,* 86, 331—348.

Günther, J. (1970) On the organisation of the exteroceptive afferences in the body segments of the earthworm. *Verhandlungen der Deutschen Zoologischen Gesellschaft,* 64, 261—265.

Günther, J. (1971) Mikroanatomie des Bauchmarks von *Lumbricus terrestris* L. (Annelida, Oligochaeta). *Zeitschrift für Morphologie der Tiere,* 70, 141—182.

Hartmann, H. B. and Boettiger, E. G. (1967) The functional organization of the propus-dactylus organ in *Cancer irroratus* Say. *Comparative Biochemistry and Physiology,* 22, 651—663.

Haskell, P. T. (1959) Hair receptors in locusts. *Nature,* 183 1106—1107.

Hepburn, H. R. Joffe, I., Green, N. and Nelson, K. J. (1975) Mechanical properties of a crab shell. *Comparative Biochemistry and Physiology,* 50A, 551—554.

Hernandez-Nicaise, M-L. (1974) Ultrastructural evidence for a sensory motoneuron in ctenophora. *Tissue and Cell,* 6, 43—58.

Hertwig, R. (1880) Ueber den Bau der Ctenophoren. *Jenaische Zeitschrift für Naturwissenschaften,* 14, 393—457.

Horch, K. W. and Salmon, M. (1972) Production, perception and reception of acoustic stimuli by semiterrestrial crabs (Ocypode & Uca, family Ocypodidae). *Forma et Functio,* 1, 1—25.

Horridge, G. A. (1963) Proprioceptors, bristle receptors, efferent sensory impulses, neurofibrils and number of axons in the parapodial nerve of the polychaete *Harmothoë. Proceedings of the Royal Society B,* 157, 199—222.

Horridge, G. A. (1966) Non-motile sensory cilia and neuromuscular junctions in a ctenophore independent effector organ. *Proceedings of the Royal Society B,* 162, 333—350.

Horridge, G. A. and Boulton, P. S. (1967) Prey detection by Chaetognatha via a vibration sense. *Proceedings of the Royal Society B,* **168**, 413–419.

Howse, P. E. (1968) The fine structure and functional organization of chordotonal organs. *Symposium of the Zoological Society of London,* **23**, 167–198.

Hughes, G. M. and Wiersma, C. A. G. (1960) Neuronal pathways and synaptic connections in the abdominal nerve cord of the crayfish. *Journal of Experimental Biology,* **37**, 291–307.

Janse, C. (1974) A neurophysiological study of the peripheral tactile system of the pond snail, *Lymnaea stagnalis* (L.) *Netherlands Journal of Zoology,* **24**, 93–161.

Josephson, R. K. (1961) The response of a hydroid to weak water-borne disturbances. *Journal of Experimental Biology,* **38**, 17–27.

Kaston, B. J. (1935) The slit sense organs of spiders. *Journal of Morphology,* **58**, 189–210.

Kennedy, D. (1974) Connections among neurons of different types in crustacean nervous systems. *The Neurosciences. IIIrd study programme,* (eds.) Schmitt, F. O. and Worden, F. G. pp. 379–388. MIT Press, Cambridge, Mass.

Kennedy, D., Evoy, W. H. and Fields, H. L. (1966) The unit basis of some crustacean reflexes. *Symposia of the Society for Experimental Biology,* **20**, 75–109.

Kinosita, H. and Kamada, T. (1939) Movement of abfrontal cilia of *Mytilus. Japanese Journal of Zoology,* **8**, 291–310.

Knapp, M. F. and Mill, P. J. (1971) The fine structure of ciliated sensory cells in the epidermis of the earthworm *Lumbricus terrestris. Tissue and Cell,* **3**, 623–636.

Kristan, W. B. Jr. (1974) Neural control of swimming in the leech. *American Zoologist,* **14**, 991–1001.

Laverack, M. S. (1962) Responses of cuticular sense organs of the lobster, *Homarus vulgaris* (Crustacea). II. Hair fan organs as pressure receptors. *Comparative Biochemistry and Physiology,* **6**, 137–145.

Laverack, M. S. (1963) Responses of cuticular sense organs of the lobster, *Homarus vulgaris* (Crustacea). III. Activity invoked in sense organs of the carapace. *Comparative Biochemistry and Physiology,* **10**, 261–272.

Laverack, M. S. (1964) The antennular sense organs of *Panulirus argus. Comparative Biochemistry and Physiology,* **13**, 301–321.

Laverack, M. S. (1968) On the receptors of marine invertebrates. *Oceanography & Marine Biology, An Annual Review,* (ed.) H. Barnes, **6**, pp. 249–324, Allen & Unwin, London.

Laverack, M. S. (1969) Mechanoreceptors, photoreceptors and rapid conduction pathways in the leech *Hirudo medicinalis. Journal of Experimental Biology,* **50**, 129–140.

Laverack, M. S. and Bailey, D. F. (1963) Movement receptors in *Buccinum undatum. Comparative Biochemistry and Physiology,* **8** 289–298.

Lissmann, H. W. (1950) Proprioceptors. *Symposia of the Society for Experimental Biology,* **4**, 34–59.

Lowe, D. G., Mill, P. J. and Knapp, M. F. (1973) The fine structure of the PD proprioceptor of *Cancer pagurus*. II. The position sensitive cells. *Proceedings of the Royal Society B*, **184**, 199–205.

McIndoo, N. E. (1911) The lyriform organs and tactile hairs of araneids. *Proceedings of the Academy of Natural Science, Philadelphia*, **63**, 375–418.

Markl, H. (1966) Schwerkraftdressuren an Honigbienen. II. Bie Rolle der Schwererezeptorischen borstenfelder verscheidener gelenke für die Schwerekompassiorientierung. *Zeitschrift für vergleichende Physiologie*, **53**, 353–371.

Masson, C. (1972) Organisation sensorielle des principales articulations de l'antenne de la Fourni *Campanotus vagus* Scop. (Hymenoptera, Formicidae). *Zeitschrift für Morphologie der Tiere*, **73**, 343–359.

Mellon, de F. (1963) Electrical responses from dually innervated tactile receptors on the thorax of the crayfish. *Journal of Experimental Biology*, **40**, 137–148.

Mellon, de F. and Kennedy, D. (1964) Impulse origin and propagation in a bipolar sensory neuron. *Journal of General Physiology*, **47**, 487–499.

Mill, P. J. & Lowe, D. G. (1973) The fine structure of the PD proprioceptors of *Cancer pagurus*. 1. The receptor strand and the movement sensitive cells. *Proceedings of the Royal Society B*, **174**, 179–197.

Moran, D. T. (1971) Loss of the sensory process of an insect receptor at ecdysis. *Nature*, **234**, 476–477.

Moran, D. T., Chapman, K. M. and Ellis, R. A. (1971) The fine structure of cockroach campaniform sensilla. *Journal of Cell Biology*, **48**, 155–173.

Moritz, K. and Storch, V. (1971) Elektronenmikroskopische untersuchung eines Mechanorezeptors von everterbraten (Priapuliden, Oligochaeten). *Zeitschrift für Zellforschung*, **117**, 226–234.

Murphy, R. K. and Mendenhall, B. (1973) Localization of receptors controlling orientation to prey by the back swimmer *Notonecta undulata*. *Journal of Comparative Physiology*, **84**, 19–30.

Nadol, J. B. Jr. and de Lorenzo, A. J. D. (1969) Observations on the organization of the dendritic processes and receptor terminations in the abdominal muscle receptor organ of *Homarus*. *Journal of Comparative Neurology*, **137**, 19–58.

Nakajima, S. and Onodera, K. (1969) Membrane properties of the stretch receptor neurone of crayfish with particular references to mechanisms of sensory adaptation. *Journal of Physiology*, **200**, 161–185.

Nicholls, J. G. and Baylor, D. A. (1968) Specific modalities and receptive fields of sensory neurons in the CNS of the leech. *Journal of Neurophysiology*, **31**, 740–756.

Nicklaus, R. (1965) Die Erregung einzelner Fadenhaare von *Periplaneta americana* in Abhangigkeit von der Grösse und Richtung der Auslenkung. *Zeitschrift für vergleichende Physiologie*, **50**, 331–362.

Ockelmann, C. W. and Vahl, O. (1970) On the biology of the polychaete *Glycera alba*, especially its burrowing and feeding. *Ophelia*, **8**, 275–294.

Olivo, R. F. (1970) Mechanoreceptor function in the razor clam; sensory aspects of the foot withdrawal reflex. *Comparative Biochemistry and Physiology*, **35**, 761–786.

Pabst, H. and Kennedy, D. (1967) Cutaneous mechanoreceptors influencing motor output in the crayfish abdomen. *Zeitschrift für vergleichende Physiologie*, **57**, 190–208.

Pilgrim, R. L. C. (1960) Muscle receptor organs in some decapod crustacea. *Comparative Biochemistry and Physiology*, **1**, 248–257.

Pringle, J. S. W. (1938a) Proprioception in insects. I. A new type of mechanical receptor from the palps of the cockroach. *Journal of Experimental Biology*, **15**, 101–113.

Pringle, J. W. S. (1938b) Proprioception in insects. II. The action of the campaniform sensilla on the legs. *Journal of Experimental Biology*, **15**, 114–131.

Pringle, J. W. S. (1955) The function of the lyriform organs of arachnids. *Journal of Experimental Biology*, **32**, 270–278.

Pringle, J. W. S. (1961) Proprioception in arthropods. In *The Cell and the Organism*. (eds.) Ramsay, J. A. and Wigglesworth, V. B., pp. 256–282, Cambridge University Press.

Prosser, C. L. (1935) Impulses in the segmental nerves of the earthworm. *Journal of Experimental Biology*, **12**, 95–104.

Rice, M. J., Galun, R. and Finlayson, L. H. (1973) Mechanotransduction in insect neurons. *Nature, New Biology*, **241**, 286–288.

Salpeter, M. M. and Walcott, C. (1960) An electron microscopical study of a vibration receptor in the spider. *Experimental Neurology*, **2**, 232–250.

Sanjeeva-Reddy, P. (1971) Function of the supernumerary sense cells and the relationship between modality of adequate stimulus and innervation pattern of the scorpion hair sensillum. *Journal of Experimental Biology*, **54**, 233–238.

Santer, R. M. and Laverack, M. S. (1971) Sensory innervation of the tentacles of the Polychaete, *Sabella pavonina*. *Zeitschrift für Zellforschung*, **122**, 160–171.

Schmidt, K. (1973) Comparative morphology of insect mechanoreceptors. *Verhandlungen der Deutschen Zoologischen Gesellschaft*, **66**, 15–25.

Schmidt, K. (1974) Die Mechanorezeptoren im Pedicellus der Eintagsfliegen (Insecta, Ephemeroptera). *Zeitschrift für Morphologie der Tiere*, **78**, 193–220.

Schöne, H. and Steinbrecht, R. A. (1968) Fine structure of statocyst receptors of *Astacus fluviatilis*. *Nature*, **220**, 184–186.

Shelton, R. G. J. and Laverack, M. S. (1968) Observations on a redescribed crustacean cuticular sense organ. *Comparative Biochemistry and Physiology*, **25**, 1049–1059.

Shelton, R. G. J. and Laverack, M. S. (1970) Receptor hair structure and function in the lobster *Homarus gammarus* (L.). *Journal of Experimental Marine Biology and Ecology*, **4**, 201–210.

Shepheard, P. R. B. (1973) Musculature and innervation of the neck of the desert locust, *Schistocerca gregaria* (Forskál). *Journal of Morphology*, **139**, 439–464.

Shepheard, P. R. B. (1974) Control of head movements in the locust, *Schistocerca gregaria*. *Journal of Experimental Biology*, **60**, 735–767.

Sherrington, C. S. (1906) On the proprioceptive system, especially in its reflex aspect. *Brain*, **29**, 467–482.

Sleigh, M. A. (ed.) (1974) *Cilia and Flagella*. Academic Press, London.

Sleigh, M. A. and McDonald, A. (ed.) (1972) The effects of pressure on organisms. *Symposia of the Society for Experimental Biology*, **26**, pp. 516, Cambridge University Press, Cambridge.

Slifer, E. H. and Sekhon, S. S. (1973) Sense organs on the antennal flagellum of two species of Embioptera (Insecta). *Journal of Morphology*, **139**, 211–226.

Smith, D. S. (1969) The fine structure of haltere sensilla in the blowfly *Calliphora erythrocephala* (Meig.) with scanning electron microscopic observations on the haltere surface. *Tissue and Cell*, **1**, 443–484.

Smith, P. H. and Page, C. H. (1974) Nerve cord sheath receptors activate the large fibre system in the leech. *Journal of Comparative Physiology*, **90**, 311–320.

Spencer, H. J. (1974) Analysis of the electrophysiological responses of the trochanteral hair receptors of the cockroach. *Journal of Experimental Biology*, **60**, 223–240.

Storch, V. and Moritz, K. (1971) Zur Feinstruktur der Sinnesorgane von *Lineus ruber* O. F. Müller (Nemertini, Heteronemertini). *Zeitschrift für Zellforschung*, **117**, 212–225.

Taylor, R. C. (1967) Functional properties of the chordotonal organ in the antennal flagellum of a hermit crab. *Comparative Biochemistry and Physiology*, **20**, 719–729.

Thurm, U. (1965) An Insect mechanoreceptor. 1. Fine structure and adequate stimulus. *Cold Spring Harbor Symposia on Quantitative Biology*, **30**, 75–82.

Thurm, U. (1968) Steps in the transducer process of mechanoreceptors. *Symposia of the Zoological Society of London*, **23**, 199–216.

Tucker, J. B. (1968) Fine structure and function of the cytopharyngeal basket in the ciliate *Nassula*. *Journal of Cell Science*, **3**, 493–514.

Tucker, J. B. (1971) Development and deployment of cilia, basal bodies, and other microtubular organelles in the cortex of the ciliate, *Nassula*. *Journal of Cell Science*, **9**, 539–567.

Walcott, C. (1969) A spider's vibration receptor: its anatomy and physiology. *American Zoologist*, **9**, 133–144.

Wales, W., Clarac, F., Dando, M. R. and Laverack, M. S. (1970) Innervation of the receptors present at the various joints of the pereiopods and third maxilliped of *Homarus gammarus* (L.) and other Macruran Decapods. *Zeitschrift für vergleichende Physiologie*, **68**, 345–384.

Wendler, G. (1966) The co-ordination of walking movements in arthropods. *Symposia of the Society for Experimental Biology*, **20**, 229–250.

Whitear, M. (1962) The fine structure of crustacean proprioceptors. 1. The chordotonal organs in the legs of the shore crab, *Carcinus maenas.* *Philosophical Transactions of the Royal Society B*, **245**, 291—325.

Wiersma, C. A. G. (1959) Movement receptors in decapod crustacea. *Journal of the Marine Biological Association of the United Kingdom*, **38**, 143—152.

Wiese, K. (1972) Das mechanorezeptorische Beuteortungssystem von *Notonecta.* 1. Die Funktion de tarsalen Scolopidial organs. *Journal of Comparative Physiology*, **78**, 83—102.

Wigglesworth, V. B. (1961) The epidermal cell. In *The Cell and the Organism,* (eds.) Ramsay, J. A. and Wigglesworth, V. B. pp. 127—143.

Wilson, A. H. and Sherman, R. G. (1975) Mapping of neuron somata in the thoracic nerve cord of the lobster using cobalt chloride. *Comparative Biochemistry and Physiology*, **50A**, 47—50.

Wolken, J. J. (1961) *Euglena. Institute of Microbiology, Rutgers University.* pp. 173.

Wyse, G. (1971) Receptor organization and function in *Limulus* chelae. *Zeitschrift für vergleichende Physiologie*, **73**, 249—273.

2

Crustacean abdominal and thoracic muscle receptor organs

H. L. FIELDS

2.1   Introduction

As is the case with most arthropods, the decapod crustaceans are highly mobile. Many species of decapods must move through aquatic and terrestrial environments which differ markedly in viscosity and density. These environmental differences add to the complexity of neural control of even simple movements. In addition to the special problems created by the external environment, precise control of the position of multiply-jointed body parts must be accomplished in the face of such variables as muscle fatigue, contraction of syngergist and antagonist muscles and applied disturbances such as might be encountered, for example, in lifting a large struggling prey. Although there may be more than one possible approach to the solution of the problem of precise motor control, the relatively small number of neurons available to arthropods demands that a relatively simple approach be taken. In fact, present knowledge indicates that the Crustacea possess a simple yet effective device to control muscle length in the face of variable load: the muscle receptor organ. This device is part of a control system whose principle of operation is similar to the vertebrate length control servo, which employs the muscle spindle.

Our knowledge of the physiology of muscle receptors and the control system of which they are a part is most complete for the abdominal muscle receptor organ of decapod crustacea. This chapter describes what is known about the anatomy and physiology of this crustacean muscle receptor organ (MRO).

2.2   Occurrence and Location

Receptors associated with the dorsal musculature of the abdomen have been found in most species of decapod Crustacea (Pilgrim, 1960). Table 2.1 shows that most groups examined, with the notable exception of Brachyura-Brachygnatha, possess MROs. These receptors are similar in their structure and their response to applied stretch in the numerous species examined.

The MROs occur in pairs on either side of the midline in each abdominal segment (Fig. 2.1). Similar receptors are found in thoracic segments, but their occurrence is variable and seems to be related to the degree of thoracic flexibility. Thus, in decapods with unfused,

Table 2.1   Crustacea in which abdominal muscle receptor organs have been determined. Symbols in the third column indicate that the paper referred to includes examination which is physiological (p), anatomical (a), or histological (h). (From Pilgrim, 1960)

Species	Type of organ	Method of examination
EUCARIDA—DECAPODA		
Natantia		
Caridea		
Pandalidae:		
Pandalus danae Stimpson	RM_1 RM_2	(p) (a)
Pandalopsis dispar Rathbun	RM_1 RM_2	(p) (a)
Cragonidae:		
Paracrangon echinata Dana	RM_1 RM_2	(p) (a)
Crago franciscorum	RM_1	(p)
Hippolytidae:		
Eualus sp. (Stimpson)	RM_1	(p)
Palaemonidae:		
Leander serratus (Pennant)	RM_1 RM_2	(a) (h)

Continued

Table 2.1—Continued

Species	Type of organ	Method of examination
Reptantia—Palinura		
Scyllaridea		
Palinuridae:		
Palinurus vulgaris L.	RM_1 RM_2	(a) (h)
Panulirus interruptus Randall	RM_1 RM_2	(p) (a)
Astacura		
Nephropsidea		
Astacidae:		
Astacus fluviatilis L.	RM_1 RM_2	(p) (a) (h)
A. leptodactylus Eschscholtz	RM_1 RM_2	(p)
A. trowbridgei Stimpson	RM_1	(p)
Procambarus clarkii (Girard)	RM_1 RM_2	(p) (a)
		(p)
P. alleni (Faxon)	RM_1 RM_2	(p) (a)
Orconectes virilis (Hagen)	RM_1 RM_2	(p)
Nephropsidae:		
Homarus vulgaris L.	RM_1 RM_2	(a) (h)
H. americanus M—Edwards	RM_1 RM_2	(p) (a)
Anomura		
Galatheidea		
Galatheidae:		
Munida quadrispina Benedict	RM_1 RM_2	(p) (a)
Thalassinidea		
Callianassidae:		
Upogebia pugettensis (Dana)	RM_1 RM_2	(p) (a)
Callianassa gigas Dana	RM_1 RM_2	(p) (a)
Paguridea		
Paguridae:		
Pagurus aleuticus (Benedict)	RM_1 RM_2	(p) (a)
P. alaskensis (Benedict)	RM_1 RM_2	(p) (a)
P. kennerlyi (Stimpson)	RM_1 RM_2	(p) (a)
P. calidus (Risso)	RM_1 RM_2	(a) (h)
P. striatus (de Latreille)	RM_1 RM_2	(a) (h)
Eupagurus prideauxi (Leach)	RM_1 RM_2	(a)
Brachyura-Brachygnatha		
Cancridae:		
Cancer magister Dana	Absent	(p) (a)
HOPLOCARIDA—STOMATOPODA		
Squillidae:		
Squilla mantis (Rondel)	RM_1 RM_2	(a) (h)

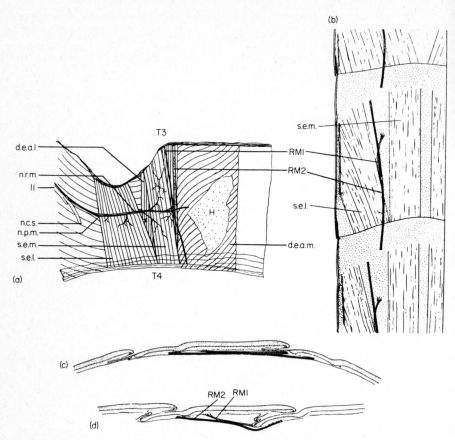

Fig. 2.1   The muscle receptor organs in the abdomen of crayfish and lobster. Representative left hemisegments. (a) Superficial view of the left extensor musculature of 3rd abdominal segment of *Procambarus clarkii.* (b) Looking dorsalward, with the flexor and deep extensor musculature removed in 2nd abdominal segment of *Homarus gammarus.* Note that the receptor muscles in *Homarus* lie between the medial and lateral heads of the superficial extensors, whereas in crayfish the receptor muscles are medial to the medial head. (c) Diagrammatic sketch of saggital sections through the abdomen in plane of the receptor muscles in *Homarus.* Top—the abdomen fully flexed; bottom—abdomen extended. RM1, RM2—Receptor muscles; d.e.a.m.—medial head of deep extensor; d.e.a.1.—lateral head of deep extensor; n.r.m.—dorsal nerve; n.p.m.—nerve to the deep extensors; n.c.s.—superficial nerve; H—patch of hypodermis; II—second nerve root; T3—tergite of 3rd abdominal segment; T4—tergite of 4th abdominal segment. ((a) from Fields, 1966; (b, c) from Alexandrowicz, 1951.)

mobile thoracic segments, MROs are found which are similar to those in abdominal segments (Alexandrowicz, 1954); whereas, in the fused thoracic segments, receptors are either absent or appear to be incomplete (Alexandrowicz, 1967). These will be discussed in a later section of this chapter (Section 2.7).

Most of the descriptive information given in this chapter derives from observations of the abdominal MROs of species of Palinura and Astacura.

2.3   Anatomy and physiology of the extensor musculature

To appreciate the anatomy and function of the MROs, knowledge of neighbouring extensor musculature is crucial. Some of the moto-neurons to these extensor muscles also supply the muscles of the MROs, and it is the contraction of these extensor muscles that is regulated in part by the MROs.

2.3.1   Fast and slow muscle fibre types

In contrast to vertebrate skeletal muscles, which respond to impulses in their motor nerves with a brief twitch, crustacean muscle responses to motor nerve impulses fall into two broad and somewhat overlapping categories: fast (or phasic) and slow (or tonic). (e.g. Atwood, 1963; Atwood and Dorai Raj, 1964; Kennedy and Takeda, 1965a, b). Crustacean fast muscle is distinguished histologically from slow, by its shorter sarcomere length and more even distribution of myofibrils when viewed in cross-section (Fig. 2.2c, d). It is capable of rapid twitches, lasting generally less than 100 ms, in response to single nerve stimuli. Generally, fast muscle cannot sustain constant tension for long periods without fatigue. Intracellular recording from fast muscle reveals that, similar to the membrane of vertebrate skeletal muscle, it is capable of producing electrogenic responses to threshold depolarization (Fig. 2.2f). As in the vertebrate neuro-muscular junction, these electrogenic responses generally appear to arise from the transmitter-induced depolarization. They differ, however, from the action potential of nerve and vertebrate muscle in that they may be graded and non-propagated. Maximal twitch tension in fast crustacean muscles is usually associated with such electrogenic responses.

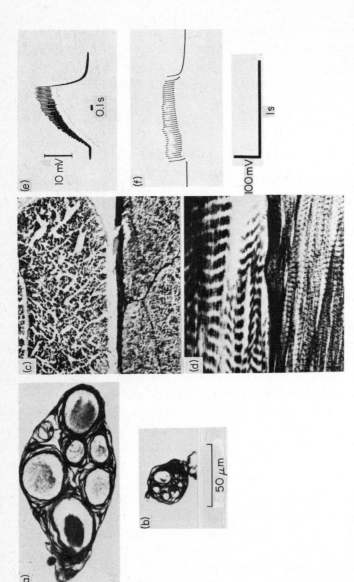

Fig. 2.2   Comparison of crustacean fast and slow muscles. The example used is the abdominal flexor system and these photographs demonstrate general contrasting features. (a) Cross-section of the branch of the third nerve root of an abdominal ganglion which supplies the fast (deep) flexor muscles. (b) Nerve branch supplying the tonic (superficial) flexors at the same magnification as in (a), showing the smaller diameter of the tonic motor axons. (c) Cross-section of slow flexor (above) and fast flexor (below) muscles, demonstrating coarse, clumped appearance of myofibrils in the former (d) Longtitudinal section of the slow flexor (above) and fast flexor (below) muscles, demonstrating the longer sarcomere length in the former (e) Intracellular recording from a slow abdominal flexor muscle fibre showing facilitation of excitatory junctional potentials. Stimulation of motor nerve at 40 Hz. Note summation and marked facilitation at this frequency. (f) Recording from a fast extensor muscle fibre, stimulating its nerve also at 40 Hz. Note lack of facilitation and minimal summation at this frequency. Also note difference in time and voltage scale. (a–d, from Kennedy and Takeda, 1965a; e, from Kennedy and Takeda, 1965b.)

In contrast, crustacean slow muscles have long sarcomeres and the myofibrils in cross-section appear to be clumped (Fig. 2.2c, d). Single stimuli to their motor nerves often produce little or no tension. When tension is produced by a train of impulses, its onset is gradual and it may be maintained for several seconds following the end of the stimulus train. Intracellular recordings from slow muscle fibres reveals a depolarization with a slower rise and decay time and generally lower amplitude than that of fast muscle. Electrogenic responses are rarely observed in slow muscles. However, with high frequency stimulation of their motor nerve, a rapid and sometimes marked increase in amplitude (facilitation) is commonly observed (Fig. 10.2e) and is correlated with a proportional tension increase.

2.3.2   Superficial and deep extensor musculature

(Kennedy and Takeda, 1965a, b; Fields, Evoy and Kennedy, 1967; Pilgrim and Wiersma, 1963)

The abdominal musculature exemplifies a functional differentiation between tonic and phasic systems. The tonic muscles, both extensors and flexors, are arranged in distinct and separate superficial sheets only a few fibres thick. In contrast, the phasic extensors and flexors are large muscles, comprising the bulk of the abdominal mass, and they are located deep to the tonic muscles. Fig. 2.3 illustrates the arrangement of the extensor musculature. It can be seen that both MROs are located superficially with the tonic muscles. The deep musculature consists of two major divisions, the lateral (d.e.a.l.), consisting of fibres orientated parallel to the long axis of the abdomen, and medial (d.e.a.m.), whose fibres have a spiral arrangement. D.e.a.l. and d.e.a.m. have the histological and physiological properties of fast muscles (Abbott and Parnas, 1965). The superficial extensor muscles run from the tergite in the segment of origin posteriorly to insert on the articular membrane or the anterior tergal ridge of the neighbouring caudal segment. These superficial muscles have the histological and physiological properties of tonic muscles.

Thus the abdominal extensor musculature can conveniently be divided into a massive deep phasic system and a delicate superficial tonic system. The phasic system is brought into action only during the brief stereotyped escape response. The tonic system is capable of smoothly adjusting the abdomen through a continuous range of positions (e.g. Kennedy, Evoy and Fields, 1966).

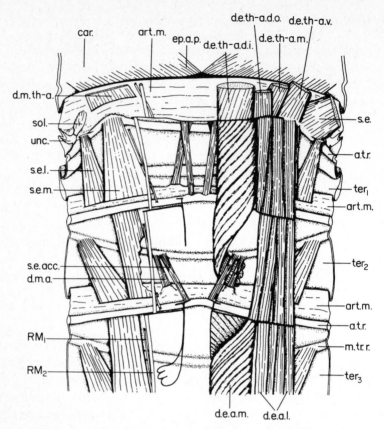

Fig. 2.3 View dorsally from ventral dissection of anterior abdomen and thoraco-abdominal joint in crayfish. On the left the deep extensor muscles have been removed. On the right the deep extensor muscle (median head) has been removed entirely from the second abdominal segment; the origin and insertion areas of the corresponding head on the left side of the figure are shown as thick lines.

a.t.r.—anterior tergal ridge (abdominal); a.tr.—anterior transverse muscle; art.m.—articular membrane; car.—carapace; d.e.a.l.—deep extensor muscle (abdominal), lateral head; d.e.a.m.—deep extensor muscle (abdominal), medial head; d.e.th-a.d.i.—deep extensor muscle (thoraco-abdominal), dorsal head—inner slip; d.e.th-a.d.o.—deep extensor muscle (thoraco-abdominal), dorsal head, outer slip; d.e.th-a.m.—deep extensor muscle (thoraco-abdominal), medial head; d.e.th-a.v.—deep extensor muscle (thoraco-abdominal), ventral head; d.m.a.—dorsal membrane muscle (abdominal); d.m.th-a.—dorsal membrane muscle (thoraco-abdominal); m.tr.r.—medial transverse ridge (of tergite); RM_1—slowly-adapting receptor muscle (abdominal); RM_2—fast-adapting receptor muscle (abdominal); s.e.—superficial extensor muscle (thoraco-abdominal); s.e.acc.—superficial extensor muscle (abdominal), accessory head; s.e.l.—superficial extensor muscle (abdominal), lateral head; s.e.m.—superficial extensor muscle (abdominal), medial head; sol.—solea (of secula); ter_{1-3}—first-third tergites (of abdomen); unc.—uncus (of secula); (From Pilgrim and Wiersma, 1963).

2.4   Anatomy of the receptor organs

2.4.1   Location

The gross anatomy and histology of the muscle receptor organs were first described for the lobsters *Homarus vulgaris* and *Palinurus vulgaris* by Alexandrowicz (1951). As pointed out above, in most decapod crustaceans that have been studied, two pairs of receptors are found on each side of each abdominal segment. In the lobster the MROs lie together, between and parallel with the two heads of the superficial extensor muscle whereas, in the crayfish, the organs lie medial to the medial head (Fig. 2.3, s.e.m.) (Florey and Florey, 1955; Wiersma, Furshpan and Florey, 1953).

2.4.2   Receptor muscles

The lateral receptor muscle (RM1) originates behind the anterior tergal ridge mediad to the superficial extensor's medial head and inserts on the anterior tergal ridge of the next posterior segment (Fig. 2.3). In the more posterior segments the origin is more posterior within the segment and thus the receptor muscle is shorter.

The medial receptor muscle (RM2) is more ventral, originating on the posterior face of the transverse inscription immediately dorso-mediad to the lateral head of the deep extensor muscle. Its insertion is actually lateral to that of RM1 on the anterior face of the next posterior transverse inscription. The deep, but not the superficial, extensors are also attached to these inscriptions.

On a given side in one segment, the two receptors differ from one another in size, structure and innervation pattern. Each MRO consists of a thin muscle bundle, distinct from the adjacent extensor musculature, and a neuron with a peripheral cell body the dendrites of which penetrate the muscle belly, generally towards the caudal end of the segment (Fig. 2.4). The lateral receptor muscle (RM1) is consistently shorter, thinner and more taut at a given degree of abdominal flexion, than the medial one (RM2).

In both receptor muscles of the lobster there is an intercalated tendinous region where the dendrites of the sensory neurons penetrate. Such tendinous zones are not obvious in the crayfish MRO, although electron-microscopic studies have demonstrated a corresponding region of increased connective tissue and relative muscle

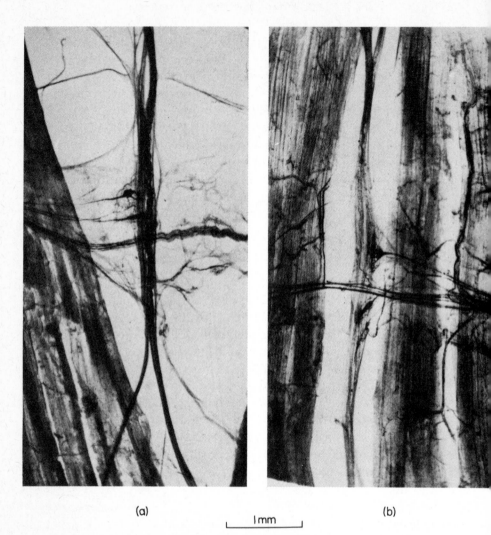

(a) |—— 1mm ——| (b)

Fig. 2.4   Muscle receptor organs in (a) 2nd and (b) 1st abdominal segments of *Homarus gammarus*. The MROs are displaced medially from the edge of s.e.l. (Fig. 2.1b) for clarity. The dorsal nerve can be seen crossing at right angles to the receptor muscles and sending a branch to innervate the receptor cell and muscles. (From Alexandrowicz, 1951.)

exclusion (Bodian and Bergman, 1962; Nadol and DeLorenzo, 1968). This intercalated connective tissue region is more prominent in RM2.

The histological features of the two muscles present a sharp contrast. In the thin lateral muscle (RM1) the fibrils are larger, more regularly arranged, and have a longer sarcomere length than those in the thick medial muscle (RM2). These histological features are paralleled by physiological differences between the receptor muscles similar to those described above (Section 2.31; Fig. 2.2). As will be discussed below, the thin receptor muscle (RM1) is a tonic muscle innervated by branches of the motoneurons supplying the tonic superficial extensors; while the thicker medial receptor muscle (RM2) is a phasic muscle supplied by branches of motoneurons innervating the fast deep extensor musculature.

RM2 contracts in a twitch-like manner to single stimuli. This rapid development of tension is usually effective in causing a sensory discharge in RM2, whereas gradual tension development often fails to cause a sensory discharge. Complete fusion of twitches in RM2 occurs only at rates of stimulation greater than 50 per second. Single impulses in the motor nerve at lower frequencies may evoke transient changes in the sensory discharge of the receptor (Section 2.6.1; Fig. 2.11b).

On stimulation of the dorsal nerve (Fig. 2.1; n.r.m.), which contains motoneurons to both RMs, the thin, lateral RM1 contracts slowly, in a graded manner, and effective contraction usually requires trains of impulses (Section 2.6.1; Fig. 2.11a). Trains of impulses in the motor nerve to RM1 produce a rise in the afferent discharge frequency which reaches a maximum directly related to stimulus frequency. Impulses in the motor nerve at frequencies as low as one per second can eventually summate to produce an afferent discharge in the slowly adapting neuron (Section 2.6.1, Fig. 2.11a); complete fusion of contractions occurs at a frequency of five per second.

Intracellular recording from the muscle elements of the receptors was undertaken by Kuffler (1954). He found only one type of junctional event in RM1: a junctional potential of 5–15 mV recorded upon nerve stimulation, regardless of the site of penetration. In RM2, two types of junctional event were recorded. There are excitatory junctional potentials of similar shape and amplitude to, but faster time course than, those observed in the slow bundle. In addition, a secondary electrogenic response was noted. This electrogenic response, which is characteristic of crustacean fast muscle, is found to be associated with firing of the sensory neuron. The muscle spike in RM2 is neither propagated, nor is it all-or-none, but rather it is a local event, the amplitude of which varies with electrode placement and with the pattern of motor nerve stimulation (Kuffler, 1954; Furshpan, 1955).

Thus, RM1 and RM2 fall into the categories of slow and fast muscles respectively, which is consistent with the histological features mentioned above. These receptor muscle properties are paralleled by the responses of the sensory neurons to which they are attached. This will be discussed in Section 2.5.

2.4.3   Innervation

Two studies (Alexandrowicz, 1951; Florey and Florey, 1955) have provided most of the information concerning innervation of the MROs. Both efferent and afferent run together in the dorsal nerve (Fig. 2.1, n.r.m.). The axons of the sensory neurons are the largest diameter fibres in this nerve trunk.

Three types of efferent elements supply each muscle receptor organ (Fig. 2.5): a motoneuron to the receptor muscle, a fibre which terminates on the receptor neuron itself, and an inhibitor to the receptor muscle. There are two large diameter motor fibres in the lobster, one to each RM (Alexandrowicz, 1951), the one to RM2 being generally the thicker. In addition, branches of motor nerves to adjacent dorsal abdominal muscles innervate the RMs near their attachments (not shown in Fig. 2.5). In *Astacus* only one large motor fibre occurs and it innervates RM2. RM1 is innervated by several small diameter fibres which appear to be shared with adjacent abdominal muscles. Kuffler (1954) also noted that, in crayfish, whereas the RM1 receives innervation along its entire length, RM2 is not innervated in the intercalated connective tissue region around the nerve cell dendrites.

Of the efferent motor fibres to RM1 at least two are excitatory (Fields *et al.* 1967). Inhibitory motor innervation of RM1 is suggested in the lobster because it is innervated by a branch of one of the accessory nerves (Section 2.6.2) which are known to be inhibitory. There is no *direct* physiological evidence for inhibitory innervation of RM1. RM2 regularly receives two excitatory motor nerve fibres (Fields, 1966; Alexandrowicz, 1967) and has been shown to receive an inhibitory motoneuron, which may be a branch of the accessory nerve (Kosaka, 1969).

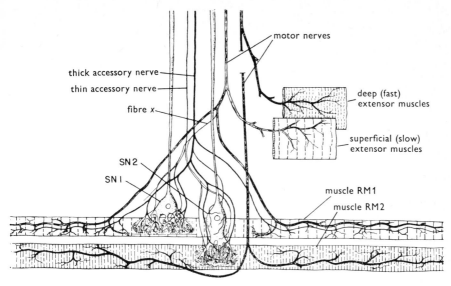

Fig. 2.5 Semi-diagrammatic drawing of the muscle receptor organs and their innervation in *Homarus gammarus*. Not all efferent components are shown in this drawing. Note sharing of motor innervation between extensor muscles and receptor muscles. RM1, RM2, receptor muscles; SN1, SN2, receptor sensory neurons. (From Alexandrowicz, 1967.)

2.5 Receptor neurons: anatomy and physiology

2.5.1 Morphology of receptor cells

The receptor neurons are multipolar cells with pyramidal to fusiform shaped somata (Fig. 2.6). The terminal dendritic tree of each cell is embedded in a region of connective tissue, which divides the RMs into two contracting regions. Electronmicroscopy has revealed that the proximal dendrites are ensheathed by glial (Schwann) cell processes which in turn are embedded in connective tissue (Nadol and DeLorenzo, 1968). The terminal dendrites lose their glial cell sheath and directly contact connective tissue or sarcolemma.

The cell associated with the lateral muscle (SN1) is orientated obliquely to RM1 and its dendritic system is spread out as the dendrites penetrate the muscle. The main dendritic trunks branch in a T-shape, running forward and backward parallel to the muscle fibres for about 500 μm along the length of RM1 (Fig. 2.6). This is not the case for the neuron associated with the medial muscle (SN2), which is orientated more perpendicularly to its muscle. Its terminal dendrites

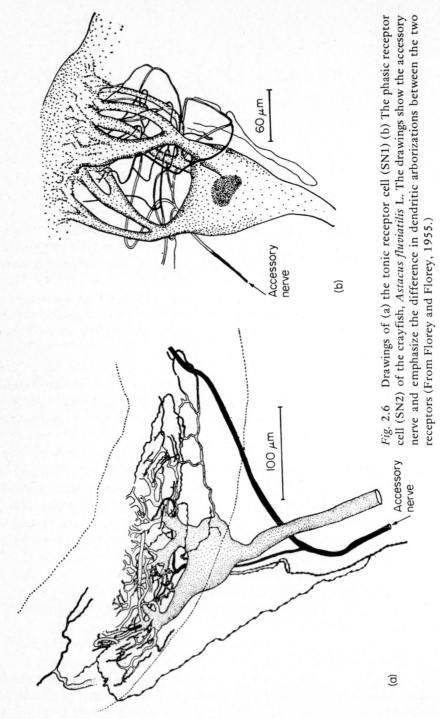

Fig. 2.6 Drawings of (a) the tonic receptor cell (SN1) (b) The phasic receptor cell (SN2) of the crayfish, *Astacus fluviatilis* L. The drawings show the accessory nerve and emphasize the difference in dendritic arborizations between the two receptors (From Florey and Florey, 1955.)

are bushy and short, are confined to a smaller area of the muscle, and do not have any particular orientation with respect to RM2 (Florey and Florey, 1955). This difference in dendritic orientation was postulated by the Floreys to account for the difference in response to stretch of the two receptor cells described below (Section 2.5.2). Thus a slight change of length in RMs would stretch the dendrites of the lateral nerve cell (SN1) which are taut initially, but a greater change would be required to place comparable stretch on the SN2 dendrites.

2.5.2   Response to muscle stretch

The earliest physiological study (Wiersma *et al* 1953) confirmed Alexandrowicz's hypothesis (Alexandrowicz, 1951), based on anatomical considerations, that the response properties of the two MROs on the same side in each segment were different. It was found that trains of impulses of two distinct amplitudes can be recorded from the distal end of the dorsal nerve in a stretched segment, and that a characteristic discharge pattern is associated with each. The most striking difference in the behaviour of the two receptors is in their response to maintained stretch. One of the fibres, usually the one producing lower amplitude spikes, maintains a rhythmic discharge for up to $3\frac{1}{2}$ hours, while the other gives a phasic pattern rarely persisting as long as 60 seconds (Fig. 2.7).

The tonic receptor is often active in the absence of applied stretch, and the threshold for activating it is much lower than for the rapidly adapting receptor. By selective destruction, it was shown that the neuron associated with the thin, lateral, coarsely-striated RM1 invariably gave rise to the low threshold, slowly adapting discharge; while the medial, thick, finely-striated RM2 was always associated with the rapidly adapting, high threshold response.

By clamping the muscle around the region of dendritic penetration and destroying the rest of the muscle, Wiersma *et al,* (1953) were able to show that the threshold for excitation and the rate of adaptation of the receptor are properties of the neurons themselves and are not related to the series-elastic properties of the muscle bundles. What, then, is the effect of these series-elastic properties on the relation between stimulus and response in this receptor?

The earliest systematic study of the mechanical properties of the MROs was undertaken by Krnjevic and van Gelder (1960, 1961). By

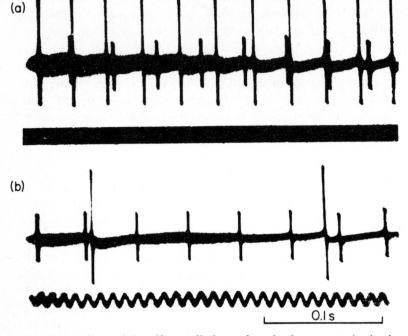

Fig. 2.7   Recordings of the afferent discharge from both receptors in the dorsal nerve of the 2nd abdominal segment of *Homarus americanus*. (a) Immediately after stretching both receptor muscles. (b) 20 seconds later, demonstrating the more rapid adaptation rate in the receptor with the higher spike amplitude. Time marker, 100 Hz. (From Kuffler, 1954.)

measuring length, tension and discharge frequency of the MROs they found a linear relationship between tension and discharge frequency, but a logarithmic relationship between receptor muscle length and discharge frequency (Chapter 15). They also found that, if the receptor muscle was suddenly stretched and held at a constant length, the tension transiently rose to an early peak and then declined to a lower constant level. The early peak tension was greater for more rapid stretches. At small initial lengths the relationship between length and tension was nearly linear. However, with increasing length, tension and impulse frequency in the receptor rose exponentially. Little qualitative difference was found between the two receptor muscles. RM2, however, required a length change twelve times greater to produce the same tension increment per unit cross-sectional area.

In a more detailed study of the same problem, Wendler (1963) observed that length change in the central dendritic region of the receptor muscle did not precisely follow the length change in the whole receptor muscle. He found that, if the receptor muscle was quickly stretched and held at a constant length, the central region of the RM, around the dendrites of the sensory cell, lengthened, even though the RM as a whole remained at constant length. Thus the length of the central dendritic region, which contains more connective tissue, parallels the receptor discharge frequency more closely than does the length of the whole receptor muscle. Other experiments demonstrated that this central region is stiffer than other regions of the receptor muscle. Since it comprises, at most, 10 per cent of the entire receptor muscle length, it would not contribute significantly to the measured mechanical properties of the muscle, but since the receptor cell dendrites are embedded in this region, it would exert great influence on the firing pattern of the sensory neuron.

Wendler also showed that, at constant length, both the tension of the receptor muscle and the discharge frequency of the receptor neuron decay with time, but that the discharge frequency declines more rapidly. This supports the hypothesis of Wiersma *et al.* (1953) that the time course of adaptation is determined partly by the property of the nerve cell and partly by the contractile and series elastic properties of the receptor muscle.

2.5.3 Generator potential

Eyzaguirre and Kuffler (1955a, b), recording with intracellular microelectrodes from the soma of the sensory neuron, showed that stretch applied to the muscle results in a depolarization, the generator potential, that follows the time course of the applied stretch. Since the recorded potentials are smaller when recordings are made further away from the dendrites, and since only the dendrites seem to be significantly sensitive to stretch, they postulated that stretch deformation of the dendrites results in a generator potential which spreads electrotonically to the soma. They found that the generator potential is graded, reflecting rate and degree of stretch. Later studies by Terzuolo and Washizu (1960, 1962) revealed that, over certain ranges of length change, there is a direct relationship between the logarithm of receptor muscle length and the generator potential, and that there is a direct linear relationship between generator potential and impulse frequency in the neuron. They also demonstrated a decrease in soma membrane resistance concomitant with the generator potential and showed that the equilibrium potential for this conductance change was near zero membrane potential. Depolarization produced by passing current through an intracellular microelectrode resulted in repetitive firing of the cell. This artificial depolarization was shown to sum with stretch-induced depolarization to increase the discharge frequency of the cell (Terzuolo and Washizu, 1962).

2.5.4   Impulse initiation

The depolarization produced by dendritic deformation generally leads to impulse initiation (Fig. 2.8). Although the site of impulse initiation in the phasic receptor is unknown, Edwards and Ottoson (1958) showed that, for the crayfish tonic receptor, impulses arise at a point on the axon 200–500 μm distant from the soma. Grampp (1966) confirmed this observation in the lobster and showed that the axon hillock, cell soma and dendrites are also capable of generating active depolarization. The resting potentials of SN1 and SN2 do not differ significantly and, with injected current, impulses arise at the same threshold level in both cells.

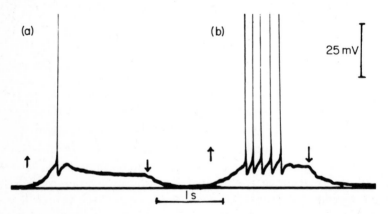

Fig. 2.8   Intracellular recording from a tonic receptor. Two stretches, indicated by ↑↓, are applied to RM1. (a) First stretch is just above threshold, producing a single action potential. (b) Second stretch is greater and produces a larger depolarization (generator potential) and a more prolonged discharge. (From Eyzaguirre and Kuffler, 1955a.)

2.5.5 Adaptation to maintained stretch

The most striking difference in behaviour of the two MROs is in their rate of adaptation to maintained stretch of their RMs. As discussed above (Section 2.5.2), the difference in viscoelastic properties of the two RMs cannot completely account for this difference in adaptation rate.

Eyzaguirre and Kuffler (1955) felt that the difference could be at least partially accounted for by the more rapid decay of the generator potential during prolonged steady stretch in the fast-adapting cell. However, Wendler (1963) found that, even with maintained anodal polarization of the tonic receptor cell, the impulse frequency still decays with time. Earlier, Terzuolo and Washizu (1960) had commented briefly that the phasic cell shows marked adaptation when activated by intracellular current pulses. Using intracellular microelectrodes for current passing and recording in the same cell, Nakajima (1964) confirmed Wendler's finding that adaptation of firing rate occurred during maintained depolarization (Fig. 2.9). He further showed that, if impulses are chemically blocked, the generator potential follows an *identical* time course in both fast and slow receptors, demonstrating that although the decay of firing rate with time results in part from a decay in the generator potential, the *difference* in adaptation rates between the two receptors with maintained depolarization is due to the difference in the accommodation rates of the two receptors. Thus the difference in decay of firing frequency with time during maintained stretch is a property of the electrically excitable component of the membrane of the neurons and not of that component which produces the generator potential.

The difference in adaptation between the two receptors can be accounted for in a highly localized region of the tonic receptor's axon. Nakajima and Onodera (1969) showed that, if current is passed locally at several regions along the axon of SN1, the ability to sustain a tonic spike discharge was localized to a patch of membrane near the zone of normal spike initiation: on the axon 200–500 μm from the soma (Fig. 2.10). Further out on the axon, SN1 accommodated to maintained current as rapidly as SN2. Thus the difference in behaviour of the two neurons can be narrowed down to their spike-initiating zones, since the generator potentials have an identical time course when the RM is bypassed, and both receptor cells have axons which rapidly accommodate.

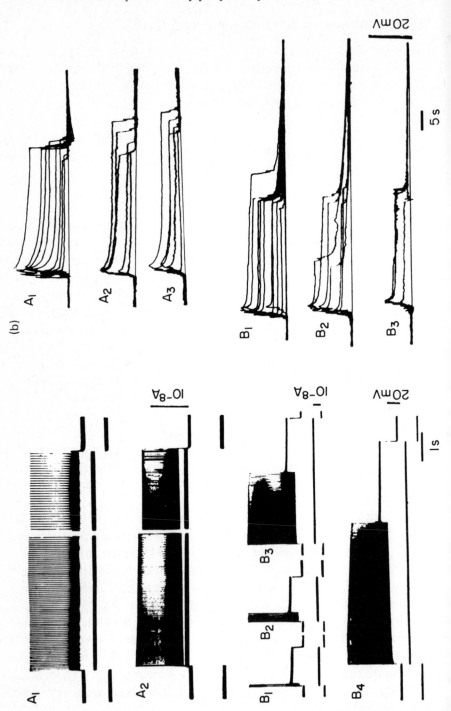

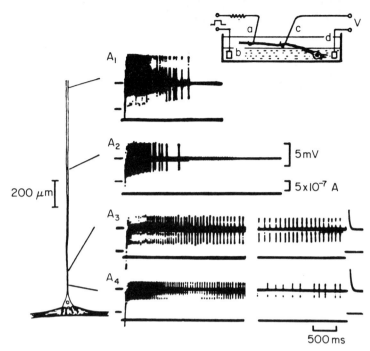

Fig. 2.10   Action potentials of a slowly-adapting cell obtained by stimulation and recording through extracellular electrodes. The arrangement is diagrammed in the inset. Traces $A_1 - A_4$ are recordings between c and d. Downward deflection indicates c positive. Lower trace is stimulating current monitor. Note that only the region near A_3 is capable of a sustained discharge to constant current application. (From Nakajima and Onodera, 1969.)

Fig. 2.9(a)   Time course of repetitive firing during the application of constant depolarizing current showing the responses to currents of different amplitudes. (a) Slowly-adapting cell. In A_1 the current was applied for 140 seconds; in A_2 a larger current was applied for 26 seconds. (b) Rapidly-adapting cell. The current increases from B_1 to B_4. At B_4 the maximum duration of repetitive firing was attained. Stronger currents than that in B_4 curtailed the duration of firing. Upper traces, action potentials; lower traces, current. (b) Time course of the generator potential induced by various degrees of maintained stretch in receptor cells treated with tetrodotoxin (10^{-7} g ml^{-1}). Pen recordings. The sudden onset of the generator potentials corresponds to the beginning of stretch; the sudden fall to the release of stretch. Some irregularities at the beginning and the end of stretches are artifacts due to the mechanical property of the stretcher. (A) Three slowly-adapting cells. (B) Three rapidly-adapting cells. (From Nakajima, 1964.)

Before the differences in decay of spike frequency with maintained depolarization can be fully accounted for, the possible mechanisms of accommodation must be examined in this system. In this regard, our knowledge of the tonic receptor neuron is the greater. Eyzaguirre and Kuffler, (1955b) have shown that SN1 exhibits both post-tetanic depression and post-tetanic hyperpolarization (PTH). Nakajima and Takahashi (1966) showed that in SN1 the post-spike hyperpolarization can be separated into two components: one, an early, brief hyperpolarization following each impulse, which is associated with an increase in potassium conductance; the other, a more prolonged post-tetanic hyperpolarization. This prolonged PTH does not reverse with hyperpolarization of the cell, is not associated with a change in membrane conductance, is enhanced by intracellular Na^+ iontophoresis and is abolished when sodium is replaced with lithium or when 2,4 dinitrophenol, which uncouples oxidative phosphorylation, is placed in the bathing medium. They concluded that the prolonged PTH is due to an electrogenic sodium pump which is activated by Na^+ entering the cell with each impulse. They did not find this PTH in the rapidly adapting neuron SN2.

Sokolove and Cooke (1971) provided further evidence on this point by showing that the degree of PTH is dependent on the number of impulses in the tetanus and that PTH can be inhibited by removal of external potassium, by application of strophanthidin, a pump inhibitor, or by cooling. They also showed that the decay of impulse frequency with maintained stretch, PTH, and post-spike depression are abolished concomitantly when the above manipulations are carried out, thus providing strong evidence of a causal relationship between the electro genic pump and spike frequency decay in the MRO.

2.6   Efferent control of the receptor

2.6.1   Effect of contraction of the receptor muscle

As discussed in previous sections, the MRO responds to deformation of its dendrites. Obviously, this can be produced either by imposed stretch of the receptor muscle or by active isometric contraction of the receptor muscle. As a result of either of these processes, tension appears across the dendritic region of the receptor, deforms the receptor dendrites, producing a depolarization which spreads electronically across the soma region to the spike initiating zone, resulting in receptor discharge (Kuffler and Eyzguirre, 1955a, b).

By combining our knowledge of the contractile properties of the receptor muscles with what we know about the electrical properties of the receptor cells, we can predict that stimulation of the motor nerves supplying the RM's will result in receptor discharge.

Single stimuli to the motor nerve supplying RM1 may result in a train of impulses lasting several seconds in the tonic MRO (Fig. 2.11a). If the initial tension on the RM1 is greater, then a greater discharge in the receptor is produced by the same stimulus to the motor nerve. Higher impulse frequencies in the motor nerve also result in increased frequencies of MRO discharge.

The additive effect of passive stretch and active contraction of the RM on the discharge frequency is also characteristic of the fast receptor, MRO2. MRO2 is most sensitive to rapid stretch or the twitch contractions of RM2. Thus it can signal the first derivative of RM length. Due to its rapid rate of adaptation and the brevity of the RM2 twitch, MRO2 can give distinguishable responses to stimuli in the motoneuron to RM2 at rates of up to 40 Hz (Fig. 2.11b) (Kuffler, 1954).

In summary, MRO1 discharge frequency is a function of at least two variables: length of RM1 (i.e. segment length) and impulse frequency in the motoneurons supplying RM1. MRO2 frequency depends on segment length and its first time derivative as well as on the frequency of impulses in the motoneurons supplying RM2. It must be emphasized that the average frequencies of firing of the MROs cannot provide unambiguous information about segmental length.

2.6.2 The accessory nerve and receptor inhibition

In addition to the motor innervation of the receptor muscles Alexandrowicz (1951) noted another efferent neural component which appeared to contact the receptor cells directly. Three such efferent fibres have been shown, in lobster (Alexandrowicz, 1951, 1967) and in some species of crayfish (Jansen, Nja, Ormstad and Walløe, 1971), to terminate around the dendrites and some of the

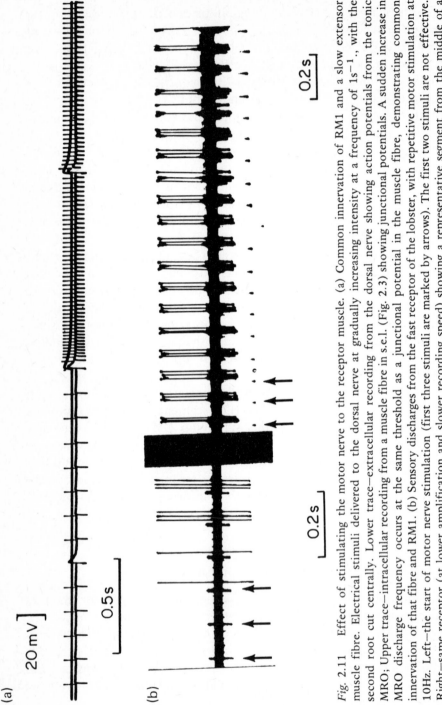

Fig. 2.11   Effect of stimulating the motor nerve to the receptor muscle. (a) Common innervation of RM1 and a slow extensor muscle fibre. Electrical stimuli delivered to the dorsal nerve at gradually increasing intensity at a frequency of $1s^{-1}$., with the second root cut centrally. Lower trace—extracellular recording from the dorsal nerve showing action potentials from the tonic MRO; Upper trace—intracellular recording from a muscle fibre in s.e.l. (Fig. 2.3) showing junctional potentials. A sudden increase in MRO discharge frequency occurs at the same threshold as a junctional potential in the muscle fibre, demonstrating common innervation of that fibre and RM1. (b) Sensory discharges from the fast receptor of the lobster, with repetitive motor stimulation at 10Hz. Left—the start of motor nerve stimulation (first three stimuli are marked by arrows). The first two stimuli are not effective. Right—same receptor (at lower amplification and slower recording speed) showing a representative segment from the middle of a continuous record with bursts of spikes accompanying each twitch. ((a) from Fields, 1966; (b) from Kuffler, 1954.)

sensory neurons (Fig. 2.5). These fibres are called accessory nerves (ANs).

Since the tonic and phasic MROs of the 8th thoracic segment are widely separated (see Section 2.7, Fig. 2.13) and are innervated by branches of the same AN, Eyzaguirre and Kuffler (1954) were able to stimulate the AN branch to MRO2 while recording in SN1. This allowed them to activate AN antidromically to SN1 without activating the motor nerves or the sensory axon of SN1. They found that such stimulation blocks impulse initiation (Fig. 2.12a). The inhibition resulting from accessory nerve stimulation is associated with a conductance increase, the inhibitory post-synaptic potential (IPSP), which rises to a maximum in 2 ms and decays in a further 30 ms (Fig. 2.12b).

The accessory nerve induced conductance change can cause either depolarization, hyperpolarization or no change, depending on the initial level of membrane potential (Fig. 2.12a, b). This conductance increase stabilizes the membrane at a level close to the resting potential.

In practice the major effect of the accessory nerve IPSP is to reduce the generator potential and thereby to prevent threshold depolarization in the spike initiation zone. Two mechanisms are involved: first, the increased membrane conductance lowers the space constant and diminishes electrotonic spread of the generator potential and second, if hyperpolarization is produced by the inhibitory transmitter, it will spread electrotonically to the spike initiating zone and be subtracted from the depolarizing generator potential.

It has been shown that the IPSP evoked by accessory nerve stimulation results from an increase in K^+ conductance (Edwards and Kuffler, 1959; Hagiwara, Kusano and Saito, 1960). The effect of inhibitory nerve stimulation is closely mimicked by application of gamma-aminobutyric acid (Hagiwara *et al.*, 1960), a putative inhibitory transmitter.

No excitatory neurons have been found to make direct contact with the receptor neuron. However, as mentioned earlier, it has been established that some decapod species may have two or three peripheral inhibitors acting on the MROs (Burgen and Kuffler, 1957; Jansen *et al.*, 1971). It should be emphasized, however, that even in species where this is true, one of the accessory nerves has a much greater inhibitory effect. The fibre which produces the largest IPSP

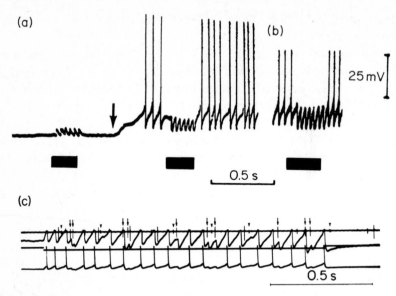

Fig. 2.12   Effect on receptor of stimulating the accessory nerve.

(a, b) Intracellular recording from the tonic receptor in the 3rd abdominal segment of the lobster. Solid bars under the trace indicate stimulation of the accessory nerve (AN) at 30 Hz. (a) With receptor muscle relaxed (left), AN stimulation produces depolarizing post-synaptic potentials. At the arrow, stretch is applied to RM producing a depolarizing generator potential and receptor discharge. AN stimulation now produces hyperpolarizing potentials and interrupts receptor discharge. (b) Increased stretch further depolarizes the receptor and the AN induced hyperpolarizations are larger. (c) Demonstration of action of more than one AN in the tonic receptor of *Astacus fluviatilis*. Bottom trace, membrane potential of receptor; top trace, spike activity in the dorsal nerve recorded *en passant*. Large accessory spikes marked with arrows, small accessory spikes marked with arrow-heads; second trace, higher amplification of receptor membrane potential. The receptor spikes are too large and too faint to reproduce. Note large accessory IPSPs preceeded by large accessory spikes in nerve record and small accessory IPSPs preceeded by small accessory spikes. In this case AN activity was reflexly produced by passing depolarizing current through the receptor (duration indicated by third trace). ((a, b) from Kuffler and Eyzaguirre, 1955; (c) from Jansen, Nja, Ormstad and Walløe, 1971.)

in MRO1, also has the highest amplitude spike as recorded with extracellular electrodes and the fastest conduction velocity. This fibre is undoubtedly the thick accessory nerve described by Alexandrowicz (1951) and is identical to the efferent element found by Eckert (1961) to be driven by stretch of the MRO. Jansen *et al.*, (1971) found two other inhibitory fibres to MRO1 in the crayfish.

One of these has a conduction velocity only half as great as that of the thick AN and produces an IPSP of only 1/5 the amplitude of those elicited by the thick AN (Fig. 2.12c). This is probably the thin AN described by Alexandrowicz. Finally, in some preparations, a third spike is visible above the noise level which is conducted at half the velocity of the thin AN and produces only a very tiny IPSP. The IPSP reversal potentials are identical for all the IPSP's produced by the three accessory nerves.

Electron-microscopic studies have shown that synapses of the accessory nerve are present in all regions of the sensory neuron except the axon (Peterson and Pepe, 1961). This is consistent with Kuffler and Eyzaguirre's (1955) conclusion that the inhibitory conductance increase involves both the soma and dendritic regions.

Although, for mainly technical reasons, most of the studies on inhibition have been carried out using the tonic MRO, the phasic receptor is also effectively inhibited by activity in AN, and there is no reason to suppose that the inhibitory mechanism involved is different from that for the tonic receptor.

It should be mentioned that an increase in frequency of regularly spaced impulses in the accessory nerve may actually *increase* ongoing activity in the tonic MRO (Perkel *et al.*, 1964). The functional significance of this finding is yet to be determined. Reflex relationships of the inhibitory nerve will be discussed in Section 2.8.3.

2.7   Thoracic MROs

In addition to the MROs that are located in each abdominal segment, similar appearing receptors are found in association with the thoracic musculature (Alexandrowicz, 1952). Information about thoracic MROs is available for several species. Although these receptors are highly variable in their location and response characteristics, a few generalizations are warranted.

The MRO's of the posterior thoracic segments are similar in appearance and response pattern to those found in the abdomen (Wiersma and Pilgrim, 1961). They are attached to a free-lying receptor muscle adjacent to the dorsal musculature (Fig. 2.13). Of the two receptors on each side of the animal, associated with the muscles bending the thoraco-abdominal joint one has a tonic, the other a phasic response to flexion of the abdomen and they are thus

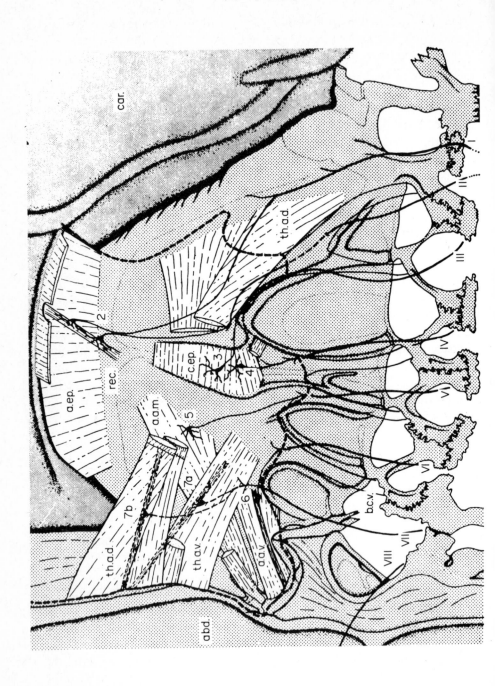

analogous to the abdominal MRO's. The tonic receptor lies lateral to the phasic receptor. The next set of MRO's anterior to those of the thoraco-abdominal joint are associated with muscles that abduct the abdomen and are activated when those muscles are stretched. In the lobster both tonic and phasic receptors are found in this location, but in *Procambarus* only the tonic receptor is present.

Anterior to these posterior thoracic MROs are a group of peripheral sensory cells which, although clearly associated with the thoracic musculature, differ markedly in structure from the abdominal MROs in that they are not attached to a separate, distinct receptor muscle and do not receive obvious efferent innervation. These are the N-cells that Alexandrowicz (1952) found in lobster and Wiersma and Pilgrim (1961) reported in both *Procambarus clarkii* and *Panulirus interruptus*. They found that all of these receptors respond tonically when activated, and no receptors with phasic responses were found in this location. Both Alexandrowicz (1956) and Wiersma and Pilgrim (1961) proposed that these anterior thoracic MROs or N-cells are derived from receptors identical to the tonic abdominal MRO. Arguments adduced in support of this proposal include the regularity of segmental innervation, the fact that an N-cell never occurs paired with a tonic receptor, although in several segments of *Leander* it is paired with a phasic receptor (Alexandrowicz, 1956), and that electrical recording reveals a slowly adapting discharge pattern (Wiersma and Pilgrim, 1961). The absence of phasic receptors and the lack of a separate receptor muscle in the N-cells has been attributed to the lower mobility of the thoracic exoskeleton. In support of this argument are the observations in *Squilla* (Alexandrowicz, 1954). In this species the fifth to eighth thoracic segments are mobile and contain complete sets of MROs

---

Fig. 2.13   Location and innervation of thoracic stretch receptors and N-cells in *Procambarus clarkii*. Semi-diagrammatic drawing of a flattened view of the left side of the thorax as seen from the inside of a parasaggital preparation. a.a.m. and a.a.v.—ventral heads of abductor abdominis muscle; a.ep.—attractor epimeralis; c.ep.—contractor epimeralis; rec.—receptor slip of attractor epimeralis; th.a.d. and th.a.v.—dorsal and ventral heads of thoraco-abdominal extensor muscle; I-VIII—second roots of thoracic ganglia; 1, SR-α; 2, SR-β; 3, SR-γ; 4, SR-δ; 5, SR-ϵ; 6, SR$_1$-TH7/8; 7a, SR$_1$-TH/Ab; 7b, SR$_2$-TH/Ab; abd.—first segment of abdomen; b.c.v.—branchiocardiac vein of the fourth leg; car.—carapace. SR α, β, γ, δ, ϵ are also known as N-cells. (From Wiersma and Pilgrim, 1961.)

identical to those in abdominal segments; the fourth thoracic segment has two receptor cells but only one RM; the third segment has one RM and one receptor cell; while the second segment has a receptor cell but no RM (Pilgrim, 1964).

The function of these thoracic receptors is unknown at present and thus no further description of them will be given in this chapter. However, the anatomical similarity of the thoracic and the abdominal MROs may indicate that they mediate similar reflex actions and perhaps have a common evolutionary predecessor.

2.8   Central connections and reflex relationships of the MROs

2.8.1   Central projections of the MRO afferent discharge

Central processes of the MRO
In work with the embryonic lobster, Alexandrowicz (1951), supplementing earlier work by Allen (1894), was able to establish that central processes of both sensory neurons enter the second root of the ganglion of the next anterior segment and simply bifurcate without apparent branching. One process ascends to the brain and the other descends to the sixth abdominal ganglion. These processes form a tract ventral to the medial giant fibre, which runs through all thoracic and abdominal ganglia.

Central nervous system response to MRO
The earliest physiological studies on this problem were undertaken by Wiersma (1958). He isolated single fibres in the circumoesophageal connectives of *Procambarus* and found some which responded tonically to abdominal flexion. One fibre was found to respond to flexion of each of the joints between tergites in the abdomen. He felt that these fibres were primary sensory fibres because their pattern of response was indistinguishable from that of the receptor neurons, they were more resistant to hypoxia than interneurons, and they had a conduction velocity of about $8 \, \mathrm{ms}^{-1}$ (indicating a large diameter fibre such as the MRO axon). Histologically, these fibres are among the largest in the central nervous system.

Wiersma and Bush (1963) subsequently investigated single units in the cord between thorax and abdomen. As would be expected, the descending fibres were from the thoraco-abdominal joint and the first intra-abdominal joint, while the ascending fibres originated from the more caudal MROs. No rapidly adapting MROs were found anterior to the one responding to flexion at the thoraco-abdominal joint, although the thorax does contain several slowly adapting receptors, as mentioned in Section 2.7.

In addition to the primary MRO fibres an interneuron was found which collects input from homolateral slowly adapting MROs (Hughes and Wiersma, 1960). The integrative site for this interneuron was shown to be in 6th abdominal ganglion as earlier histological and physiological evidence had suggested (Allen, 1894; Alexandrowicz, 1951; Wiersma, 1958). Wiersma and Bush (1963) demonstrated that this interneuron collects from all slowly adapting MROs of the same side. Thus there is at least one histologically established site of convergence for the MRO input.

2.8.2   Efferent elements of the abdominal ganglia

In each half segment of the crayfish abdomen, the entire ganglionic outflow to extensor and flexor musculature consists of 28—30 efferent elements. The largest number, ten, supply the massive fast flexors, contraction of which produces the rapid escape reflex. At least one of the efferent elements to the fast flexors is a peripheral inhibitor. The fast extensor muscles are supplied by four excitatory motoneurons and one inhibitor. The tonic (i.e. superficial) flexor and extensor muscles are both supplied by five excitatory motoneurons and one inhibitor. Each of the twelve efferent elements to the tonic musculature has characteristic properties and can be identified physiologically in different individuals (Kennedy and Takeda, 1965; Kennedy *et al.*, 1965). As discussed above, axonal branches of certain tonic extensor motoneurons supply RM1. The remaining efferent elements are inhibitory fibres (accessory nerves) supplying the receptor cells. From one to three are present in each half segment depending on the species and segmental level.

2.8.3   Reflex connections of the MRO

Of the above efferent elements, impulses in the tonic MRO have been shown to have central reflex connections with, at most, only three or four. One superficial extensor motoneuron (SEMN) is driven, as well as one or two accessory nerves. In some preparations one of the SEMNs is inhibited. Impulses in the phasic MRO have only been shown to excite the accessory nerves. It seems likely, however, that other reflex connections exist which are as yet undiscovered, especially for the phasic MRO.

Effect of MRO 1 on tonic extensor motoneurons: Myotatic Reflex
With simultaneous monitoring of the electrical activity in the superficial extensor muscles and impulses in the MROs, it has been shown that the tonic, but not the phasic, receptor provides excitatory drive to at least one of the SEMNs (Fig. 2.14a) (Fields and Kennedy, 1965). The slope of this relationship is constant in a given preparation and is shown in Fig. 2.15. Receptor firing rates of 30 Hz in intact, unrestrained preparations are not uncommon, which demonstrates that activity of a single tonic MRO in its normal physiological range of frequencies is sufficient to drive a tonic extensor motoneuron. Under some conditions a single impulse in the tonic receptor can evoke an impulse in a tonic extensor motoneuron (Fig. 2.14b) (Fields, 1966).

Using the technique of *en passant* recording of the dorsal nerve, n.r.m., while monitoring junctional potentials in extensor muscle fibres, it is possible to identify the action potential of each SEMN.

Fig. 2.14   Responses to stretching the slow muscle receptor organ, recorded *en passant* in the superficial division of the dorsal nerve (upper trace) and intracellularly in a slow extensor muscle fibre (lower trace). Increased frequency of the afferent impulses from the slow receptor (large spikes) is associated with reflex discharge of one of the slow extensor motor neurons (intracellularly recorded excitatory junctional potentials and associated extracellular action potentials). Augmented output is also seen in the accessory nerve (downward arrows). Upward arrows indicate impulses in slow extensor inhibitor. 10 mV. Calibration refers to intracellular trace. (b) Central interaction of tactile and stretch receptor input to tonic motoneurons. A nerve containing only tactile afferent fibres (n.c.s.) was electrically stimulated (dots) at gradually increasing intensity. RM_1 was stretched slightly to produce a steady background discharge in the tonic MRO. n—a spike in the tonic MRO; m—an extracellular muscle potential, associated with the intracellular junctional potential recorded in a slow extensor muscle fibre (lower trace). Stimulus intensity was greater in (c) than in (b) and consequently each MRO impulse was more effective in eliciting a motoneuron impulse. In (d), the stimulus intensity was equal to that in (b) but the MRO was discharging at a high frequency. ((a) from Kennedy, Evoy and Fields, 1966; (b) from Fields, 1966.)

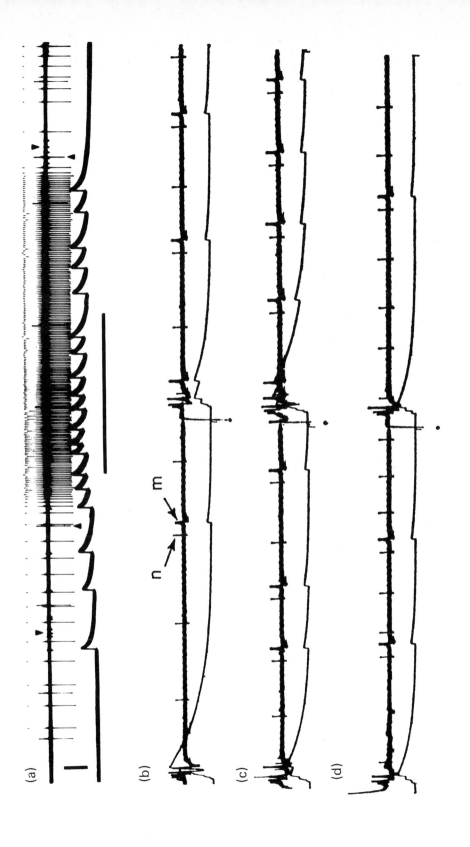

(a)

(b)

n
m

(c)

(d)

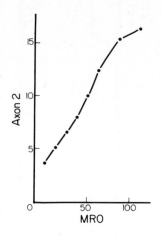

Fig. 2.15   Input/output curve for the connection between the tonic MRO afferent and slow extensor motoneuron 2. The central process of the MRO afferent was isolated from the ventral nerve cord and stimulated electrically; output frequencies were averaged over 1 second periods, after a steady value had been reached following the onset of stimulation. (From Fields, Evoy and Kennedy, 1967.)

The small number of motoneurons in n.r.m., and the fact that each SEMN has a characteristic spike amplitude and firing pattern has made it possible to establish that there is a constant relationship of motoneuron size to function. The SEMNs can thus be individually recognized in different preparations and have been numbered one to six in order of increasing size. SEMN5 is the inhibitor, the others are excitatory motoneurons.

Impulses in MRO1 generally produce slow extensor junctional potentials of only one amplitude (Fig. 2.14a). This junctional potential was shown to be associated with SEMN2 (Fields, *et al.,* 1967). SEMN2 is unique in that it supplies practically 100 per cent of the tonic extensor muscle fibres and produces relatively large junctional potentials in most. Since the tension developed is proportional to both the number of muscle fibres and the degree of depolarization of each, discharge of MRO1 sufficient to drive SEMN2 is likely to produce significant tension. In spite of the fact that SEMN2 innervates almost all of the slow extensor muscle fibres it does not innervate either receptor muscle in its segment. MRO1 has no demonstrable excitatory effect on any contralateral SEMN, but weakly excites at least one SEMN in the adjacent segments ipsilaterally.

Thus the tonic MRO — SEMN2 connection provides a myotatic reflex in each segment which would oppose any applied force lengthening that segment.

In addition to the above effects of MRO1 on SEMN2, Sokolove (1973) has presented evidence that during 'voluntary' extension, MRO1 discharge may excite other slow extensor motoneurons. These were not identified, however, and the possibility that receptors other than MRO1 might be involved was not ruled out.

In favorable preparations SEMN1 can be clearly identified. It has been shown to be weakly inhibited by discharge of MRO1. SEMN1 usually supplies less than 40 per cent of the fibres in the lateral head of the superficial extensor muscle (s.e.1.) and generally produces very small junctional potentials in the fibres it does innervate. There is some indirect evidence that SEMN1 is weakly excitatory to the tonic receptor muscle. The functional significance of SEMN1 is unclear at present (Fields *et al.*, 1967).

Elements not affected by MRO discharge

No other segmental motoneurons have been found to be excited or inhibited by either MRO. Although most abdominal reflexes have been shown to be reciprocal at the segmental level, no inhibition of flexor motoneurons by the MRO could be demonstrated, even though this was carefully looked for by Kennedy *et al.*, 1966. Similarly, the peripheral inhibitory motoneurons to extensors and flexors are not affected by MRO discharge (Fields, 1966).

Effect of MRO1 on the accessory nerve: self-inhibition

Eckert (1961a, b) was the earliest investigator to examine the reflex activity elicited by the MROs. He observed that flexion of the abdomen increases the spike frequency in two efferent elements, provided that the tonic MROs are intact. The lower amplitude efferent element described by Eckert is probably SEMN2. The other flexion-activated element had higher amplitude action potentials, responded phasically, and was shown to inhibit MRO discharge. This latter element is, by definition, the accessory nerve (AN). Thus MRO1 inhibits itself by this reflex activation of AN.

As mentioned in Section 2.6.2, in some species there are as many as two inhibitory nerves besides the thick AN. There is the thin AN

(Fig. 2.5) which produces a smaller inhibitory effect than the thick and a very small fibre producing a still smaller effect. The thin AN is activated by impulses in MRO1, although the gain of the MRO1-thin AN reflex is lower than that for the MRO1-thick AN reflex (Jansen *et al.*, 1971). Both thick and thin ANs are excited by MRO1 in the contralateral hemisegment. In contrast to the two larger ANs, the smallest inhibitory fibre is usually spontaneously active and cannot be excited by ipsilateral MRO1 activity. However, as with the thick and thin AN, it can be excited by stretch of the contralateral MRO1. Because the inhibitory effect of the two smaller diameter inhibitory fibres is so small, little is known of their function. Further discussion of the role of the accessory nerve will, accordingly, focus on the function of the thick AN. Subsequent use of the term accessory nerve (AN) in this chapter refers exclusively to the thick accessory nerve.

Eckert demonstrated that the tonic MRO is more effective than the phasic in activating AN, though AN is effective in inhibiting both tonic and phasic receptors. This is consistent with the anatomical observation that at least the thick AN branches to supply both receptor cells (Florey and Florey, 1955). Eckert (1961a) demonstrated that each MRO excites the AN in neighbouring segments and that summation of this excitation occurs when two or more MROs discharge simultaneously. Fields *et al.* (1967) further showed that the MRO-AN reflex is strongly biased in an anterior direction. For example, a given frequency of discharge in the MRO of abdominal segment three produces a higher frequency of firing in the AN of abdominal segment two then in the AN of its own segment and has a very small effect on the AN of the posterior adjacent segment.

This anterior bias of the MRO-AN reflex is in marked contrast to the MRO-SEMN2 'resistance' reflex, which is strongest in the same segment ipsilaterally and falls off sharply in neighbouring segments. The functional significance of this anterior bias was elucidated by Page and Sokolove (1972), who recorded MRO and AN discharge during unrestrained abdominal movements. They observed that during active tonic flexion all centrally originating drive to the tonic extensors is removed. However, flexion stretches the tonic MRO and thus provides excitatory drive to both AN and SEMN2. Page and Sokolove noted that most tail flexions begin with flexion of the most caudal segments and proceed anteriorly. During these flexions high frequency discharge of the AN was observed. This AN discharge inhibits the tonic MRO and prevents activation of the MRO-SEMN2 reflex. Thus the anterior bias of the MRO-AN reflex serves as an intersegmental mechanism to block the MRO-SEMN2 reflex which would otherwise oppose voluntary flexion. This gives some functional meaning to the central inhibitory modulation of MRO activity.

Eckert (1961) studied the central organization of the MRO-AN reflex by observing the effect of sectioning the ventral nerve cord on activation of the accessory nerve by the MROs. He found that sectioning just anterior to the second abdominal ganglion (cut 1, Fig. 2.16) diminished activation of the contralateral accessory nerve by MROs in the same segment. Cutting posterior to the third abdominal ganglion (cut 2) had no additional effect. He then cut posterior to the second abdominal ganglion (cut 3), thus isolating it, and found that the response of the contralateral accessory nerve to MRO activity was completely abolished. He concluded that the central processes of the receptors do not make effective synaptic contact with the inhibitor in the ganglion of entry but do so in the adjacent ganglia, and that MRO reflex inhibition does not require either the brain or the sixth abdominal ganglion for effective operation.

In summary, activity in the tonic MRO gives rise to two segmental reflexes which are, to a large extent, of opposite sign. The functional significance of these reflexes has been elucidated by studies of relatively intact, behaving crayfish.

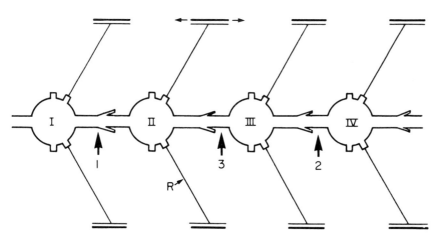

Fig. 2.16 Diagram of the first four abdominal ganglia in a crayfish. Large arrow heads (1, 2, 3) indicate sites of nerve cord sectioning; small arrow heads, the MRO's stimulated; R, the recording site on an accessory nerve. Details of the effects of sectioning the nerve cord are given on p. 101.

2.8.4 The phasic MRO

The sensitivity of the phasic MRO to rate of tension development and the fact that its RM is fast and shares its motoneuron supply with the fast extensors strongly suggest that its reflex role is limited to regulation of the twitch muscles of the abdomen. To date, however, no central reflex action for the phasic MRO has been found other than activation of the inhibitory AN in the same and adjacent segments as described above. Because of the brevity of phasic events and their all-or-none nature, the fast system is intrinsically more difficult to study in intact animals, and efforts to determine the role of the phasic MRO in movement control have been unsuccessful. Thus, the discussion that follows refers of necessity only to the tonic or postural control system of the crayfish abdomen.

2.9   Behavioural studies

2.9.1   Control of receptor muscles

Before the question of the functional significance of the MRO can be approached it is crucial to understand just what situations produce activity in the receptor. Although, as discussed above, impulses are generated by dendritic deformation, this may come about through either imposed tail flexion or activity in the motoneurons supplying the RMs. Before the reflex actions of the tonic MRO were known, Eckert (1961b) observed that, in relatively intact preparations, the receptor muscles actively contracted during tail extension, but produced no sensory discharge unless shortening was prevented. Thus, in the intact animal the RMs were contracting isotonically during extension and there was no tension increase to deform the receptor dendrites. On the other hand, if the tail was flexed manually during an active extension a brisk receptor discharge ensued.

2.9.2   Discharge of MRO1 in relation to 'voluntary' extension

Further observations on intact crayfish, held by carapace clamp but with the abdomen unrestrained in air, revealed periodic fluctuations in discharge of the tonic MRO (Fields, 1966). Two observations were made in this preparation that indicate the functional significance of the MRO. First, an increased discharge in the tonic MRO is often correlated with the onset of abdominal extension. Discharge in the tonic receptor is usually maximal during the first few degrees of extension and diminishes as the movement progresses. Second, if flexion is experimentally imposed upon an active 'voluntary' extension, the crayfish generally re-extends to its original position. Monitoring of MRO discharge under these circumstances revealed that it is maximal for maximal flexion and gradually diminishes as the original tail position is attained.

2.9.3 The adequate stimulus to MRO1

These results clarify the meaning of the tonic receptor discharge. At no time is it a simple function of segment length. Its output is a signal that the actively contracting receptor muscle is developing tension faster than it can be unloaded by the adjacent tonic extensors contracting in parallel. If the tonic extensor muscles contract faster, thus shortening the segment as fast or faster than the tonic RM can shorten, no MRO discharge will be seen. This suggests that the MRO operates as a detector, signaling deviation of segment length from some 'desired' or 'expected', centrally programmed position.

This view is greatly strengthened by the work of Sokolove (1973), again using the unrestrained abdomen. He observed that, when extension is initiated, it eventuates in complete extension, if the abdomen is unrestrained and fully immersed in water. In this situation there is usually no discharge from MRO1. If the abdomen is stopped by an obstruction during extension, the steady state receptor discharge is a function of the *difference* between the position at which it is stopped and full extension. Full extension presumably is the desired 'position'. This relationship is independent of the velocity and force of extension at the time of obstruction, indicating that segment length or abdominal position is the output which is controlled.

Thus, when taken as a unit, the MRO-SEMN2 reflex forms part of a negative feedback control system for regulating muscle length. The discharge in the tonic receptor is the feedback signal in this system. Any factor which slows the shortening of the segment during active extension (e.g. gravity or extensor fatigue) constitutes a disturbance. In a functional sense, it is these disturbances that constitute the adequate stimulus for discharge of the tonic receptor.

The motoneuron(s) supplying the receptor muscle can be considered to supply a control signal for segment length. Thus the system has a centrally determined set-point which is translated at the segmental level into impulses per second in the motoneuron or motoneurons supplying RM1. Figure 2.17 is a block diagram summarizing the elements of this postural control system. By altering the drive on SEMN4 (the main motoneuron supply to the tonic receptor muscle) the set point can be varied, allowing the servo to operate over a wide range of abdominal positions.

2.10   Command interneurons, central programmes and proprioception

2.10.1   Command interneurons

Although we have outlined above a fairly complete description of the segmental elements for postural control, summarized in Fig. 2.17, and have shown that the tonic MRO is often brought into action during tail extension, we have said little about the central neural mechanisms which activate this segmental apparatus. Our knowledge of these mechanisms, though far from complete, is sufficient to provide further understanding of postural control at the single neuron level. Furthermore, analysis of this system has provided insights into the relationship between reflex movements and those that are centrally programmed.

In the crayfish, single interneurons have been isolated which, when stimulated, elicit complex, coordinated sequential behaviours (Wiersma and Ikeda, 1964; Evoy and Kennedy, 1967; Atwood and Wiersma, 1967).

It should be emphasized that only a small fraction of the thousands of interneurons in the crayfish central nervous system are capable of eliciting these complex behaviours. Such interneurons are called *command interneurons* and it is assumed that they make direct or indirect connections with many ganglionic motoneurons, inhibitors and interneurons. The particular set of command fibres activated and the temporal properties of their selection is termed motor tape, programme or score.

For abdominal movements three major types of command fibre have been found: extension, flexion and suppression (Evoy and Kennedy, 1967). Each command fibre seems to elicit a slightly

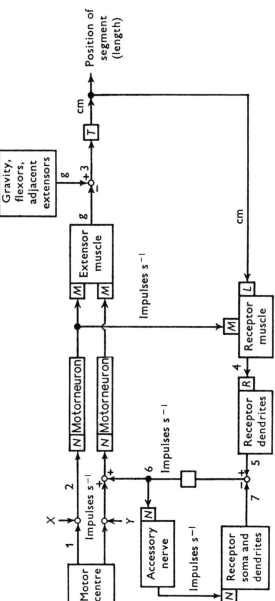

Fig. 2.17 Block diagram of the servo-control system for regulation of the tonic contraction. Segment length is the output. Large rectangles represent structures named within; small squares represent transducing functions; cm, *length*; g, *tension*; N—neuro-neural synapse; M—neuromuscular junction; T—converts tension to length change (i.e. series and parallel visco-elastic properties of the slow extensor muscles); L—converts length change to tension (visco-elastic properties of RM_1); R—converts dendritic deformation to membrane depolarization (receptor potentials). x, y, other inputs. Open circles represent summing points and closed circles represent 'pick-off' points for feedback loops. 1—Reference input, related to 'desired' position (i.e. set point); 2—actuating signal (excitatory input to motorneuron supplying RM_1); 3—net tension in extensors; 4, by altering the mechanical properties of RM_1, i.e. its length-tension curve, activity in the motor nerve to RM_1 determines the discharge rate of the MRO at any given length. RM_1 length equals segment length. If segment length is greater than 'desired', RM_1 tension produces a generator potential 5—which exceeds threshold for MRO discharge, 6, 7—inhibitory junctional potential set up by accessory nerve activity (From Fields, 1966).

different ganglionic outflow pattern. One command might elicit extension of caudal segments, another, extension of rostral segments. Thus abdominal position depends on which command or combination of command fibres is brought into play (Kennedy, Evoy, Dane and Hanawalt, 1967).

Although our knowledge of the mechanism of selection of a particular command fibre for a given movement is rudimentary, we do have a fairly detailed description, at the single neuron level, of the mechanism of tonic abdominal extension in the crayfish.

Each command interneuron makes synaptic contact with one or more of the efferent elements (Fig. 2.18). When the command fibre is brought into action, activation or inhibition of at least one and usually several efferent elements necessarily follows. The exact pattern of motor elements excited and inhibited is quite variable. It depends on which command fibres are activated and their frequency of firing (Evoy and Kennedy, 1967).

Certain general organizational rules are followed. First, there is reciprocity of activated elements. Stimulation of an extension command fibre usually activates extensor excitor motoneurons, but other actions observed might include activation of the flexor inhibitor and/or inhibition of flexor excitor motoneurons and the extensor inhibitor. Second, output is invariably symmetrical. Finally, different command fibres differ in the degree to which they excite or suppress a given ganglionic element, but it is always in the appropriate direction for the movement elicited.

2.10.2. Two types of postural extension command

It is clear that two major categories of extension command are possible: those which provide strong drive to SEMN4 (the motoneuron supplying the tonic RM) and those which provide little or no drive to SEMN4. That both types of command exist and are employed by intact, unrestrained crayfish has been clearly demonstrated (Fig. 2.19) (Fields, Evoy and Kennedy, 1967).

If a set amount of extension is desired, SEMN4 (and/or SEMN1) will be activated by command fibre activity (Fig. 2.19a). The degree of activity in SEMN2 will depend on the net load encountered by the muscles in each segment. Thus a load-cancelling function requiring a feedback loop through MRO1 is built into certain motor programmes

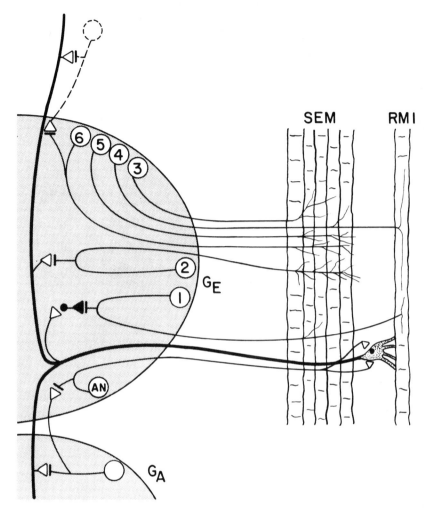

Fig. 2.18   Simplified diagram of segmental elements of the tonic extension control system. Numbers refer to superficial extensor motoneurons. AN is the accessory nerve. All the peripheral axons shown run together in the dorsal nerve (n.r.m.) before it branches to supply the superficial extensor muscle (S.E.M.). Presynaptic terminals are indicated by triangles (∇—excitatory; $\blacktriangledown$—inhibitory). Connections between MRO1 and other neuronal elements have not been shown to be monosynaptic. An inhibitory interneuron is shown interposed between MRO1 and SEMN1. The MRO1 to AN connection is not in the ganglion of entry (G_E) but in adjacent ganglia (G_A). The connection of MRO1 to SEMN6 is suggested but not established (see text). Connections between other segmental receptors and efferent elements are not shown, nor are descending and ascending interneurons which may be involved in postural control.

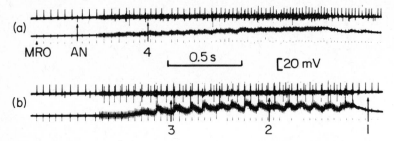

Fig. 2.19   Two types of extension command. Selective actions of central command fibres upon slow extensor motoneurons. Upper traces, *en passant* records from the dorsal nerve; lower traces, intracellular records from a slow extensor muscle fibre. Artifacts mark the beginning and end of 100 Hz stimulation of two different command fibres (a and b) isolated from the connective between 5th and 6th abdominal ganglia. In (a) primarily axon 4 is driven, and there is strong reafferent excitation of the MRO. In (b) there is strong drive on axons 2 and 3, together with a larger unit which is not identified and the discharge frequency of the MRO is not changed. AN—spike in the accessory nerve; 1, 2, 3, 4—motor axons. (From Fields, Evoy and Kennedy, 1967.)

for tonic extension. Those programmes which activate SEMN4 embody load-cancellation.

In contrast, other command fibres exist which can produce extensions with minimal proprioceptive feedback, either by removing the MRO from the circuit with AN inhibition or by providing an extension command using elements that do not cause contraction of the receptor muscle. Such commands will produce an unloading of the RM, thus providing a relatively pure tension command regardless of load (Fig. 2.19b).

It is clear that the crayfish often employs commands which activate the receptor muscle, since bursts of MRO activity may be observed in intact animals during even small amplitude spontaneous tail extensions (Fields, 1966). Command fibres which appear to activate only SEMN4 have also been observed in the intact animal (Fields *et al.*, 1967).

To determine the factors which select the command fibre array is obviously an enormous task. However, it is of fundamental importance for a complete description of the neural basis of behaviour. One illustrative example of command selection is given below. Stimulation of hairs on the telson is a consistently effective method of producing activity in the slow extensor motoneuron innervating RM1

(SEMN4), with relatively little activity in the other slow extensor motoneurons. In the intact animal such stimulation usually produces extension of the tail and the crayfish walks away from the stimulus. In this case, activation of an abdominal extension command fibre with load cancellation is only part of a very complex motor programme resulting from the 'decision' to move away from the stimulus. Since it is virtually impossible for the crayfish to walk forward with its tail flexed it must maintain the extended position during this behaviour. A common way for the crayfish to do this is by maintaining tension on RM1 by activating SEMN4. In this case the motor programme is triggered by telson stimulation but, since the behaviour far outlasts the stimulus, other factors must account for the sustained activity in the appropriate command fibres. How such activity is sustained is one of the more intriguing problems of neurobiology.

Thus the postural control system serves as an example of the interaction of central motor programmes with environmental disturbances. If the extension programme provides strong excitatory input to SEMN4, then the activity of SEMN2 will depend on both the initial segment length and load as well as on direct input from the command fibre. If the central programme does not employ SEMN4, then there is no load cancellation and although the precise array of motoneurons activated and tension developed is more predictable, the final position is less certain.

2.11 Summary

Thus we see that understanding the role of a particular proprioceptive organ depends on an accurate description of the behavioural situations in which it discharges. It has been possible to determine this for the tonic MRO because its activity may be recorded in relatively unrestrained, neurally intact preparations. Because of the small number of neurons involved in the segmental reflexes mediated by the MRO a relatively complete description of this load-cancelling servo-device has been possible.

There is much accumulated evidence that the vertebrate muscle spindle functions in a similar manner (Stein, 1974). For example, during stepping movements in the cat, alpha and gamma motoneurons are activated together (Severin, 1967). Because of the muscle spindle-mediated myotatic reflex in the contracting muscle, such movements must be considered to be 'servo-assisted'. The degree of activity in the alpha motoneurons (supplying the parallel extrafusal fibres) will depend on the initial position as well as on the load encountered during shortening of the muscle.

In conclusion, the details of one solution of the problem of position control in the face of variable load have been presented. The solution employed appears to be a general one seen in both crustacean and vertebrate systems.

Acknowledgements

The author would like to acknowledge the assistance of Dr S. D. Anderson and Ms Annette Lowe in the preparation of this manuscript. The work was partially supported by P.H.S. Grant NS70777—01.

References

Abbott, B. C. and Parnas, I. (1965) Electrical and mechanical responses in deep abdominal extensor muscles of crayfish and lobster. *Journal of General Physiology*, **48**, 919—931.

Alexandrowicz, J. S. (1951) Muscle receptor organs in the abdomen of *Homarus vulgaris* and *Palinurus vulgaris*. *Quarterly Journal of Microscopical Science*, **92**, 163—199.

Alexandrowicz, J. S. (1952) Receptor elements in the thoracic muscles of *Homarus vulgaris* and *Palinurus vulgaris*. *Quarterly Journal of Microscopical Science*, **93**, 315—346.

Alexandrowicz, J. S. (1954) Notes on the nervous system in the Stomatopoda IV. Muscle receptor organs. *Pubblicazioni della Stazione zoologica di Napoli*, **25**, 94—111.

Alexandrowicz, J. S. (1956) Receptor elements in the muscles of *Leander serratus*. *Journal of the Marine Biological Association of the UK*, **35**, 129—144.

Alexandrowicz, J. S. (1967) Receptor organs in thoracic and abdominal muscles of crustacea. *Biological Review*, **42**, 288—326.

Allen, E. D. (1894) Studies on the nervous system of Crustacea, I. Some nerve elements of the embryonic lobster. *Quarterly Journal of Microscopical Science*, **36**, 461—482.

Atwood, H. L. (1963) Differences in muscle fiber properties as a factor in 'fast' and 'slow' contraction in *Carcinus*. *Comparative Biochemistry and Physiology*, **10**, 17—32.

Atwood, H. L. and Dorai Raj, B. S. (1964) Tension development and membrane responses in phasic and tonic muscle fibers of a crab. *Journal of Cellular and Comparative Physiology*, **64**, 55—72.

Atwood, H. L. and Wiersma, C. A. G. (1967) Command interneurons in the crayfish central nervous system. *Journal of Experimental Biology*, **46**, 249–261.

Bodian, D. and Bergman, R. A. (1962) Muscle receptor organs of crayfish: functional-anatomical correlations. *Bulletin of the Johns Hopkins Hospital*, **110**, 78–106.

Burgen, A. S. V. and Kuffler, S. W. (1957) Two inhibitory fibres forming synapses with a single nerve cell in the lobster. *Nature*, **180**, 1490–1491.

Eckert, R. O. (1961a) Reflex relationships of the abdominal stretch receptors of the crayfish, I. Feedback inhibition of the receptors. *Journal of Cellular and Comparative Physiology*, **57**, 149–162.

Eckert, R. O. (1961b) Reflex relationships of the abdominal stretch receptors of the crayfish, II. Stretch receptor involvement during the swimming reflex. *Journal of Cellular and Comparative Physiology*, **57**, 163–197.

Edwards, C. and Hagiwara, S. (1959) Potassium ions and the inhibitory process in the crayfish stretch receptor. *Journal General Physiology*, **43**, 315–321.

Edwards, C. and Kuffler, S. W. (1959) The blocking effect of α-aminobutyric acid (GABA) and the action of related compounds on single nerve cells. *Journal of Neurochemistry*, **4**, 19–30.

Edwards, C. and Ottoson, D. (1958) The site of impulse initiation in a nerve cell of crustacean stretch receptor. *Journal of Physiology*, **143**, 138–148.

Evoy, W. H. and Kennedy, D. (1967) The central nervous organization underlying control of antagonistic muscles in the crayfish. I. Types of command fibers. *Journal of Experimental Zoology*, **165**, 223–238.

Eyzaguirre, C. and Kuffler, S. W. (1954) Inhibitory activity in single cell synapses. *Biological Bulletin*, **107**, 310.

Eyzaguirre, C. and Kuffler, S. W. (1955a) Processes of excitation in the dendrites and in the soma of single isolated sensory nerve cells of the lobster and crayfish. *Journal of General Physiology*, **39**, 87–119.

Eyzaguirre, C. and Kuffler, S. W. (1955b) Further study of soma, dendrite and axon excitation in single neurons. *Journal of General Physiology*, **39**, 121–153.

Fields, H. L. (1966) Proprioceptive control of posture in the crayfish abdomen. *Journal of Experimental Biology*, **44**, 455–468.

Fields, H. L., Evoy, W. H. and Kennedy, D. (1967) Reflex role played by efferent control of an invertebrate stretch receptor. *Journal of Neurophysiology*, **30**, 859–874.

Fields, H. L. and Kennedy, D. (1965) Functional role of muscle receptor organs in crayfish. *Nature*, **206**, 1235–1237.

Florey, E. and Florey, E. (1955) Microanatomy of the abdominal stretch receptors of the crayfish (*Astacus fluviatilis* L.). *Journal of General Physiology*, **39**, 69–85.

Furshpan, E. J. (1955) Studies on certain sensory and motor systems of decapod crustaceans. *Ph.D. Thesis*. California Institute of Technology, Pasadena, California, U.S.A.

Grampp, W. (1966) The impulse activity in different parts of the slowly adapting stretch receptor neuron of the lobster. *Acta Physiological Scandanavica*, **66, Suppl, 262**, 1—36.

Hagiwara, S., Kusano, K. and Saito, S. (1960) Membrane changes in crayfish stretch receptor neuron during synaptic inhibition and under action of gamma-aminobutyric acid. *Journal of Neurophysiology*, **23**, 505—515.

Hughes, G. M. and Wiersma, C. A. G. (1960) Neuronal pathways and synaptic connections in the abdominal cord of the crayfish. *Journal of Experimental Biology*, **37**, 291—307.

Jansen, J. R. S., Nja, A., Ormstad, K. and Walløe, L. (1971) On the innervation of the slowly adapting stretch receptor of the crayfish abdomen. An electrophysiological approach. *Acta Physiological Scandanavica*, **81**, 273—285.

Jasper, H. H. and Pezard, A. (1934) Relation entre la rapidité d'un muscle strié et sa structure histologique. *Comptes rendus hebdom cidaire des Séances de l'Académie des Sciènces*, **198**, 499—501.

Kennedy, D., Evoy, W. H. and Fields, H. L. (1966) The unit basis of some crustacean reflexes. *Symposia of the Society for Experimental Biology*, **20**, 75—109.

Kennedy, D., Evoy, W. H., Dane, B. and Hanawalt, J. T. (1967) The central nervous organization underlying control of antagonistic muscles in the crayfish. II Coding of position by command fibers. *Journal of Experimental Zoology*, **165**, 239—248.

Kennedy, D. and Takeda, K. (1965a) Reflex control of abdominal flexor muscles in the crayfish, I. the twitch system. *Journal of Experimental Biology*, **43**, 211—227.

Kennedy, D. and Takeda, K. (1965b) Reflex control of abdominal flexor muscles in the crayfish, II. the tonic system. *Journal of Experimental Biology*, **43**, 229—246.

Kosaka, R. (1969) Electrophysiological and electron microscopic studies on the neuromuscular junction of the crayfish stretch receptors. *Japanese Journal of Physiology*, **19**, 160—175.

Krnjević, K. and van Gelder, N. M. (1960) The effects of stretch on the tension and rate of discharge of crayfish stretch receptors. *Journal of Physiology*, **154**, 27.

Krnjević, K. and van Gelder, N. M. (1961) Tension changes in crayfish stretch receptors. *Journal of Physiology*, **159**, 310—325.

Kuffler, S. W. (1954) Mechanisms of activation and motor control of stretch receptors in lobster and crayfish. *Journal of Neurophysiology*, **17**, 558—574.

Kuffler, S. W. and Edwards, C. (1958) Mechanism of gamma aminobutyric acid (GABA) action and its relation to synaptic inhibition. *Journal of Neurophysiology*, **21**, 586—610.

Kuffler, S. W. and Eyzaguirre, C. (1955) Synaptic inhibition in an isolated neve cell. *Journal of General Physiology*, **39**, 155—184.

Nadol, J. B., Jr. and DeLorenzo, A. J. D. (1968) Observations on the abdominal stretch receptor and the fine structure of associated axo-dendritic synapses and neuromuscular junctions in *Homarus. Journal of Comparative Neurology*, 132, 419–444.

Nakajima, S. (1964) Adaptation in stretch receptor neurons of crayfish. *Science*, 146, 1168–1170.

Nakajima, S. and Onodera, K. (1969) Membrane properties of the stretch receptor neurons of crayfish with particular reference to mechanisms of sensory adaptation. *Journal of Physiology*, 200, 161–185.

Nakajima, S. and Takahashi, K. (1966) Post-tetanic hyperpolarization and electrogenic sodium pump in stretch receptor neurone of crayfish. *Journal of Physiology*, 187, 105–127.

Page, C. H. and Sokolove, P. G. (1972) Crayfish muscle receptor organ: role in regulation of postural flexion. *Science*, 175, 647–650.

Perkel, D. H., Schulman, J. H., Bullock, T. H., Moore, G. P. and Segundo, J. P. (1964) Pacemaker neurons: effects of regularly spaced synaptic input. *Science*, 145, 61–63.

Peterson, R. P. and Pepe, F. A. (1961) The fine structure of inhibitory synapses in the crayfish. *Journal of Biophysics and Biochemical Cytology*, 11, 157.

Pilgrim, R. L. C. (1960) Muscle receptor organs in some decapod Crustacea. *Comparative Biochemistry and Physiology*, 1, 248–257.

Pilgrim, R. L. C. (1964) Stretch receptor organs in *Squilla mantis* latr. (Crustacea: Stomatopoda). *Journal of Experimental Biology*, 41, 793–804.

Pilgrim, R. L. C. and Wiersma, C. A. G. (1963) Observation on the skeleton and somatic musculature of the abdomen and thorax of *Procambarus clarkii* (Girard) with notes on the thorax of *Panulirus interruptus* (Randall) *Journal of Morphology*, 113, 453–487.

Severin, F. V., Orlovskii, G. N. and Shik, M. L. (1967) Work of the muscle receptors during controlled locomotion. *Biophysics*, 12, 575–586.

Sokolove, P. G. (1973) Crayfish stretch receptor and motor unit behavior during abdominal extensions. *Journal of Comparative Physiology*, 84, 251–266.

Sokolove, P. G. and Cooke, I. M. (1971) Inhibition of impulse activity in a sensory neuron by an electrogenic pump. *Journal General Physiology* 125–163.

Stein, R. B. (1974) Peripheral control of movement. *Physiological Review*, 54, 215–243.

Terzuolo, C. A. and Washizu, Y. (1960) A study of the generator potential of the crayfish stretch receptor. *Physiologist*, 3, 162.

Terzuolo, C. A. and Washizu, Y. (1962) Relation between stimulus strength, generator potential and impulse frequency in stretch receptor of crustacea. *Journal of Neurophysiology*, 25, 56–66.

Wendler, L. (1963) Über die Wirkungskette zwischen Reiz und Erregung (Versuche an den abdominalen Streckreceptoren des Flusskrebses). *Z. vergl. Physiol.* 47, 280–315.

Wiersma, C. A. G. (1958) On the functional connections of single units in the central nervous system of the crayfish *Procambarus clarkii* (Girard). *Journal of Comparative Neurology*, **110**, 421.

Wiersma, C. A. G. and Bush, B. M. H. (1963) Functional neuronal connections between the thoracic and abdominal cords of the crayfish *Procambarus clarkii* (Girard). *Journal of Comparative Neurology*, **121**, 207–235.

Wiersma, C. A. G., Furshpan, E. and Florey, E. (1953) Physiological and pharmacological observations on muscle receptor organs of the crayfish, *Cambarus clarkii* (Girard). *Journal of Experimental Biology,* **30**, 136–150.

Wiersma, C. A. G. and Ikeda, K. (1964) Interneurons commanding swimmeret movements in the crayfish, *Procambarus clarkii* (Girard). *Comparative Biochemistry and Physiology*, **12**, 509–525.

Wiersma, C. A. G. and Pilgrim, R. L. C. (1961) Thoracic stretch receptors in crayfish and rock lobster. *Comparative Biochemistry and Physiology*, **2**, 51–64.

3 Non-impulsive thoracic-coxal receptors in crustaceans

B. M. H. BUSH

3.1. Anatomy

This chapter deals with a group of muscular and connective strands innervated by a small number of large diameter sensory nerve fibres. They lie predominantly within the segmental thoracic compartments at the bases of the pereiopods, though most of them span the most proximal, thoracic-coxal, limb joint.

Thoracic-coxal receptor organs have been described only in the Decapoda (Alexandrowicz and Whitear, 1957; Alexandrowicz, 1958, 1967), but comparable structures may well be present in other groups of Crustacea. Within the decapods they differ somewhat from one 'tribe' to another (Figs. 3.1 and 3.2). The following basic components of the receptor complex of each leg, however, are common to all decapods. (a) The *muscle receptor* comprises a long thin, almost cylindrical, bundle of muscle fibres enclosed in an elastic connective tissue sheath, innervated by two large and one or two small diameter sensory fibres at its proximal end and by one or two fine motor axons which run along its length. (b) The *levator* and *depressor receptors* are two very thin elastic strands which originate in the thorax and end distally in the coxopodite, among the insertions of the basipodite (limb) levator and depressor muscles,

115

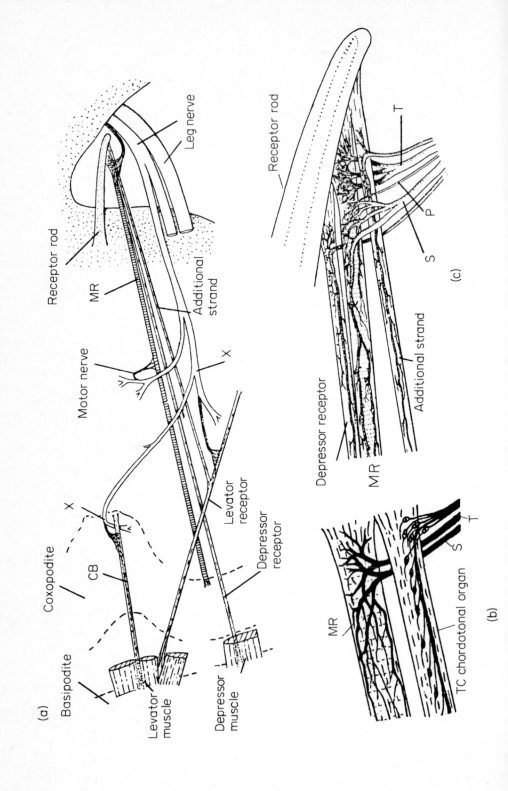

(a)

Basipodite

Coxopodite

CB

X

Receptor rod

MR

Motor nerve

Additional strand

X

Levator receptor

Depressor receptor

Levator muscle

Depressor muscle

Leg nerve

(b)

MR

TC chordotonal organ

S

T

(c)

Receptor rod

Depressor receptor

MR

Additional strand

S

P

T

respectively. Except in Brachyura, one or both are joined by an *additional elastic strand* (Fig. 3.1b, c). Each of these three or four strands is innervated by from one to three stout fibres and one or more fine fibres, but they lack any receptor muscle or motor supply. (c) A *chordotonal organ* (Chapter 6) is also present in the Astacura, running parallel with the muscle receptor from its origin alongside it (Fig. 3.1a) to its insertion in the coxa slightly distal to that of the muscle receptor.

3.1.1   The muscle receptor

At its proximal end this receptor organ commonly attaches poster-iorly on the mesial rim of the endophragmal septum separating thoracic compartments. However, in *Palinurus, Homarus*, and the cheliped segment in brachyurans, it originates on a special projection in this region termed a 'receptor rod'. It runs ventrolaterally to insert inside the anterior proximal rim of the coxa or, in the four walking leg segments of brachyurans, distally on the coxal promoter tendon. In all cases it is functionally in parallel with the limb promotor muscle so that it is *stretched by 'remotion'* of the coxopodite (i.e. backward movement of the leg) and, due to its elasticity, it shortens with 'promotion'.

The sensory innervation of the muscle receptor also differs character-istically between the decapodan tribes (Figs. 3.1a, b; 3.2a). The two large fibres are most clearly differentiated in the Brachyura (Fig. 3.2a) (Alexandrowicz and Whitear, 1957). In this tribe, the 'S fibre' bifurcates peripherally to supply the two connective tissue *S*trands which flank the receptor muscle proximally (Fig. 3.4a), while the '*T* fibre' divides into several smaller branches which enter a short proximal *T*endon of the receptor muscle itself. Proximal to their branch points they run together as roughly uniform cylinders of 40–60 μm diameter (Fig. 3.4b) in adult *Carcinus*, for 1 mm to 6 mm (posterior leg), to their respective segmental thoracic ganglia. A third,

---

Fig. 3.1   (a, c) Anatomy of the thoracic-coxal receptors in *Palinurus vulgaris* (Palinura; posterior view); CB—coxo-basal chordotonal organ; MR—thoracic-coxal muscle receptor; P,S,T—afferent fibres. (b) Diagram showing the sensory innervation of the muscle receptor in Astacura. (From Alexandrowicz, 1967.)

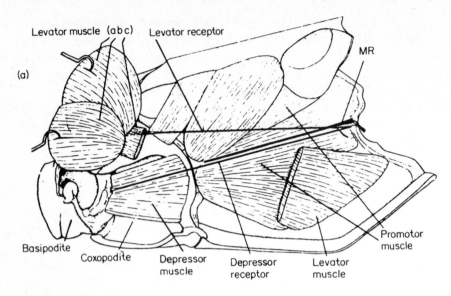

(a)

Levator muscle (abc) Levator receptor

MR

Basipodite Coxopodite Depressor muscle Depressor receptor Levator muscle Promotor muscle

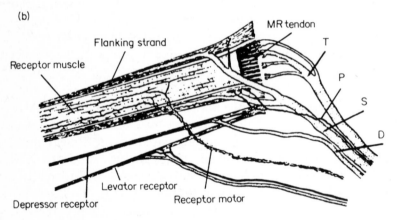

(b)

Receptor muscle Flanking strand MR tendon T P S D

Depressor receptor Levator receptor Receptor motor

Fig. 3.2 Anatomy of the thoracic-coxal receptors in *Carcinus maenas* (Brachyura; posterior views). MR—thoracic-coxal muscle receptor; D,P,S,T— afferent fibres. (From Alexandrowicz and Whitear, 1957.)

smaller 'P' fibre $(5-10\,\mu\mathrm{m}$ diameter), which innervates the ventral flanking strand only, accompanies them.

The muscle receptor in *Eupagurus* (Anomura) resembles the brachyuran plan (Alexandrowicz, 1958), but that of the Astacura (*Homarus* and *Astacus*) (Alexandrowicz and Whitear, 1957) and Palinura (rock-lobsters) (Alexandrowicz, 1967) lacks any flanking strands, and the two large sensory neurons branch profusely in the proximal region of the receptor muscle (Fig. 3.1a). The prawn, *Leander* (Natantia), exhibits both bifurcating and multiply branching forms of 'S fibre' endings (Cannone and Bush, unpublished observations), suggesting a polymorphic evolutionary origin to the 'higher' forms.

3.1.2 Central morphology of the sensory neurons

In contrast to the astacuran chordotonal organ and most other crustacean mechanoreceptors, the sensory cell bodies of the muscle receptor and the levator and depressor receptors lie *centrally* in the segmental thoracic ganglia (Alexandrowicz and Whitear, 1957). This has recently been shown unequivocally for the muscle and depressor receptor neurons in the posterior leg segment of *Carcinus*, by cobalt iontophoresis via the cut ends of their afferent fibres (Fig. 3.3a–d). S, T and D neurons branch characteristically within the ganglion, and their somata occur consistently in similar, well separated positions (Fig. 3.3a, b). Their general appearance in these preparations resembles that of the ipsilateral promotor motoneurons of the same segment, (Fig. 3.3c) and their respective arborization fields overlap extensively. Indeed one could conceive that they might represent motoneurons modified in evolution for a proprioceptive function!

3.1.3 Ultrastructure of the muscle receptor

The fine structure of the sensory dendrites of the thoracic-coxal receptors has been investigated in *Carcinus*, *Pagurus* and *Astacus* by Whitear (1965), and in *Cancer* by Krauhs and Mirolli (1975). At the ultrastructural level there are no significant differences between these decapods, or between the dendritic endings of the different receptors (excluding the chordotonal organ—Chapter 9). Figure 3.4 illustrates the major features.

Between the ganglion and the receptor, each *nerve fibre* is encased in several layers of sheath cells, closely packed on the inside (Fig. 3.4b), which persist into the connective tissue of the organ (Fig. 3.4a). Each of the relatively long *primary dendritic branches* of both S and T fibres give off numerous short *secondary branches* roughly at right angles. These penetrate between 'string cells' and enter an

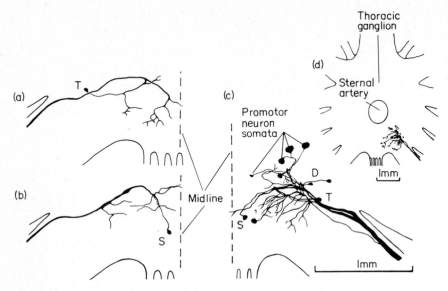

Fig. 3.3   Central morphology of (a, b) individual T and S sensory neurons and (c, d) combined S, T, D and promotor neurons in the posterior thoracic ganglion of *Carcinus maenas*, following 'back-filling' with cobalt chloride, (Luff, Godden and Bush, unpublished observations.)

adjacent longitudinally oriented 'vacuolated string', turning to run lengthways in one or both directions within the string (Fig. 3.4d–f). The naked, unsheathed character of these terminal 'dendrite fingers', about 0.1 μm in diameter and several μm long, gives the appearance of vacuoles when cut in cross section. The strings themselves consist of an extracellular amorphous connective tissue substance containing bundles of collagen-like fibrils, which also resemble (Krauhs and Mirolli, 1975; Whitear, personal communication) invertebrate elastic fibres (Elder, 1973). Each string is encased in and presumably secreted by, one or more string cells.

The dorsal and ventral *flanking strands* of the muscle receptor in *Carcinus* are separated proximally from the connective tissue sheath surrounding the muscle but merge with it more distally, well before the main S fibre branches terminate (Fig. 3.4a). Each strand contains about eight vacuolated strings, which run most of the length of the S dendrites, merging into the general connective tissue of the strands proximally and distally. The P fibre enters some of the same vacuolated strings as the ventral branch of S and, therefore, is presumably stimulated by the same mechanism. The strings of the T fibre are shorter and more numerous, lying within the short proximal 'tendon' of the receptor muscle. Proximally they are attached via connective tissue to the hypoderm of the endosternite; distally they connect with the tapered proximal ends of the muscle fibres (Fig. 3.4g). Thus the vacuolated strings in which the *T fibre terminates* are clearly *in series* with the receptor muscle, while those of the *S fibre* are *in parallel* with it (Whitear, 1965).

The muscle fibres of the crab receptor broadly resemble slow, tonic, crustacean skeletal muscle in fine structure (Fig. 3.4c) (Bush and Luff, unpublished observations). Two fine (5—15 μm diameter) motor axons leave the promotor muscle nerve opposite the receptor and join it between the flanking strands (Fig. 3.2b), to run along the surface for almost its whole length, giving off numerous branches which penetrate the fibres at frequent intervals (Fig. 3.4c).

3.2 Afferent nerve response to receptor stretch

The most striking feature of the responses of all the thoracic-coxal receptor neurons with central somata is their lack of propagated, all-or-none impulses (Ripley, Bush and Roberts, 1968). Extra- or intracellular electrodes record only slow, graded, usually non-overshooting, depolarizing potential changes in response to stretching the receptor muscle, either directly or by remotion of the leg at the thoracic-coxal joint. These responses have the characteristics of receptor potentials (Fig. 3.5b—f), with dynamic and static components related to the corresponding phases of the stretch stimulus, and graded with rate (Fig. 3.5c) and amplitude (Fig. 3.5d) of stretch. Simultaneous monitoring at two or more points along the nerve shows that they are conducted decrementally (Fig. 3.5e, f), without generating impulses, throughout the length of the afferent nerve. That these 'non-impulsive' afferent responses are indeed biologically functional is demonstrated by the resulting reflex excitation of impulses in motoneurons innervating the coxal promotor muscle, with which the receptor muscle lies in parallel (Fig. 3.5a) (Bush and Roberts, 1968).

Essentially similar non-impulsive responses have been recorded in the afferent fibre of the depressor receptor (Fig. 3.5d) and the one levator receptor fibre tested in *Carcinus*, and in the sensory fibres of the muscle receptor in a number of decapod crustaceans, including *Leander* (Natantia), *Astacus* and *Homarus* (Astacura) (Fig. 3.5i, j), *Panulirus* (Palinura), *Eupagurus* and *Birgus* (Anomura) and *Carcinus*,

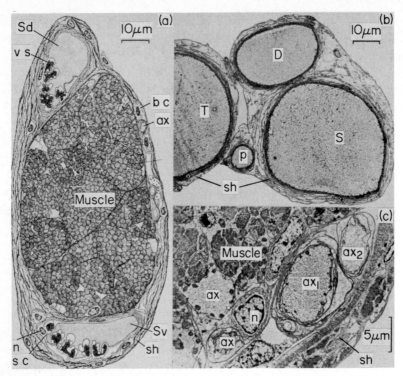

Fig. 3.4   *Fine structure of the muscle receptor in Carcinus maenas (a–g) and Cancer antennarius (h–j).* (a) Drawing from electronmicrographs of a transverse section of the muscle receptor through the dorsal and ventral flanking strands (anterior to the left). (b) Transverse electronmicrograph through the sensory nerve near the muscle receptor. (c) Transverse section through the edge of the receptor muscle. (d) Stereogram drawing showing relationship of sensory dendrite to the connective tissue. (e) Transverse section of vacuolated strings and dendrite fingers of S fibre. (f) Longitudinal section of string and fingers of T fibre. (g) Transverse section of junction between vacuolated string and receptor muscle (arrow). (h) Transverse section of dendrite fingers in relaxed receptor and (i) stretched receptor. (j) Unstained transverse section from a receptor incubated in LaCl$_3$, showing lanthanum particles (black dots) in extracellular spaces between string cells and dendrite branchlet, and in the string around and between fingers.

ax	— motor axon;	sh	— sheath
b c	— blood cell	s c	— string cell
d	— dendrite	v s	— vacuolated string
df	— dendrite finger	D,P,S,T	— afferent nerve fibres
m	— mitochondrion	Sd, Sv	— dorsal, ventral branch
n	— nucleus		of S fibre

(a, d–g from Whitear, 1965; b, c from Bush and Luff, unpublished observations; h–j from Krauhs and Mirolli, 1975.)

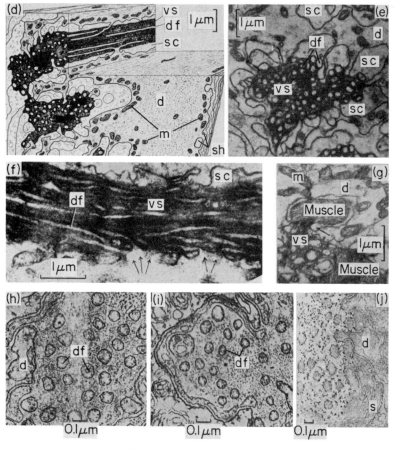

Fig. 3.4   d, e, f, g, h, i, j.

Cancer (Fig. 3.5f), *Maia*, *Portunus* and *Potamon* (Brachyura) (Bush and Roberts, 1968; Mirolli, 1976; Bush and Cannone, unpublished observations). The afferent fibres vary in length with species and limb segment, from 1—2 mm in the prawn and anterior pereiopod segments of small specimens of other species to 10—20 mm in the posterior leg segments of large crabs; and even these latter, relatively long nerves lacked impulses (Fig. 3.5f). The only partial exception encountered was the occurrence in some specimens of *Carcinus*, *Portunus*, *Potamon* and *Birgus* of occasional graded, active membrane responses, similar to those seen in many crustacean muscle fibres (Atwood, 1967), superimposed upon the basic receptor potentials (Fig. 3.5g, h). Their

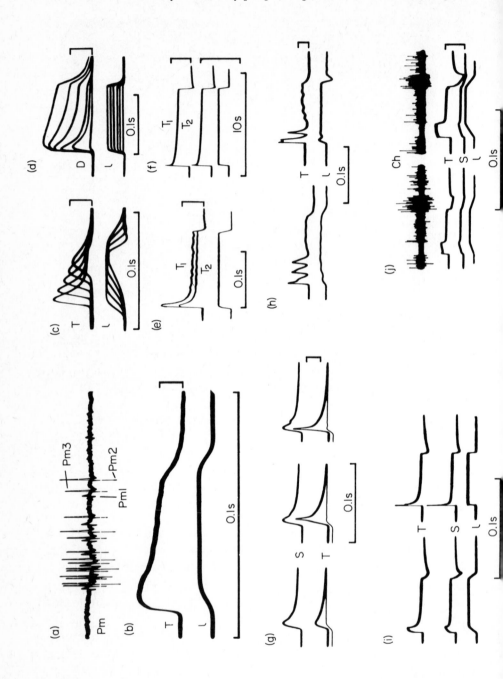

evidently rare occurrence suggests that they may not be particularly important physiologically.

In contrast to the foregoing non-impulsive sensory neurons with central somata, the *thoracic-coxal chordotonal organ* of the astacurans does exhibit afferent impulses (Bush and Cannone, unpublished observations). Moreover these peripheral bipolar neurons respond only to *relaxation* of the receptor strand, being inhibited by strand stretch (Fig. 3.5j). This receptor thus appears to complement the muscle receptor with which it lies in parallel.

The crustacean thoracic-coxal muscle receptor was the first reported instance of a mechanoreceptor lacking afferent impulses. Similar decremental afferent conduction has since been reported in four 'giant' (40−50 μm diameter) sensory fibres of a *non-muscular stretch receptor* at the base of the *uropod* in *Emerita* (Anomura), whose somata lie within the last abdominal ganglion (Paul, 1972). Their graded receptor potentials also elicit appropriate resistance reflexes, which in this case involve coordinated excitation and inhibition of antagonistic muscles. Possibly this receptor (with its accompanying 'muscular strand') represents part of a homologous series including the thoracic-coxal receptors. Non-impulsive conduction has also been observed in several arthropod (and vertebrate) *visual receptors*, including the barnacle lateral ocellus, whose decrementally conducting sensory *axons* may be up to 11 mm long but only 5−20 μm in diameter (Shaw, 1972). '*Non-spiking*' *interneurons* have also been reported recently, in the motor control

---

Fig. 3.5 Basic evidence for slow, graded, non-impulsive, decrementally conducted afferent responses to receptor stretch. (a−e *Carcinus*), (f) *Cancer*, (g) *Potamon*, (h) *Portunus* (i) *Astacus*, (j) *Homarus*. (a) Promotor stretch reflex and (b) concurrent T fibre response. (c) Graded velocity. (d) Graded amplitude stretch of *depressor* receptor. (e, f) Decremental conduction in T fibre as seen by two electrodes separated by 2 mm (e, nerve length 4 mm) and 7 mm (f, nerve length 12 mm). (g, h) Graded active membrane responses. (i) Astacuran S and T responses to two velocities of stretch. (j) Chordotonal organ nerve responses to strand release, and concomitant S and T fibre responses, to two different stretch amplitudes.

l−length monitor traces (increase upwards); S, T−afferent fibres (intracellular, depolarization upwards); Ch−chordotonal organ afferent nerve; Pm−promotor motor nerve; Pm1, Pm2, Pm3−promotoneurons 1, 2, 3. (cf. Fig. 3.14b−d).

Calibrations: 20 mV (S, T) (From Bush and Roberts, 1968; 1971; and Bush and Cannone, unpublished observations.)

systems for ventilation in the lobster (Mendelson, 1971) and walking in the cockroach (Pearson and Fourtner, 1975). Thus signal transmission within neurons entirely lacking impulses may be more widespread than hitherto believed.

3.3   Properties of the afferent fibres

Consistent with the evidently passive, decremental nature of conduction in the S and T fibres, their *voltage/current relationships* are usually almost linear (i.e. ohmic) over a wide range, usually with little or no rectification (Fig. 3.6a–c; Roberts and Bush, 1971). Some S and T fibres show a small graded active response to depolarizing current, with a variable threshold around 20–30 mV depolarized (Fig. 3.6b), but more often such local membrane responses are lacking (Fig. 3.6c).

3.3.1   Cable constants

Cable constant determinations for these fibres have proved somewhat unreliable, owing in part to the complex nature of the resting potential (Section 3.3.2). Since any damage to the fibres, due for instance to multiple impalements, would tend to depress their membrane resistance and hence their apparent length and time constants, the higher values obtained for these constants may be nearer the true values. Assuming this to be so, the following were the 'best' estimates, based on the mean of the three highest values for these constants in *Carcinus* S and T fibres of about 50 μm diameter and 4.5 mm length, with resting membrane potentials around 60 mV and input resistance of about 8×10^5 Ω. The highest individual value is given in brackets. *Length constant* (measured from pairs of stretch-evoked receptor potentials recorded simultaneously at two points about 2 mm apart), S 7.0 mm (11.0 mm); T 14.5 mm (16.0 mm). *Time constant* S 8.0 ms (11.0 ms); T 5.3 ms (7.3 ms). The surprisingly high length constants, which yield electrotonic length values of 0.8 (S) and 0.3 (T), were obtained in fresh, well perfused, *in situ* preparations (Cannone, 1974), cf. the much lower values (S 5.0 mm; T 3.3 mm) previously reported from isolated preparations by Roberts and Bush (1971).

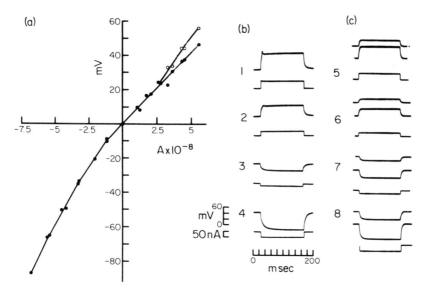

Fig. 3.6 Electrical properties of receptor afferent fibre membrane in *Carcinus*. (a) Voltage/current relations of an S fibre at steady state (●————●) and at peak of initial transient (○————○). (b) Sample of two depolarizing and two hyperpolarizing pulses, used in making graph. (c) Similar responses recorded at two points 2 mm apart in another S fibre lacking active responses (nerve length 5.5 mm). The relatively large electrotonic decrement with distance in this situation (cf. Fig. 3.5e, f), is probably due to the high resting conductance of the peripheral end (transducer region) of the sensory fibre membrane (see Section 3.3.2).

Calibrations: 20 mV, 20 ms/division. (From Roberts and Bush, 1971; and unpublished observations.)

3.3.2   The resting membrane potential

The resting membrane potential of S and T fibres of unstretched crab muscle receptors is commonly in the range −50 mV to −70 mV. Though potassium dependent, the observed slope of only 30 mV per ten-fold change in external potassium concentration indicates that other factors are involved (Roberts and Bush, 1971). Recent evidence suggests that an electrogenic sodium pump contributes to the resting potentials of S and T fibres (Mirolli, 1974, and unpublished observations). Cooling and ouabain depolarize them, while warming hyperpolarizes them (Fig. 3.7a−c). Replacing all external Na^+ with

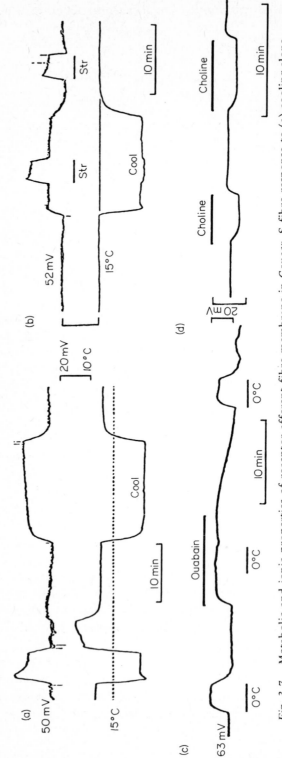

Fig. 3.7 Metabolic and ionic properties of receptor afferent fibre membrane in *Cancer*. S fibre response to (a) cooling alone, (b) cooling and receptor stretch (str), (c) cooling and 10^{-3} mM ouabain, (d) choline replacing external sodium. Lower traces in (a) and (b) show temperature of bathing solution (cooling downwards); all other traces show S fibre intracellular membrane potential (depolarization upwards).

choline initially hyperpolarizes the fibres (d) and doubles their input resistance, indicating that, even with the receptor fully relaxed, the normal resting sodium flux is large. Mirolli also observed a standing potential gradient of about 6 mV/mm along the S fibre of unstretched receptors, with the distal end more depolarized. This he attributes to spatial separation of two current sources: an inward sodium current peripherally due to the permanently 'leaky' sensory terminals, and an outward current in the rest of the sensory fibre due to the sodium pump. The lack of any significant effect of cooling on the stretch-evoked receptor potentials (Fig. 3.7b) is consistent with such a spatial separation, as also is the presence of mitochondria in all segments of the afferent fibres except the terminal dendrite fingers (Whitear, 1965; Krauhs and Mirolli, 1975).

Mirolli proposes an equivalent circuit model comprising a constant voltage source (or voltage clamp) peripherally and a constant current source (the sodium pump) proximally. The former will tend to reduce the time constant of the fibre so permitting transmission of faster transients along the fibre, while the latter will enable the voltage clamp (or short circuit) condition to exist at the receptor endings by maintaining a large potential difference between these and the rest of the cell. At the same time the pump may help to boost the afferent signal by increasing the driving potential, to an extent proportional to the membrane resistance of the fibre. A high membrane resistance would normally imply a large membrane time constant, and hence a slow time course of the afferent signal, but this constraint is removed by the high resting conductance peripherally. The overall effect of this system is to allow the cell 'to use the precision of analogue signaling over substantially longer distances without losing the information embodied in the time course of the stimulus' (Mirolli, unpublished observations; see Krauhs and Mirolli, 1975).

3.3.3 Sensory transduction and its ionic basis

Sustained stretch of the receptor muscle in *Cancer* is characterized at the ultrastructural level by a reduction (of about 25 per cent from minimum to maximum length) in the cross-sectional diameter of the fine terminal dendrite fingers of the S fibre (Fig. 3.4h cf. i; Krauhs and Mirolli, 1975). This is consistent with the hypothesis that the dendrite fingers constitute the main, if not the sole, site of mechano-electric transduction. That the extracellular space around these terminal fingers is readily accessible to ions, as this hypothesis requires, is shown by the aggregation of lanthanum particles in this space (Fig. 3.4j).

Constant current pulses injected into either sensory fibre result in decreasing potential excursions with increasing receptor length (Bush, Cannone and Godden, unpublished observations) (Fig. 3.15c). This indicates a decreasing membrane resistance, presumably at the sensory terminals, with increasing stretch. That the primary transduction event is indeed a membrane permeability change is supported by voltage clamp experiments on the T fibre (Bush, Godden and Macdonald, 1975b). The stretch-evoked receptor currents closely resemble the corresponding unclamped receptor potentials. At normal resting potentials they are inward going, and they reverse to outward currents at membrane potentials around +20 to +30 mV, suggesting that the receptor current is predominantly carried by sodium ions (cf. Roberts and Bush, 1971). However, since the receptor potentials are only reduced to about 20—30 per cent of their former value when all external sodium is replaced by Tris or choline, and the residual response then reverses at about −10 to −20 mV, other ions (probably including calcium) evidently contribute to the receptor currents.

In S and T fibres displaying a graded spike-like component superimposed upon the basic receptor potential (Fig. 3.5g, h), tetrodotoxin abolishes only these 'spiky' components, without affecting the underlying response, indicating a small degree of depolarization-sensitive sodium activation in these afferent fibres. Where such secondary membrane responses are absent, the receptor potential is unaffected by tetrodotoxin.

3.4   Receptor potentials and the parameters of stretch

Characteristic S and T fibre responses to ramp-function and 'trapezoid' receptor length changes are illustrated in Fig. 3.8a—d (see Bush and Roberts, 1971). With slow ramps the initial response ('a' in A) can sometimes be resolved into an α component, probably representing an active membrane response of the afferent fibre (cf. Fig. 3.5g) and a β component, which reflects the initial mechanical

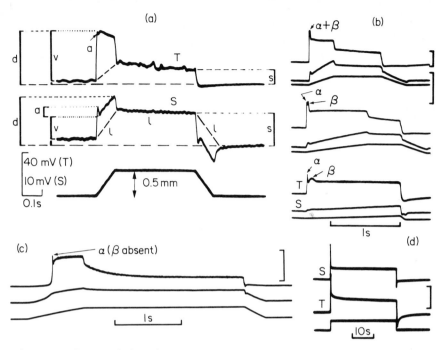

Fig. 3.8 Characteristics of T and S fibres receptor potentials. (a–c) *Carcinus,* (d) *Potamon.* Initial lengths of de-efferented receptor muscle cf. *in situ* lengths: (a, b, d) mid-length; (c) minimum. Components: *d*—dynamic, *s*—static; *l*—length, *v*—velocity, *a*—'acceleration' (= α + β).

Lower traces—length monitor traces (increase upwards); S, T—afferent fibre membrane potentials (intracellular, depolarization upwards).

Calibrations (b–d): 20 mV (S, T). (After Bush and Roberts, 1971; Cannone, 1974; Bush and Cannone, unpublished observations.)

condition of the receptor muscle and can thus be regarded as the 'true' initial response of the receptor to the onset of stretch (Fig. 3.8b, cf. Fig. 3.8c). (see also Cannone, 1974, pp 40–42; Figs. 14–17).

3.4.1   Tension changes with receptor stretch

The 'tension response' to trapezoid stretch, recorded at the distal end of the receptor, is remarkably similar in shape to the receptor potentials (Fig. 3.9a) (Bush and Godden, 1974). Not surprisingly it has features of both S and T receptor potentials, since the mechanical linkages involved include both the receptor muscle and its roughly tubular connective tissue sheath in series, respectively, with the T and S fibre. The initial tension transient, like the β component in the T fibre response, varies with receptor length and interval since the previous stretch, as well as the degree of any contractile response, suggesting that it may depend partly upon the formation of cross-bridges between thick and thin muscle filaments (Hill, 1968; cf. Lannergren, 1971).

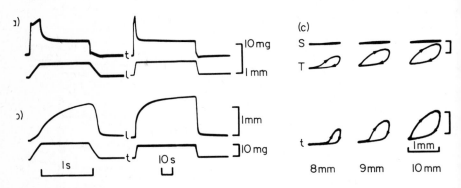

Fig. 3.9   (a) Tension and (b) length changes in response to ramp-function length (a) and tension (b) increments in the receptor muscle in *Carcinus*. (c) x-y displays showing hysteresis in S and T membrane potentials and tension (recorded at different times) with 1 Hz sinusoidal length change of 1 mm amplitude. Initial length of de-efferented receptor muscle cf. *in situ* lengths: (a, b) mid-length, (c) as indicated (min 8 mm, max 12 mm).

l, t—length, tension monitor traces (increase upwards); S, T—afferent fibres (intracellular, depolarization upwards).

Calibrations (c): 20 mV (S, T); 20 mg (t). (From Bush, Godden and Macdonald, 1975a; and unpublished observations.)

A more direct analysis of the relationships between receptor potentials and tension may become possible with servo-controlled tension rather than length changes (Bush, Godden and Macdonald, 1975a). The slow length changes recorded under such 'tension clamp' conditions (Fig. 3.9b) emphasise the highly visco-elastic behaviour of the receptor muscle. Further evidence for the close relationship between T fibre response and tension is seen in X-Y displays during low frequency sinusoidal oscillation of the muscle receptor (Fig. 3.9c). Pronounced length-dependent hysteresis is apparent in both the tension and the T fibre membrane potential changes.

3.4.2   Stretch velocity

The amplitude and shape of the dynamic components of both S and T fibre receptor potentials vary systematically with velocity of stretch (Fig. 3.10). The T fibre shows a much higher velocity sensitivity than the S fibre, its dynamic amplitude varying almost linearly with log velocity over most of the physiological range (Bush and Roberts, 1971). The accompanying tension responses appear rather intermediate in velocity sensitivity (Bush and Godden, unpublished observations).

Sinusoidal changes in length also evoke characteristic S and T response waveforms, and further illustrate the T fibre's marked dynamic sensitivity. Frequency-response curves display a steeper slope for the T than the S fibre over an equivalent velocity range to that studied with ramp stretches.

3.4.3   Receptor length

The form of the S and T fibre responses, like the underlying tension changes, depends upon the receptor's resting length, particularly at the shorter lengths and in the absence of efferent activity (Fig. 3.11). Moreover the responses to successive trapezoid stretches at short intervals show considerable 'hysteresis', the responses obtained at successively decreasing lengths being smaller than those at comparable increasing lengths. This is particularly evident in the T fibre's dynamic and initial membrane potentials, whereas the S fibre's response is more consistent, less hysteretic, and nearly linearly related to receptor length. The tension responses again appear more or less intermediate in these respects.

3.4.4   Time dependent changes

The interval between successive stretches also greatly affects the responses, particularly at short resting lengths (Bush, Cannone and Godden, unpublished observations). For example, the duration of the T fibre's dynamic response to a ramp stretch of a relatively slack receptor increases progressively with increasing inter-stimulus interval. The time constant of this 'recovery' process varies inversely as the resting length, being about 2—3 minutes with the de-efferented receptor at minimum *in situ* length. Such variability in the T fibre

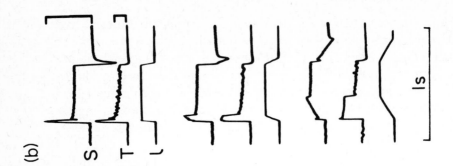

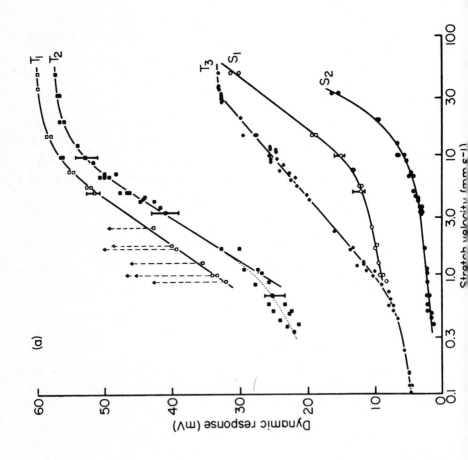

response to a 'constant' stimulus, greatly influences its effective information content, as witness the concomitant variation in the promotor stretch reflex.

3.5 Role of the receptor muscle

Morphologically, the T fibre in Brachyura terminates in series with the receptor muscle, whereas the S fibre endings appear to lie in parallel with its proximal region (Fig. 3.2b). That this model is functionally accurate is supported by observations on receptors in which the receptor muscle is ruptured within its sheath near the proximal end (Cannone, 1974; Bush and Godden, unpublished observations). The T fibre membrane potential in such preparations varies little, if at all, with receptor length within its normal length range, while the S fibre potential retains its linear dependence upon length, generally with an increased slope. Clearly therefore, efferent activation of the intact receptor muscle is likely to have a marked influence upon the T fibre's response, though its effect on the S fibre in crabs is less predictable.

3.5.1 Effects of receptor motor stimulation

Repeated microelectrode penetrations at any point along the receptor muscle in *Carcinus* reveal approximately 10 to 20 electrically distinct muscle fibres, with membrane potentials ranging from about -40 to -70 mV (Bush and Roberts, 1971; Cannone, 1974). Motor nerve stimulation elicits depolarizing junctional potentials of $1-12$ mV unfacilitated amplitude and widely differing time courses, mostly about 10x slower than those of promotor muscle

---

Fig. 3.10   Stretch velocity effects on dynamic responses of T and S fibres. (a) Peak amplitude of dynamic component ('d' in Fig. 3.8) *vs* stretch velocity; triangles in T_1 represent peaks of initial transients, squares the beginnings of the true velocity components. (b) Sample records from S_1, T_1 series.

l—length monitor traces (increase upwards); S, T—afferent fibres (intracellular, depolarization upwards).

Calibrations: 20 mV (S, T). (From Bush and Roberts, 1971).

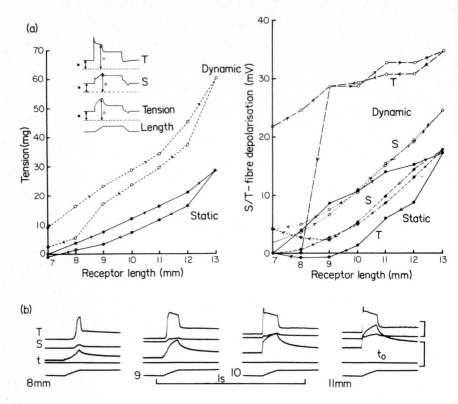

Fig. 3.11   Receptor length effects on afferent responses and tension. (a) Dynamic and static receptor tension and S and T amplitudes (as indicated in inset) to constant trapezoid stretches applied at 10 second intervals (*in situ* length range: 8–12 mm). (b) Sample positive dynamic components from which curves in (a) were plotted.

l, t—length, tension monitor traces (increase upwards); S, T—afferent fibres (intracellular, depolarization upwards).

Calibrations: 20 mV (S, T); 20 mg (t). (From Bush and Godden, unpublished observations.)

fibres. Facilitation and summation with repetitive stimulation can lead to high levels of sustained depolarization, particularly at high frequencies. Under isometric conditions this results in slow tension development, the rate and extent of which increases with stimulation frequency (Fig. 3.12a, b). Around or below minimum *in situ* length, however, when the resting receptor appears slack, the development of tension may be delayed (Fig. 3.12a). Optimal length for tetanic

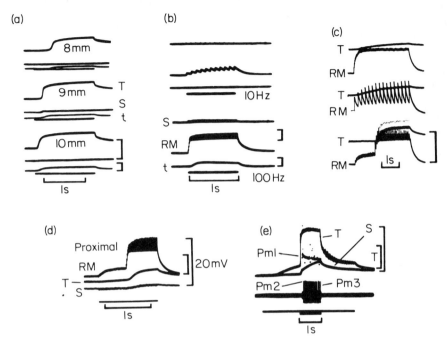

Fig. 3.12   Receptor efferent stimulation effects.

(a) Isometric tension (t) and S and T fibre responses to 100 Hz trains (bars) at three receptor lengths as indicated (range *in situ*: 8–12 mm). (b) Tension, e.j.p.s., and S fibre responses at mid-length to 10 Hz and 100 Hz trains. (c) Dual excitatory innervation shown by e.j.p.s. and T responses with low intensity, high frequency (top); higher intensity, low frequency (middle); and smoothly increasing intensity (bottom). (cf. Fig. 1e in Bush and Cannone, 1974.) (d) Increasing intensity stimulation of promotor nerve proximal to receptor motor nerve branch point. (e) High frequency train (100 Hz) (thick bar on bottom trace), superimposed on low frequency efferent stimulation (10 Hz), evoke reflex and small S response in addition to large T response.

t—tension monitor traces (increase upwards); S, T—afferent fibres (intracellular, depolarization upwards). RM—receptor muscle fibre (intracellular); Pm1, Pm2, Pm3—promotoneurons 1, 2, 3.

Calibrations: 20 mV (S, T, RM); 20 mg (t). (From Bush and Godden, unpublished observations; Cannone, 1974.)

tension is between mid- and maximal *in situ* lengths (Bush and Godden, unpublished observations).

Isometric contraction of the receptor muscle results in depolarization of the T fibre, but commonly has little or no effect on the S fibre (Fig. 3.12a, b) (Bush and Godden, 1974). There is an almost linear relation between isometric tension and T depolarization, both during tension development at one frequency and at different frequencies. The T fibre thus appears to be responding primarily to receptor tension under isometric conditions.

Graduation of efferent stimulation intensity reveals two excitatory motor units innervating the receptor muscle (Bush and Cannone, 1974). This can be seen in the successive recruitment of two amplitudes of depolarizing, excitatory junctional potentials and, at sufficiently high frequencies, of two increments in T fibre depolarization (Fig. 3.12c). The smaller but not the larger amplitude e.j.p.s. can also be evoked by stimulation of the promotor nerve distal to the point at which the receptor motor nerve branches off it (cf. Fig. 3.12d), indicating that the smaller unit is shared with the promotor muscle. This inference is supported by serial transverse electronmicrographs, which clearly show the smaller of the two receptor efferents to branch at this point (Cannone, 1974).

In some preparations, high frequency stimulation of both receptor efferents does lead to a small depolarization of the S fibre in addition to a large T fibre response (Fig. 3.12d, e). Possibly in these cases the receptor muscle contracts more strongly distally, so as to stretch the S fibre terminals in parallel with its proximal end. Alternatively, if the receptor muscle were to contract more strongly proximally, this could lead to a reduced S response to a simultaneous stretch stimulus at the same time as the T response is enhanced by the efferent stimulus. Reciprocal variations of S and T responses have been noted previously (Bush and Roberts, 1971), and it seems possible that one role of the dual motor innervation of the receptor muscle might be to influence S and T responses differentially in this sort of way.

3.5.2   Interaction of receptor contraction and stretch

As expected from the foregoing observations, efferent activation of the receptor muscle can have a pronounced effect upon the afferent responses to receptor stretch. At any given length, the tension increments, and hence the T fibre depolarizations, in response to concurrent efferent stimulation and receptor stretch, summate (Fig. 3.13a, b; Bush and Godden, 1974). A more striking effect is seen at short lengths, when efferent stimulation preceding a ramp stretch, even by as much as 10–20 seconds or more, produces marked enhancement of the receptor potentials, especially the T fibre dynamic duration, and tension response. Very brief 'conditioning

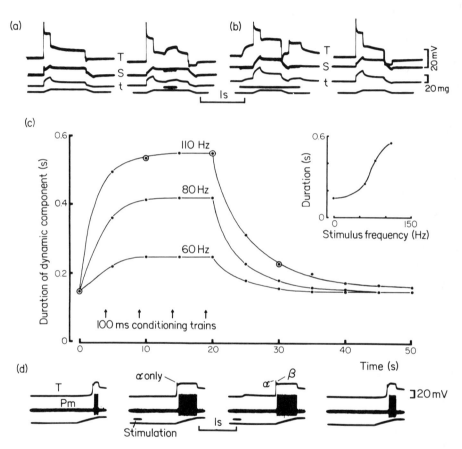

Fig. 3.13   Receptor motor effects on stretch-evoked responses.

(a, b) Summation of responses to trapezoid stretch and 100 Hz motor stimulation (bars) at initial lengths of 10 mm (a) and 11 mm (b). (c) Time course of changes in T fibre dynamic component duration with and then without preceding 100 msec. receptor motor conditioning trains at three frequencies; inset shows plateau levels. (d) Sample responses (at times indicated by circles on top curve); note small T response to conditioning train only in third record. Initial receptor length in c, d approximately minimum *in situ* length. (After Bush and Godden, 1974; Cannone, 1974.)

trains' (e.g. 100 ms) can be effective at shorter intervals, and successive pre-stretch stimuli summate progressively (Fig. 3.13c, d; Cannone, 1974). After cessation of a series of conditioning stimuli, the responses rapidly revert to their pre-conditioned levels (Fig. 3.13c, d). That this decline in the response is a consequence of the continuing passive stretches is shown by the complimentary effect, in which the response 'recovers' if left long enough without stretching. Possibly stretch disrupts some mechanical link in the muscle (cross-bridges?) which, however, re-form spontaneously if left undisturbed, or more rapidly if contraction is stimulated (cf. Hill, 1968).

Functional significance of the efferent control
The foregoing observations suggest that an important function of efferent activation of the receptor muscle is to maintain the sensitivity of the muscle receptor, in particular the T fibre, under conditions of varying length. One aspect of this concerns the 'hysteresis' in the T fibre response to a constant stretch stimulus at different initial lengths. Such hysteresis clearly constitutes a severe source of 'ambiguity' in the afferent response. However, receptor efferent stimulation can reduce or, at adequate frequencies, virtually eliminate the hysteresis in the T fibre response (Bush and Godden, unpublished observations). That is, it is serving to maintain the sensitivity, or 'gain', of the receptor at a steady level. It does not increase the receptor gain, since the T (and S) fibre receptor potentials, in response to a given passive stretch, do not increase during concurrent efferent stimulation. In contrast, however, the *negative dynamic component* of the T fibre, which is greatly enhanced during receptor contraction (Fig. 3.13b). Thus a further, possibly ancilliary, role for the efferent innervation, may be to allow effective signalling of those movements which result in receptor shortening, or unloading of the receptor muscle.

3.6   The promotor stretch reflex

The thoracic-coxal receptor muscle lies in parallel with, and in Brachyura among, the fibres of the coxal promotor muscle. By analogy with mammalian spindle terminology this can therefore be considered as its 'extrafusal' muscle. Passive remotion of the coxa in *Carcinus*, which stretches the receptor muscle, excites several motoneurons innervating the promotor muscle (Bush and Roberts,

1968). This promotor stretch reflex (or 'resistance reflex') is mediated by the muscle receptor, since cutting its afferent nerve abolishes the reflex response to coxal movement.

The major features of this reflex and the methods used in its analysis are summarized in Figs. 3.14 and 3.15. Here the reflex responses of only the three most commonly evoked promotoneurons are illustrated, Pm1, Pm2 and Pm3, in order of increasing thresholds for reflex recruitment. However, a total of up to nine occasionally respond, though more than four or five is unusual in these highly dissected preparations (Cannone, 1974).

3.6.1 Afferent input to the reflex

The main conclusion from these experiments is that the T fibre provides the dominant afferent input to the promotor stretch reflex, while the reflex role of the S fibre remains enigmatic (Bush and Cannone, 1973). The main evidence for this is: (a) The instantaneous frequencies of all responding promotoneurons follow the waveforms of the corresponding components of the T fibre receptor potentials much more closely than those of the S fibre (Fig. 3.14b–e; Fig. 3.13d) (b) Much greater depolarizations of the S than the T fibre are required to produce comparable effects (Fig. 3.15d, e). (c) Hyperpolarizing (or cutting) T, but not S, inhibits the stretch reflex (Fig. 3.15f, g) (d) Small amplitude vibration of the muscle receptor is a potent stimulus to the promotor reflex and the T fibre but not the S fibre and (e) Adequate isometric contraction of the receptor muscle, evoked by receptor efferent stimulation, leads to T fibre depolarization which, if sufficient, elicits reflex promotoneuron discharge in the absence of any S fibre response (Fig. 3.17c).

3.6.2 Positive feedback to the receptor muscle

In addition to exciting motoneurons to the 'extrafusal', promotor muscle, stretching the muscle receptor also reflexly activates one of the two motoneurons innervating the receptor muscle (Fig. 3.16a) (Bush and Cannone, 1974). At high velocities of stretch, producing large T fibre dynamic components, the second, smaller receptor efferent may also be transiently activated (Fig. 3.16b). Strong depolarizing currents injected into the T fibre also excite one or, at high levels, both receptor efferents, and the current and stretch-evoked responses sum as in the promotor reflex (Fig. 3.16c–e).

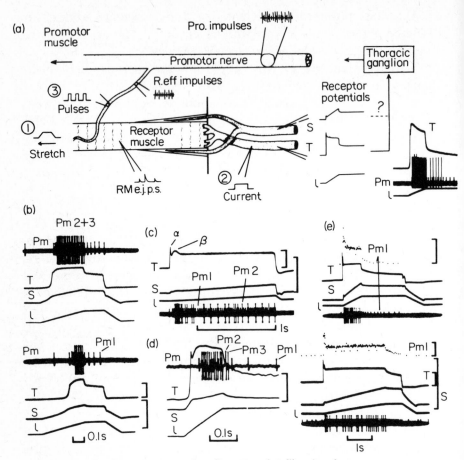

Fig. 3.14   The promotor stretch reflex: T and S fibre involvement.
(a) Diagram showing three methods of stimulating promotoneuron discharge (1–3), and the major types of recording made in analysing the reflex. (b) Constant velocity stretch at two relatively slack lengths (longer in upper record). (c–e) Different velocities of stretch (note different time scales). Instantaneous frequency displays in E are for Pm1. Note reflex response to β but not α component in (c). Pm2 and Pm3 cannot be discriminated owing to their similar spike amplitudes. *Calibrations*: 20 mV (S, T); 100 Hz (Pm1). (After Bush and Cannone, 1973; Cannone, 1974.)

Depolarization of the S fibre, on the other hand, even with very large currents, does not activate either of the receptor efferents. Again, hyperpolarization of the T fibre but not the S fibre blocks the receptor efferent reflex. In this case, therefore, the T fibre provides the sole afferent input to the receptor reflex.

Compared to the 'extrafusal' stretch reflex, the receptor stretch reflex is clearly not a very powerful one. Thus the threshold T fibre depolarizations for recruiting each receptor motoneuron are much higher than, and the slopes of the curves relating reflex frequency with T fibre membrane potential are much lower than, those of the three lowest threshold promotoneurons (Fig. 3.16f; Cannone, 1974). The higher threshold receptor motor unit, Rm2, with the smaller impulses, is probably the one which is shared with the promotor muscle, since it is the smaller of the two which is seen to branch in electronmicrographs. Its effect on the promotor muscle, however, is unknown.

As already seen, both receptor efferent axons produce depolarizing, *excitatory* junctional potentials in the receptor muscle fibres which they innervate (Fig. 3.12c, d; Bush and Cannone, 1974). The relative amplitudes of e.j.p.s. from the two axons differ in different muscle fibres, and may even be reciprocal in neighbouring fibres. Both, however, result in depolarization of the T fibre, though by different relative amounts in different preparations (Cannone, 1974). There seems no doubt, therefore, that the receptor reflex is a positive feedback, self-excitatory reflex. A possible role for this reflex may be to maintain the gain of the sensory system, particularly that of the T fibre, in normal, non-isometric conditions when the 'extrafusal' stretch reflex tends to unload the receptor. The slow contractile response of the receptor muscle, particularly with the low motor frequencies encountered in the reflex, presumably obviates any possibility of oscillation, of the kind normally inherent in positive feedback systems.

3.6.3   Synaptic transmission at the afferent synapses

Micro-assays of transmitter-synthesizing enzymes in *Carcinus* (Bush, Emson and Joseph, 1975; Emson, Bush and Joseph, 1976) and *Homarus* (Barker, Herbert, Hildebrand and Kravitz, 1972), have shown relatively high concentrations of choline acetyltransferase in the thoracic-coxal receptor (and other) sensory nerves. This suggests that acetylcholine may mediate synaptic transmission at the central synapses of the S and T fibres, in keeping with other sensory nerves in Crustacea (Florey, 1973). Whether the promotor stretch reflex depends exclusively upon chemical, cholinergic afferent transmission, however, is uncertain. The inhibitory effect of T fibre

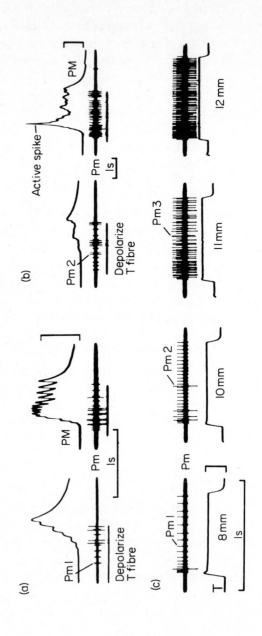

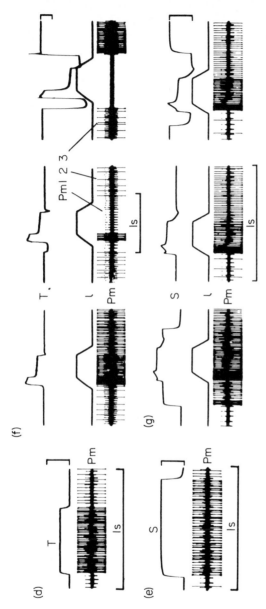

Fig. 3.15   Promotor responses to current pulses injected into T and S fibres.

(a, b) Intracellular records from two promotor muscle fibres with promotor nerve response: (a) responded only to Pm1, (b) only to Pm2. (c) Responses to constant current pulses (13 nA) into the T fibre at four receptor lengths. (d, e) Depolarization of either S or T fibre evokes promotor discharge. (f, g) Hyperpolarization of T but not S inhibits the stretch-evoked reflex; depolarization sums with it in either (only S shown).

Calibrations: 20 mV (PM; S, T). (After Bush and Cannone, 1973; Cannone, 1974.)

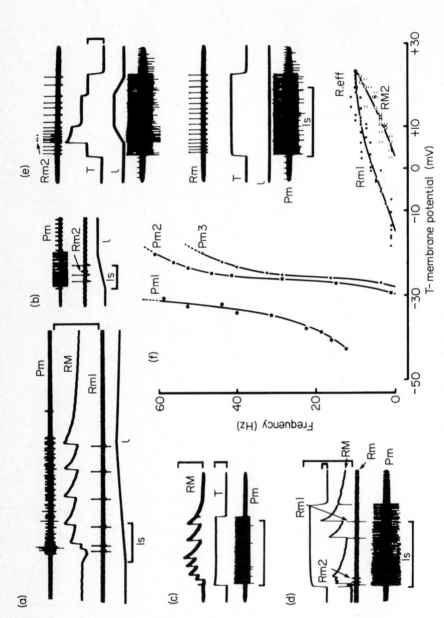

Fig. 3.16   Positive feedback reflex excitation of receptor muscle.
(a, b) Receptor stretch reflexly excites one or both receptor efferents, and sums with current injected into the T fibre (e); current alone, if strong enough, can excite both efferents (d, cf. c). Receptor efferent thresholds are higher, sensitivities lower than for promotoneurons Pm 1–3 (f).

hyperpolarization would be easier to reconcile with an electrical synapse, though the possibility of both electrical and chemical transmission being involved is not excluded (cf. the crayfish 'escape reflex', Zucker, 1973).

Measurements of reflex transmission delays are not very helpful, particularly with decremental conduction and peripheral recording. Despite the rather long delays encountered (4—10 ms), the precise promotoneuron following of the T fibre potentials does suggest a monosynaptic pathway, but intraganglionic recording will be necessary to test this (cf. Burrows, 1975).

3.7   Behavioural role of the thoracic-coxal receptors

Being the only known stretch receptor at this joint (except in Astacura), the crustacean T-C muscle receptor undoubtedly plays a major role in the proprioceptive regulation of the thoracic-coxal joint and hence of the whole limb, and complements the astacuran TC chordotonal organ. The strong promotor stretch reflex clearly implicates this receptor in a servo-control capacity, comparable to other muscle receptors. As yet there is no direct evidence on its normal behaviour *in situ*. However, recordings from 'lively', dissected preparations often show spontaneous bursts of activity occurring more of less *synchronously* in the larger receptor efferent and several promotoneurons (Fig. 3.17a, b; Cannone, 1974). With the receptor clamped isometrically the former leads to T depolarization (Fig. 3.17b), though whether this would also result in the intact animal is doubtful. What evidence there is, suggests that 'co-activation' of receptor and promotor units may be common, so that 'servo-assistance' of movement and posture, as proposed for mammalian muscle spindles (see Matthews, 1972), appears a likely role. Nevertheless, the results exemplified in Figs. 3.12e, 3.17c, suggest that the T-C muscle receptor could also function in a 'follow-up servo' mode as postulated for the crayfish abdominal MRO (Chapter 2), or so as to 're-set' a postural base-level as suggested for the myochordotonal organ (Evoy and Cohen, 1971; Chapter 6).

The function of the levator and depressor receptors is enigmatic. Alexandrowicz (1958) hypothesized that they will contribute to the proprioceptive regulation of the depressor and levator muscles, respectively, which certainly seems probable. To what extent they

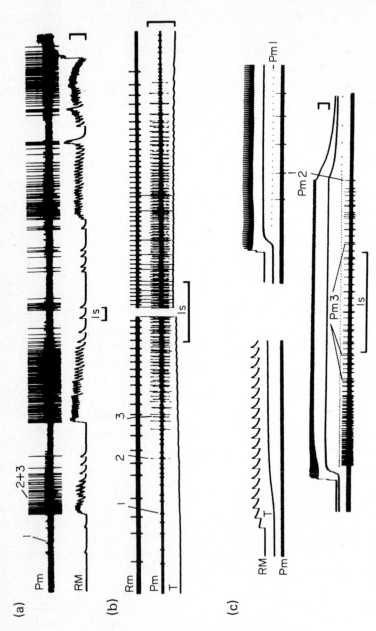

Fig. 3.17   Central and reflexly evoked promotor and concomitant receptor activation.

(a, b) Spontaneous bursts of activity in receptor efferent (Rm1), seen as receptor e.j.p.s. in (a), occur roughly synchronously with promotoneuron bursts. (c) Three frequencies of receptor motor stimulations eliciting e.j.p.s., T depolarization and reflex promotoneuron discharge.

Calibrations: 20 mV (RM; T). (After Cannone, 1974; Bush and Cannone, unpublished observations.)

contribute to the resistance reflexes mediated (primarily?) by the CB chordotonal organ (Bush, 1965), however, is unknown.

Other intra-segmental reflexes involving the T-C muscle receptor, and also various inter-segmental reflexes such as that from the CP chordotonal organs to a T-C receptor motoneurone in the crayfish (Moody, 1970), probably contribute to intra- and inter-segmental coordination within the limb. It may be significant that stretching of the T-C muscle receptor was not found to influence the myochordotonal motoneurons, although CB shortening had a strong reflex excitatory action on the excitatory myochordotonal unit (Bush and Clarac, 1975).

Acknowledgements

I thank Drs. Cannone, Godden and Mirolli for their comments on this chapter and for permission to use their unpublished data, including material from Cannone's Ph.D. thesis. Work in the author's laboratory was supported by the M.R.C. and the S.R.C.

References

Alexandrowicz, J. S. (1958) Further observations on proprioceptors in Crustacea and a hypothesis about their function. *Journal of the Marine Biological Association, U.K.,* **37**, 379–396.

Alexandrowicz, J. S. (1967) Receptor organs in the coxal region of *Palinurus vulgaris. Journal of the Marine Biological Association, U.K.,* **47**, 415–432.

Alexandrowicz, J. S. and Whitear, M. (1957) Receptor elements in the coxal region of Decapoda Crustacea. *Journal of the Marine Biological Association, U.K.,* **36**, 603–628.

Atwood, H. (1967) Crustacean neuromuscular mechanisms. *American Zoologist,* **7**, 527–551.

Barker, D. L., Herbert, E., Hildebrand, J. G. and Kravitz, E. A. (1972) Acetylcholine and lobster sensory neurons. *Journal of Phsyiology* **226**, 205–229.

Burrows, M. (1975) Monosynaptic connexions between wing stretch receptors and flight motoneurons of the locust. *Journal of Experimental Biology,* **62**, 189–219.

Bush, B. M. H. (1965) Leg reflexes from chordotonal organs in the crab, *Carcinus maenas. Comparative Biochemistry and Physiology,* **15**, 567–587.

Bush, B. M. H. and Cannone, A. J. (1973) A stretch reflex in crabs evoked by muscle receptor potentials in non-impulsive afferents. *Journal of Physiology,* **232**, 95–97P.

Bush, B. M. H. and Cannone, A. J. (1974) A positive feed-back reflex to a crustacean muscle receptor. *Journal of Physiology*, **236**, 37—39P.

Bush, B. M. H. and Clarac, F. (1975) Intersegmental reflex excitation of leg muscles and myochordotonal efferents in decapod Crustacea. *Journal of Physiology*, **246**, 58—60P.

Bush, B. M. H., Emson, P. C. and Joseph, M. H. (1975) Transmitter enzymes and amino acid levels in sensory and motor nerves of the shore crab. *Journal of Physiology*, **245**, 6—7P.

Bush, B. M. H. and Godden, D. H. (1974) Tension changes underlying receptor potentials in non-impulsive crab muscle receptors. *Journal of Physiology*, **242**, 80—82P.

Bush, B. M. H., Godden, D. H. and Macdonald, G. A. (1975a) A simple and inexpensive servo system for the control of length or tension of small muscles or stretch receptors. *Journal of Physiology*, **245**, 1—3P.

Bush, B. M. H., Godden, D. H. and Macdonald, G. A. (1975b) Voltage clamping of non-impulsive afferents of the crab thoracic-coxal muscle receptor. *Journal of Physiology*, **245**, 3—5P.

Bush, B. M. H. and Roberts, A. (1968) Resistance reflexes from a crab muscle receptor without impulses. *Nature*, **218**, 1171—1173.

Bush, B. M. H. and Roberts, A. (1971) Coxal muscle receptors in the crab: The receptor potentials of S and T fibres in response to ramp stretches. *Journal of Experimental Biology*, **55**, 813—832.

Cannone, A. J. (1974) Analysis of a crab sketch reflex mediated by non-impulsive muscle receptor afferents. *Ph.D. Thesis.* University of Bristol, U.K.

Elder, H. Y. (1973) Distribution and functions of elastic fibres in the invertebrates. *Biological Bulletin*, **144**, 43—63.

Emson, P. C., Bush, B. M. H. and Joseph, M. H. (1976) Transmitter metabolizing enzymes and free amino acid levels in sensory and motor nerves and ganglia of the shore crab (*Carcinus maenas*). *Journal of Neurochemistry*, **26**, (in press).

Evoy, W. H. and Cohen, M. J. (1971) Central and peripheral control of Arthropod movements. *Advances in Comparative Physiology and Biochemistry*, **4**, 225—266.

Florey, E. (1973) Acetylcholine as sensory transmitter in Crustacea. *Journal of Comparative Physiology*, **83**, 1—16.

Hill, D. K. (1968) Tension due to interaction between the sliding filaments in resting striated muscle. The effect of stimulation. *Journal of Physiology* **199**, 637—684.

Krauhs, J. M. and Mirolli, M. (1975) Morphological changes associated with stretch in a mechano-receptor. *Journal of Neurocytology*, **4**, 231—246.

Lännergren, J. (1971) The effect of low-level activation on the mechanical properties of isolated frog muscle fibers. *Journal of General Physiology*, **58**, 145—162.

Mathews, P. B. C. (1972) Mammalian muscle receptors and their central actions. Edward Arnold, London.

Mendelson, M. (1971) Oscillator neurons in crustacean ganglia. *Science*, **171**, 1170–1173.

Mirolli, M. (1974) Evidence for a metabolically dependent electrogenic process contributing to the resting potential of a crustacean stretch receptor. *The Physiologist*, **17**, *289*.

Moody, C. (1970) A proximally directed intersegmental reflex in a walking leg of the crayfish. *American Zoologist*, **10**, 501.

Paul, D. H. (1972) Decremental conduction over 'giant' afferent processes in an arthropod. *Science*, **176**, 680–682.

Pearson, K. G. and Fourtner, C. R. (1975) Nonspiking interneurons in the walking system of the cockroach. *Journal of Neurophysiology*, **38**, 33–52.

Ripley, S. H., Bush, B. M. H. and Roberts, A. (1968) Crab muscle receptor which responds without impulses. *Nature*, **218**, 1170–1171.

Roberts, A. and Bush, B. M. H. (1971) Coxal muscle receptors in the crab: the receptor current and some properties of the receptor nerve fibres. *Journal of Experimental Biology*, **54**, 515–524.

Shaw, S. R. (1972) Decremental conduction of the visual signal in barnacle lateral eye. *Journal of Physiology*, **220**, 145–175.

Whitear, M. (1965) The fine structure of crustacean proprioceptors. II. The thoracico-coxal organs in *Carcinus, Pagurus*, and *Astacus. Philosophical Transactions of the Royal Society, Series B*, **248**, 437–456.

Zucker, S. (1973) Crayfish escape behavior and central synapses. I. Neural circuit exciting lateral giant fibre. *Journal of Neurophysiology*, **35**, 599–620.

4

Abdominal and thoracic receptors in insects, centipedes and scorpions

L. H. FINLAYSON

4.1   Peripheral neurons in insects

The neurons that are involved in proprioception in arthropods are of
two types, the type I or uniterminal neuron of chordotonal organs
and the type II or multiterminal neuron of other kinds of stretch
receptor (Finlayson, 1968). Uniterminal neurons characteristically
have a single terminal process which contains the organelles
characteristic of a cilium (Chapter 9). Multiterminal neurons characteristically are multipolar and have no ciliary structures within them.
Some type II neurons have but a single sensory process (bipolar) but
they are presumably also multiterminal. The term multiterminal
refers to the many fine endings along the lengths of the sensory
processes that have been revealed by the electron microscope
(Osborne, 1963a, 1964; Osborne and Finlayson, 1965; Whitear,
1965). The neurons of the lyriform organ of Arachnida are not
obviously ciliary in structure but they are uniterminal (Salpeter and
Walcott, 1960) and may be derived from ciliary neurons. The only
receptors innervated by uniterminal (type I) neurons which will be
considered in this chapter are the chordotonal organs of the insect
abdomen. The multiterminal neurons that will be described, have for
the most part not been shown to be proprioceptors, but it may be

inferred from their location in the body that their function is likely to be proprioceptive. However, certain multiterminal neurons have been shown to respond to stretch, in particular those associated with connective tissue strands and muscles. As well as those multiterminal neurons associated with connective tissue strands and muscles there are a number in the body of insects associated with various other tissues (Finlayson, 1968). It seems likely that the orientated stretch receptors have evolved from multiterminal neurons on the body wall or attached to various tissues. It is appropriate, therefore, to consider first of all those multiterminal neurons in insects which are not associated with a special accessory structure and then to consider those sensilla which have an accessory component such as a connective tissue strand or a special muscle fibre.

4.1.1   Multiterminal neurons without accessory structures

Subepidermal neurons
The presence of a system of subepidermal multipolar neurons in insects has been known for a long time (Viallanes, 1882). Frequently this system has been called a subepidermal plexus but the work of Osborne (1963b) on the larva of the blowfly *Phormia* showed that there is no fusion of the processes from subepidermal neurons and no evidence of synaptic associations. The description of the system given by Osborne (1963b) is the most comprehensive available and it will be used as a basis for a comparative description of the system in insects (Fig. 4.1).

Although the subepidermal system of neurons at first sight may seem to be random in its distribution, in fact the position of the neurons is remarkably constant. In the blowfly larva the number of neurons in each segment is always the same; 24 in the prothorax, 28 in both the mesothorax and the metathorax and 30 in each abdominal segment. The majority of dendrites ramify under the epidermis but Osborne (1963b) noted that some terminate on muscles. Also according to Osborne (1963b) the whole body wall of the blowfly larva is covered by a fine meshwork of dendritic processes from these multiterminal neurons. In a later electron microscope study of the terminations of these neurons, Osborne (1964) found that they send fine processes through the basement membrane and into invaginations of the epidermal cells. The ultimate terminations which are in such intimate contact with the epidermal

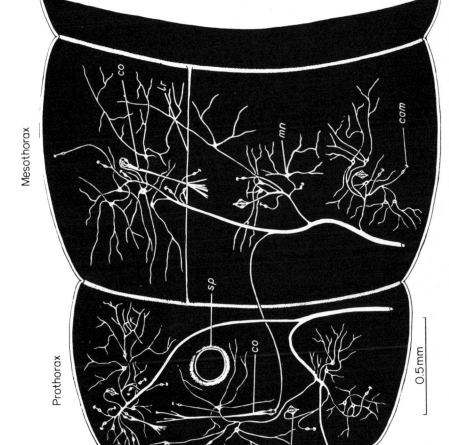

Fig. 4.1   Lateral view (from inside) of the right half of the pro- and mesothorax of a *Phormia* larva, to show the topography of the sensory neurons and sensilla. cam—campaniform sensillum; cs—cuticular sensillum (possibly a chemoreceptor); co—chordotonal organ; lr—longitudinal stretch receptor; mn—multipolar neuron; sp—spiracle. (From Osborne, 1963b.)

cells, are 'naked' in the sense that they are no longer sheathed with the neurilemma (Schwann) cells that cover all the rest of the cell body and its processes. The only physiological investigation of the function of subepidermal neurons is a preliminary report by Finlayson and Osborne (in Finlayson, 1968) of the activity recorded from the two most dorsal subepidermal neurons of an abdominal segment of the blowfly larva. These neurons fire continuously and can be stimulated to increase their discharge frequency by pulling on the cell body. In this respect, they react in a way typical of a neuron in a stretch receptor. It is likely that they are stimulated by distortions of the body wall but conclusive proof of this function has yet to be obtained. Their main function may be as exteroreceptors, in that they will respond to deformations of the body wall caused by external agencies, but they must also respond to alterations in the shape of the body brought about by the activity of the muscular system.

Neurons situated more deeply in the body
Apart from muscle receptor organs and connective-tissue strand organs, which will be dealt with later, there are in the insects a number of multiterminal peripheral neurons situated more deeply in the body than the subepidermal plexus. Their dendrites terminate on a variety of tissues. Although the cells resemble the N-cells of Crustacea in that they are not associated with a special receptor muscle or any orientated connective tissue component (Chapter 2), there is no reason to suppose that they are derived from more highly organized receptors, as Wiersma and Pilgrim (1961) suggest for the N-cells in Crustacea. In the blowfly larva (Osborne, 1963b; Whitten, 1963) and in the tsetse larva (Finlayson, 1968), a group of neurons of this type has been described in the lateral region of each abdominal segment. The basic number of the cells is three and the topographical distribution in the blowfly larva is shown in Fig. 4.2. The most anterior of these neurons (mn1) is situated in a connective tissue capsule, attached to the posterior edge of the transverse intersegmental muscle. Its axon runs in a fine nerve which ultimately joins one of the ventral unpaired nerves of the median nervous system. A dendritic process runs along the edge of the transverse intersegmental muscle and enters a strand of connective tissue that anchors the main longitudinal tracheal trunk to the body wall. A second sensory process runs into a nerve branch supplying the

Fig. 4.2   Drawing of the musculature of the lateral region of the second abdominal segment of a *Phormia* larva to show the topography of neurons *mn*1, *mn*2 and *mn*3 and the distribution of their processes. ib—imaginal bud; *mn*1, *mn*2, *mn*3—multipolar neurones. (From Osborne, 1963b).

ventro-lateral muscle. This branch was not seen in the tsetse larva (Finlayson, 1968). The second neuron (mn2) lies near the motor nerve into which the sensory process from cell mn1 runs. One process from mn2 innervates the rudimentary spiracular trachea, a second innervates the epithelium of a trachea, a third runs into the motor branch adjoining the capsule of the cell body, a fourth runs towards cell mn1 and a fifth, which is probably the axon, enters the main segmental nerve.

Neuron mn3 lies within the nerve that innervates the lateral transverse muscles; sometimes a fourth cell mn4 lies adjacent to mn3. One process from each of these cells always runs in the nerve towards the central nervous system and is presumably the axon. One process from cell mn3 leaves the main nerve and runs over the epithelium of a trachea. The other processes innervate the nerve trunk and its branches to the lateral transverse muscles. Neuron mn4 often appears to be associated with a strand of connective tissue which attaches the nerve to adjacent muscles. In the first abdominal segment mn3 and mn4 are always present and both are situated in a single capsule attached to the nerve. The dendritic processes always innervate the epithelium of a trachea. Neuron mn4 was not found in the tsetse larva but this species was not subjected to the extremely intensive examination that Osborne (1963b) gave to the blowfly larva. The spiracular neuron of the adult tsetse fly may be the homologue of neuron mn2 of the larva (Finlayson, 1966). In adult flies, and in other insects, a similar system of subepidermal neurons exists, but their detailed topography has not been worked out as completely as in the blowfly larva.

Neurons on or in nerves
Neurons situated on the course of nerves or inside nerves have been found in the tsetse fly (Finlayson, 1966, 1968), the blowfly (Gelperin, 1971) and the stick insect (Finlayson and Osborne, 1968). One of the processes of the spiracular neuron of the tsetse adult passes to the spiracular muscle (Finlayson, 1966). It has at least three other processes whose destinations are unknown. Some of these neurons are definitely secretory in function (Finlayson and Osborne, 1968; Gelperin, 1971). In the stick insect, a group of about four neurons on the nerve linking the lateral branch of the median unpaired system to the main segmental nerve system of the succeeding segment, are all neurosecretory in nature (Finlayson and

Osborne, 1968; Fig. 4.3). Although they are electrically active, they do not respond to stretching (Finlayson and Orchard, unpublished observations). The other neurons of the system are not neurosecretory (Fifield and Finlayson, unpublished observations). Thus neurons 9 and 10, which are situated on the long nerve that runs to the dorsal muscle and the dorsal longitudinal stretch receptor (Finlayson and Osborne, 1968) respond to stretch by increasing their frequency of firing (Finlayson and Orchard, unpublished observations). Other neurons which are not neurosecretory and which are electrically active, fire regularly and are apparently unaffected by mechanical stimulation. Preparations of isolated nerves containing neurons taken from the abdomen of the tsetse fly are also electrically active but are apparently unaffected by stretching the nerves (Finlayson and Anderson, unpublished observations).

Gelperin (1971) described neurons within nerves in the abdomen of the blowfly *Phormia*. Within the first and second lateral branches of Gelperin's 'median abdominal nerve', which is the main abdominal nerve running to the posterior end of the abdomen and giving off branches to the 3rd–10th abdominal segments, he found a total population of about four to eight cells. He also found cell bodies in the branches of the other large nerves which run from the thoracic ganglionic mass into the abdomen. These nerves he calls the 'medial accessory abdominal nerve' and the 'lateral accessory abdominal nerve'. The 'lateral accessory abdominal nerve' contains part of the segmental system of the last thoracic segment and the first and second abdominal segment while the 'medial accessory abdominal nerve' contains the segmental nerve of the first abdominal segment and possibly part of the second. The main nerve is his 'median abdominal nerve', which serves the bulk of the abdomen. Gelperin (1971) attempted to eliminate orientated stretch receptors from taking part in the monitoring of food intake by removing all of the muscles and other tissues from abdominal *sternites* 2 and 3, but scraping the sternites would have no effect on the *dorsal* longitudinal stretch receptors which could be monitoring distension of the abdomen. However, he showed convincingly, by recording by means of suction electrodes placed in the vicinity of neurons in nerves, that at least some of them respond to stretch. He also showed that they alter the frequency of their discharge as the crop contracts and expands. The nervous system of the blowfly adult has many cross connections between the main nerves (median abdominal, lateral

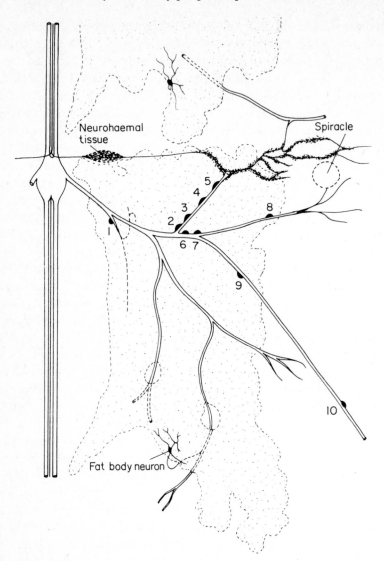

Fig. 4.3   Diagram of the innervation of one side of an abdominal segment of the stick insect (*Carausius*) showing neurons 1–10 on the course of the nerves. (After Finlayson and Osborne, 1968.)

accessory abdominal and medial accessory abdominal) in the abdomen and Gelperin's hypothesis is that this network of nerves is stretched by extension of the crop when the blowfly feeds. Those neurons within the nerves which are mechanoreceptive and not neurosecretory respond to this stretch. He described the mechano-receptive neurons as 'nerve cord stretch receptors', analogous to those in the crayfish ventral nerve cord (Hughes and Wiersma, 1960; Kennedy, Evoy and Fields, 1966). This is an inappropriate term since these neurons are widely distributed in the peripheral nerves of flies, including the tsetse fly (Finlayson, unpublished observations) and have not been demonstrated in an insect nerve cord. Gelperin (1971) could invoke an increase in discharge in the co-called 'nerve-cord neurons' when they were completely isolated from the rest of the nervous system and from the rest of the tissues of the blowfly. It seems highly likely, therefore, that they function as proprioceptors. However, there are similar neurons in the stick insect and, no doubt, in other insects, which do not appear to be situated in an appropriate position to subserve such a function. Getting and Steinhardt (1972) have shown behaviourly and electrophysiologically that the input from abdominal or foregut stretch receptors has no influence on the threshold for proboscis extension, initiated by stimulation of labellar chemoreceptors. However, they also showed behaviourally that ingestion of food raises the threshold for stimulation of feeding behaviour via tarsal chemoreceptors, confirming the role of proprio-ceptors in part of the control system (see Chapter 5).

4.1.2   Multiterminal neurons attached to accessory structures

The accessory component of this type of stretch receptor is a strand or tube of connective tissue, or a special receptor muscle, or a combination of connective tissue and muscle (Finlayson, 1968). The simplest type is a strand of connective tissue to which the neuron is attached (Figs. 4.4, 4.5), as in cockroaches (Dictyoptera), dragonflies (Odonata), stoneflies (Plecoptera), earwigs (Dermaptera), mayflies (Ephemeroptera), and beetles (Coleoptera) (Table 4.1). In three orders of exopterygote insects, the Phasmida, Orthoptera and Hemiptera (Heteroptera), stretch receptor neurons and their con-nective tissue accessory component are attached to ordinary muscles of the body, although in the Orthoptera (Slifer and Finlayson, 1956) the particular muscle fibre to which the neuron is attached is modified, in that its striations are not so conspicuous as those of the normal fibres of the rest of the muscle band. In the three

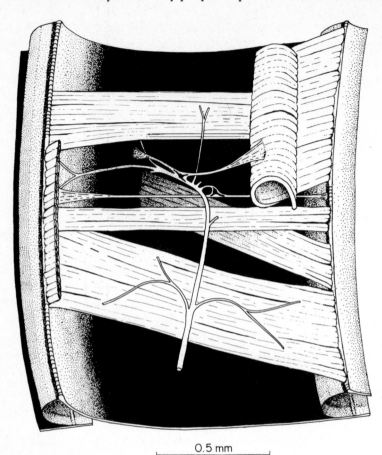

0.5 mm

Fig. 4.4   Right side of the second abdominal segment of *Forficula* (Dermaptera) with various muscles cut or removed, to reveal the two stretch receptors. (From Osborne and Finlayson, 1962.)

endopterygote groups, Neuroptera, Trichoptera and Lepidoptera there is a special receptor muscle completely independent of the ordinary muscles of the body (Fig. 4.6). Similarly, in the centipedes and scorpions there are special receptor muscles (Fig. 4.7; Rilling, 1960; Bowerman, 1972a).

Response to stretch of neurons on connective tissue strands
The connective tissue strand receptor of the dragonfly larva responds to stretch by increasing the frequency of its discharge (Finlayson and Lowenstein, 1958; Lowenstein and Finlayson, 1960). The response

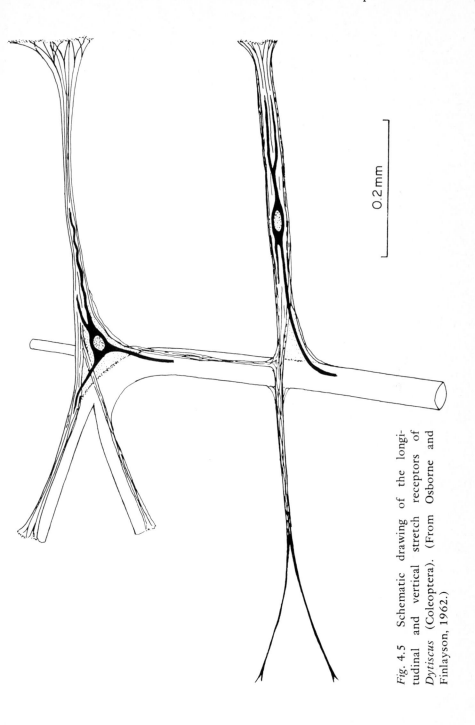

Fig. 4.5 Schematic drawing of the longitudinal and vertical stretch receptors of *Dytiscus* (Coleoptera). (From Osborne and Finlayson, 1962.)

Table 4.1   Stretch receptors in the abdomen of insects

Group	Genus	Stages examined			Dorsal longitudinal				Vertical	Authors
		Adult	Pupa	Larva	Connective tissue	Muscle	Whole segment	Part segment		
Odonata	*Aesbna*	–	+	+	+	–	–	+	+	F and L, (1958)
Ephemeroptera	*Ephemera*	+	+	+	+	–	–	+	+	O and F, (1962)
Plecoptera	*Dictyopterygella**	–		+	+	–	+	+ (in thorax only)	+	O and F, (1962)
Dictyoptera	*Periplaneta*	+		–	+	–	–	+	+	F and L, (1958)
	Blaberus	+		–	+	–	–	+	+	O and F, (1962)
Orthoptera	*Locusta* *Schistocerca* *Romalea* *Dissosteira* *Melanoplus* *Phoetaliotes*	+		–	+	+ (1)	+	–	+	S and F, (1956) O and F, (1962)
Phasmida	*Carausius*	+		–	–	+ (2)		+	?	S and F, (1956)

Order	Genus										Reference
Dermaptera	*Forficula*	+	+	+	–		+	–		+	O and F, (1962)
Hemiptera	*Rhodnius*	+	+	+	+		+(3)	–	+	–	A, (1972)
Coleoptera	*Dytiscus*	+	–	+	–		–	+	+	+	O and F, (1962)
Neuroptera	*Sialis*	–	–	+	+		+	+	–	+(4)	O and F, (1962)
Trichoptera	*Limnophilus*	+	+	–	–		+	+	–	–	O and F, (1962)
	Phryganea	+	+	+*	+		+	+	–	–	
Lepidoptera	*Antheraea*	+	+	+	+		+	+	–	+	
	Hyalophora	+	+	+	+		+	+	–	–	F and L, (1957, 1958)
	Samia	+	+	+	+		+	+	–	–	
	Actias	+		+*							
Hymenoptera	*Apis*	+	–	–	+		–	+	–	–	F and L, (1957, 1958)
Diptera	*Phormia*	–	–	+*(5)	+		–	+	+	+	O (1963b)
	Glossina	+	–	+*	+		–	+	–	–	F, (1972—larval tsetse, adult unpublished observations)
					(Larva)		(Larva)	(Adult)	(Larva)		

(1) On special fibre in dorsal longitudinal muscle band
(2) On dorsal longitudinal muscle band
(3) On dorsal longitudinal muscle band
(4) No connective tissue strand, probably reduced
(5) Ventral longitudinal receptor also
* Identified in thorax also

S = Slifer
F = Finlayson
L = Lowenstein
O = Osborne
A = Anwyl

of the receptor to increasing amplitude of stretch is linear up to a maximum level, beyond which the phenomenon of over stretch occurs and the frequency of firing drops (Finlayson and Lowenstein, 1958). When the receptor is stretched within its physiological range and maintained in the new position, the new rate of firing falls very slowly. The connective tissue strand receptor of the dragonfly is, therefore, a typical tonic receptor showing slow adaptation. It can, however, also monitor phasic changes in length (see Section 4.5.2). Similar phasic-tonic responses are shown by the dorsal longitudinal receptor of the cockroach *Blaberus* (Lowenstein and Finlayson, 1960).

4.2   Stretch receptors in insects

4.2.1   Distribution and anatomy

The distribution and topography of stretch receptors in the thorax and abdomen of insects is summarized in Table 4.1. In every species of insect in which multiterminal stretch receptors have been found there is always one in a dorsal longitudinal position. Although the thorax has not been so intensively studied, dorsal longitudinal stretch receptors have been found in the thorax of adult stick insects (Slifer and Finlayson, 1956) stonefly larvae (Osborne and Finlayson, 1962) moth larvae (Lowenstein and Finlayson, 1960) and blowfly larvae (Osborne, 1963b). The stick insect, being apterous, is not typical of adult insects and the evolutionary fate of the series of dorsal longitudinal receptors in the thorax of winged insects is not known. Gettrup (1962, 1963) has described stretch receptors in the thorax of the desert locust (*Schistocerca gregaria*) and Wilson and Gettrup (1963) suggest that they may be homologous with the dorsal abdominal series. In many insects, there is in addition another stretch receptor lying approximately at right angles to the dorsal longitudinal receptor, named by Finlayson and Lowenstein (1958) the *vertical receptor*. A vertical receptor is present in all the exopterygote insects which have been examined, with the exception of *Rhodnius* (Hemiptera) (Anwyl, 1972) and it also occurs in the endopterygote order Coleoptera (Osborne and Finlayson, 1962). Dorsal longitudinal stretch receptors have been identified in fourteen orders of insects (Slifer and Finlayson, 1956; Lowenstein and Finlayson, 1958; Osborne and Finlayson, 1962; Osborne, 1963b;

Anwyl, 1972), and it is reasonable to assume that they are universally present throughout the Class. The association of the multiterminal neuron with a receptor muscle occurs in both exopterygote and endopterygote groups, and in both there are receptors which consist only of a connective tissue strand plus a neuron. The connective tissue component of a stretch receptor consists either of a ribbon-like strand of connective tissue fibres as in the cockroach and the dragonfly larva or a tube of closely compacted fibres as in the bee and dipterous larvae (blowfly, tsetse fly). In the Trichoptera and Lepidoptera, a similar tubular connective tissue component occurs but in these orders it is attached to a special receptor muscle fibre. The presence of multiterminal neurons in or on the course of nerves has been described in the previous section (Section 4.1). Stretch receptors may have evolved from neurons of that type or from neurons of the subepidermal plexus; there is evidence in favour of both hypotheses.

In the stoneflies (Plecoptera) the neuron of the vertical receptor may be situated on the tergal epidermis and send its single process into a connective tissue strand slung between the tergal epidermis and the main tergal nerve where it passes below the second and third dorsal muscle bands. It would appear that the vertical receptor in the stoneflies has evolved from a neuron on the body wall which has sent its process along a strand of connective tissue. Connective tissue links without innervation are common in insects; the whole interior of the body is enmeshed in connective strands slung between the various organs. The neuron of the longitudinal receptor of the stonefly larva, on the other hand, sends its single process towards the epidermis in the region of the posterior intersegmental fold. It could be argued, therefore, that this neuron is one of the deeper system which sends a process towards the body wall. The stoneflies are unique among insects in possessing bipolar neurons in both longitudinal and vertical stretch receptors. In the Ephemeroptera (mayflies) there is a bipolar neuron in the longitudinal receptor, again with the process running towards the intersegmental fold, in this case, the anterior fold. Supporting the hypothesis that the stretch receptor neurons are derived from neurons of the deeper system is the arrangement seen in the larva of the alder fly *Sialis*, (Neuroptera). The vertical receptor is slung between a branch of the tergal nerve innervating the dorsal epidermis and two smaller longitudinal muscles above the main 2nd and 3rd dorsal muscles bands (Fig. 4.6). It is extremely simple in

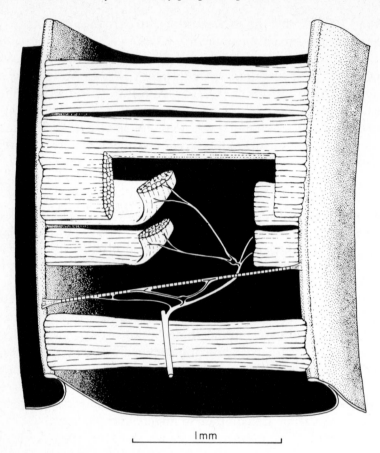

Fig. 4.6   Right side of the fifth abdominal segment of the larva of *Sialis* (Neuroptera) with the third dorsal muscle band removed to show the location of the receptors. Motor and sensory innervation are shown. (From Osborne and Finlayson, 1962.)

structure, consisting of a neuron anchored to the nerve in which its axon runs, and with no accessory connective tissue fibres. The longitudinal receptor of *Sialis* consists of a receptor muscle and a multiterminal neuron but, as in the vertical receptor, there is no sign of connective tissue fibres in the profusion that they occur in the other groups of insects. In the cockroach *Periplaneta*, both receptors are of the connective tissue strand type and they are linked together by a connective tissue junction. However, in another cockroach, *Blaberus*, there is no such connection between the two strands.

  In those insects which possess a muscle receptor organ, or in which the dorsal longitudinal stretch receptor neuron is associated with one of the ordinary muscles of the body, there is considerable variety in their anatomy, although they are relatively few in number. In the locust (Orthoptera) the neuron is attached to one of the dorsal longitudinal muscle bands of the abdomen. It varies in position, normally being at the edge of the band but occasionally being nearer the middle. Its connective tissue component is firmly attached to the muscle band and the fibre to which the neuron is applied appears to differ from the normal fibres in having fainter striations. As in most insects which have a dorsal longitudinal muscle receptor organ, there is no vertical receptor. In the assassin bug *Rhodnius* the arrangement resembles that seen in the locust but the neuron is not attached to a particular muscle fibre. Its processes ramify over several fibres (Anwyl, 1972). The cell body of the neuron in *Rhodnius* is usually situated within the dorsal motor nerve, close to the fork at which this nerve branches from the tergal nerve. Occasionally the neuron may lie in the tergal nerve close to the fork. The dendrites from the neuron innervate three of the eight muscles. In *Rhodnius*, as in *Sialis*, it would appear that the stretch receptor neuron is one of the deeper series which has sent its processes on to a dorsal longitudinal muscle. In *Sialis* there is no evidence that the dendrites of the vertical receptor ramify over the two muscle fibres to which they are attached.

In the Trichoptera and in the Lepidoptera there is a dorsal longitudinal muscle receptor organ and no vertical receptor. In both these orders, as might be expected from their close phylogenetic relationship, the muscle receptor organ is similar. It consists of a single muscle fibre to which is attached a tube of connective tissue. The detailed structure of the lepidopteran muscle receptor organ has been worked out (Osborne and Finlayson, 1965) and will be described in the Section 4.5.

Only in the larva of the blowfly has a ventral longitudinal stretch receptor been found in an insect (Osborne, 1963b). Finlayson, (1972) did not find a corresponding receptor in the tsetse larva.

4.2.2   Ultrastructure

Lepidopteran muscle receptor organ
In the caterpillar larva of a lepidopteran such as the giant silk moth *Antheraea pernyi* there is a pair of receptors in the meso- and metathorax and abdominal segments 1 to 9. Each receptor consists

of a modified muscle fibre that lies just above the dorsal longitudinal muscles (Figs. 4.7, 4.8, 4.9). It is attached to the intersegmental folds at each end of the segment by connective tissue fibres. The stretch receptor can be located by reference to the arrangement of the dorsal longitudinal muscle fibres. Each dorsal band is composed of three sub-divisions (Fig. 4.7). The main segmental nerve sends major branches through the dorsal muscle band in two regions; through the narrow gaps between the sub-divisions. Soon after the nerve passes through the first of these fissures it gives rise to the motor innervation of the muscle receptor organ, and a branch which runs to the body wall receives the axon of the stretch receptor neuron (Figs. 4.7, 4.9). There are about a dozen motor endings on the receptor muscle fibre (Fig. 4.8). Running along the dorsal edge of the stretch

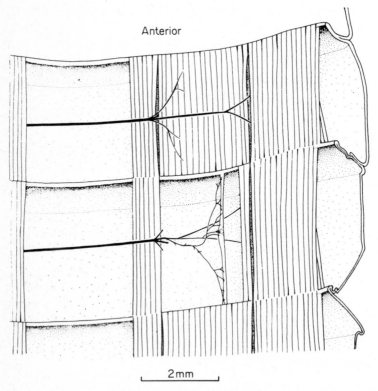

Anterior

2mm

Fig. 4.7   Diagram of the dorsal longitudinal muscle band of 3rd and 4th abdominal segments of a moth pupa, with part of the band in 4th segment removed to show the location and innervation of the MRO. (From Finlayson and Lowenstein, 1958.)

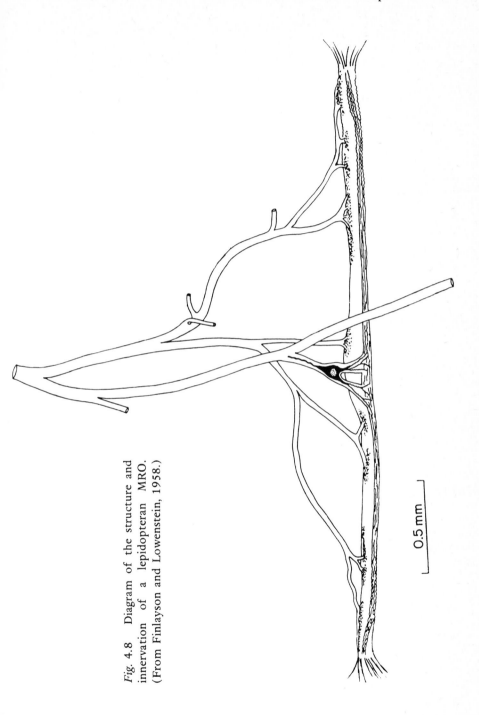

Fig. 4.8 Diagram of the structure and innervation of a lepidopteran MRO. (From Finlayson and Lowenstein, 1958.)

0.5 mm

Fig. 4.9   Photograph of a dissection of the dorsal muscle band of the caterpillar of *Antheraea pernyi* showing the MRO. (Long arrow)—sensory neuron; (short arrows)—motor innervation.

receptor is a tract which is richly supplied with connective tissue fibres. This tract stains with methylene blue and would appear to be the equivalent of the connective fibre component of the simpler type of stretch receptor, which lacks a muscular component. In the earlier studies of stretch receptors (Finlayson and Lowenstein, 1958), it was assumed that the dendrites of the sensory neuron passed into this fibre tract. Later studies, using the electron microscope (Osborne and Finlayson, 1965), showed that interpretation to be not strictly accurate (see p. 11). In the central region of the muscle receptor organ there is a swelling in which lies a giant nucleus, measuring as much as 680 μm in length (Finlayson and Mowat, 1963). In the same region, but lying closer to the fibre tract, is another giant nucleus of lesser size. It is usually very elongated and measurements from specimens fixed in Bouin give an average size of 410 x 33 μm for larvae and 428 x 21 μm for adults (Fig. 4.10b). In light microscopy, the striations of the receptor muscle are much less pronounced than those of the ordinary body muscles. The central giant nucleus lies in a core of cytoplasm in which striations cannot be seen. A possible interpretation of the structure of the stretch receptor is that it consists of three large cells, the sensory neuron, a single muscle cell

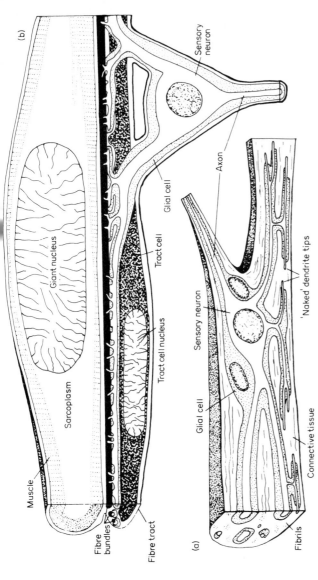

Fig. 4.10   Schematic drawings to show the main ultrastructural features of (a) the cockroach and (b) the moth stretch receptors. Only the regions in the vicinity of the sensory neuron are shown. (a) In the cockroach the neuron and glial cell sheath are embedded in the connective tissue strand. The tips of the dendrites are without the glial cell sheath, and are buried in the matrix of the connective tissue. No connections are observed between the naked dendrite tips and the connective tissue fibrils. (b) In the moth the dendritic terminals are also naked and are associated with dense accumulations of connective tissue fibres, the fibre bundles. In the central region the muscle fibre is filled with clear sarcoplasm, the sarcoplasmic core, and the giant nucleus. The fibre tract contains the main dendrites and the tract cell, whose nucleus is shown. (From Osborne, 1970.)

with a very large nucleus and a single cell which produces the fibre tract. This interpretation was borne out by studies with the electron microscope (Osborne and Finlayson, 1965). Although the central region does not appear to be striated when viewed with the light microscope, the electron microscope reveals features characteristic of striated muscle, although in the region of the giant nucleus the precise organization has broken down (Figs. 4.11, 4.12). The electron microscope reveals that there are muscle fibrils in the

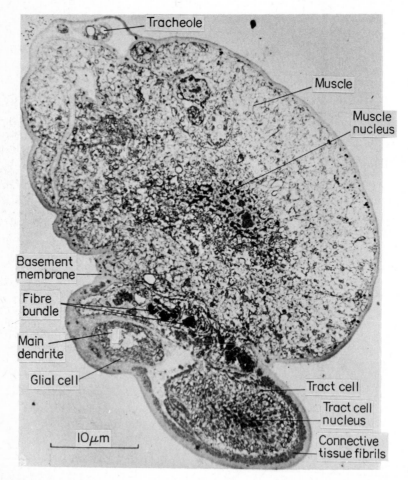

Fig. 4.11   Low-power electron micrograph of a transverse section through the central region of the receptor. Both the muscle nucleus and tract cell nucleus can be seen. The main dendrite is at this level, not inside the fibre tract. (From Osborne and Finlayson, 1965.)

cytoplasm of the central nuclear region but the amount is small. It is not surprising, therefore, that these fibrils could not be seen with the light microscope, especially as the muscle of the stretch receptor is not so conspicuously striated as ordinary muscle. The transverse tubular system characteristic of striated muscles is also present in the central region (Fig. 4.12). The giant nucleus of the central swelling, which we now know to be a hypertrophied muscle nucleus, has a greatly folded nuclear membrane. The muscle cell is innervated by motor endings distributed along its length, including the central region (Fig. 4.8). The fibre tract is tubular in structure with connective tissue fibres in the wall. Studies with the electron microscope (Osborne and Finlayson, 1965) revealed the presence of bundles of connective tissue fibres where the tract meets the muscle fibre (Figs. 4.11, 4.12). These dense bundles are not visible in preparations made for the light microscope. They run longitudinally for most of the length of the receptor, sometimes fusing together and changing diameter in an irregular fashion. Towards the end of the receptor the fibrils become fewer, until they finally lose their distinction from the rest of the fibrous masses which form the inner ring of the wall of the fibre tract. In addition to being larger than the inner ring of fibrils, with which they are probably homologous, they show a tendency for dense accumulations of fibrils to adhere so closely that deeply staining compound fibres are formed which, in cross section (Fig. 4.13), look like dark spots, or irregular dark areas in a lighter background. The degree of condensation of fibrils varies within the bundles of a single receptor. In addition to the fibre tract giant nucleus the core of the fibre tract consists of cytoplasm, containing mitochondria and a well developed endoplasmic reticulum. In the central region of the receptor the dendrites of the sensory neuron do not run in the fibre tract but alongside it in association with the fibre bundles (Figs. 4.10, 4.11, 4.12). The dendrite sheath gradually merges with the fibre tract and the dendrites then run in the fibre tract itself. It was also discovered that the endings of the dendrites from the sensory neurons did not pass along the core of the fibre tract, as had been assumed in the earlier studies; rather, throughout the entire length of the dendrites, side branches are given off which send minute processes into the connective tissue fibre bundles (Figs. 4.10, 4.12, 4.13). Each dendritic branch is wrapped in glial membranes but the tips of the final terminations are unwrapped. The resulting naked endings are closely applied to the fibre bundles, or penetrate them.

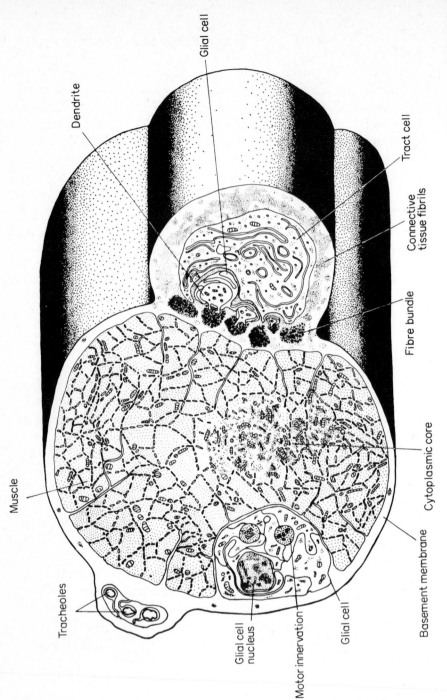

Fig. 4.12 Reconstruction based on electron micrographs of the general structure of the lepidopteran muscle receptor organ. Note the muscle with the cytoplasmic core and motor innervation. Attached to the edge of the muscle fibre is a tube of connective tissue, the fibre tract, containing the tract cell and dendrites enveloped by the glial cell. Processes from the dendrites penetrate between the dense connective tissue fibrils of the fibre bundles. (From Osborne and Finlayson, 1965.)

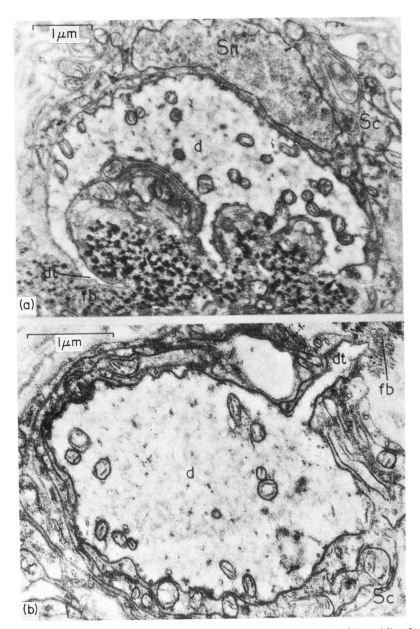

Fig. 4.13   Electron micrographs of sections through large dendrites (d) of a lepidopteran MRO showing unsheathed terminations (dt) penetrating fibre bundles (fb). Sc—Schwann (glial) cell; Sn—Schwann (glial) cell nucleus. (From Osborne and Finlayson, 1965.)

Towards the end of the receptor the fibre tract decreases in diameter and the fibre bundles and dendritic terminations become fewer. Finally the bundles and dendrites disappear altogether and the much diminished fibre tract alone remains. A similar association of dendritic terminations and connective tissue bundles was found by Whitear (1965) in the thoracico-coxal organ of Crustacea.

The cockroach stretch receptor

In the cockroach (Osborne, 1963a) the receptor is much simpler in structure (Fig. 4.10a), consisting simply of a multiterminal neuron ensheathed by glial cells and embedded in a strand of connective tissue. As in the moth the tips of the dendrites are unsheathed, but unlike the moth they lie in the matrix of connective tissue and are not associated with special fibre bundles. Connective tissue fibrils are present but there are no intimate connections between them and the naked dendrites.

In neither the cockroach nor the lepidopteran receptor is there any indication of special structures in the tips of the dendrites. The neurons, therefore, may be truly described as multiterminal, as opposed to the uniterminal ciliary type of neuron found in many mechanoreceptors in insects.

4.2.3   Physiology of the lepidopteran muscle receptor organ

There appears to be a basic minimum resting discharge which is evoked when an isolated caterpillar MRO is put under slight tension (Finlayson and Lowenstein, 1958). This basic discharge frequency in high Na^+/low Mg^{++} saline is frequently around 20 to 30 impulses per second but in a saline which more nearly resembles the lepidopteran haemolymph (Weevers, 1966a) the resting discharge is lower in frequency. Weevers (1966a, b) does not indicate whether the completely relaxed receptor in lepidopteran saline continues to fire, but in view of the fact that this saline is much less excitatory than high Na^+/low Mg^{++} saline it is unlikely that the completely relaxed receptor will fire at all. The MRO is a tonic receptor maintaining a resting discharge frequency for long periods (Finlayson and Lowenstein, 1958; Weevers, 1966b). The impulse frequency, after adaptation to different lengths of stretch, is shown in Fig. 4.14 (Weevers, 1966b). In this respect, therefore, the lepidopteran receptor acts as a

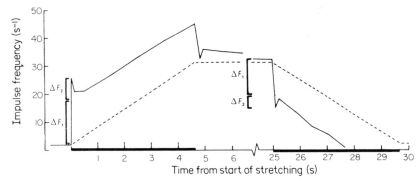

Fig. 4.14   The complex response of a caterpillar MRO to slow constant-velocity stretching and relaxing (————) compared with the changes in adapted impulse frequency (—————) over the same length range, plotted on the same time-scale. The heavy line above the abscissa signifies stretching; that below signifies releasing. The dashed line is the position response; ΔF_1, the movement response; ΔF_2 the acceleration response. (From Weevers, 1966b.)

classic tonic mechanoreceptor. Analysis of records from receptors which have been subjected to stretch and then maintained at their new length show also a phasic response in the form of a higher frequency discharge for a short period at the greater length, followed by a subsequent drop to the new level of tonic discharge frequency. Weevers (1966b) has analysed the complex response of an MRO to slow, constant velocity stretching and relaxing (Fig. 4.14). He recognizes three components in the response of the receptor: (1) the *position* response, as already described, (2) the *movement* response, which is a frequency of firing higher than the frequency shown by the receptor at the same length but in a static condition and (3) the *acceleration* response, which is represented by an overshoot of increased frequency at the beginning of movement and on relaxation. As the velocity of stretch is increased, both the movement response and the acceleration response increase also. A decrease in velocity results in a corresponding reduction in discharge frequency, or a temporary cessation of impulses if the deceleration is sufficiently rapid (Lowenstein and Finlayson, 1960; Weevers, 1966b). The silent period that follows the cessation of stretching may be a consequence of the receptor's inability to return to the resting level as rapidly as the length change is imposed on it. There may be a lag because of the viscous property of the muscle and other components. When an MRO is subjected to sinusoidal stretch and relaxation, as the frequency of sinusoidal stimulation increases the number of impulses which can be produced during the stretching phase becomes fewer and fewer, until the receptor neuron may be firing only one impulse per stretch. Also, as the frequency of sinusoidal stimulation increases, the silent period following relaxation becomes increasingly longer. It is possible for this receptor to follow sinusoidal stimulation at least to a frequency of about 5 Hz (Lowenstein and Finlayson, 1960). Interpretation of the results of sinusoidal stimulation are complicated and the isolated receptor is undoubtedly in a very unnatural situation. It is not possible, therefore, to predict from such experiments the upper frequency of phasic stimulation to which the muscle receptor organs can respond. It is clear, however, that the lepidopteran MRO can signal both maintained changes in its length and changes in frequency of repetitive stretches. It can also, by means of the acceleration response, monitor the velocity of stretch. When the number of impulses drops to one, or very few, the precision of the receptor will undoubtedly be less and it will be signalling only frequency of repetitive stretching and relaxation, which in the whole insect is likely to be induced by respiratory movements.

4.2.4   Comparison of stretch receptors in insects

In those insects which possess only a dorsal longitudinal receptor (Table 4.1), the dendrites ramify over the entire length of the organ, in contrast to the arrangement in insects with a vertical receptor (Osborne and Finlayson, 1962). This arrangement probably increases the sensitivity of the longitudinal receptors and eliminates the need for a vertical receptor. Such an interpretation throws doubt on Finlayson and Lowenstein's (1958) hypothesis that these two receptors are antagonistic. They may, in fact, be complementary, functioning over different ranges. When the longitudinal receptor is slightly extended, the vertical receptor may be unaffected, but as extension is increased the vertical receptor may be pulled into a more longitudinal position in order to monitor the upper end of the range of extension.

The muscular component present in neuropteran, trichopteran and lepidopteran longitudinal receptors probably further improves the efficiency by allowing reflex adjustment of length. It is significant that only the Neuroptera of these three orders possesses a vertical receptor. The neuropteran vertical receptor, however, has no orientated strand and may be degenerate (Osborne and Finlayson, 1962).

The evidence from comparative anatomy favours the hypothesis that the trends for the dorsal longitudinal receptor to increase in length until it spans the entire segment, for its neuron to send dendrites along a greater length of the organ and for a muscular component to be incorporated, have resulted during evolution in the reduction and loss of the vertical receptor.

4.3 Structure and physiology of muscle receptor organs in centipedes

All the stretch receptors described so far in centipedes are muscle receptor organs in which specialized muscle fibres are innervated by a number of sensory neurons (Osborne, 1961; Rilling, 1960; Varma, 1972). In the geophilid centipedes such as *Haplophilus* the body segments are of equal length but there is a secondary joint near the anterior end, giving extra flexibility to the body. The skeletal muscles and the stretch receptors are correspondingly segmentally arranged (Fig. 4.15) (Osborne, 1961). In *Lithobius*, which is intermediate in form between the burrowing geophilid centipedes and the very fast running fleet centipedes like *Scolopendra*, the body segments have alternate long and short tergites (tergite heteronomy) (Fig. 4.16). Almost all the dorsal muscles form bisegmental units running over a long and short segment. The receptor muscles are also arranged bisegmentally (Fig. 4.17) (Osborne, 1961; Rilling, 1960). In the fleet centipedes, such as *Scolopendra*, where tergite heteronomy is evident but not so pronounced, the receptor muscles and some of the skeletal muscles from adjacent long and short segments are fused to form bisegmental units (Osborne, 1961; Varma, 1972). Although the receptor muscle spans two segments, it is attached midway, near the anterior edge of the short tergite (Osborne, 1961). According to Varma (1972), it is attached for a considerable part of its length to the dorsal integument as it traverses the short tergite. In both *Lithobius* and *Scolopendra* the MRO is forked at its posterior end (Figs. 4.16, 4.17).

The MRO's of centipedes are innervated by motor axons and by sensory neurons, the latter occurring singly or in groups. In *Lithobius*, 13 sensory neurons were described by Rilling (1960) and

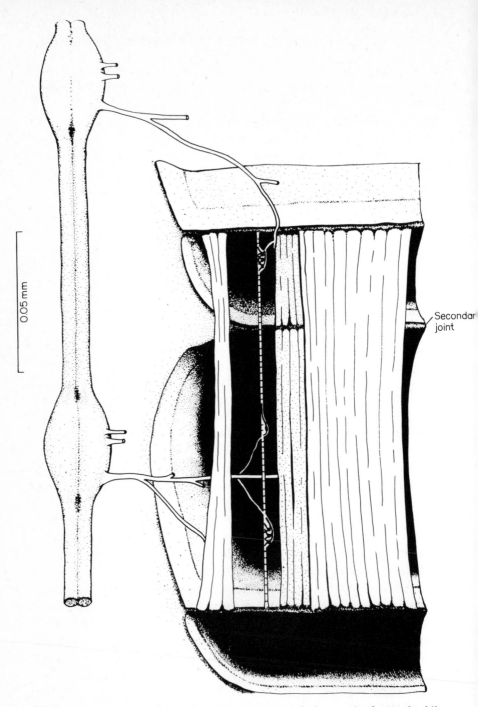

Fig. 4.15   Diagram of muscle receptor organ of the centipede *Haplophilus*, lying above the dorsal musculature. Nerve cord displaced to the left. (From Osborne, 1961.)

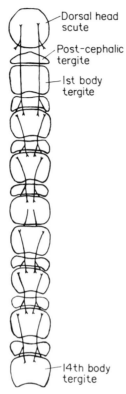

Fig. 4.16 Diagram of dorsal head scute, post-cephalic tergite and the following 14 tergites of the centipede *Lithobius*, to show the relationship between the topography of the muscle receptor organs and the arrangement of the tergites. (From Osborne, 1961.)

the same number was found by Varma (1972) in *Scolopendra*. Both bipolar and multipolar neurons are present (Varma, 1972). In *Haplophilus* Osborne (1961) found seven neurons on each receptor muscle arranged in two groups of three and a solitary neuron (Fig. 4.15). The solitary neuron is situated near the centre of the receptor muscle and has one long main dendrite that runs for a distance along the muscle. One group of neurons is situated near the anterior end of the receptor muscle and the other group near the posterior end (Osborne, 1961).

There are two regions of the centipede body in which the arrangement of muscle receptor organs differs from those described. The first of these is the head and anterior region (Fig. 4.17). In *Lithobius* each receptor muscle is attached anteriorly to the dorsal

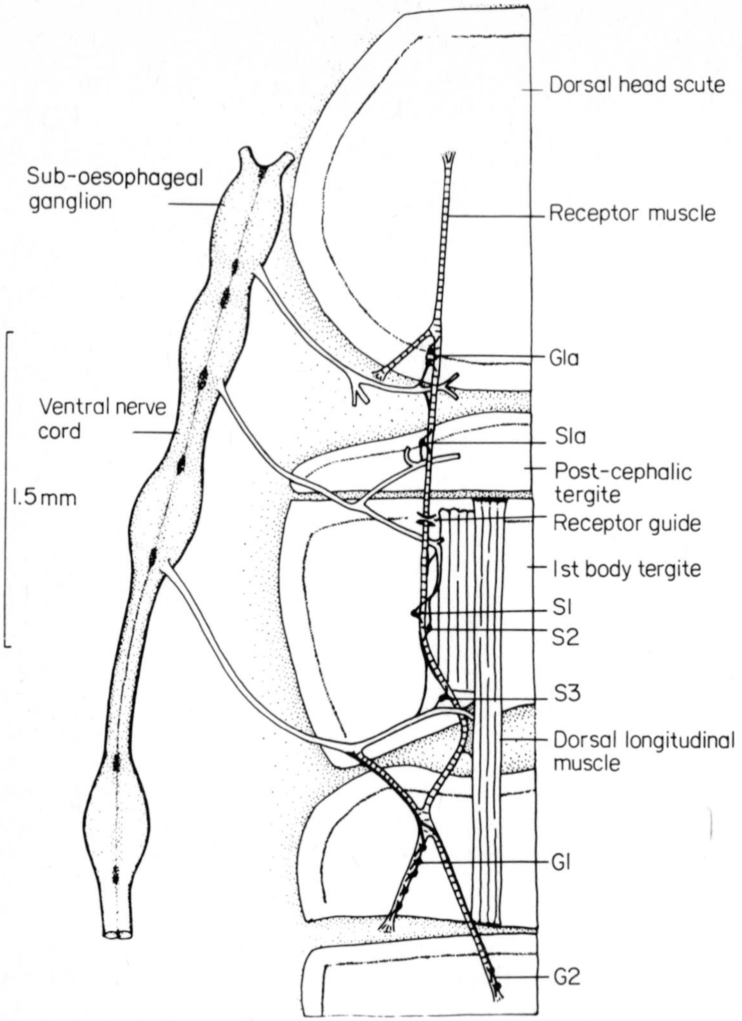

Fig. 4.17   Diagram of anterior region of the centipede *Lithobius* showing the complex muscle receptor organ which runs from the dorsal head scute to the 3rd body tergite. G1, G1a, G2—groups of sensory neurons; S1, S1a, S2, S3—sensory neurons. (From Osborne, 1961.)

head scute and posteriorly to the anterior end of the third body tergite and to the intersegmental membrane between the second and third tergites. It is also attached, as are the body receptors, to a dorsal longitudinal muscle (Osborne, 1961). At the anterior end of the first tergite, the receptor muscle runs through a small collar-like upfolding of the epidermis, inside which it is free to slide longitudinally. The term 'receptor guide' was proposed by Osborne for this structure. A side branch from the receptor muscle is anchored to the intersegmental membrane between the dorsal head scute and the post-cephalic tergite. According to Varma (1972) a similar arrangement exists in *Scolopendra.* The other part of the centipede body which differs from the basic plan is the region of the two consecutive long tergites (7 and 8). In *Scolopendra*, the muscle receptor organ of this region is attached anteriorly to tergite 7 and posteriorly to the anterior region of tergite 8, the posterior region of tergite 9 and the anterior region of tergite 10, (Varma, 1972). In *Lithobius*, according to Rilling (1960) and Osborne (1961) the equivalent structures are two receptors (Fig. 4.16). A short receptor muscle is attached anteriorly to the epidermis of tergite 7 and posteriorly to the epidermis at the anterior border of tergite 8. The receptor spanning tergites 8 and 9 has the normal topography. Receptor muscles of the other body receptors diverge anteriorly from the dorsal midline, but those of tergite 7 in *Lithobius* converge anteriorly (Osborne, 1961). The head receptor on the other hand runs more or less parallel with the longitudinal axis of the body (Osborne, 1961; Rilling, 1960). There are only two sensory neurons on the stretch receptor of tergite 7 in *Lithobius* (Osborne, 1961).

The only physiological work done on centipede receptors is reported by Osborne (1961). Recordings were obtained from S3 and G1 (Fig. 4.17). S3 was isolated by severing the axons of G1. Stretching the receptor produces an increase in discharge frequency; relaxation results in post excitatory inhibition (Section 6.5.1) followed by a gradual return to a resting discharge. S3 thus shows the typical response of a slowly adapting tonic mechanoreceptor. The effect of sinusoidal stimulation is to produce a burst of impulses only during the stretch phase of the cycle. No clear evidence that S3 responds to movement velocity as well as displacement was obtained. A recording from G1, obtained by cutting the axon of S3, showed that at least some of the units in this group are slowly adapting.

Manton (1958) has correlated chilopod body design with the locomotory activity of the group and Osborne (1961) discusses the role of muscle receptor organs in the light of her hypotheses. The geophilid centipedes, such as *Haplophilus*, are borrowing forms and have very flexible bodies. Each body segment has a large and a small tergite and sternite, and thus each segment has two telescopic joints which enable the animal to twist and turn in an extreme manner. They also enable it to extend and shorten the body to a considerable degree. The short legs anchor the body and much of the burrowing activity is carried out in a way similar to the earthworm. *Lithobius*, on the other hand, having longer legs and a reduction of lateral undulation is better adapted for running. If lateral undulations were large, the legs would tend to become entangled and stumbling would occur. The reduction of body undulations is achieved by heteronomy of tergites and the bisegmental arrangement of the tergal muscu-lature. In fleet centipedes, only the joints of the posterior end of the long tergites flex conspicuously when the animal is running. Since pronounced flexure of all inter-tergal joints can take place (e.g. when the animal turns or is cleaning itself), the control of inter-tergal flexures when running must be a function of the tergal muscles. The body receptors of *Scolopendra* and *Lithobius* diverge anteriorly. This would appear to improve the ability of the receptors in monitoring lateral movements. It is probable, therefore, that the amplitude of lateral body undulations during running is regulated by afferent discharges from the muscle receptor organs. The progressive develop-ment of tergite heteronomy, correlated with the bi-segmental arrangement of muscles and muscle receptor organs, in the fleet centipedes, *Scolopendra* and *Lithobius*, has the effect of halving the number of inter-tergal joints. This arrangement produces a more rigid body, which is better able to reduce lateral undulations. In *Scutigera*, the fastest running type of fleet centipede, each alternate tergite has become so reduced that in effect the number of inter-tergal joints is halved.

In fleet centipedes the head is moved from side to side during locomotion. Manton (1958) states that the extra rigidity of the joint between the two long tergites (7 and 8) in both *Scolopendra* and *Lithobius* is effective in damping down lateral body undulations instigated by movement of the head. It is interesting to note that in *Lithobius* the receptors associated with the joint between tergites 7 and 8 are less complex in terms of sensory units than the rest of the body receptors. In *Scutigera*, tergites 7 and 8 are fused and it is unlikely therefore that there will be stretch receptors associated with these tergites. Head movement in *Lithobius* is undoubtedly moni-tored by the head receptors (Fig. 4.9). As each head receptor muscle is free to slide through its receptor guide, head movement may cause neurons as far back as S3 to discharge. This would relay information into the ganglion of the second body segment and possibly initiate reflex responses before lateral undulations became prominent in that region.

In centipedes, the axons from some neurons on the receptor muscles enter the ganglion in the preceding segment. Afferent fibres from the abdominal muscle receptor organs in decapod Crustacea also enter the ganglion of the preceding segment (Chapter 2).

It is noteworthy that the centipedes have a number of sensory units on their muscle receptor organs, whereas the insects never have more than one neuron associated with each receptor muscle or connective tissue strand. The abundance of neurons in the former may be correlated with the high degree of coordination needed between the body segments of chilopods during locomotion.

4.4 Structure and physiology of muscle receptor organs in scorpions

4.4.1 Structure

Bowerman (1972a) describes the topography and structure of muscle receptor organs in the postabdomen (tail) of the scorpion *Centruroides gracilis.* This is the first description of such structures in scorpions. The paired MRO's are located in the ventral region of segments one to four. Each receptor muscle is attached directly to the anterior edge of one segment and indirectly by a long tendon to the anterior edge of the succeeding segment (Fig. 4.18). The receptor muscles are composed of two distinct morphological types of fibre, a medial bundle of larger diameter fibres with longer sarcomeres and a lateral column of smaller diameter fibres with shorter sarcomeres (Fig. 4.19). The MRO's lie in parallel with the extensor muscles and so are stretched by flexion at the posterior joint. When the succeeding segment is raised, the tendon which is attached to its anterio-ventral rim, is pulled, thereby stretching the receptor muscle. The sensory cells which innervate the receptor muscle are located in nerves and there are two discrete populations, each in a separate nerve (Fig. 4.19). Group 1 lies on a nerve which runs parallel with the receptor muscle and is attached to it by connective tissue. It contains about a dozen cell bodies. The second group of sensory cells is located in a smaller nerve from the ganglion, which also contains the motor nerves of the extensor and of the receptor muscles.

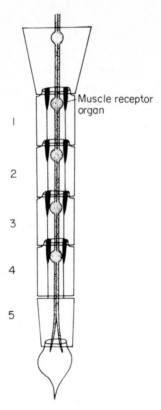

Fig. 4.18   Diagram of scorpion postabdomen with each joint in an extended position. The paired muscle receptor organs are located in segments 1–4, as are the postabdominal ganglia. (From Bowerman, 1972a.)

4.4.2   Physiology

The dozen or so cells of the group 1 population (Fig. 4.19) can be distinguished by their differing directional sensitivity. Both tonic and phasic responses occur, tonic units responding either to stretch or to relaxation. Only a single tonic stretch unit was found consistently in each preparation (Fig. 4.20) but several units were found to respond to relaxation. The single tonic stretch (elongation) sensitive unit is silent at positions in the relaxed half of the range. When a new position is being imposed by stretch, the frequency of discharge of the tonic stretch unit increases, then shows rapid adaptation, followed by a slow decline (Fig. 4.21). When a new position is imposed by relaxation, the tonic stretch unit is silent for a period

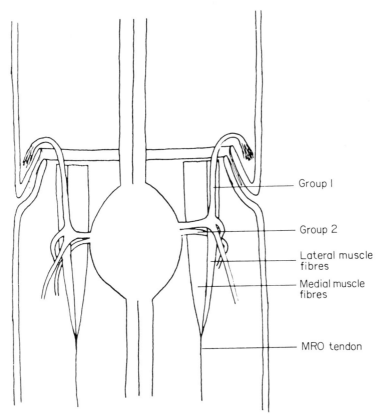

Group I

Group 2

Lateral muscle fibres

Medial muscle fibres

MRO tendon

Fig. 4.19   Diagram of a postabdominal ganglion and muscle receptor organs of a scorpion, showing the position of the two groups of sensory neurons and the lateral and medial muscle fibres. (From Bowerman, 1972a.)

before low frequency discharge returns. This unit shows, therefore, a movement response similar to that of the lepidopteran MRO. If activity were exclusively a function of receptor length the discharge would begin and end at approximately the same position during stretching and relaxing. The unidirectional phasic activity extends the range of sensitivity of the muscle receptor organs during stretching, and greatly reduces it during relaxing. There are about six tonic relaxation units and the responses of four of them are shown in Fig. 4.20. The discharge frequency of each tonic relaxation unit increases progressively the more the MRO is relaxed. A phasic sensitivity similar to that of the tonic stretch unit, but opposite in direc-tionality, is exhibited by these tonic relaxation units. One phasic

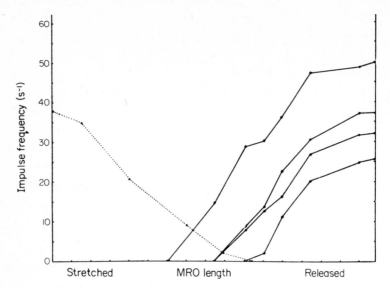

Fig. 4.20   Relationship between discharge frequency and MRO length for tonic Group 1 neurons of scorpion MRO. (------) stretch-sensitive unit; (————) release-sensitive units. (From Bowerman, 1972a.)

stretch unit was identified, which fired only if the velocity of stretch reached a certain level. The frequency of the discharge appears to be directly related to the velocity of stretch. Three phasic relaxation units were identified. All three units fired during the fastest release under the experimental conditions, but at lower velocities the phasic

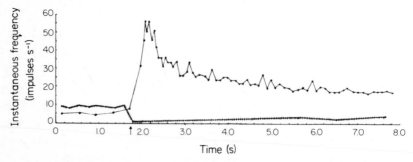

Fig. 4.21   Movement sensitivity of tonic stretch-sensitive neuron in Group 1 of scorpion MRO. The discharge frequency subsequent to movement of the MRO to a set length is dependent on whether the position is reached by releasing (+++++) or by stretching (————). The arrow indicates the time at which the receptor length was reset. (Instantaneous frequency = reciprocal of interspike interval). (From Bowerman, 1972a.)

units dropped out in succession, indicating that there is range fractionation (Chapter 6) of the response of these units (Bowerman, 1972a).

The second group (group 2) consists of six stretch-sensitive units. They are all phasic-tonic, with responses similar to that of the lepidopteran MRO (Fig. 4.22; Bowerman, 1972a).

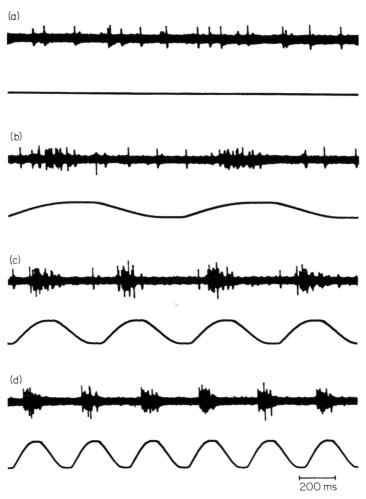

Fig. 4.22   Response of Group 2 neurons of scorpion MRO to sinusoidal stretch-release stimulation. (a) Resting discharge; (b) 1 Hz; (c) 2 Hz and (d) 3 Hz. The receptor tendon was coupled to a loudspeaker driven by a low-frequency sine-wave generator. Lower trace monitors MRO length, with upward deflection indicating stretch. (From Bowerman, 1972a.)

Nothing is known of the detailed structure of the sensory units in the scorpion, but they are presumably multiterminal neurons. Bowerman (1972a) suggests that the group 2 terminations are probably more intimately associated with the muscle component of the receptor than those of the group 1 neurons. Group 1 neurons appeared to be less specific in their response than group 2 neurons. Thus they respond to stretch and release and also to flexion, extension and rotation of the anterior joints; whereas group 2 units do not respond to such movements of the joint at the anterior end of the segment. Thus the group 2 population, specifically monitor position and movement of the posterior joint in the dorso-ventral plane (Bowerman, 1972a).

4.5   Stimulus transduction in insect stretch receptors

Two types of insect stretch receptor have been studied with the electron microscope, the cockroach receptor, which is a connective tissue strand type (Osborne, 1963a) and the lepidopteran MRO (Osborne and Finlayson, 1965). In the cockroach receptor (Fig. 4.10a) the terminations of the neuron lie in the connective tissue matrix and Osborne suggested that the transduction mechanism is compression of the naked tips brought about by the reduction in diameter that must accompany elongation of the strand. In the lepidopteran receptor the unsheathed terminations lie among the special connective tissue fibre bundles. As Osborne (1970) pointed out, the fibres of the lepidopteran receptor are likely to be collagenous in nature, similar to those described by Whitear (1965) in the thoracico-coxal MRO of the crab. He also pointed out that, paradoxically, insect receptor strands are highly elastic although vertebrate collagen is notoriously inelastic, and suggested that the fibrils of the insect stretch receptors may have a different molecular structure from vertebrate collagen and be at least partially elastic. However, this hypothesis need not be invoked unless each fibril extends the full length of the receptor. If the fibres run for only part of the length they could slip past each other when the strand is stretched, and thereby allow pressure or a shearing force to be exerted on the tips of the dendrites which lie among them. Whichever hypothesis more nearly describes the real morphological relationships of the dendritic tips and the connective tissue components of insect stretch receptors, it seems clear that transduction takes place by deformation of the unsheathed terminations. Similar naked terminals have been shown to be the sensitive regions in vertebrate muscle spindle and Pacinian corpuscle (see Osborne, 1970). No accessory structures have been seen in the tips of the naked dendrites which would serve in the transduction process but the original investigations of the cockroach and the lepidopteran receptor were carried out at a relatively early stage in electron microscopy and it would be worthwhile re-examining them to see if there are accessory structures like the neurotubular cytoskeleton of the LC3 multiterminal neuron and the LR7 uniterminal neuron of the cibarium and mouthparts of the tsetse fly. It is suggested (Rice, Galun and Finlayson, 1973) that the effective stimulus is a very slight bend which stretches one side of the tip over the neurotubular cytoskeleton so as to increase membrane capacitance or membrane ionic conductance.

4.6 Proprioceptive functions of stretch receptors in the control of posture and locomotion

4.6.1 Insects

Apart from speculations by Finlayson and Lowenstein (1958) on the possible function of the longitudinal and vertical stretch receptors and the ventro-lateral chordotonal organ of the dragonfly larva, the only study of the proprioceptive function of stretch receptors in the abdomen of insects was that carried out by Weevers (1965, 1966c) in the caterpillar. Recordings of membrane potentials from muscles show that muscle reflexes controlled by stretching or relaxing MROs exist in the caterpillar (Weevers, 1966c). The most typical response is the excitation of a muscle by stretching the MRO of the same segment. Such excitation occurs in ipsilateral and contralateral muscles of the same segment but inhibition of a contralateral muscle may occur (Weevers, 1966c), indicating that the responses of individual muscles in the caterpillar may vary. When a single MRO is stretched, at least 32 motor units show reflex responses, the vast majority of muscles being excited. This is not surprising because the musculature of the caterpillar is exceedingly complex and is functioning to maintain the hydrostatic skeleton as well as taking part in locomotory activities. The response of two muscle groups to stretching the ipsilateral MRO are shown in Fig. 4.23 (Weevers, 1966c). The muscle in Fig. 4.23a

(a)

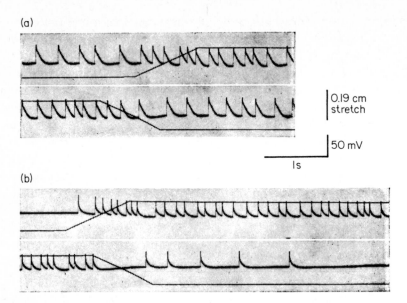

0.19 cm
stretch

50 mV

Is

(b)

Fig. 4.23   Oscillograph records of action potentials in two muscle groups of a caterpillar, showing the effect of stretching the ipsilateral MRO. (a) was taken from group f; (b) from group G (Fig. 4.24). The second beam shows the movement of the stretching forceps. (From Weevers, 1966c.)

was already active, but stretching of the MRO increased the frequency of muscle action potentials, while relaxation led to a decrease in their frequency. The muscle in Fig. 4.23b was not spontaneously active, but became active when the MRO was stretched, and lapsed into inactivity again when the MRO was released. Muscles in other segments may also respond to stretching of the MRO (Fig. 4.24), but such intersegmental reflexes are typically less intense than the intrasegmental ones. Inhibitory reflexes are confined to muscles other than the longitudinal intersegmental and long diagonal muscles. One group of integumentary muscles in particular is strongly inhibited by stretch of either the ipsilateral or contralateral intrasegmental MRO (Fig. 4.25). Reflex responses of muscles have a phasic component also, which mirrors the phasic response of the receptor neurons (Fig. 4.26). Displacement, movement and acceleration components are all seen in the reflex action of the muscle. When two receptors are stretched simultaneously the ipsilateral and contralateral reflexes reinforce each other to give a response larger than the sum of the responses to stretching the

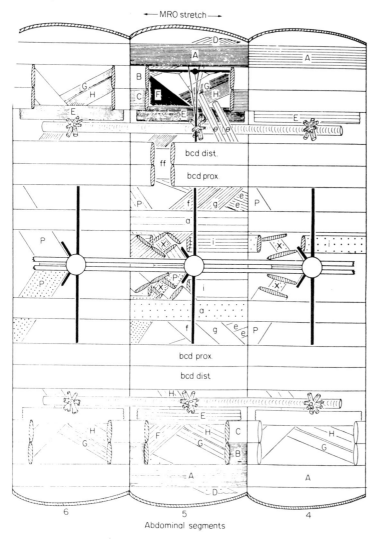

Fig. 4.24 Diagram of three adjacent segments (4 to 6) of a caterpillar showing the kind (excitatory or inhibitory) and approximate intensity of reflex effects following stretching of the receptor on the left side of segment 5. Closely spaced lines indicate intense excitation and closely spaced dots intense inhibition. (From Weevers, 1966c.)

receptors singly. The reflex fields of neighbouring receptors overlap and spatial facilitation produces a disproportionate increase in the overall response when two receptors are stimulated simultaneously. Weevers (1966c) recorded muscle activity from many groups of

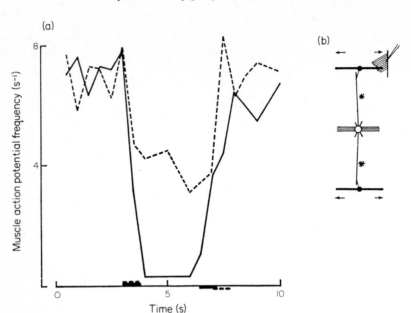

Fig. 4.25 (a) An inhibitory stretch reflex in muscle group L of a caterpillar. (————) shows the response to stretching the ipsilateral MRO by 0.15 cm; (— — —) shows the effect of subjecting the contralateral receptor to a similar stimulus. The full line above and below the abscissa shows the time of stretching and of releasing the ipsilateral MRO; the dashed line, the time of stretching and releasing the contralateral MRO. (The average of two such responses). (b) The location of muscle group L. *—the position of the spiracles. (From Weevers, 1966c.)

muscles in the caterpillar in response to stretching MROs. In Fig. 4.24, which summarizes the results he obtained, the most intense excitatory and inhibitory reflexes are shown by the most closely spaced lines and dots respectively. Muscles which are not shaded show no change in discharge frequency when the MRO is stretched. Weevers (1966c) found evidence that the moto neurons of nerve 2 in the abdominal ganglion of silk moth caterpillars make synaptic contacts only in the ganglion anterior to the one in which the axons leave the CNS, confirming earlier observations by Von Holst (1934). Therefore, the reflex pathway for muscles innervated by nerve 2 involves synaptic connections in the ganglion of the segment anterior to the stimulated receptor and responding muscles. The simplest possible arrangement of the central connections of an MRO to produce the reflex muscle activity found by Weevers is shown in

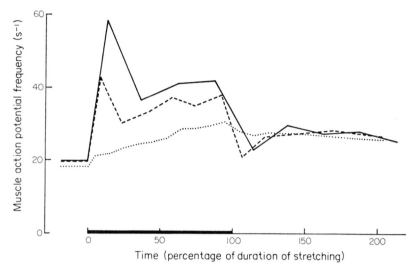

Fig. 4.26   The stretch reflex in muscle group E (Fig. 4.24) of a caterpillar. The ipsilateral MRO in the same segment was stretched by 0.19 cm at three different rates. (. . . . . .) 0.008 cm s^{-1}; (-----) 0.18 cm s^{-1}; (————) 0.36 cm s^{-1}. The tonic discharge in group E was intensified experimentally. Each curve is the average of two responses. (From Weevers, 1966c.)

Fig. 4.27. The muscles most strongly excited are those which lie functionally in parallel with the stretched sense organ. It is concluded that a major function of the caterpillar MRO is to mediate the negative feedback reflex tending to stabilize body position independent of load.

The caterpillar moves by peristaltic waves passing forward as the caterpillar progresses. The peristaltic movement results in the lifting and advancing of each segment in turn. If the posterior end of the caterpillar is held stationary (Weevers, 1965) the peristaltic wave ceases and then recommences at the posterior end. In the normal course of events this response would result in the posterior end of the caterpillar being moved forward thereby unloading the stretched MROs of the segments involved. In view of the very extensive reflex effects of MRO stretch, Weevers (1965) considered the possibility that a chain of segmental reflexes, activated by the MROs, might alone be sufficient to coordinate a locomotor wave. However, some muscles which take an active part in locomotion do not show any reflex change in response to stretching a single MRO. It is unlikely, therefore, that any locomotory wave controlled solely by the MROs

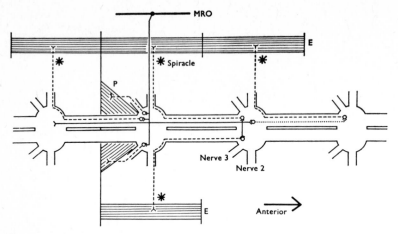

Fig. 4.27   The simplest possible central connexions of a single MRO of a caterpillar which could produce a stretch reflex with the observed spread. For simplicity only one muscle innervated by nerve 2 and one innervated by nerve 3 is shown in each half segment. (————) indicates the ramifications of the MRO sensory axon; (— — — —) signify motor axons; (*) show the position of the spiracles. E, P,—muscle groups (Fig. 4.24). (From Weevers, 1966c.)

could be as complex as the pattern of activity shown by the muscles during normal peristaltic locomotion. Furthermore, the kinds of reflex effect resulting from MRO stretch are very uniform, the majority being excitatory and occurring with more or less the same latency. Since the reflex fields of adjacent receptors overlap, particular patterns of MRO excitation involving several receptors might have reflex effects not predictable by algebraic addition of the responses to stretch of each individual receptor. Finally, there might be special functions, different in kind, from those served by the stretch reflexes; functions only fulfilled by the MROs during peristaltic locomotion. Weevers stretched one receptor during the passage of a locomotory wave in the pinned-out preparation, at the same time recording from a muscle which in the inactive animal was reflexly excited by the same stimulus. Discharge frequency increased during the period of stretch, confirming that the stretch reflexes acted during peristaltic locomotion. MRO denervation was also tried in order to see whether it produced any effects which could be interpreted as the result of interrupting a chain of segmental reflexes necessary or important in allowing the conduction of a wave along the body (Weevers, 1965). Neither stimulation of an MRO nor elimination of an MRO had any pronounced effect on the transmission of wave conduction in the caterpillar, confirming the conclusion of von Holst (1934) that wave conduction in the caterpillar is primarily a central nervous process. Weevers (1965) concluded that chain reflexes mediated by MROs are unlikely to play more than a modifying part in the conduction of the peristaltic locomotory wave, but he pointed out the need for caution in interpreting the importance of MROs on the basis of the experiments he was able to carry out and he advocated further study involving total MRO inactivation in the intact animal or an investigation of CNS mechanisms involved in locomotion.

4.6.2   Scorpions

Bowerman (1972b) investigated reflexes in the four main muscle groups and the receptor muscle of the post-abdominal segments of the scorpion. Unilateral stimulation of the MRO by stretching the receptor tendon results in an increased motor discharge to the extensor ventro-lateral and dorso-lateral muscles and depression of activity to the flexor muscles. Release of the MRO enhanced flexor discharge and depressed motor output to the extensor ventro-lateral and dorso-lateral muscles. Reflex motor activity in the nerve of the receptor muscle was similar to that in the nerve supplying the extensor muscles. Ipsilateral MRO stretch excited receptor muscle motor neurons and release terminated excitation. Bowerman (1972b) did not differentiate between motor output to the medial and the lateral muscle components of the MRO. Stimulation of the MROs on both sides of a segment resulted in greater reflex output than unilateral stimulation. In order to establish the relative contribution of each MRO group of sensory neurons the group 1 nerve was sectioned close to the receptor. All reflex motor activity in the nerve to the main muscles, evoked by MRO stimulation, was unaffected by this operation, indicating that the neurons of group 2 are primarily responsible for modulating reflex motor output. Selective elimination of the group 2 neurons was not accomplished because of the shortness of the nerve and so the possibility of a group 1 contribution to the segmental reflexes has not been entirely ruled out.

4.6.3   Millipedes

Athough nothing is known about the proprioceptive sense organs in millipedes, the use of proprioceptive information in the control of locomotion in *Trigoniulus* (Barnwell, 1965) and in *Schizophyllum* has been analysed in a detail (Burger, 1971; Burger and Mittelstaedt, 1972) that has not been applied to the other arthropods discussed in this chapter; although studies by Grosslight and Harrison (1961) indicate that a similar mechanism is probably involved in the mealworm (*Tenebrio*) larva.

A millipede can maintain a straight course without the need for any external guiding stimuli. This means that any accidental bends that take place between segments during locomotion must be corrected immediately so that the many segments of the elongated body are kept in a straight line. This is the first type of negative feedback which must be assumed if the millipede in normal unimpeded locomotion is to keep its flexible body in as nearly a straight line as possible.

The other situation in which a millipede frequently finds itself in nature, is being diverted from a straight path by an obstacle. When this happens the animal is forced to turn but as soon as its antennae have passed the obstacle it makes a compensatory turn in the opposite direction to return its body to the original orientation. The angle of the compensatory turn is determined by the angle of the preceding induced turn (Chapter 14). A similar compensatory turn takes place if the diversion is induced by a visual stimulus.

In making a turn, successive segments of the multi-segmented body are involved in a repetitive activity. In order to make a compensatory turn, the millipede must store information about the angle of bend and number of segments involved so that the correct path can be restored. Burger (1971) investigated the sensory cues used by the millipede in this control system and although she was primarily interested in the storage of information, her experiments give some information about the proprioceptive system involved. The experimental arrangement consisted of an alleyway along which the millipede was constrained to run. This alleyway consisted of two straight parts which could be set at a variable angle to one another or moved into a new angle while the animal was in motion.

When a millipede is induced to go round a bend and then, while it is turning, the rear segments are brought quickly into line with the front end the angle of the compensatory turn that is subsequently made is smaller than it would have been if the animal had been allowed to complete the induced turn. This means that the size of the compensatory turn is proportional to the number of segments that have passed the bend.

When a compensatory turn is interrupted by quickly pushing the rear end of the animal into line with the front end which had already passed the compensatory turning point, the front end immediately begins an additional turn in the same direction as the compensatory turn. The amount of the additional turn increases in proportion to the number of segments prevented from making the turn by the experimental displacement. The store of proprioceptive information must obviously be unloaded by the animal bending at successive segmental joints; it cannot be unloaded by an imposed 'passive' movement of the whole body into the final compensatory orientation. If two bends are imposed on the animal the compensatory bend assumed when it becomes free to turn, takes both induced bends into consideration and the resultant compensatory bend is appropriate to restore the millipede to its original course. If the second induced turn is in the compensatory direction the millipede continues in a straight line when it emerges from the alleyway but when its body is completely free it may then make a compensatory turn which will depend on how much information is left in store. If nothing is left in the store, it will not compensate but if there is still information about the first induced bend, the store must be unloaded by the animal making a further compensatory turn in the opposite direction to the first induced turn. For example, if it is induced to turn through 75° in one direction, then induced to turn 30° in the opposite (compensatory) direction, when the whole body becomes completely free, it will turn in the compensatory direction through 45°.

Although nothing is known of the receptors involved in the control of orientation in millipedes, they are obviously segmental, rather than multisegmental units, because these behavioural experiments show that the numbers and relative positions of segments are recorded by the nervous system. It is unlikely that the precision required in this system could be achieved by external receptors alone and so it is probable that millipedes have internal stretch receptors, arranged in segmental units. The fact that successive *passive* bendings induced by an experimenter can cause compensatory turning, indicates a sensory system that is not intimately linked with the effector system. Evidence that body receptors rather than leg receptors are involved was obtained by Burger (1971) in experiments in which the induced turns were imposed by dragging the millipedes round bends. The animals subsequently performed appropriate compensatory turns. In such a situation any proprioceptive information from the limbs would probably be chaotic and certainly not directly related to the angle of bend. Not only are leg receptors unlikely to be used but, because of the precise monitoring of angular displacement required, it is likely that the stretch receptors are MRO's rather than of the connective tissue strand type.

4.7   Chordotonal organs

Studies on chordotonal organs have been confined almost exclusively to organs with many scolopales, such as the complex organs of the leg of insects (Usherwood, Runion and Campbell, 1968; Young, 1970) and similar complex organs in the Crustacea (see Chapters 6, 7 and 9). Little is known about simple chordotonal sensilla in which the sensory element consists of but a single neuron or a very small number of neurons. They have been described in every abdominal segment of insects; in which they occupy constant positions. The distribution of chordotonal sensilla in the first seven or eight abdominal segments in ten orders of insects is shown in Table 4.2. So far they have always been found in mid-ventral, ventro-lateral, lateral or dorsal positions, or in all these positions. It is probably safe to assume that, in those insects in which chordotonal sensilla have been described in only one of these positions in abdominal segments, the others have not been looked for or have been overlooked. In those insects in which a thorough search has been made for simple chordotonal sensilla, large numbers have been found. Slifer (1936) found 76 pairs of chordotonal organs in *Melanoplus* (Orthoptera) but these also include complex organs. Hertweck (1931), found 45 pairs of simple chordotonal sensilla in the larva of *Drosophila* (Diptera). The only thorough investigation of the physiology of a simple strand type chordotonal sensillum in the abdomen of an insect has been carried out by Orchard (1975b). He studied the ventro-lateral chordotonal sensillum of the stick insect *Carausius morosus* (Phasmida), which occurs in pairs, one on each side of the mid-ventral line of abdominal segments 2–7. There are two

Table 4.2 The distribution of connective chordotonal organs* in the first seven or eight abdominal segments in various orders of insects (Orchard, 1975a).

| Order | Genus | Chordotonal organs | | | |
		Mid-ventral	Ventro-lateral	Lateral	Dorsal
Ephemeroptera	*Ephemera* larva[1]			+(3)+(1)	
Odonata	*Aeshma* larva[2]			+(3)	
Orthoptera	*Melanoplus*[3]	+(1–5)	+(1–6)	+(2)seg.1	
Phasmida	*Carausius*[4]	+(1)	+(2)		
Dictyoptera	*Blaberus*[4,5,6]	+(4–7)		+(7)	
	Periplaneta[5,6]	+(3–5)			
Hemiptera	*Rhodnius*[7]	+(1–2)	+(1–2)		
Lepidoptera	*Tortrix* larva[8]			+(4)	
Diptera	*Drosophila*[10]	+(1)	+(1)	+(5)	+(1)
	Chironomus[8]			+(3)	+(2)+(2)
	Corethra[8]			+(3)	
	Culex[8] larva			+(?)	
	Glossina[9]			+(?)	
	Phormia[11]	+(1)	+(1)	+(5)	+(1)
	Ptychoptera[8]			+(?)	
	Tabanus[8]			+(3)	
Hymenoptera	*Nematus*[8] larva			+(7)	
Coleoptera	*Dytiscus*[8]			+(4)	
	Ergates[12] larva			+(2)	
	Monohammus[12]			+(2)	

*Discounting tympanal organs; +, present; (1–7), denotes number of scolopales; (?), details unknown. [1]Osborne and Finlayson (1962); [2]Mill (1965); [3]Slifer (1936); [4]Orchard (quoted in text and unpublished observations); [5]Kehler and Rowe (1968); [6]Kehler, Smalley and Rowe (1970); [7]Anwyl (1972); [8]Graber (1882); [9]Finlayson, 1972; [10]Hertweck (1931); [11]Osborne (1963b); [12]Hess (1917).

scolopales in each organ and, correspondingly, two sensory neurons. Orchard (1975b), points out that the so-called ventral phasic receptors of the cockroach *Blaberus discoidalis* (Dictyoptera) studied by Kehler, Smalley and Rowe (1970) and described as a new type of receptor are probably the mid-ventral chordotonal organs. They differ from the ventro-lateral organ of *Carausius* in containing up to 6 scolopales. The chordotonal organ of *Carausius* has no resting discharge and though it responds readily to vibration (Fig. 4.28), Orchard could not find a correlation between frequency of stimulation and frequency of response. It responds to the compression phase of the ventilatory cycle (Fig. 4.28) although, as with the

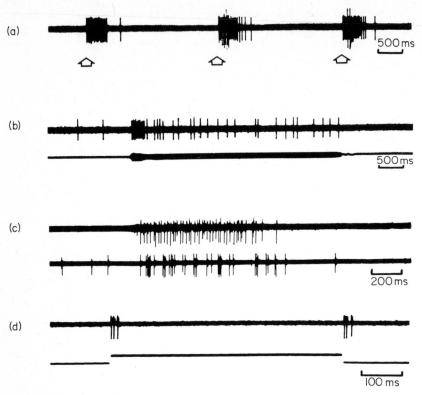

Fig. 4.28   Oscillograph records of responses of the ventro-lateral chordotonal organ of the stick-insect (*Carausius*): (a) to vibrational stimuli produced by tapping the table (indicated by arrows). (Note the phasic response and the presence of 2 units), (b) To 100 Hz vibration produced by holding a vibrating probe in the saline approximately 1 mm above the preparation (probe record on lower trace). (c) Efferent activity in the left segmental nerve from the 2nd abdominal ganglion (upper trace), and afferent activity from the left chordotonal organ in the 4th abdominal segment (lower trace) during an expiratory burst. (d) Response to displacement of the body wall by 50 μm (lower trace). Upward deflection represents stretching of the attachment strand. Note the phasic response to application and release. (From Orchard, 1975b.)

'ventral phasic receptor' of Kehler *et al.* (1970), the response is variable, occurring in some preparations but not in others. The variability may be due to the abnormal state of the animal after being opened up. Direct mechanical stimulation of the sensillum by distortion of the body wall, which induces stretching of the attachment strand, elicits a rapidly-adapting phasic discharge. Upon release, a further rapidly-adapting phasic response occurs (Fig. 4.25). There is no tonic response. The critical velocity to which the receptor will respond was found to be 4 mm s^{-1}. The frequency of impulses and the number of impulses per burst increase with increasing velocity of stimulation, whilst the latency decreases. At a constant velocity of 6 mm s^{-1}, with gradually increasing amplitude of stretch there is no apparent change in response; the number of impulses per burst, impulse frequency and latency all remain comparatively constant. With constant velocities greater than 20 mm s^{-1}, however, an increase in amplitude of stretch is accompanied by an increase in the number of impulses per burst, although the impulse frequency remains constant. The response to relaxation is similar to the response to stretch. In an attempt to determine whether the relaxation response is produced by the same neuron as the stretch response, the effect of shortening the pulse interval was observed. The relaxation response is abolished when a stimulus duration of 7 ms is used, but there is no effect on the stretch response. This suggests that the same cells respond both to stretch and to relaxation.

The ventro-lateral chordotonal organ of *Carausius* contains two scolopales associated with two sensory neurons, suggesting that each scolopale encloses only one dendrite. In this respect it differs from the chordotonal organs at the joints of decapod crustacean legs (Bush, 1965b) and some insect chordotonal organs (see Howse, 1968), in which each scolopale encloses two or more dendrites. The 'abdominal vibration receptors' of the cockroach (Florentine, 1967, 1968) are probably chordotonal organs also. Their lateral position, as located electrophysiologically, and their highly phasic response to mechanical stimuli, suggests that they are lateral chordotonal organs. The ventro-lateral chordotonal organs of *Carausius* are phasic and respond only to changes in position (movement). The rate of movement appears to be coded in terms of impulse frequency, with the frequency response being constant for any amplitude at a constant velocity. There is also an increase in the number of impulses in certain other conditions. Thus an increase in amplitude of square wave stimulation, or an increase in amplitude at constant velocities of more than 20 mm s^{-1}, produce an increase in the number of impulses. At amplitudes of 100 μm, an increase in velocity also produces an increase in the number of impulses. This would, at first sight, appear to be contradictory, but it must be remembered that the CNS is obtaining additional information from other sense organs and so may be able to interpret a 'number of impulses' code. The rate of movement coded by frequency could be the important factor although there may be information concerning the amount of movement coded by number of impulses. A simpler explanation may be that the receptor is used merely to record events. A decrease in latency with an increase in the rate of displacement has been reported in other sense organs (see Florentine, 1967, 1968) but it is difficult to ascertain whether the animal would use this information. The ventro-lateral organ of *Carausius* resembles the velocity sensitive, phasic movement fibres found in the CB and MC joint chordotonal organs of *Carcinus* (Bush, 1965a, b). These fibres have higher thresholds and, generally, higher saturation velocities than other movement fibres and usually no spontaneous resting or position discharge (Bush, 1965b) (Section 6). Unlike the chordotonal organs found in the decapod crustacean limb and the insect limb, the ventro-lateral chordotonal organ of *Carausius* contains no tonic units. It is possible that the insect abdomen has evolved a proprioceptive system with a variety of receptors (stretch receptors, chordotonal organs, multipolar neurons and campaniform organs), all of which combine to give a wide range of unit types similar to those found in the chordotonal organs of the decapod limb. There are no campaniform organs as such in Crustacea (Chapter 1) and no multiterminal neurons have been described in the legs of Crustacea (Finlayson, 1968). The receptor responds in a similar manner to stretch and release as do the mid-ventral chordotonal organs of the cockroach (Kehler *et al*., 1970) and the cuticular stress detectors in crustacean limbs (Clarac, Wales and Laverack, 1971) (Section 7.3). It seems that in *Carausius* the same scolopidia are involved in the stretch and the relaxation responses. In support of this hypothesis (a) response to relaxation is abolished by shortening the pulse width and (b) although two scolopidia are present, both stretch and relaxation responses have been shown to consist of impulses from *two* fibres.

Proprioceptive function

Contraction of the longitudinal and vertical muscles of an abdominal segment of *Carausius* produces the compression phase of ventilation and stimulates the chordotonal organs which respond throughout the entire period of compression. The minute contractions produced during the incompletely fused tetanus (Burke, 1954) during contraction of the muscles are sufficient to stimulate the receptor during the whole of the compression phase. It seems likely that the ventro-lateral chordotonal organs of *Carausius* provide information about the activity of either longitudinal or vertical muscles during the expiratory phase of ventilation. Inspiration may be monitored by lateral chordotonal organs. It was postulated by Finlayson and Lowenstein (1958) that the 'oblique receptor' in the dragonfly larva, later identified as a lateral chordotonal organ (Mill, 1965), may be stretched during inspiration and relaxed during expiration. A system of abdominal chordotonal organs in the abdomen of insects may provide information concerning the entire cycle of ventilation with differing sensitivities and/or the differing positions of the pairs of chordotonal organs allowing each phase to be monitored.

Acknowledgements

I wish to thank Drs Osborne and Orchard for reading the manuscript and the former for supplying original drawings; also Miss Fifield for redrawing Fig. 4.3.

References

Anwyl, R. (1972) The structure and properties of an abdominal stretch receptor in *Rhodnius prolixus*. *Journal of Insect Physiology*, **18**, 2143–2154.

Barnwell, F. H. (1965) An angle sense in the orientation of a millipede. *Biological Bulletin*, **128**, 33–50.

Bowerman, R. F. (1972a) A muscle receptor organ in the scorpion post-abdomen. I. The sensory system. *Journal of Comparative Physiology*, **81**, 133–146.

Bowerman, R. F. (1972b) A muscle receptor organ in the scorpion post-abdomen. II. Reflexes evoked by MRO stretch and release. *Journal of Comparative Physiology*, **81**, 147–157.

Burger, M.-L. (1971) Zum Mechanismus der Gegenwendung nach mechanisch aufgezwungener Richtungsänderung bei *Schizophyllum sabulosum* (Julidae, Diplopoda). *Zeitschrift für vergleichende Physiologie*, **71**, 219–254.

Burger, M.-L. and Mittelstaedt, H. (1972) Course control by stored proprioceptive information in millipedes. *Proceedings of the 3rd International Symposium of Biocybernetics. Leipzig. August 1971, Biocybernetics Volume IV. Drischel, H. and Dettmar, P. (eds.) Fischer-Verlag Jena.*

Burke, W. (1954) An organ for proprioception and vibration sense in *Carcinus maenas*. *Journal of Experimental Biology*, **31**, 127–137.

Bush, B. M. H. (1965a) Proprioception by chordotonal organs in the mesocarpopodite and carpo-propopodite joints of *Carcinus maenas* legs. *Comparative Biochemistry and Physiology*, **14**, 185–199.

Bush, B. M. H. (1965b) Proprioception by the coxo-basal chordotonal organ, CB, in legs of the crab, *Carcinus maenas. Journal of experimental Biology*, **42**, 285–297.

Clarac, F., Wales, W., and Laverack, M. S. (1971) Stress detection at the autotomy plane in the decapod Crustacea. II. The function of receptors associated with the article of the basi-ischiopodite. *Zeitschrift für vergleichende Physiologie*, **73**, 343–407.

Finlayson, L. H. (1966) Sensory innervation of the spiracular muscle in the tsetse fly (*Glossina morsitans*) and the larva of the wax moth (*Galleria mellonella*). *Journal of Insect Physiology*, **12**, 1451–1454.

Finlayson, L. H. (1968) Proprioceptors in the invertebrates. *Symposia of the Zoological Society of London*, **23**, 217–249.

Finlayson, L. H. (1972) Chemoreceptors, cuticular mechanoreceptors, and peripheral multiterminal neurones in the larva of the tsetse fly (*Glossina*). *Journal of Insect Physiology*, **18**, 2265–2276.

Finlayson, L. H. and Lowenstein, O. (1955) A proprioceptor in the body musculature of Lepidoptera. *Nature*, **176**, 1031.

Finlayson, L. H. and Lowenstein, O. (1958) The structure and function of abdominal stretch receptors in insects. *Proceedings of the Royal Society of London* (B), **148**, 433–449.

Finlayson, L. H. and Mowat, D. J. (1963) Variations in histology of abdominal stretch receptors of saturniid moths during development. *Quarterly Journal of Microscopical Science*, **104**, 243–251.

Finlayson, L. H. and Osborne, M. P. (1968) Peripheral neurosecretory cells in the stick insect (*Carausius morosus*) and the blowfly larva (*Phormia terrae-novae*). *Journal of Insect Physiology*, **14**, 1793–1801.

Florentine, G. J. (1967) An abdominal receptor of the American cockroach *Periplaneta americana* (L.) and its response to airborne sound. *Journal of Insect Physiology*, **13**, 215–218.

Florentine, G. J. (1968) Response characteristics and probable behavioural roles for abdominal vibration receptors of some cockroaches. *Journal of Insect Physiology*, **14**, 1577–1588.

Gelperin, A. (1971) Abdominal sensory neurons providing negative feedback to the feeding behaviour of the blowfly. *Zeitschrift für vergleichende Physiologie*, **72**, 17–31.

Getting, P. A. and Steinhardt, R. A. (1972) The interaction of external and internal receptors on the feeding behaviour of the blowfly, *Phormia regina*. *Journal of Insect Physiology*, **18**, 1673–1681.

Gettrup, E. (1962) Thoracic proprioceptors in the flight system of locusts. *Nature*, **193**, 498–499.

Gettrup, E. (1963) Phasic stimulation of a thoracic stretch receptor in locusts. *Journal of Experimental Biology*, **40**, 323–333.

Graber, U. (1882) Die chordotonalen Sinnesorgane und das Gehör der Insekten. 1. Morphologischer Theil. *Archiv für mikroskopische Anatomie*, **20**, 506–640.

Grosslight, J. H. and Harrison, P. C. (1961) Variability of response in a determined turning sequence in the meal worm (*Tenebrio molitor*): an experimental test of alternative hypotheses. *Animal Behaviour*, 9, 100–103.

Hess, H. N. (1917) The chordotonal organs and pleural discs of cerambycid larvae. *Annals of the Entomological Society of America.* 10, 63–74.

Hertweck, H. (1931) Anatomie und Variabilität des Nervensystems und der Sinnesorgane von *Drosophila melanogaster* (Meigen). *Zeitschrift für wissenschaftliche Zoologie*, 139, 559–663.

Holst, E. von (1934) Motorische und tonische Erregung und ihr Bahrenverlauf bei Lepidopteren-larven. *Zeitschrift für vergleichende Physiologie*, 21, 395–414.

Howse, P. E. (1968) The fine structure and functional organization of chordotonal organs. *Symposia of the Zoological Society of London*, 23, 167–198.

Hughes, G. M. and Wiersma, C. A. G. (1960) Neuronal pathways and synaptic connexions in the abdominal cord of the crayfish. *Journal of Experimental Biology*, 37, 291–307.

Kehler, J. G. and Rowe, E. C. (1968) Two new types of mechanoreceptor in the cockroach abdomen. *American Zoologist*, 8, 773.

Kehler, J. G., Smalley, K. N. and Rowe, E. C. (1970) Ventral phasic mechanoreceptors in the cockroach abdomen. *Journal of Insect Physiology*, 16, 483–497.

Kennedy, D., Evoy, W. H. and Fields, H. L. (1966) The unit basis of some crustacean reflexes. *Symposia of the Society for Experimental Biology*, 20, 75–109.

Lowenstein, O. and Finlayson, L. H. (1960) The response of the abdominal stretch receptor of an insect to phasic stimulation. *Comparative Biochemistry and Physiology*, 1, 56–61.

Manton, S. M. (1958b) Habits of life and evolution of body design in Arthropoda. *Journal of the Linnaean Society of London, Zoology*, 44, 58–72.

Mill, P. J. (1965) An anatomical study of the abdominal nervous and muscular systems of dragonfly (Aeschnidae) nymphs. *Proceedings of the Zoological Society of London*, 145, 57–73.

Orchard, I. (1975a) The physiology of peripheral neurones in the stick insect *Carausius morosus*. *Ph.D. Thesis*, University of Birmingham, U.K.

Orchard, I. (1975b) Structure and properties of an abdominal chordotonal organ in the stick insect (*Carausius morosus*) and the cockroach (*Blaberus discoidalis*). *Journal of Insect Physiology*, 21, 1491–1499.

Osborne, M. P. (1961) Studies on the sensory nervous system of insects and centipedes. *Ph.D. Thesis*, University of Birmingham, U.K.

Osborne, M. P. (1963a) An electron microscope study of an abdominal stretch receptor of the cockroach. *Journal of Insect Physiology*, 9, 237–245.

Osborne, M. P. (1963b) The sensory neurons and sensilla in the abdomen and thorax of the blowfly larva. *Quarterly Journal of Microscopical Science*, **104**, 227–241.

Osborne, M. P. (1964) Sensory nerve terminations in the epidermis of the blowfly larva. *Nature*, **201**, 526.

Osborne, M. P. (1970) Structure and function of neuromuscular junctions and stretch receptors. In *Insect Ultrastructure* ed. A. C. Neville, Blackwell, Oxford. *Symposia of Royal Entomological Society of London*, No. 5, pp. 77–100.

Osborne, M. P. and Finlayson, L. H. (1962) The structure and topography of stretch receptors in representatives of seven orders of insects. *Quarterly Journal of Microscopical Science*, **103**, 227–242.

Osborne, M. P. and Finlayson, L. H. (1965) An electron microscope study of the stretch receptor of *Antheraea pernyi* (Lepidoptera, Saturniidae). *Journal of Insect Physiology*, **11**, 703–710.

Rice, M. J., Galun, Rachel, and Finlayson, L. H. (1973) Mechanotransduction in insect neurons. *Nature, New Biology*, **241**, 286–288.

Rilling, G. (1960) Zur Anatomie des braunen Steinlaufers *Lithobius forficatus* L. (Chilopoda), Skeletmuskelsystem, peripheres Nervensystem und Sinnesorgane des Rumpfes. *Zoologische Jahrbucher (Anatomie)*, **78**, 39–128.

Salpeter, M. M. and Walcott, C. (1960) An electron microscope study of a vibration receptor in the spider. *Experimental Neurology*, **2**, 232–250.

Slifer, E. H. (1936) The scoloparia of *Melanoplus differentialis* (Orthoptera, Acrididae) *Entomological News*, **47**, 174–180.

Slifer, E. H. and Finlayson, L. H. (1956) Muscle receptor organs in grasshoppers and locusts (Orthoptera, Acrididae). *Quarterly Journal of Microscopical Science,* **97**, 617–620.

Usherwood, P. N. R., Runion, H. I. and Campbell, J. I. (1968) Structure and physiology of a chordotonal organ in the locust leg. *Journal of Experimental Biology*, **48**, 305–323.

Varma, L. (1972) Muscle receptor organs of the centipede *Scolopendra morsitans* (L). *Zoologischer Anzeiger*, **188**, 400–407.

Viallanos, H. (1882) Recherches sur l'histologie des insectes. *Annales des Sciences Naturelles* (6), **14**, 1–348.

Weevers, R. de G. (1965) Proprioceptive reflexes and the co-ordination of locomotion in the caterpillar of *Antheraea pernyi* (Lepidoptera). In *The physiology of the insect central nervous system*. Treherne, J. E. and Beament, J. W. L. (eds.) Academic Press, London and New York. 113–124.

Weevers, R. de G. (1966a) A lepidopteran saline: effects of inorganic cation concentration on sensory, reflex and motor responses in a herbivorous insect. *Journal of Experimental Biology*, **44**, 163–175.

Weevers, R. de G. (1966b) The physiology of a lepidopteran muscle receptor. I. The sensory response to stretch. *Journal of Experimental Biology,* **44,** 177–194.

Weevers, R. de G. (1966c) The physiology of a lepidopteran muscle receptor. III. The stretch reflex. *Journal of Experimental Biology,* **45,** 229–249.

Whitear, M. (1965) The fine structure of crustacean proprioceptors. II. The thoracico-coxal organs in *Carcinus, Pagurus* and *Astacus. Philosophical Transactions of the Royal Society* (B), **248,** 437–456.

Whitten, J. M. (1963) Observations on the cyclorrhaphan larval peripheral nervous system: muscle and tracheal receptor organs and independent type II neurons associated with the lateral segmental nerves. *Annals of the Entomological Society of America,* **56,** 755–763.

Wilson, D. M. and Gettrup, E. (1963) A stretch reflex controlling wingbeat frequency in grasshoppers. *Journal of Experimental Biology,* **42,** 521–535.

Wiersma, C. A. G. and Pilgrim, R. L. C. (1961) Thoracic stretch receptors in crayfish and rocklobster. *Comparative Biochemistry and Physiology,* **2,** 51–64.

Young, D. (1970) The structure and function of a connective chordotonal organ in the cockroach leg. *Philosophical Transactions of the Royal Society of London,* B. **256,** 401–428.

5 Receptors of the mouthparts and gut of arthropods

W. WALES

5.1 Introduction

The receptors of the mouthparts and gut of arthropods have recently received considerable attention for several reasons. Firstly, the crustacean mouthpart appendages are believed to have evolved from the same ancestral type of appendage as the walking leg and comparative studies of the proprioceptor system may be expected to elucidate how different limbs have become specialized for their particular roles. It may also be possible, through comparative studies, to recognize both primitive features and functional specializations; e.g. the CAP organs, which are absent in brachyurans (Laverack and Dando, 1968; Wales, 1972; Wales, Clarac, Dando and Laverack, 1970; Wales and Laverack, 1972a). Secondly, the crustacean mandible is of particular interest since its proprioceptors can be seen to play an important role in modifying a central programme of cyclical activity (Macmillan, Wales and Laverack, 1976. Wales, Macmillan and Laverack, 1976b). Thirdly, both the insect and crustacean foregut are controlled by a largely autonomous stomatogastric nervous system. This is composed of ganglia with small numbers of neurons and offers a variety of systems of greater complexity than the cardiac ganglion for the analysis of neuronal interactions (Maynard,

213

1966, 1969, 1971). The study of feeding behaviour and its control is of course interesting in itself.

'Mouthparts' is something of a blunderbus term in that it includes different numbers and types of appendages when applied to different animals. The crustaceans and, more particularly, the insects have been extremely successful in adapting to a wide range of ecological niches and, in doing so, have developed mouthparts specialized to their mode of life. In this chapter the term 'mouthparts' includes both the segmental appendages, such as mandibles and maxillae, and the nonsegmental appendages; the labrum, paragnatha, hypopharynx and epipharynx.

5.2　Crustacean mouthparts

The term mouthparts applied to the Crustacea in the classical sense (Borradaile, 1917), encompasses a larger and functionally more diverse group of appendages than in the insects. It includes all appendages from the mandibles to the 3rd maxillipeds, inclusive.

Our knowledge of crustacean mouthpart proprioceptors is almost exclusively restricted to the Astacura and Palinura. Ong (1969) described the structure of mandibular sensory receptors in the calanoid copepod *Gladioferens pectinatus* but does not suggest which, if any, of the structures described may be proprioceptive in function. Wetzel (1935), in his description of the amphipod *Caprella* describes the sensory endings in the mandible, but none falls into a known category of proprioceptor. Clearly there is a need for further investigation of the lower crustaceans.

In the decapods the proprioceptor systems are based upon the connective chordotonal organ innervated by the type I bipolar cells, and upon the type II multipolar cells associated with accessory connective tissue strands or muscles. The 3rd maxilliped of *Homarus gammarus* has connective chordotonal organs at all but the pro-dactylopodite and thorax-coxopodite joints (Fig. 5.1). The number of chordotonal organs is less than occurs in the walking legs, there being one less at all joints except for the coxo-basipodite joint, which has a single connective chordotonal organ in each case (Wales *et al.*, 1970). The TC MRO is present in the 3rd maxilliped but it has not yet been studied in detail. At the mero-carpopodite and carpo-propodite joints, CAP sensilla are associated with the connective chordotonal organ and these may also prove to be of a proprio-

(a)   (b)

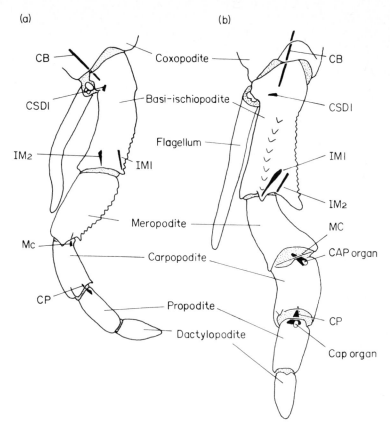

CB

CSDI

IM2

Mc

CP

Coxopodite

Basi-ischiopodite

Flagellum

IMI

Meropodite

Carpopodite

Propodite

Dactylopodite

CB

CSDI

IMI

IM2

MC

CAP organ

CP

Cap organ

Fig. 5.1   Posterior (a) and ventral (b) views of the 3rd maxilliped of *Homarus gammarus* showing the location of the CAP organs and the joint chordotonal organs. The latter are shown in transparency. The joint chordotonal organs, CB, IM1, IMa, MC and CP, are named from the joint which they monitor. (After Wales, Clarac, Dando and Laverack, 1970.)

ceptive nature. In the basi-ischiopodite, CSD organs are present. These organs are modified connective chordotonal organs which have become associated with specialized cuticular structures. The role of CSD organs in the 3rd maxilliped has not been demonstrated but it seems probable that, as in the walking legs, they are monitoring the total load on the limb. Although the 3rd maxilliped does not bear the body weight, it is subjected to considerable loads during feeding, when pulling on the food to help sever it as the mandibles bite. The proprioceptors of the 1st and 2nd maxillipeds and of the two pairs of maxillae are not known, with the exception of an unusual and interesting receptor at the base of the scaphognathite (Pasztor, 1969).

The mandible is the most anterior of the crustacean mouthpart appendages and, in *Homarus gammarus*, is a large and accessible structure (Wales, Macmillan and Laverack, 1976a) with a rich sensory innervation. Two proprioceptors have been described from the lobster mandible; one is a loose collection of sensory neurons in the posterior stomach nerve (Dando and Laverack, 1969; Wales and Ferrero, 1976) and the other, an unusual muscle receptor organ (Wales and Laverack, 1972a, b). However, the finding of Ferrero and Wales (1976) that electrical stimulation of most mandibular nerves excites the mandible common inhibitor system, suggests that there are many other proprioceptive inputs.

The mandibular muscle receptor organ (MRO) (Wales and Laverack, 1972a) consists of a ribbon-like muscle, innervated at its ventral insertion by 10—20 multiterminal sensory neurons (Fig. 5.2). These neurons may be bipolar or multipolar but in all cases they are multiterminal. The dendritic processes branch dichotomously and diverge to innervate the base of the receptor muscle. Although the dendritic processes have not been observed within the muscle insertion it seems probable that the number of terminals produced by each cell is small. The mandibular MRO shows some differences to the TC MRO's which occur at the corresponding joint of the walking leg. Thus the mandibular MRO is innervated by a large number of neurons with peripherally located somata, whereas the TC MRO's are innervated by a small number of neurons with centrally located somata. Also the mandibular MRO is innervated distally with a large number of morphologically similar endings, while the TC MRO is innervated proximally with two dissimilar types of ending. Moreover, the two MRO's are stretched by quite different movements. It is, therefore, most unlikely that these MRO's are homologous structures.

The mandibular MRO of *Homarus gammarus* responds to receptor muscle stretch and to mandible opening with increased activity (Wales and Laverack, 1972b). The units of the MRO fire at a rate related to the degree of passive muscle stretch but are also velocity sensitive. Fig. 5.3 shows the response of a typical unit to mandible opening at three different velocities during passive receptor muscle stretch. The unit exhibits a 'tension peak' at high rates of stretch but responds tonically at lower velocities, with a frequency related to the

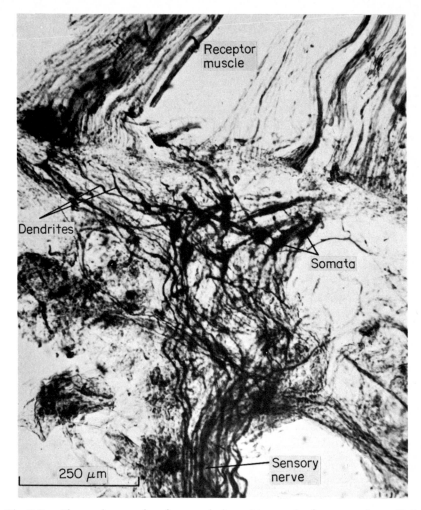

Fig. 5.2   Photomicrograph of a methylene blue stained preparation of the mandibular MRO from *Homarus gammarus*. The nerve trunk and sensory neurons are normally closely applied to the receptor muscle at its base and they have been separated in this preparation to allow the stain to penetrate to the somata and their dendritic processes. Some bundles of small axons pass into the hypodermis. (From Wales and Laverack, 1972a.)

degree of stretch. Although the majority of units are of this type, some aberrant units were found which exhibit their maximum activity at an intermediate degree of stretch and fire at a reduced frequency to higher stretch.

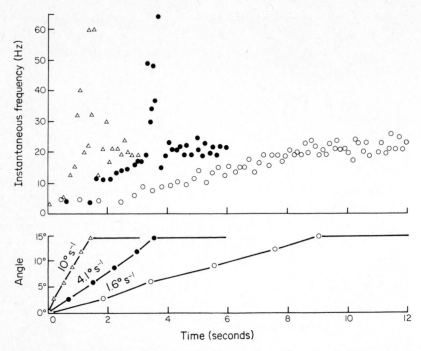

Fig. 5.3   Graph showing the response of a single tonic unit of the mandibular MRO of *Homarus gammarus* to mandible opening at different angular velocities. The instantaneous frequency of a single impulse is the reciprocal of the interval between that impulse and the preceding one. The response is plotted for three different angular velocities: $o-1.6^\circ$ s^{-1}; $\bullet-4.1^\circ$ s^{-1}; $\Delta-10^\circ$ s^{-1}. The angle of mandible opening is measured from the closed position. (From Wales and Laverack, 1972b.)

The role of the mandibular MRO has not yet been determined but it is interesting to note that it is the one mandibular proprioceptor which does *not* activate the mandible common inhibitor neurons (Wales and Ferrero, 1976). Current theories of MRO function indicate that it may be part of an error detecting servo-system. The mandible of *Homarus* has a large powerful musculature (Wales *et al.*, 1976a) which produces the force necessary to break hard substances such as cuticle. When biting on hard substances there is a danger that sudden collapse of the food will produce mandible movement and, due to the high muscle tensions, could result in self-injury. It is probable that the mandibular MRO is monitoring and controlling muscle length and hence mandible position. The velocity sensitive component of the sensory neuron response (Fig. 5.3) could be extremely valuable in this respect.

The second mandibular proprioceptor is associated with the posterior stomach nerve (Wales and Ferrero, 1975). This nerve is a branch of the outer mandibular nerve, which has a large number of neurons located in or close to it where it overlies the lateral articulation of the mandible. The more dorsal cells are gut proprioceptors (Section 5.5.1) but the remaining cells, which are bipolar or tripolar, innervate the hypodermis of that region (Dando and Laverack, 1969). The neuronal processes divide dichotomously and the endings may innervate widely separated regions. These sensory cells respond to mandibular movement in a phaso-tonic manner and are both movement and position sensitive (Wales and Ferrero, 1976).

The next major group of receptors is the peri-oesophageal complex, termed 'mouthpart receptors' (MPR) by Dando and Laverack (1968). The MPR system is an elaborate system of strands and neurons, which respond to a number of movements in the region of the mouth. This system has so far been described for astacuran and brachyuran decapods. The description given here is for *Homarus gammarus* (Laverack and Dando, 1968) and *Nephrops norvegicus* (Moulins, Dando and Laverack, 1970), which are basically similar.

The MPR system is bilaterally symmetrical and consists of three pairs of receptors associated with a pair of main strands which lie either side of the oesophagus. The main strand is attached to the mandible and to the oesophagus along most of its length. It is inelastic but seems to be under constant tension. MPR1 lies at the anterior end of the main strand and is innervated by the inferior oesophageal nerve from the commisural ganglion. It has four or five multiterminal sensory cells which innervate a smaller discrete elastic strand attached to the main strand. Some branches also innervate the surrounding connective tissue.

MPR1 contains some regularly firing units and up to three neurons may fire in this manner over a long period without change. The receptors also respond phasically to movement of the main strand or to mandible movements. The phasic units adapt very rapidly and respond only during movement. The receptor activity increases with mandible closing, the tonic units firing at a rate dependent on mandible position.

MPR2 and MPR3 are located at the posterior end of the main strand and are innervated by a branch of the paragnathal nerve from the sub-oesophageal ganglion. MPR2 is a less discrete structure than MPR1 and MPR3 and has five—ten multiterminal sensory neurons.

MPR3 is associated with a more discrete elastic strand which attaches to the main strand and inserts close to the base of the paragnatha. MPR2 and MPR3 also respond to movements and to stretching of the main strand. The responses are more complex as units respond to both mandible opening and closing. The units are largely phasic or phaso-tonic and they adapt when position is maintained. Tonic units are occasionally found, some of which fire over restricted ranges of stretch as described for the mandibular MRO above.

The main strand undergoes restricted movement and would seem to play a stretch regulating role as well as transmitting the stimuli. The other ends of the receptor strands are attached to mobile structures and thus the three receptors receive some dissimilar input. MPR1 responds to oesophageal movements via the main strands and to mandibular movements. It also responds to movements of the mouth, labrum and paragnatha. MPR2 responds predominantly to oesophageal movements but also to mandible movements via the main strand. MPR3 responds more dramatically to mandibular movement and to paragnathal movements. MPR2 and MPR3 also respond to labrum movements to a lesser degree, and to mouth movements.

The MPR receptors respond to a large variety of movements associated with feeding but do so to different degrees and in some cases asynchronously. This proprioceptive information also reaches the CNS by different pathways and it is probable that the CNS can identify specific movements from the nature of the sensory input of the three receptors. The function of this receptor system is unknown but Laverack and Dando (1968) postulate that MPR1 and MPR3 may be involved in mandible resistance reflexes and that all three MPR's may be involved in the control of oesophageal and foregut activity. Maynard (1966) has shown that input via the stomatogastric nerve can prime the activity of stomatogastric neurons (Section 5.5.1) and such activity can be due to sensory input.

5.3　Insect mouthparts

Cephalization is further developed in insects than in other arthropods and, as a result, the mouthparts form a more discrete group of appendages. The insects have a smaller number of mouthparts; a labrum, one pair of mandibles, one pair of maxillae with palps, and a labium bearing a pair of palps. These mouthparts may differ considerably in form from species to species.

Some extensive investigations of mouthpart sense organs have been published. Corbière-Tichané (1971a-c, 1973), in the cave-living coleopteran larvae of *Speophyes lucidulus*, and Richard (1951), in the termite *Calotermes flavicollis*, have described the full range of mouthpart sense organs. Thomas (1966), in her detailed study of the mouthparts of the locust, *Schistocerca gregaria*, describes only the external cuticular structures, and the occurrence of internal stretch receptors or chordotonal organs is unknown. Other authors have restricted their investigation to particular mouthpart appendages (Moulins, 1969, 1971; Peters, 1962; Sturckow, Adams and Wilcox, 1967) or to a specific type of sense organ (MacFarlane, 1953; Moulins, 1966, 1974; Pringle, 1938; Slifer, 1936).

The proprioceptive systems of insects are largely based upon sensilla associated with the cuticle. According to Dethier (1963) insects have five types of proprioceptive sense organs; hair plates, campaniform sensilla, chordotonal organs, stretch receptors and statocyst-like organs. Only the latter have not been described for insect mouthparts.

A variety of sensilla occur on the mouthparts of *Schistocerea* (Thomas, 1966) the larva of *Speophyes* (Corbière-Tichané, 1971a, 1973) and *Calotermes* (Richards, 1951). Hair-like mechanoreceptor sensilla are found on all the locust mouthparts, the greatest variety being found on the clypeo-labrum. Sensory sensilla are widespread on most surfaces of the appendages, showing more dense grouping in some regions than others. The termite, *Calotermes*, and the larvae of *Speophyes* have fewer sensilla on their mouthparts but, as with the locust, these are widespread and of varied morphology.

In locusts, hair plates occur on the maxillary and labial palps and on the labium at the junction of the prementum and the membrane proximal to it. Due to the proximity of the mouthparts to each other, much of the information from other tactile hairs will be from contact between appendages. We must consider that under these conditions a more diffuse group of mechanosensory sensilla may provide valuable proprioceptive information.

More pertinent to proprioception in the mouthparts, perhaps, are the campaniform sensilla. These are found on most of the locust mouthparts though they are absent from the clypeo-labrum and very few occur on the labium of *Schistocerea*. They do, however, occur on the labial palp. They are also found on the hypopharynx of *Schistocerca* (Thomas, 1966), *Forficula* (Moulins, 1969) and *Blabera*

(Moulins, 1971). In many cases the campaniform sensilla are found in groups and they tend to occur on the internal surfaces of the appendages. In the larvae of *Speophyes* the campaniform sensilla are small in number and widespread, not congregated into groups (Fig. 5.4). Pringle (1938) demonstrated the proprioceptive nature of the campaniform sensilla in the maxillary palps of the cockroach *Periplaneta*.

Chordotonal organs have been found in the mouthparts of all insects where they have been sought. Their presence has been demonstrated in the orthopteran *Melanoplus* (McFarlane, 1953; Slifer, 1936), the larva of *Speophyes* (Corbière-Tichané, 1971a, 1973), *Calotermes* (Richards, 1951) and the homopteran *Brevicoryne* (Wensler, 1974). In *Melanoplus* and *Calotermes*, chordotonal organs are described in the labium and maxilla only, whereas in the larva of *Speophyes* they also occur in the mandibles (Fig. 5.4). The chordotonal organs of the mandible attach to the medial or biting surface and a particularly large group innervate the lacinia of the maxillae. Chordotonal organs have also been described in the mandibles of *Brevicoryne*, by Wensler (1974), where they are believed to monitor bending of the mandibular stylets.

The final group of mouthpart proprioceptors are the stretch receptors. These are sensory structures composed of multiterminal sensory neurons, the terminal dendritic processes of which are devoid of a glial sheath and are embedded in connective tissue. Stretch receptors have been described primarily in association with the preoral cavity (cibarium) (Moulins, 1966, 1974; Rice, 1970) although several authors have also described multiterminal neurons in the labellar nerve of flies (Peters, 1962; Peters and Richter, 1963; Sturckow *et al.*, 1967) and elsewhere in the labium, labrum and hyhopharynx (Richard, 1951). The multiterminal neurons in the labellar nerve have been shown to be proprioceptors rather than interneurons as was first thought (Sturckow *et al.*, 1967). These neurons respond to proboscis movements and deflection of taste hairs on the labellum.

In *Blaberus craniifer* there are four stretch receptors associated with the preoral cavity, one in the epipharynx and three in the hypopharynx (Fig. 5.5). These receptors are complex, having up to fourteen multiterminal cells with a variety of morphological properties. The dendritic endings of the cells lie in connective tissue strands and many of the neuron somata lie in discrete clusters or

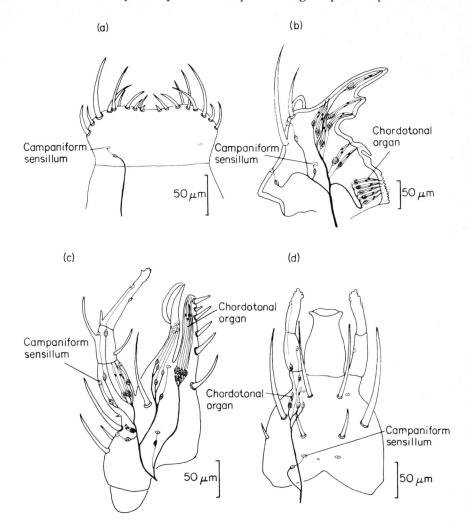

Fig. 5.4   The distribution of campaniform sensilla and chordotonal organs in the mouthparts of the larvae of *Speophyes lucidulus*.. (a) the labrum; (b) mandible; (c) the maxilla; (d) the labium. The sensilla and their innervation are shown in transparency. Only half of the innervation is shown for the labrum and labium. (After Corbière-Tichané, 1973.)

pseudoganglia (Fig. 5.5). Moulins (1974) describes four types of sensory cell;

Type A: These have dendrites in the receptor strand and in the sub-epidermal connective tissue. This is their first description in

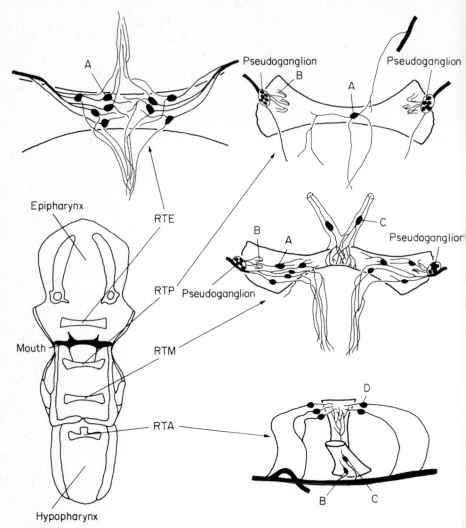

Fig. 5.5   The stretch receptors associated with the preoral cavity of the cockroach *Blaberus craniifer*. There are four stretch receptors, RTE, RTP, RTM and RTA, which are innervated by four types of cell, A, B, C and D. (See p.1.1. for description). (After Moulins, 1974.)

insect proprioceptors but similar cells have been described for the crustacean perioesophageal complex (Section 5.2).

Type B: Bipolar uniterminal or multiterminal neurons with a small number of terminals, the somata of which are grouped together to form pseudoganglia.

Type C: Bipolar multiterminal neurons with a large number of terminals. This type is very similar to the sensory neurons of the crustacean abdominal MRO's.

Type D: Bipolar neurons with an unbranched dendrite.

The four stretch receptors are each unique in the types of sensory cells which innervate them (Fig. 5.5). The epipharyngeal tension receptor (RTE) has only Type A cells. The posterior hypopharyngeal tension receptor (RTP) has two types of cells, A and B. The medial and anterior hypopharyngeal tension receptors (RTM and RTA) each have three types of cells. The former has types A, B and C; the latter B, C and D.

The stretch receptors also receive an accessory innervation by fibres containing dense-cored vesicles. The efferent fibres do not synapse directly onto the sensory cells but, none-the-less, may effect a motor control. Maddrell (1966) has shown that neural control of the mechanical properties of cuticle can occur, and Moulins (1974) proposes that the efferent control of these sense cells may be mediated by modification of the mechanical properties of the system.

In the tsetse fly and the blowfly, the preoral cavity forms the cibarial pump and this is innervated by three pairs of multiterminal neurons. Rice (1970) has shown that these neurons respond to simulated cibarial pumping and the activity of the neurons differs with the region of the anterior wall stimulated. Rice concludes that these neurons provide a means of accurately measuring the volume of food ingested and the receptor input may thus influence satiation behaviour and the manufacture of digestive enzymes.

5.4   Arachnid mouthparts

The arachnids have two pre-ambullatory appendages which are used in feeding, and these differ in size and function throughout the class. In the scorpions the pedipalps are large and chelate whereas the chelicerae are small. The pedipalps are used in prey capture and the chelicerae to break down the food. The sun spiders (Solifugae) on the other hand have very large chelicerae and the pedipalps are leg-like, with terminal adhesive organs used for prey capture. The spiders (Araneidae) have chelicerae of moderate size but the pedipalps of the male are modified to form copulatory organs. In spite of this complexity the chelicerae and pedipalps will be considered to be mouthparts for the purpose of this review. The arachnids have no mandibles, although *Limulus* (Xiphosura) does break up its food with mandible-like processes (the gnathobases) which occur on the basal segment of each walking leg.

The spiders have slit sensilla, which correspond functionally to the campaniform sensilla of insects, widespread over the body surface. Slit sensilla can occur singly, or in groups to form lyriform organs. Barth and Libera (1970) have performed an extensive survey of the slit sensilla on the spider *Cupieunius salei* and have shown them to be widespread over the chelicerae and pedipalps as well as the legs (Section 8.3.1). Lyriform organs also occur close to most of the joints of these appendages (Fig. 5.6). The proprioceptive nature of lyriform organs was demonstrated by Pringle (1955).

Internal proprioceptors have also been demonstrated anatomically and electrophysiologically in the pedipalps of scorpions and spiders. At the patella-tibia joint of the pedipalp in the scorpion *Centruroides vittatus* there are two proprioceptive organs (Bowerman and Larimer, 1973). These receptors consist of small groups of ten or less, bipolar neurons. The terminal insertions of the dendrites appear to be associated with the articular membrane of the joint. Both groups of neurons are position sensitive and the tonic units respond either to extension or to flexion of the joint. Both receptors have approximately equal numbers of extension and flexion units.

In the tarantula, *Eurypelma heutzi,* two groups of bipolar sensory neurons are also described for the femur-patella joint of the pedipalp (Rathmayer, 1967). Group one has nine to eleven bipolar neurons; group two has four neurons. The majority of the receptor cells are extension sensitive and are of a phaso-tonic nature. They are largely position sensitive, but at least three respond to movement and these are purely phasic.

5.5　Crustacean gut

The gut of Malacostracans is usually straight. It consists of a foregut, which may only be a simple tubular oesophagus but is commonly enlarged to form a titurating stomach, the walls of which have developed opposing chitinous denticles or ossicles. This is particularly true of the larger decapods. The midgut varies greatly in size and bears cecae, one pair of which is modified to form the digestive

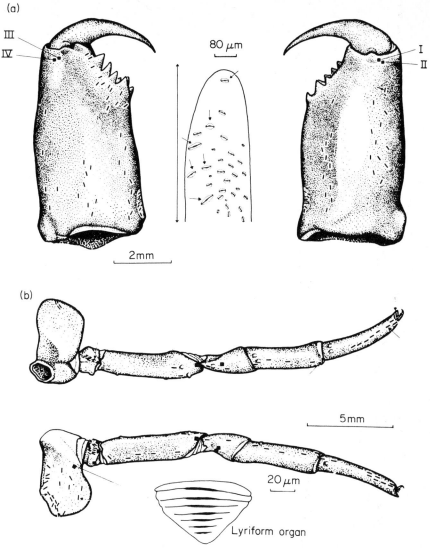

Fig. 5.6   The distribution of slit sensilla (−) and lyriform organs (■) on (a) the chelicerae and (b) the pedipalps of the spider *Cupiennius salei*. (After Barth and Libera, 1970.)

gland. The posterior part of the gut is an intestine formed entirely or partly by the hindgut and which terminates via the rectum. The foregut and hindgut are cuticularized. The following description refers to the decapod crustaceans.

5.5.1   Control of foregut movement

The foregut of decapods is innervated by the stomatogastric system (see Bullock and Horridge, 1965). The titurating stomach undergoes rhythmical movements the motor pattern of which originates in the stomatogastric ganglion (Maynard, 1966, 1969, 1971). This ganglion is largely autonomous, but recordings from intact, free-moving animals suggest that it is under the control of modulating command interneurons (Morris and Maynard, 1970). It has been shown that stimulation of the stomatogastric nerve (Maynard, 1966), or of a sensory input (Dando and Laverack, 1969; Dando, Chanussot and Nagy, 1974) will produce modifications of the patterned output of the stomatogastric ganglion. Moreover, Dando and Selverston (1972) have shown the presence of command fibres in the inferior ventricular nerve which innervate cells of the stomatogastric ganglion via the stomatogastric nerve. The cells of the stomatogastric ganglion do not receive direct sensory input, and modulation of the output normally occurs via the stomatogastric nerve from the CNS. Stimulation of the stomatogastric nerve (Maynard, 1966) will also excite other neurons of unknown properties.

Proprioceptors are widespread on the foregut (Dando and Laverack, 1969; Dando and Maynard, 1974; Larimer and Kennedy, 1966; Orlov, 1926a, b). Dando and Maynard (1974) give an extensive description of foregut receptors and review the sensory innervation of the foregut. At the opening of the mouth into the oesophagus lies the MPR system, described above (see Section 5.2). The function of this receptor system has yet to be determined and its input may be important to the function of the stomatogastric system (Laverack and Dando, 1968). Multiterminal cells occur on the oesophagus (Orlov, 1926a, b; Dando and Maynard, 1974). These cells may be bipolar, but are frequently multipolar, and are probably proprioceptors. Dando and Maynard describe uniterminal cells on the oesophagus and cardiac sac. It is probable that these are chemoreceptive in nature, although they are not associated with cuticular sensilla. Unipolar cells also occur on the pterocardiac and pyloric ossicles and it is likely that these are proprioceptive, monitoring movement of the ossicles.

A number of small multiterminal neurons were found on the muscles of the cardiac stomach lateral wall and the anterior pyloric muscles by Dando and Maynard (1974), who suggest that these may constitute muscle receptor organs. As the function of these muscles is not purely to set the level of activity in the sense organ the term MRO is used here in a loose sense (Section 2.3). These neurons are most probably monitoring muscle movement, which correlates with stomach movements produced by these muscles. It has yet to be demonstrated that their association with the muscle has any greater functional significance.

The posterior stomach nerve of decapods (see Section 5.2) also contains neurons of an important foregut proprioceptor. This group of cells has been named the posterior stomach nerve (Dando and Laverack, 1969), the posterior stomach receptor (Dando *et al.*, 1974) and the gastric mill receptor (Dando and Maynard, 1974). In the posterior stomach nerve, where it ascends the lateral wall of the carapace, lies a group of up to 80 bipolar neurons, the distal processes of which innervate the gastric mill. The neurons are mainly bipolar with long dendritic processes, the terminal regions of which undergo multiple bifurcations. The neurons of this proprioceptor respond to movements of the gastric mill with phasic and tonic components (Fig. 5.7). They are sensitive to both the rate and the magnitude of the displacement. Stimulation of the posterior stomach nerve produces changes in the stomatogastric output (Dando and Laverack, 1969; Dando *et al.*, 1974).

Another interesting proprioceptor is described by Larimer and Kennedy (1966), in the stomatogastric ganglion of crayfish. This multiterminal neuron is the only known sensory neuron in the stomatogastric ganglion. It innervates the cardiac ossicle region of the

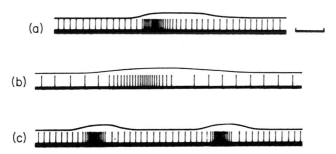

Fig. 5.7   Intracellular recordings from a cell in the posterior stomach nerve of *Homarus gammarus*, showing the response to forward movements of the urocardiac ossicle (upward movement of top trace). Time calibration for (a) and (c) is 1 s; for (b) it is 0.5 s. (From Dando and Laverack, 1969.)

gastric mill and responds to movement of this region. The unit is basically tonic but has a velocity sensitive component.

Dando and Laverack (1969), suggest that the cells of the posterior stomach nerve and of the stomatogastric sensory neuron will provide positive feedback, via interneurons, to the stomatogastric ganglion, thus priming the system. Inhibition of the gastric mill movements could be effected by the pyloric sensory cells (Orlov, 1926b) sending information about the passage of food to the midgut. Bethe (see Bullock and Horridge, 1965) demonstrated that brainless crabs would continue to eat in excess, suggesting that the brain is also involved in the inhibition of stomatogastric activity.

5.5.2   Midgut receptors

At the junction of the foregut and midgut is a small number of large multiterminal neurons (Dando and Maynard, 1974; Orlov, 1926b), which are probably proprioceptors.

5.5.3   Control of hindgut movement

Movements of the hindgut and anus of *Homarus gammarus* are described by Winlow and Laverack (1972a, b). The hindgut movements are primarily neurogenic, although localized movements occur in the denervated hindgut. No physiological evidence was found for the occurrence of hindgut proprioceptors. However, a pair of bipolar multiterminal neurons innervate the soft cuticle of the anus. These neurons respond to anal opening and in some preparations other neurons were detected physiologically which responded to anal closure (Fig. 5.8) (Winlow and Laverack, 1970, 1972a).

5.6   Insect gut

The insect gut consists basically of a foregut divided into anterior pharynx, oesophagus, crop and narrow proventriculus; a midgut termed the ventriculus; and a hindgut consisting of an anterior intestine and a posterior proctodeum. Both the foregut and hindgut are lined with cuticle (Barnes, 1968). The foregut and hindgut are innervated by quite separate parts of the central nervous system. The foregut is innervated by the stomodeal nervous system, arising from the sub-oesophageal ganglion, and the hindgut by the proctodeal nerves.

Fig. 5.8   The response of the anal receptors of *Homarus gammarus* to anal movements. Upper trace—activity in the teased anal nerve; lower trace—the anal movements (downwards denotes opening). The movements were produced by stimulation of the ventral nerve cord (indicated by a horizontal bar). (From Winlow and Laverack, 1972a.)

5.6.1   Control of foregut movement

Foregut peristaltic moments are under neural control. Clarke and Grenville (1960) showed that gut movements in *Schistocerca* cease on ablation of the ventricular ganglia, whereas the hypocerebral ganglion appears to play only a modifying role. In *Acheta* (Möhl, 1972) removal of both the ventricular and hypocerebral ganglia fails to abolish gut movements. In the totally denervated foregut, after removal of the ganglion and the oesophageal nerves, contractions of a limited nature will occur. Thus, in *Acheta*, the transmission of the gut wave appears to be partly myogenic and partly neurogenic, with the probability of a wave travelling the length of the gut being much greater with the oesophageal nerves and the ganglia intact. The neural coordinating mechanism is located in the oesophageal *nerves* and the ganglia in this insect play a stabilizing role. Möhl was able to demonstrate a correlation between the motor activity in the branches of the oesophageal nerve and gut movement.

The foregut of insects receives an abundant sensory innervation and multiterminal sensory cells are widespread (Dando, Chanussot and Dando, 1968; Clarke and Langley, 1963; Gelperin, 1967; Langley, 1965; Möhl, 1969, 1972). There is also indirect evidence of gut proprioceptors (Clarke and Langley, 1963; Dethier and Boden-stein, 1958; Dethier and Gelperin, 1967; Gelperin, 1966a, b, 1972; Langley, 1966). Möhl, 1972 has shown clearly that the foregut proprioceptors of *Acheta* respond to gut movements (Fig. 5.9a, b) and that they are particularly responsive to cyclical movements but do not seem to give information regarding maintained tension. The sensory input affects the motor activity in the oesophageal nerves (Fig. 5.9c, d). Stretching the gut produces a decrease in motor activity which may be preceded by a short increase at higher rates of

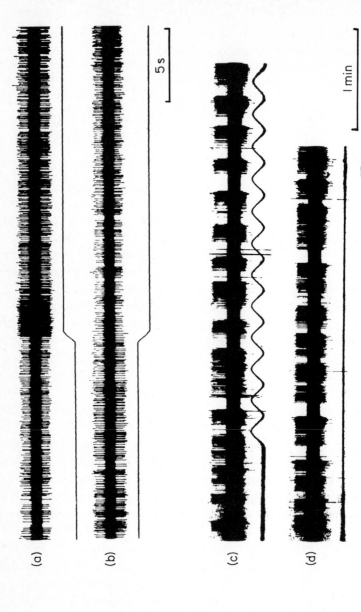

Fig. 5.9 (a, b) Sensory activity recorded from side branches of the oesophageal nerve of *Acheta*, cut close to their entry into the oesophageal nerve. (c) Motor activity in a side branch of the oesophageal nerve, cut close to its entry into the gut wall, during cyclical stretching of the foregut. Note that the motor activity is modulated by the rhythmical sensory input due to stretching. (d) Motor activity in the absence of stretch. In all cases upward movement of the lower trace indicates stretching of the foregut. (From Möhl, 1972.)

stretch. Removal of the ventricular and hypocerebral ganglia of the stomodeal nervous system produces no qualitative change.

The recurrent nerve from the hypocerebral ganglion contains two multiterminal stretch sensitive neurons which respond to expansions of the foregut lumen (Fig. 5.10) (Gelperin, 1967).

5.6.2 Control of hindgut movement

The larva of the beetle *Oryctes nasicornis* has a complex S-shaped hindgut with a terminal rectal ampulla. Nagy (1974) has shown that, as for the foregut of *Acheta*, the transmission of the peristaltic activity is partly myogenic. The denervated hindgut of *Oryctes* will continue to contract but in a poorly coordinated fashion. The motor innervation coordinates the movement, and sensory input from hindgut proprioceptors may modulate motor activity.

The presence of hindgut proprioceptors is indicated by the work of Dethier and Gelperin (1967), Gelperin (1972) and Nunez (1964). Proprioceptors of *Oryctes* hindgut are described by Nagy (1974). The hindgut is cuticularized and the posterior cuticular intima has mechanoreceptive hairs and campaniform sensilla. The latter will certainly provide proprioceptive information. Nagy has also shown the presence of multipolar cells located near the anus and the posterior end of the rectal ampulla. These monitor the volume of the rectal ampulla and the flow of faeces through the anus. Other proprioceptors occur in association with the extrinsic muscle of the rectal ampulla. These are a stretch receptor and a muscle receptor organ (Fig. 5.11) and they attach between the lateral muscles of the rectal ampulla and the wall of the ampulla. The muscle receptor organ has a single multiterminal sensory neuron; whereas the stretch receptor has two neurons, a primary cell located proximally and a smaller, secondary cell located distally. These receptors reflexly excite the muscle with which they are associated (Fig. 5.11).

5.6.3 Control of feeding

Gelperin (1971) described a small population of neurons in the first and second lateral branches of the median abdominal nerve of *Phormia* which are activated by gentle stretch. *In situ* recordings from these cells has shown them to fire at a rate dependent on crop volume. These neurons are involved in the regulation of appetite, and

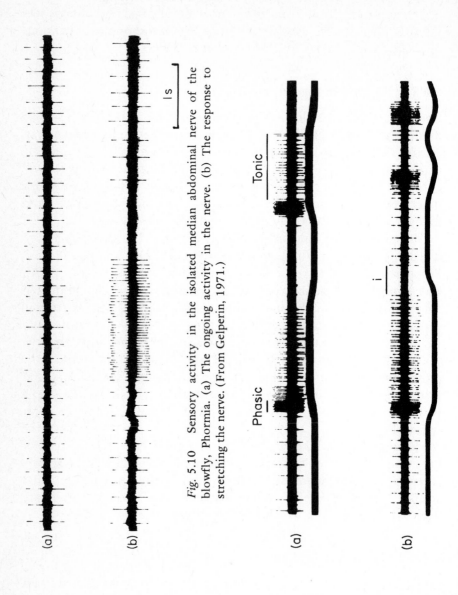

Fig. 5.10   Sensory activity in the isolated median abdominal nerve of the blowfly, Phormia. (a) The ongoing activity in the nerve. (b) The response to stretching the nerve. (From Gelperin, 1971.)

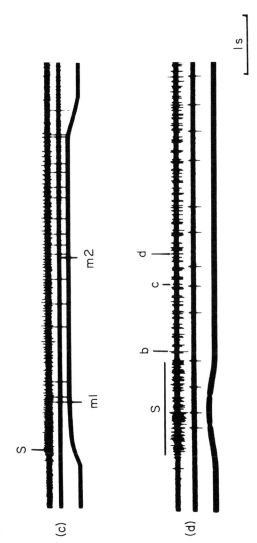

Fig. 5.11 (a, b) Afferent activity of the stretch receptor associated with the lateral muscle of the rectal ampulla, showing the phaso-tonic nature of the response. Relaxation of the strand inhibits (i) the response. Upward movement of the lower trace in A and downward movement in B denotes stretching of the receptor. (c) Reflex excitation of motor neurons to the lateral muscle of the rectal ampulla in response to excitation of the stretch receptor. (d) Reflex excitation of motoneurons of the peri-anal muscles in response to stimulation of receptors on the dorsal part of the rectal ampulla. s—sensory response; m1, m2, b, c, d—different motor units. (From Nagy, 1974.)

section of this abdominal nerve produces hyperphagy in the blowfly. Gelperin suggests that these cells interact in the CNS to reduce the probability of the animal feeding. Nagy (1974) described neurons in other abdominal nerves, but these do not respond to gut movement and their function is unknown. Section of the recurrent nerve from the hypocerebral ganglion also leads to more frequent and prolonged feeding bouts (Dethier and Bodenstein, 1958; Dethier and Gelperin, 1967). Gelperin (1967) has shown this nerve to contain stretch sensitive neurons responding to foregut movement.

Dethier and Gelperin (1967) concluded that several inhibitory inputs act on the feeding drive, but these inputs are inequal and transection of the median abdominal nerve produces a more pronounced hyperphagy than that of the recurrent nerve.

5.6.4 Other functions of gut proprioceptors

There is now considerable evidence of gut and abdominal proprioceptors being involved in the control of several endocrine systems. It appears that proprioceptive input may affect the neurondocrine control of protein metabolism (Clarke and Langley, 1963), enzyme production in the midgut (Langley, 1966) and body wall distension during feeding (Maddrell, 1963, 1966). These interesting aspects of proprioceptor function are reviewed by Finlayson (1968).

Acknowledgements

I wish to express my gratitude to Adrian Bottoms for assistance with translation and for helpful criticism of the manuscript. I also thank Roy Summers for his photographic skills in reproducing the illustrations.

References

Barnes, R. D. (1968) *Invertebrate Zoology*. Saunders, London.

Barth, F. G. and Libera, W. (1970) Ein Atlas der Spaltsinnesorgane von *Cupiennius salei* (Keys). Chelicerata (Araneae). *Zeitschrift für Morphology der Tiere*, **68**, 343–369.

Borradaile, L. A. (1917) On the function of the mouthparts of the common prawn. *Proceedings of the Cambridge Philosophical Society. Biological Sciences*, **19**, 56.

Bowerman, R. F. and Larimer, J. (1973) Structure and physiology of the patella-tibia joint receptors in scorpion pedipalps. *Comparative Biochemistry and Physiology A*, **46**, 139–152.

Bullock, T. H. and Horridge, G. A. (1965) *Structure and function in the nervous system of invertebrates.* Freeman, New York.

Clarke, K. U. and Langley, P. A. (1963) Studies on the initiation of growth and moulting in *Locusta migratoria migratorioides.* III The role of the frontal ganglion. *Journal of Insect Physiology,* 9, 411—421.

Clarke, K. U. and Grenville, H. (1960) Nervous control of movements in the foregut of *Schistocerca gregaria* Forsk. *Nature,* 186, 98—99.

Corbière-Tichané, G. (1971a) Recherches sur l'équipement sensoriel du Coleoptere cavernicole. *Thése,* Université de Provence.

Corbière-Tichané, G. (1971b) Ultrastructure de l'equipement de la mandibule chez la larve du *Speophyes lucidueus* Delar. *Zeitschrift für Zellforschung und mikroskopische Anatomie,* 112, 129—138.

Corbière-Tichané, G. (1971c) Ultrastructure des organes chordotonaux des pièces cephaliques chez la larve du *Speophyes lucidulus* Delar. *Zeitschrift für Zellforschung und mikroskopische Anatomie,* 117, 275—302.

Corbière-Tichané, G. (1973) Sur les structures sensorielles et leurs fonctions chez la larve de *Speophyes lucidulus. Annales de Spéléologie,* 28, 247—265.

Dando, J., Chanussot, B. and Dando, M. R. (1968) Le systeme nerveux stomodéal post-céphalique de *Schistocerca gregaria* Forsk. (Orthoptère) et *Blabera cranifer* Burm. (Dictyoptère). *Comptes rendus hebdomadaire des séances de l'Academie.* 267, 1852—1855.

Dando, M. R., Chanussot, B. and Nagy, F. (1974) Activation of command fibres to the stomatogastric ganglion by input from a gastric mill proprioceptor in the crab, *Cancer pagrurus. Marine Behaviour and Physiology,* 2, 197—228.

Dando, M. R. and Laverack, M. S. (1968) A mandibular proprioceptor in the lobster, *Homarus vulgaris. Separatum Experientia,* 24, 931.

Dando, M. R. and Laverack, M. S. (1969) The anatomy and physiology of the posterior stomach nerve (p.s.n.) in some decapod crustacea. *Proceedings of the Royal Society, London B.,* 171, 465—482.

Dando, M. R. and Maynard, D. M. (1974) The sensory innervation of the foregut of *Panulirus argus. Marine Behaviour and Physiology,* 2, 283—305.

Dando, M. R. and Selverston, A. I. (1972) Command fibres from the supra-oesophageal ganglion to the stomatogastric ganglion in *Panulirus argus. Journal of Comparative Physiology,* 78, 138—175.

Dethier, V. G. (1963) *The physiology of insect senses.* Methuen, London.

Dethier, V. G. and Bodenstein, D. (1958) Hunger in the blowfly. *Zeitschrift für Tierpsychologie,* 15, 129—140.

Dethier, V. G. and Gelperin, A. (1967) Hyperphagia in the blowfly. *Journal of Experimental Biology,* 47, 191—200.

Ferrero, E. and Wales, W. (1976) The mandibular common inhibitor system 1. Axon topography and the nature of coupling. (In preparation).

Finlayson, L. H. (1968) Proprioceptors in the invertebrates. *Symposia of the Zoological Society of London,* 23, 217—249.

Gelperin, A. (1966a) Control of crop emptying in the blowfly. *Journal of Insect Physiology*, 12, 331–345.

Gelperin, A. (1966b) Investigations of a foregut receptor essential to taste threshold regulation in the blowfly. *Journal of Insect Physiology*, 12, 829–841.

Gelperin, A. (1967) Stretch receptors in the foregut of the blowfly, *Science*, 157, 208–210.

Gelperin, A. (1971) Abdominal sensory neurons providing negative feedback to the feeding behaviour of the blowfly. *Zeitschrift für vergleichende Physiologie*, 72, 17–31.

Gelperin, A. (1972) Neural control systems underlying insect feeding behaviour. *American Zoologist*, 12, 489–496.

Langley, P. A. (1965) The neuroendocrine system and stomatogastric nervous system of the adult tsetse fly *Glossina morsitans*. *Proceedings of the Zoological Society of London*, 144, 415–424.

Langley, P. A. (1966) The control of digestion in the tsetse fly *Glossina morsitans*. Enzyme activity in relation to the size and nature of the meal. *Journal of Insect Physiology*, 12, 439–448.

Larimer, J. L. and Kennedy, D. (1966) Visceral afferent signals in the crayfish stomatogastric ganglion. *Journal of Experimental Biology*, 44, 345–354.

Laverack, M. S. and Dando, M. R. (1968) The anatomy and physiology of mouthpart receptors in the lobster, *Homarus vulgaris*. *Zeitschrift für vergleichende Physiologie*, 61, 176–195.

McFarlane, J. E. (1953) The morphology of the chordotonal organs of the antenna, mouthparts and legs of the lesser migratory grasshopper, *Melanoplus mexicanus mexicanus* (Saussure). *Canadian Entomologist*, 85, 81–103.

Macmillan, D., Wales, W. and Laverack, M. S. (1976) Mandibular movements and their control in *Homarus gammarus*. III Effects of load changes. (In press).

Maddrell, S. H. P. (1963) Control of ingestion in *Rhodnius prolixus*. *Nature*, 198, 210.

Maddrell, S. H. P. (1966) Nervous control of the mechanical properties of the abdominal wall at feeding in *Rhodnius*. *Journal of Experimental Biology*, 44, 59–68.

Maynard, D. M. (1966) Integration in crustacean ganglia. In: Nervous and hormonal mechanisms of integration. *Symposia of the Society for Experimental Biology*, 20, 111–149.

Maynard, D. M. (1969) In: *The interneurone*, Brazier. Mary A. B. (ed.) p. 58–70. University of California Press, Los Angeles.

Maynard, D. M. (1971) *Simpler networks. Conference on patterns of integration*. Academy of Science. New York.

Möhl, B. (1969) Nervenzellen im Bereich des Ösophagus der Grillen und ihre mögliche Funktion. *Experentia*, 25, 947.

Möhl, B. (1972) The control of foregut movements by the stomatogastric nervous system in the European house cricket *Acheta domesticus* L. *Journal of Comparative Physiology*, 80, 1–28.

Morris, J. and Maynard, D. M. (1970) Recordings from the stomatogastric nervous system in intact lobsters. *Comparative Biochemistry and Physiology A*, **33**, 969–974.

Moulins, M. (1966) Présence d'un récepteur de tension dans l'hypopharynx de *Blabera cranifera* Burm. (Insecta, Dictyoptera). *Comptes rendus hebdomadaire des séances de l'Académie des Sciences*, **262**, 2476–2479.

Moulins, M. (1969) Etude anatomique de l'hypopharynx de *Forficula auricularia* L. (Insecte, Dermaptere): Téquments, musculature, organes sensoriels et innervations. Interpretation morphologique. *Zoologische Jahrbucher. Anatomie und Ontogenie der Tiere*, **86**, 1–27.

Moulins, M. (1971) La cavite préorale de *Blabera craniifer* Burm. (Insecte, Dictyoptere) et son innervation: Etude anatomo-histologique de l'epipharynx et l'hypopharynx. *Zoologische Jahrbucher. Anatomie und Ontogenie der Tiere*, **88**, 527–586.

Moulins, M. (1974) Recepteurs de tension de la région de la bouche chez *Blaberus craniifer* Burmeister (Dictyoptera :,Blaberidae). *International Journal of Insect Morphology and Embryology*, **3**, 171–192.

Moulins, M., Dando, M. R. and Laverack, M. S. (1970) Further studies on mouthpart receptors in Decapoda Crustacea. *Zeitschrift für vergleichende Physiologie*, **69**, 225–248.

Nagy, F. (1974) Le système neuromusculaire et sensoriel de l'intestin posterieur et de la région proctodeal, chez la larvae *d'Oryctes nasicornis* L. [Col scarabeidae]. Thése *Universite de Dijon, France*.

Nunez, J. A. (1964) Trinktriebregelung bei Insekten. *Naturwissenschaften*, **17**, 419.

Ong, J. E. (1969) The fine structure of the mandibular sensory receptors in the brackish water calanoid copepod *Gladioferens pectinatus* (Brady). *Zeitschrift für Zellforschung und mikroskopische Anatomie*, **97**, 178–195.

Orlov, J. (1926a) Die Innervation des Darmes de Flusskrebses. *Zeitschrift für mikroskopisch-anatomische Forschung*, **4**, 101–148.

Orlov, J. (1926b) System nerveux intestinal de l'ecrevisse. *Bulletin de l'Institut de recherches biologiques de l'Universite de Molotov*, **5**, 1–32.

Pasztor, V. M. (1969) The neurophysiology of respiration in decapod crustacea. II The sensory system. *Canadian Journal of Zoology*, **47**, 435–441.

Peters, W. (1962) Die propriorezeptiven organe am Prosternium und an den labellen von *Calliphora erythrocephala*. *Zeitschrift für Morphologie und Okologie der Tiere*, **51**, 211–226.

Peters, W. and Richter, S. (1963) Morphological investigations on the sense organs of the labella of the blowfly, *Calliphora* MG. *Proceedings of the 16th International Congress of Zoology*, **3**, 89–92.

Pringle, J. W. S. (1938) Proprioception in insects 1. A new type of mechanical receptor from the palps of the cockroach. *Journal of Experimental Biology*, **15**, 101–113.

Pringle, J. W. S. (1955) The function of the lyriform organs of arachnids. *Journal of Experimental Biology*, **32**, 270–278.

Rathmayer, W. (1967) Elecktrophysiologische untersuchungen au Proprioceptoren im bein einer Vogelspinne (*Eurypelma heutzi* Chamb.) *Zeitschrift für vergleichende Physiologie*, **54**, 438–454.

Rice, M. J. (1970) Cibarial stretch receptors in the tsetse fly (*Glossina austeni*) and the blowfly (*Calliphora erythrocephala*). *Journal of Insect Physiology*, **16**, 277–289.

Richards, G. (1951) L'innervation et les organes sensoriel des pieces buccales du termite à cou jaune (*Calotermes flavicollis* Fab). *Annales des sciences naturelles (Zoologie)*, **13**, 397–412.

Slifer, E. H. (1936) The scoloparia of *Melanoplus differentialis* (Orthoptera Acrididae). *Entomological News*, **47**, 174–180.

Sturckow, B., Adams, J. R. and Wilcox, T. A. (1967) The neurones in the labellar nerve of the blowfly. *Zeitschrift für vergleichende Physiologie*, **54**, 268–289.

Thomas, J. G. (1966) The sense organs on the mouth parts of the desert locust (*Schistocerca gregaria*). *Journal of Zoology*, **148**, 420–448.

Wales, W. (1972) A comparative study of proprioception in the appendages of decapod crustaceans. *Ph.D. Thesis*, University of St. Andrews, Scotland, U.K.

Wales, W., Clarac, F., Dando, M. R. and Laverack, M. S. (1970) Innervation of the receptors present at the various joints of the pereiopods and third maxilliped of *Homarus gammarus* (L.) and other Macruran Decapods (Crustacea). *Zeitschrift für vergleichende Physiologie*, **68**, 345–384.

Wales, W. and Ferrero, E. (1976) The mandibular common inhibitor system. 2. Input sensitivity. (In preparation).

Wales, W., Macmillan, D. and Laverack, M. S. (1976a) Mandibular movements and their control in *Homarus gammarus*. 1. Mandible morphology. (In press).

Wales, W., Macmillan, D. and Laverack, M. S. (1976b) Mandibular movements and their control in *Homarus gammarus*. 2. The normal pattern of activity. (In press).

Wales, W. and Laverack, M. S. (1972a) The mandibular muscle receptor organ of *Homarus gammarus* (L.) (Crustacea, Decapoda). *Zeitschrift für Morphologie der Tiere*, **73**, 145–162.

Wales, W. and Laverack, M. S. (1972b) Sensory activity of the mandibular muscle receptor organ of *Homarus gammarus* (L.) 1. Response to receptor muscle stretch. *Marine Behaviour and Physiology*, **1**, 239–255.

Wensler, R. J. D. (1974) Sensory innervation monitoring movement and position in the mandibular stylets of the aphid, *Brevicoryne brassicae*. *Journal of Morphology*, **143**, 349–363.

Wetzel, A. (1935) Uber das periphere Nervensystem der Caprelliden. *Zeitschrift für Morphologie und Ökologie der Tiere*, **30**, 206–296.

Winlow, W. and Laverack, M. S. (1970) The occurrence of an anal proprioceptor in the decapod crustacea *Homarus gammarus* (L.) (Syn. H. vulgaris M. Ed.) and *Nephrops norvegicus* (Leach). *Life sciences*, **9**, 93—97.

Winlow, W. and Laverack, M. S. (1972a) The control of hindgut motility in the lobster, *Homarus gammarus* (L.) 1. Analysis of hindgut movements and receptor activity. *Marine Behaviour and Physiology*, **1**, 1—27.

Winlow, W. and Laverack, M. S. (1972b) The control of hindgut motility in the lobster, *Homarus gammarus* (L.) 2. Motor output. *Marine Behaviour and Physiology*, **1**, 29—47.

6

Chordotonal organs of crustacean appendages

P. J. MILL

6.1   Introduction

The first crustacean limb chordotonal organ was discovered by Barth (1934), who described a myochordotonal organ at the ischio-meropodite joint in a variety of decapods. However, this was largely ignored for the next two decades, until Burke (1954) described a proprioceptor spanning the distal leg joint in *Carcinus maenas*. It is now known that there is at least one chordotonal organ at most functional joints; also one or two myochordotonal organs at the ischio-meropodite joint and two cuticular stress detectors (Chapter 7) in the basi-ischiopodite region (Fig. 6.1). In recent years considerable effort has been made towards obtaining an insight into the structure and function of these receptors.

6.2   Anatomy

6.2.1   Joint chordotonal organs

The joint chordotonal organs are connective organs (Chapter 9) and typically consist of an elastic receptor strand (or sheet) which lies close to, or spans, a joint. The strand consists of a matrix of strand

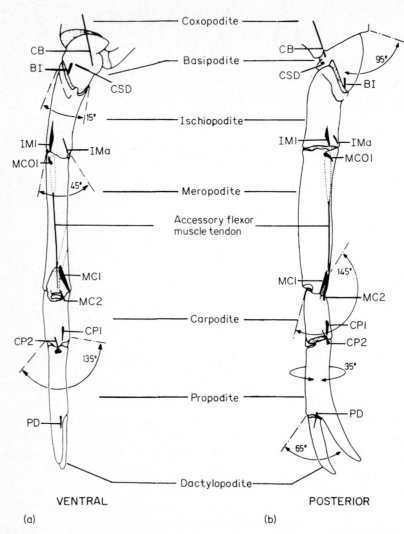

Fig. 6.1   (a) posterior and (b) ventral views of a left walking leg of *Hommarus gammarus* illustrating the position of the joint chordotonal organs and the myochordotonal organ. Details of the location of the TC chordotonal organ and TC muscle receptor organ are given in Fig. 3.1; the cuticular stress detectors in Figs 7.1, 7.2. The angle through which each joint moves is also shown. (After Wales *et al.*, 1970.)

cells and collagen surrounded by an outer layer of amorphous connective tissue. Associated with the strand are a number (20–80) of bipodal sensory cells. Some or all of the cell bodies are embedded in the strand (Fig. 6.2) and their dendrites run singly or, more usually, in pairs, parallel to the long axis of the strand. Generally, two groups of sensory cells can be distinguished; a proximal one containing the larger cells and a distal one containing the smaller cells, although some small cells are often present in the proximal group. Following the lead of Alexandrowicz and Whitear (1957), Alexandrowicz (1958) and Wiersma (1959) these receptors are named according to the joint at which they are located and whose movement and position they monitor. Thus the receptor spanning the pro-dactylopodite joint is the PD organ (Tables 6.1, 6.2).

Walking legs, claws and 3rd maxillipeds
The walking legs and claws have been investigated in the Astacura, Palinura, Brachyura and Anomura, but the 3rd maxillipeds have only been examined in the first two groups.

Thorax-coxopodite joint. A single TC organ (the elastic receptor of Alexandrowicz and Whitear, 1957) occurs at this joint in the legs and claws of the Astacura (*Homarus* and *Astacus*), but not in the Palinura, Brachyura or Anomura (Alexandrowicz and Whitear, 1957; Alexandrowicz, 1958, 1967). It runs parallel to the TC Muscle Receptor Organ (MRO) (Chapter 3), originating on the same receptor rod (pillar of the endosternite in the fifth pereiopods) as the MRO, in the thorax, and inserting about half-way along the anterior surface of the coxopodite. Some of the sensory cells at the proximal end are not embedded in the receptor strand and there is no apparent differentiation into two cell groups.

Coxo-basipodite joint. There is a single CB organ, which is essentially the same in the Astacura, Palinura, Brachyura and Anomura. Like the TC organ its receptor strand is not attached to a muscle tendon and, in this respect, these two organs differ from all of the more distal ones. The CB receptor strand originates proximally on the cuticle near the thorax-coxopodite articulation. In the pereiopods it runs to the rim of the basipodite near the insertion of the basipodite levator muscle, or between the insertions of the anterior and posterior levator muscles in the Palinura and Brachyura,

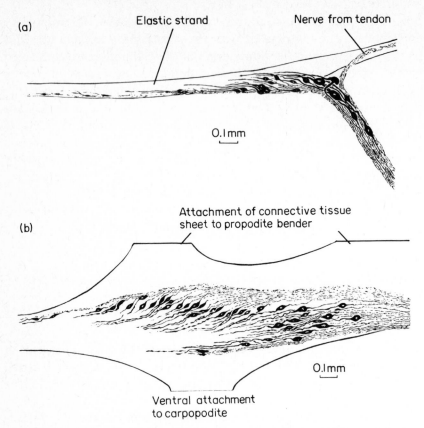

Fig. 6.2   Diagrams of (a) CP2 and (b) CP1 to show the general form of the two main types of joint chordotonal organ. In both cases the distal part of the strand containing many of the smaller sensory cell bodies is omitted. (From Whitear, 1962.)

while in the third maxillipeds it inserts near the basipodite-flagellum joint. It thus crosses the coxo-basipodite joint and is elongated by extension of the basipodite. Again like the TC organ, many of the proximal sensory cell bodies are not embedded in the receptor strand and the division into two cell groups is not clear (Alexandrowicz and Whitear, 1957; Alexandrowicz, 1958, 1967; Whitear, 1962; Wales, Clarac, Dando and Laverack, 1970).

Basi-isochiopodite joint.   This joint is fused in the appendages of the Palinura, Brachyura and Anomura and in the 3rd maxillipeds of the Astacura. However, in pereiopods 2–5 of the Astacura this joint

Table 6.1

	Pereiopods				3rd Maxillipeds	
	Astacura	Palinura	Anomura	Brachyura	Astacura	Palinura
PD	✓	✓	✓	✓	–	✓
CP1	✓	✓	✓	✓	–	–
CP2	✓†	✓†	✓†	✓	✓†	✓†?
MC1	✓	✓	✓	✓	–	–
MC2	✓†	✓†	✓†	✓	✓†	✓†?
IM1	✓	✓	✓	✓	✓	✓
IMa	✓	–	–	–	✓	?
MCO2	–	✓	✓	✓	–	–
MCO1p + MCO1m	–	–	–	✓	–	–
MCO1	✓†‡	✓†‡	✓†§‡?		?	?
BI	✓(not claw)	←———Joint fused———→			←Joint fused→	
CB	✓	✓	✓	✓	✓	?
TC	✓	–	–	–	?	?

	Antennules		Antennae		
	Palinura	Hoplocarida	Palinura	Anomura	Amphipoda
J3 ≡ CF	✓	?	} (J2–J3) {	✓	?
J2	✓	✓		*	✓
J1	✓	✓	✓	*	?

† = CAPs present
‡ = Annex sensory cells present
§ = Claw only
* = Six basal segments (no receptors described).

remains moveable and a BI organ is present. The receptor strand lies entirely within the basipodite with its proximal end attached to the ischiopodite retractor muscle tendon and it is elongated by retraction of the ischiopodite (Clarac and Masson, 1969; Wales *et al.*, 1970).

Ischio-meropodite joint. At this joint the situation is more complex. In the Astacura, Palinura, Brachyura and Anomura there is a receptor, IM1, which resembles CP1 and MC1 (see below). The flattened receptor strand forms part of a connective tissue sheet which, in the pereiopods, runs from the meropodite reductor tendon to the postaxial wall of the ischiopodite. In the 3rd maxillipeds the receptor strand originates on the meropodite productor muscle tendon and, in *Homarus*, the sheet partially spans the joint. Distally the strand inserts on the cuticle at the distal end of the ischiopodite and thus does not cross the joint. The sensory cell bodies lie in a curved row. In all cases reduction of the meropodite elongates the receptor strand (Clarac, 1968a; Clarac and Masson, 1969; Wales *et al.*, 1970; Alexandrowicz, 1972).

In the Astacura there is an additional organ, IMa (IM2 of Clarac and Masson, 1969; IM3 of Alexandrowicz, 1972; See Section 6.3 for a discussion of the terminology). It consists of a thin receptor strand in which the sensory cell bodies lie in almost a straight line. Two sensory cell groups can be distinguished in the pereiopods but not in the 3rd maxillipeds. In the pereiopods, the receptor strand runs from the main nerve trunk distally to the ventral region of the ischio-meropodite joint. The distal termination is difficult to determine but both Clarac and Masson (1969) and Alexandrowicz (1972), working on *Astacus* and *Homarus* respectively, indicate that it does not cross the joint but ends on the ischiopodite. In the 3rd maxillipeds it runs from the ventral side of the meropodite productor muscle tendon to end in the large tooth at the distal end of the ischiopodite. In the pereiopods, the strand is elongated by both reduction and production of the meropodite about a mean position; in the 3rd maxillipeds, elongation is by production alone (Wales *et al.*, 1970). In *Ligia* (Isopoda), Alexander (1969), has described two single-celled organs which he suggests are proprioceptors.

Mero-carpopodite and Carpo-propodite joints.   It is convenient to treat both of these joints together. In the pereiopods there are two receptors at each joint; MC1 and MC2 at the mero-carpopodite joint and CP1 and CP2 at the carpo-propodite joint. (Whitear, 1962; Wales *et al.*, 1970; Alexandrowicz, 1972). In the 3rd maxillipeds only MC2 and CP2 are present (Wales *et al.*, 1970).

MC1 consists of a sheet of connective tissue with one side attached to the tendon of the accessory flexor muscle (AFM) of the carpopodite, the other to the pre-axial wall of the meropodite; while distally it narrows to a strand which inserts on the rim of the meropodite. CP1 is similar to MC1 and consists of a sheet of connective tissue anchored to the tendon of the bender (productor) muscle of the propodite and to the ventral surface of the carpopodite, with a distal strand attached to the ventral rim of the carpopodite. Thus in neither case does the receptor span the joint.

Table 6.2

Organ	Pereiopods — Proximal end on tendon of	Elongated by	Across joint	3rd Maxillipeds — Proximal end on tendon of	Elongated by	Across joint
PD	Closer (Flexor)	Dactylopodite closing	√	Closer	Dactylopodite closing	√
CP1	Bender (Productor)	Propodite bending	—	—	—	—
CP2	Stretcher (Reductor)	Propodite stretching	√	Ventral bender	Propodite bending	√
MC1	Accessory Flexor	Carpopodite flexion	—	—	—	—
MC2	Flexor	Carpopodite flexion	√	Flexor	Carpopodite flexion	√
IM1	Reductor	Meropodite reduction	—	Productor	Meropodite reduction	!
IMa	Reductor	Meropodite reduction and production	—	Productor	Meropodite production	—
MCO2	—	Carpopodite flexion	√?	—	—	—
MCO1	—	Carpopodite extension	—	—	—	—
BI	Retractor	Ischiopodite retraction	—	—	—	√
CB	—	Basipodite extension	√	—	Basipodite extension	√
TC	—	Coxopodite?	√	—	—	—

Organ	Antennules — Proximal end on tendon of	Elongated by	Across joint	Antennae — Proximal end on tendon of	Elongated by	Across joint
J3	Depressor	Flagellum depression	—	St. 2 —	Flagellum and Segment 3 extension	—
J2	Depressor	Segment 3 depression	√	St. 1 —	Segment 3 extension	√
J1	—	Segment 2 elevation and depression	√	—	Segment 2 ?	√

(From various sources, as mentioned in the text).

However, their arrangement is such that flexion of the carpopodite elongates MC1 and bending of the propodite elongates CP1. The sensory cell bodies are all embedded in the connective tissue and the dendrites of the smallest ones extend into the distal strand. In *Palinurus* (Palinura) and *Pagurus* (Anomura), the large cells of the proximal group form a fairly straight row (Alexandrowicz, 1972) but in the brachyurans *Carcinus, Cancer* and *Maia* and the astacuran *Homarus*, the row of large cells is markedly curved so that the largest cells lie lateral (MC1), or ventral (CP1), with the smallest ones mesial (MC1) or dorsal (CP1) (Whitear, 1962; Alexandrowicz, 1972). Alexandrowicz suggested that these variations indicate a dependence on the topographic relationships of the strand with neighbouring structures, rather than having any functional significance.

The receptor strands of MC2 and CP2 are generally rounded but tend to spread out near their distal insertions, and they both cross their respective joints. MC2 originates on the carpopodite flexor muscle tendon and inserts on the cuticle at the proximal end of the carpopodite. In the pereiopods CP2 originates on the propodite stretcher (reductor) muscle tendon, but in the 3rd maxillipeds the origin is on the tendon of the ventral propodite bender muscle (Wales *et al.*, 1970; Alexandrowicz, 1972). An exception to this is found in the claws of *Pagurus ochotensis*, where the origin is on the carpopodite cuticle (Field, 1974). In all cases the insertion is on the cuticle at the proximal end of the propodite. MC2, like MC1, is elongated by flexion of the carpopodite. However, in the pereiopods, CP2 is elongated by the opposite direction of movement to that which elongates CP1, i.e. stretch of the propodite. In the 3rd maxillipeds, movement at this joint is complex, but elongation of CP2 is brought about by contraction of the ventral bender muscle. (Wales *et al.*, 1970).

In *Carcinus*, the sensory cell bodies of MC2 and CP2 appear similar to those of the PD organ (see below), lying more or less in a row with the most proximal, largest ones not embedded in the receptor strand (Whitear, 1962). In *Pagurus*, MC2 is similar but, in CP2 of *Pagurus* and MC2 of *Homarus*, the row of cells is curved; and in CP2 of *Homarus* and both organs of *Palinurus*, the cells lie almost transverse to the axis of the strand (less obvious in the 3rd maxillipeds) and towards its distal end, with the largest cells distal to the smallest (Wales *et al.*, 1970; Alexandrowicz, 1972).

In *Homarus* and *Nephrops* Wales *et al.*, noted that the dendrites of some of the sensory cells associated with MC2 and CP2 do not terminate in the sensory strand but form a distinct group with their dendrites running towards the cuticle to innervate the cuticular articulated pegs (CAPs) (Chapter 1). Indeed, in MC2 these cells are not embedded in the receptor strand but lie adjacent to it. Alexandrowicz (1972) thought that probably all the sensory cells of MC2 and CP2 in *Homarus* and *Palinurus* are associated with the CAPs but observed that in *Pagurus* some remain associated only with the receptor strand. There are no CAPs in brachyurans and so in the members of this group presumably all of the dendrites are associated with the receptor strand. In *Ligia* a two-celled organ, which may be a proprioceptor, is present at each of these joints (Alexander, 1969).

Pro-dactylopodite joint.   With the exception of the 3rd maxillipeds of the Astacura (*Homarus* and *Astacus*), all of the appendages have a single PD organ. The receptor strand is normally oval in section except in the pereiopods of *Homarus* where it is flattened and in the 3rd maxillipeds of *Palinurus* and *Panulirus* where it is in the form of a sheet. It is anchored proximally to the tendon of the dactylopodite closer (flexor) muscle and distally to the cuticle at the proximal end of the dactylopodite, thus spanning the joint (Burke, 1954; Whitear, 1962; Wales *et al.*, 1970; Mill and Lowe, 1972). Closing the dactylopodite causes elongation of the receptor strand (Wiersma and Boettiger, 1959; Lowe and Mill, 1972).

Two groups of sensory cells are recognizable and some of the larger ones (in the proximal group) are not embedded in the strand. The distal group consists of a row of small, bipolar cells. Within the proximal group it is possible to distinguish cells with dendrites entering the anterior surface of the strand from those with dendrites entering the dorsal surface (Wiersma and Boettiger, 1959; Hartman and Boettiger, 1967; Mill and Lowe, 1971, 1973). Furthermore, the latter are associated with a collagen rich area of the strand (Boettiger and Hartman, 1968; Mill and Lowe, 1973).

Antennae and antennules

Few studies have so far been made on the chordotonal organs of the antennae but the indications are that a single receptor is present at some or all of the basal joints. Dealing with the structure of these appendages in those animals which have been investigated, the antennae and antennules of *Panulirus* (Palinura), the antennae of *Caprella* (Peracarida–Amphipoda) and the antennules of *Squilla*

(Hoplocarida) consist of three basal segments and one (antennae) or two (antennules) flagella. The antennae of *Petrocheirus* (Anomura) have six basal segments (coxopodite to carpopodite), with a squame arising from the basipodite, and a distal flagellum.

In *Panulirus* and *Squilla* there is a single chordotonal organ (J1) at the first joint (between the two proximal segments), the receptor strand of which is attached at both ends to the hypodermis and spans the joint (Fig. 6.3). In *Panulirus*, it originates near the proximal end of the first segment and lies dorso-medially in the antennae (Hartman and Austin, 1972), but in the antennules the origin is nearer the distal end of the first segment and the strand runs ventro-laterally (Wyse and Maynard, 1965). In the antennules of *Squilla* the receptor strand originates on the dorso-lateral wall of the first segment and is, unlike in *Panulirus*, sheet-like (Sandeman, 1964). In all cases the insertion is near the proximal end of the second segment. In the antennules of *Panulirus*, Wyse and Maynard (1965) found that it was more or less possible to distinguish three groups of sensory cells (also in J2 and J3), the most distal of which is contained in an accessory strand attached near the distal end of the main strand. The main strand is elongated by passive elevation or extension of the second segment.

Spanning joints two and three of the antennae of *Panulirus* is a complex organ (J2–J3) consisting of two receptor strands lying on a sheet of connective tissue which stretches from the medial to the lateral wall of the second and third segments and has a thin distal

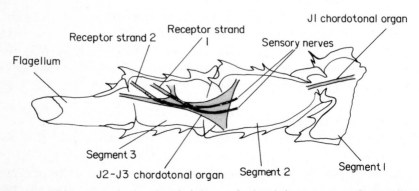

Fig. 6.3   Diagram of the basal joints of the left antenna of *Panulirus interruptus* showing the J1 and J2–J3 joint chordotonal organs. (After Hartman and Austin, 1972.)

attachment in the flagellum. One receptor strand (St.1) is attached to the sheet in segment two and inserts medially in segment three; the other (St.2) extends distally from the sheet in segment three to attach to the medial side of the same segment. Although the larger sensory cells lie proximally on the receptor strands, there is no division into groups. St.1 is elongated by extension of the third segment; St.2 by extension of both the third segment and of the flagellum (Hartman and Austin, 1972). J2—J3 also occurs in *Homarus* (Sigvardt, 1974).

In the antennules of *Panulirus* there is a separate receptor at each of the second and third joints. J2 has a main receptor strand which originates on the tendon of the depressor muscle of the third segment and inserts at the proximal end of the third segment. Like J1, it has an accessory strand at its distal termination, which contains the distal cell group. Depression of the third segment elongates the main strand (Wyse and Maynard, 1965). Similarly, in *Caprella*, J2 spans the joint, running from the tendon of the extensor muscle of segment three to insert on the cuticle of the third segment (Wetzel, 1934). In *Squilla*, J2 has a thin receptor strand associated with a tendon which runs from the medio-lateral wall of the second segment to the proximal edge of the third segment, this insertion being shared with the tendon of the extensor muscle of the third segment. The dendrites, however, run in a sheet of tissue which inserts towards the distal end of the second segment (Sandeman, 1964).

J3 in *Panulirus* has a receptor strand which originates on the flagellum depressor tendon and inserts ventrally on the third segment between the bases of the flagella. Thus it does not span the joint. As with J1 and J2 there is an accessory strand near the distal end, but here it is small and devoid of sensory cells. The main strand as a whole is elongated by depression of the outer flagellum, but the accessory strand is arranged so that the region containing the middle cell group is apparently elongated by passive elevation of this flagellum (Wyse and Maynard, 1965). A corresponding organ has not been described in *Caprella* or *Squilla* but in *Petrocheirus*, Taylor (1967a) mentions a CF organ (which may be homologous) spanning the carpopodite-flagellum joint.

Swimmerets

Two proprioceptors have been described in the swimmerets of *Homarus*. They both originate in the abdomen and span the coxopodite. One inserts on the tendon of the main return-stroke (protractor) muscle, the other on the anterior rim of the basipodite (Davis, 1968a).

6.2.2   Myochordotonal organs

In the Astacura there is a single myochordotonal organ (MCO1) in the pereiopods but in the Palinura, Brachyura and Anomura there are two (MCO1 and MCO2) (Section 6.3; Fig. 6.4). These organs are associated with the accessory flexor muscle (AFM) which has two heads (proximal and distal) joined by a long thin tendon. The proximal head originates at the extreme proximal end of the meropodite and is spindle-shaped, while the flattened distal head originates on the anterior wall towards the distal end of the meropodite. Distally, the tendon inserts onto the base of the main flexor muscle tendon.

MCO1

This myochordotonal organ was first described by Barth (1934) in a large variety of decapods, mainly brachyurans. It lies ventrally at the proximal end of the meropodite and is juxtaposed to the proximal head of the AFM. In *Palinurus*, there is a strand of connective tissue running alongside the proximal head ('flanking strand') which joins a connective tissue 'strap' in the region of the proximal tendon. This strap runs ventrally towards the cuticle, terminating near the distal end of the cuticular canals of the CAPs (Chapter 1). The more proximal sensory cells are all small and lie near the integument with their distal processes running towards the cuticular canals. The more distal cells lie near the tendon and the distal processes of some also run towards the cuticular canals, while those of others run into the strap (Alexandrowicz, 1972).

In *Astacus* and *Homarus* there is also a connective tissue strap and it is associated with the tendon near the origin of the dorsal part of the muscle, which is separate from the larger, ventral part in the Astacura (Clarac and Masson, 1969; Alexandrowicz, 1972). The arrangement of cells in *Homarus* is similar to that in *Palinurus* (Alexandrowicz, 1972). The small proximal group of cells have been termed the annex cells by Clarac and Masson (1969) (Fig. 6.4).

In *Pagurus* the strap, or the main part of it, ends in a cuticular depression (Alexandrowicz, 1972). Clarac and Masson did not note any annex cells in their description of *Eupagurus*, nor are there any

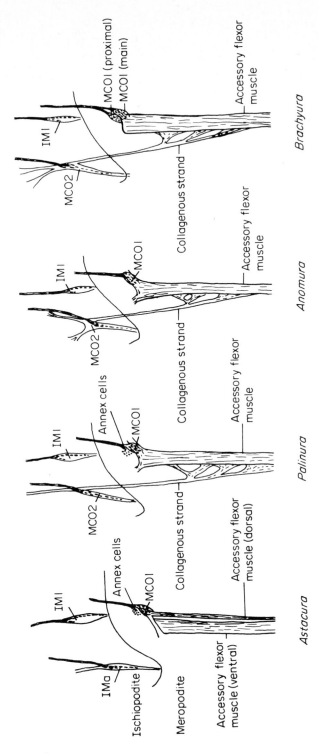

Fig. 6.4 The relationships between the IM organs, the annex cells and the proximal head of the accessory flexor muscle in Astacura, Palinura, Anomura and Brachyura. (After Clarac and Masson, 1969; Alexandrowicz, 1972.)

in the Brachyura. Their presence correlates with that of CAPs (Table 1.1).

In the Brachyura, MCO1 consists of a main part (MCO1m) and a proximal part (MCO1p) (Barth, 1934; Clarac, 1968a). According to Barth the sensory cells of MCO1m are embedded in a sheet of connective tissue which extends from the ventral part of the proximal head of the AFM to the posterior wall of the meropodite, and their dendrites run obliquely downwards to attach ventrally in a depression in the cuticle. This has been confirmed by Clarac (1968a) in *Carcinus* with the proviso that the connective tissue is restricted to the extreme proximal end. MCO1p contains the larger cells and their dendrites are also attached to the cuticular depression. Clarac observed that this part of the organ is associated with the dorsal region of the AFM. In this connection it is of interest to note that in *Carcinus* and *Cancer* (also *Palinurus* and to a less marked extent *Pagurus*) the muscle fibres of the dorsal region are of the 'fibrillen' type; those of the ventral region 'felder' (Cohen, 1963b; Clarac, 1968a; Alexandrowicz, 1972). MCO1 has also been observed in *Crangon* (Caridea) by Clarac and Masson (1969).

The dendrites of MCO1 are elongated when the proximal head of the AFM is stretched during extension of the carpopodite (Hwang, 1961).

MCO2

This organ was first described in detail by Cohen (1963a) in *Cancer magister*. It consists of a group of cells associated with a receptor strand similar to those of the joint chordotonal organs. However, the receptor strand is attached to a collagenous ligament which originates as several strands associated with connective tissue around the leg nerve in the distal ischiopodite. These join and the ligament runs distally and ventrally to make a fan-shaped insertion on the connective tissue surrounding the proximal head of the AFM (Barth, 1934; Cohen, 1963a; Alexandrowicz, 1972). Thus MCO2 is functionally connected to the AFM. Distally the receptor strand diverges from the ligament, runs ventrally and then divides into two, inserting near the ischio-meropodite joint. Cohen thought that both insertions were on the ischiopodite, an opinion shared by Alexandrowicz (1972), who demonstrated a similar structural arrangement in *Palinurus* and *Pagurus*. However, Clarac (1968a) stated that, in *Carcinus mediterraneus*, one inserted in the ischiopodite, the other (containing most of the distal sensory cells) crossing the joint to insert on the meropodite. Whitear (1962), who briefly mentions this organ in a study on *Carcinus maenas*, observed a single distal strand inserting on the meropodite; as did Gordon (1973) in *Cancer pagurus*.

The dendrites of MCO2 are elongated when the proximal head of the AFM is shortened (Section 6.4) (Cohen, 1963a).

6.2.3   Flagellum proprioceptors

In the distal region of the first segment of the antennal flagellum of *Petrocheirus* (Anomura), Taylor (1967a) described a chordotonal organ which differs from those discussed in Section 6.1.1, in that there is no receptor strand. It consists of a ring of sensory cells, the dendrites of which run distally to insert in the proximal wall of the second flagellar segment.

Proprioceptors have also been described in the antennular flagella of *Panulirus* (Laverack, 1964). They probably occur at each annulation and there appear to be two sensory cells per segment, one giving information on medial movement, the other on outward movement.

6.3   Problems of terminology and homology

The problem concerns those receptors located in the region of the ischio-meropodite joint. These have been referred to above as IM1, IMa, MCO1 and MCO2. Table 6.3 gives the names which these organs have been called by various authors and an attempt will now be made to rationalize this problem.

There is little controversy surrounding IM1. All those who have described it in the Astacura use this term (Clarac and Masson, 1969; Wales *et al.*, 1970; Alexandrowicz, 1972): Alexandrowicz has pointed out its similarities with MC1 and CP1, namely that the receptor strand is associated with a sheet of connective tissue and does not cross the joint and that the nerve cells are arranged in a curving line. In the Palinura and Brachyura, where IMa is absent, Clarac (1968a) and Clarac and Masson (1969) have used the term IM; Alexandrowicz (1972) IM1. To avoid any possible confusion with IMa it is suggested that the term IM1 always be used.

Table 6.3

Organ	Barth (1934) Mainly Brachyura	Whitear (1962) Brachyura	Cohen (1963) Brachyura	Cohen (1965) Brachyura	Clarac (1968) Brachyura	Clarac and Masson (1969) Astacura	Palinura Brachyura	Alexandrowicz (1973) Astacura	Palinura Anomura
IM1	*	*	*	*	IM	IM1	IM	IM1	IM1
IMa	–	–	–	–	–	IM2	–	IM3	–
MCO1	Myochordotonal organ	Barth's organ	*	Proximal Myochordotonal organ	MCO1 or Barth's organ	Myochordotonal organ, MCO	MCO1	Barth's organ, IM2	Barth's organ, IM2
MCO2	*	IM	Myochordotonal organ	Main Myochordotonal organ	MCO2	–	MCO2	–	IM3

* Not described
– Absent

IMa only occurs in the Astacura and it is interesting that in this group MCO2 is absent. Both Clarac and Masson (1969) and Alexandrowicz (1972) point out the probable homology of these two organs. IMa was called IM2 by Clarac (1968) and Wales *et al.* (1970). However, Alexandrowicz (1972) pointed out that MCO1 is the ischio-meropodite equivalent of the MC2, CP2 series (see p. 250) and he thus used the term IM3. Since IM2 implies a homology with MC2 and CP2 and since IM3 implies the presence of an organ named IM2, the new name IMa is proposed.

The terms MCO1 and MCO2 for the two myochordotonal organs were first used by Clarac and Masson (1969). In the Astacidea, where MCO2 is absent, Clarac and Masson used MCO to refer to MCO1, but it is suggested here that the latter term is less likely to cause confusion and should always be used when referring to the organ first described by Barth (1934). Some confusion exists in the earlier work of Cohen (1963a, 1965) who assumed that MCO2 was Barth's main myochordotonal organ and MCO1 the proximal myochordotonal organ. Whitear (1962) mentioned the presence of these organs in *Carcinus*. She referred to MCO1 as Barth's organ and to MCO2 as IM. Alexandrowicz (1972) also avoided the use of the term 'myochordotonal' and stated that MCO1 should be referred to as IM2 since it occurs at the ischio-meropodite joint and is undoubtedly homologous with MC2 and CP2 (in all three organs some of the dendrites are associated with the CAPs, when the latter are present). Similarly, since MCO2 occurs at the ischio-meropodite joint and is homologous with IMa, he argued that it also should be an IM organ, namely IM3. However, since the use of 'myochordotonal' is so well established in the literature for these two organs (e.g. Barth, 1934; Cohen, 1963a) and since they have a special relationship with the AFM and are thus functionally associated with the mero-carpopodite, not the ischio-meropodite, joint (Section 6.4), it is suggested that the terms MCO1 and MCO2 be retained.

6.4   Mechanics

With the exception of the carpo-propodite the joints of the pereiopods only allow movement in one plane; antero-posterior or dorso-ventral. The relationship of the PD organ to movement at the pro-dactylopodite joint has been investigated in *Cancer* by Lowe and Mill (1972), who showed that receptor strand length increases almost linearly with angular movement of the dactylopodite away from the fully closed position through about two-thirds of the movement arc, but that further change in angle is accompanied by a decreasing rate of increase in strand length (Fig. 6.5). It was calculated that a linear relationship should occur between the square of receptor strand length (L^2) and the cosine of the angle of the joint (cos α) and found that, using fairly large crabs, the equation $L^2 = 54.5 + 43.0 \cos \alpha$ fitted the experimental data with a correlation coefficient in excess of 0.99. Strand length depends on the size of the animal. For the above, typical values were 6.3 mm in the fully open position and 10.2 mm in the fully closed position; a length increase of 62 per cent. This gives an average length change of 30–35 μm/degree of joint movement. However, each dendrite is only a few hundred μm long and so will experience at most a length change in the order of 1–3 μm/degree. In *Carcinus*, the length change of PD has been estimated at only about 20 per cent (Burke, 1954); that of MC1 and CP1 about 70 per cent, of MC2 and CP2 about 25 per cent (Bush, 1965a) and of CB about 45 per cent (Bush, 1965b).

Cohen (1963a) described the effect on MCO2 of changes in the mero-carpopodite joint angle. The tendon of the AFM inserts distally

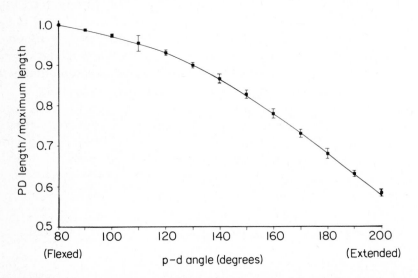

Fig. 6.5   Length of the PD receptor strand: maximum length of the strand plotted against the angle of articulation of the pro-dactylopodite joint. These results are the means for 12 legs and the vertical bars indicate ±1 s.e. (After Lowe and Mill, 1972.)

on that of the main flexor muscle. Thus, when the carpopodite is actively flexed, the AFM is shortened, and when it is extended, the AFM is elongated; a total length change of about 5.7 mm occurring. Since the angular movement is about 100°, this gives 57 µm of length change/degree of joint movement. Fig. 6.6 shows the relationship between the AFM, the collagenous strand and the receptor strand (containing the sensory cells) of MCO2. It can be seen that the dendrites are elongated during flexion and relaxed during extension of the carpopodite. The maximum increase in length of the receptor strand from the 'rest' position (45° from fully flexed) to full flexion is about 120 µm, giving 2.6 µm/degree of joint movement. Thus

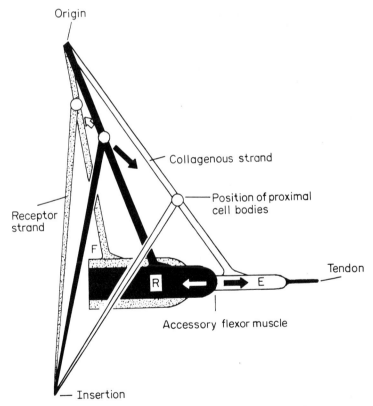

Fig. 6.6   The relationships between the proximal head of the accessory flexor muscle and the various components of MCO2 in the pereiopods of *Cancer*. R—indicates the 'rest' position; E and F—the positions produced by passive extension and flexion respectively of the carpopodite. (After Cohen, 1963a.)

there is a mechanical reduction of about 22 : 1 between change in length of the AFM and of this strand (Cohen, 1963a). However, the actual reduction as far as individual dendrites are concerned is probably much greater since they are only a few hundred μm long. The length change experienced by individual dendrites is thus of the same order of magnitude in both PD and MCO2.

In the walking legs of the Astacura, Palinura and Anomura and of the brachyuran *Dormidiopsis dormia* the carpo-propodite joint allows both antero-posterior movement (135° in *Homarus*) and rotation (35° in *Homarus*) of the propodite. Rotation does not occur in the chelae or third maxillipeds of *Homarus* but, in the latter appendages, the propodite is capable of dorso-ventral as well as antero-posterior movement (Wiersma and Ripley, 1954; Wales *et al.*, 1970).

In the antennae of *Panulirus* movement of most joints is also uniplanar. Joint 2, however, allows 70° of dorso-ventral movement as well as about 22° of movement in the medio-lateral plane. The receptor strands of the J2—J3 organ lie on the same connective tissue sheet and this raises the possibility that movement at either joint 2 or joint 3 could excite both parts of the organ. The main connective tissue sheet is shortest longitudinally when segment 3 is flexed and the flagellum extended (the 'rest' position) (Fig. 6.7). Strand 1 is elongated by extension of segment 3, while strand 2 is elongated by extension both of segment 3 and of the flagellum (Hartman and Austin, 1972). In *Homarus* the flagellum is not extended while segment 3 is flexed (Sigvardt, 1974).

6.5 Sensory responses

Activity in the limb chordotonal organs was first recorded by Burke (1954), who described units in the PD organ of *Carcinus* which were unidirectionally sensitive to movement of the dactylopodite, and others which gave a maintained discharge when the joint was held in one of its extreme positions. Wiersma and Boettiger (1959) described these in more detail and also observed units which were intermediate in their response characteristics. They noted that the sensory cells in the proximal group are movement sensitive, those of the distal group position sensitive and, furthermore, that movement units are generally larger than position units. Mendelson (1963) and Bush (1965a) have demonstrated that similar responses are obtained by joint movement, *in situ* movement of the receptor strand and

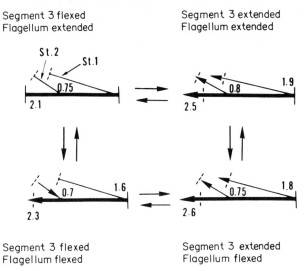

Fig. 6.7   Diagrammatic representation of the J2–J3 chordotonal organ of *Panulirus interruptus* showing the lengths attained by the main strand (horizontal line) and the two receptor strands (St.1 and St.2) resulting from different positions of segment 3 and the flagellum. (From Hartman and Austin, 1972.)

movement of the partially or completely isolated strand. Similar types of unit have been found in most of the limb chordotonal organs and myochordotonal organs of decapods. Also unidirectional movement and position units have been recorded from the pereiopods of *Ligia* (Alexander, 1969).

In the following account the terminology of Boettiger and Hartman (1968), which involves use of the terms 'elongation' and 'relaxation' for change in length of the receptor strand, will be followed. This makes comparisons between receptors easier and reference to Tables 6.2 and 6.4 will show which direction of movement at the appropriate joint causes elongation and relaxation of the various receptors.

6.5.1   Unidirectional movement units

All phasic units respond to movement at the joint in one direction only (Burke, 1954; Wiersma and Boettiger, 1959). Hence there are elongation sensitive movement cells (ESMCs) and relaxation sensitive movement cells (RSMCs). Some of the units are extremely sensitive and may respond at a velocity of less than one degree of joint movement per second. These are almost pure movement units in that they attain their maximum frequency over the whole arc of joint movement while the movement velocity is low, or is at least close to the firing threshold. However, near threshold, RSMCs tend to be less sensitive towards the fully elongated position; ESMCs towards the fully relaxed position (Wiersma and Boettiger, 1959) (Figs. 6.8, 6.9). The ESMC illustrated, reaches a saturation frequency of 24–28 Hz at a movement velocity of only about 5 mm s^{-1}. It is rather less sensitive near the relaxed end of its range and, below saturation frequency its range depends on the velocity of movement. At 0.65 mm s^{-1} (Fig. 6.9B) it only fires over about half of the movement range but, as velocity increases, it starts to fire progressively closer to the relaxed end of the range (Mill and Lowe, 1972). In the PD organ, pure movement units are located in the central zone of the movement cell group (Hartman and Boettiger, 1967; Boettiger and Hartman, 1968).

Wiersma and Boettiger also described units whose firing frequency is dependent both on strand length and on the velocity of movement (i.e. position and velocity sensitive) and the term 'differential sensitivity' was proposed for this phenomenon by Mill and Lowe (1972). In the PD organ, these velocity sensitive units occupy the proximal and distal ends of the movement cell group. RSMCs and ESMCs at the proximal end *both* show an increase in sensitivity as the fully elongated position is approached, while those at the distal end both show an increase in sensitivity as the fully relaxed position is approached (although distal ESMCs are rare) (Fig. 6.10). Proximal RSMCs and some proximal ESMCs and distal RSMCs, respond over most or all of the movement range; other proximal ESMCs and distal RSMCs only start to fire when the appropriate extreme position is approached (Hartman and Boettiger, 1967; Boettiger and Hartman, 1968). An example of a velocity sensitive ESMC is shown in Fig. 6.8b. The firing frequency is still increasing at quite high movement velocities and a plot of frequency against log rate for this unit produces a linear relationship. At all velocities it tended to show an increase in sensitivity with elongation of the strand (differential sensitivity), this becoming more marked as the velocity increased (Mill and Lowe, 1972).

Phasic units have also been described with a tonic background discharge unrelated to receptor strand length. They are generally velocity sensitive and most of the movement units of the MCO2

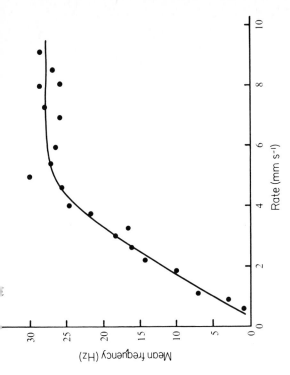

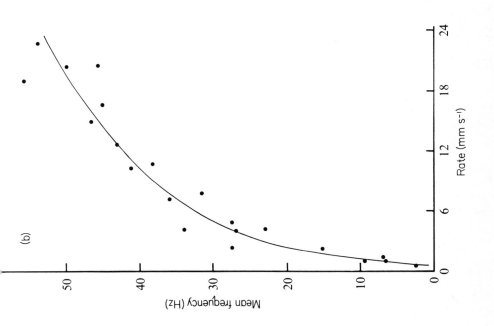

Fig. 6.8 Graphs showing the relationship between mean frequency and velocity of stretch for (a) an almost pure movement ESMC (elongation sensitive movement cell) from *Cancer pagurus* (the same unit as in Fig. 6.9B); and (b) a velocity sensitive ESMC. (From Mill and Lowe, 1972.)

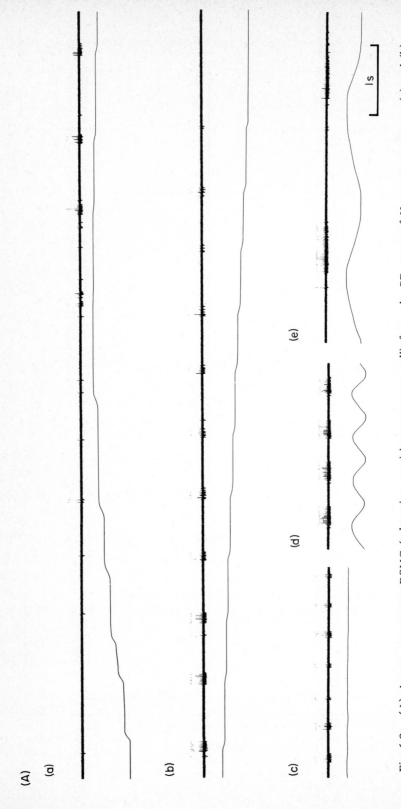

Fig. 6.9 (A) A pure movement RSMC (relaxation sensitive movement cell) from the PD organ of *Homarus gammarus*. (a) and (b) are continuous. (c) note the extreme sensitivity of this unit. Lower trace indicates elongation (upwards) and relaxation (downwards) of the receptor strand.

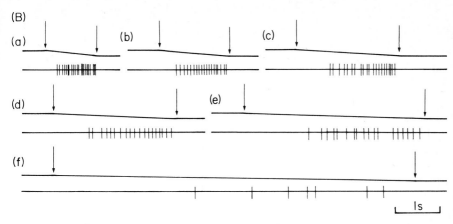

Fig. 6.9 (B) an almost pure movement ESMC (elongation sensitive movement cell) from the PD organ of *Cancer pagurus* (the same unit as in Fig. 6.8A). Lower trace indicates elongation (downwards) of the receptor strand. Arrows indicate periods of stimulation. (From Mill and Lowe, 1972.)

organ of *Cancer* are of this type (Cohen, 1963a). They have also been found occasionally in the joint chordotonal organs (Bush, 1965a; Clarac, 1968b; Mill and Lowe, 1972). (Clarac and Mill and Lowe classified them as 'intermediate' units). Movement in the direction opposite to that which causes phasic excitation produces inhibition of the tonic discharge (Mill and Lowe, 1972), except in some units in MCO1 (Clarac, 1968b), and the excitatory phasic burst is followed by a period of 'post-excitatory depression' (Cohen, 1963a; Mill and Lowe, 1972) (Fig. 6.11).

Cohen (1963a) stated that the phasic units of MCO2 respond over only part of the movement arc of the carpopodite, most having a

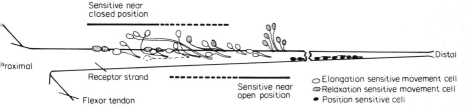

Fig. 6.10 Diagram from the anterior side to show the organization of the PD organ in *Cancer*. The lines referring to close and open joint position sensitivity refer to the movement sensitive cells. The dashed part of the lines refers to movement sensitive cells which are equally sensitive in any position. (After Hartman and Boettiger, 1967.)

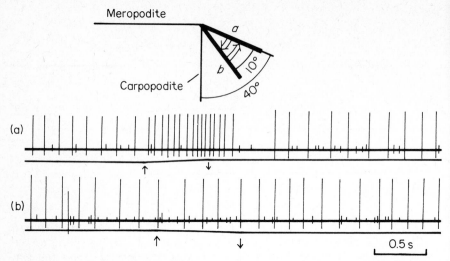

Fig. 6.11   Response of a single ESMC from the MCO2 of *Cancer*. (a) shows $10°$ flexion of the carpopodite (receptor strand elongation), (b) shows $10°$ extension of the carpopodite (receptor strand relaxation). Lower trace indicates carpopodite movement. Arrows indicate period of stimulation. Spikes redrawn. (After Cohen, 1963a.)

range of $40°-60°$, as compared with the total movement arc of $90°-100°$, and he referred to this phenomenon as 'range fractionation' (Fig. 6.12). There is considerable overlap between the ranges of these units and Clarac (1968b) noted a greater preponderance of them near the rest position of the joint. In MCO1 some of the phasic units show a maximum response as positions other than the extreme ones are approached and there is some evidence here also for range fractionation (Clarac, 1968b). It is possible that a more extensive investigation of differential sensitivity and range fractionation will reveal that they differ only in degree. Thus at low movement velocities the range of differentially sensitive units may be so curtailed as to produce range fractionation. One reason that has been suggested for range fractionation (Cohen, 1963a) is that it may be to allow high resolution to small stimulus increments, with the wide range units being relatively insensitive to avoid mechanical damage caused by considerable elongation. However, the wide-range pure movement units are often the most sensitive (Wiersma and Boettiger, 1959; Mill and Lowe, 1972). Wood (1974) observed neither differential sensitivity, nor range fractionation in the PD of *Pachygrapsus*.

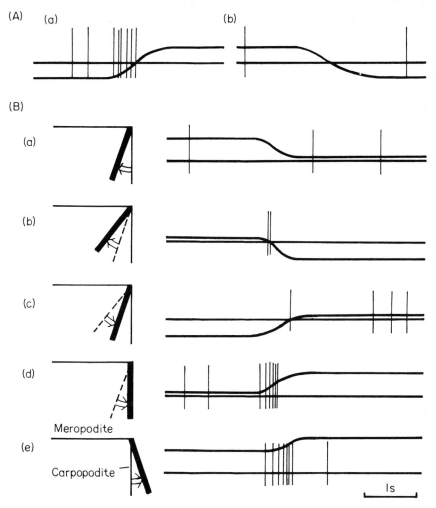

Fig. 6.12 Response of a RSMC. (A) shows the response to 10° of extension of the carpopodite (receptor strand relaxation) and 10° of flexion of the carpopodite (receptor strand elongation). (B) shows a sequence of 10° movements of the carpopodite as illustrated to the left of each trace. In A and B lower trace indicates carpopodite movement. Spikes redrawn. (After Cohen, 1963a.)

Phasic units have also been recorded from the antennular flagella of *Panulirus*, where they respond to movements away from the rest (straight) position (Laverack, 1964), and from the antennal J2–J3 organ of *Panulirus* (Hartman and Austin, 1972), but are apparently absent in the antennal flagellum chordotonal organ of *Petrocheirus*

(Taylor, 1967a, b). Although the majority of movement units in J2–J3 are clearly unidirectional, whether they be stimulated by movement at one or both joints, there are a small number in receptor strand 1 (and a single one recorded from receptor strand 2) which are unusual in that they respond to directions of movement of segment 3 and of the flagellum which apparently produce opposite length changes of the receptor strands. However, for movement of each joint separately the units are unidirectional (Hartman and Austin, 1972).

In movement units from the PD organ Wiersma and Boettiger (1959) observed that below saturation frequency the impulses are separated by intervals which appear to be multiples of the shortest interspike interval observed at saturation ('skip discharge'). Wiersma, Van Der Mark and Fiore (1970) confirmed this and further showed that the shortest interspike interval is about 20 ms, a result also obtained by Mill and Lowe (1972) and Wood (1974) (Fig. 6.13). Since such a value may be artificially produced by the influence of the 50 Hz mains frequency, Wiersma *et al.* stimulated the receptor by purely mechanical means, when they obtained similar results, but with a longer fundamental period (47 ms), possibly caused by some inherent periodicity in the mechanical system. Wood (1974) also found a few position units which exhibited a skip discharge.

6.5.2   Position units

In the joint chordotonal organs, some of the cells are almost pure position units, with little or no adaptation following movement. The majority, however, have a distinct phasic component (Wiersma and Boettiger, 1959; Wiersma, 1959; Bush, 1965a, b; Mill and Lowe, 1972; and Bush, 1965b) noted that this is particularly apparent at higher movement velocities. In the PD organ the position units are located in the distal cell group (Wiersma and Boettiger, 1959; Hartman and Boettiger, 1967). In both types the maximal firing rate of elongation sensitive position cells (ESPCs) usually occurs at the elongated end of the range; that of RSPCs at the relaxed end. In ESPCs and RSPCs with a phasic component, relaxation and elongation respectively cause phasic inhibition (Mill and Lowe, 1972) (Fig. 6.14a). In MCO1 tonic units with a marked phasic component occur which fire maximally at the rest position of the carpopodite and this maximal firing away from an extreme position has been

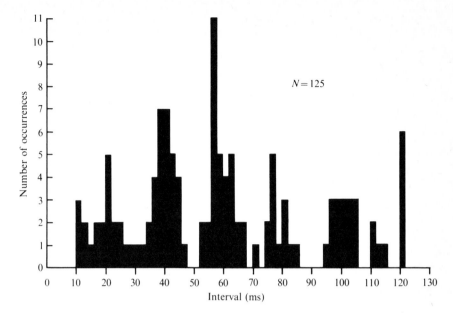

Fig. 6.13   Pulse interval histogram of the output from an ESMC from the PD organ of *Cancer*. Note the increased number around 20 ms and at multiples of this. (From Mill and Lowe, 1972.)

described in the CB organ of *Carcinus* (Bush, 1965b) and on one occasion in the PD organ of *Cancer* (Mill and Lowe, 1972), but is unusual in the joint chordotonal organs. In MCO2 most of the position units approach the 'pure position' ideal, with a small phasic component. Most of the ESPCs respond in the part of the movement arc between rest and full flexion of the carpopodite, firing maximally at the latter (Cohen, 1963a). Cohen reported an absence of RSPCs, but see Section 6.6.

In the antennal flagellar chordotonal organ of *Petrocheirus* all of the cells are position sensitive. They remain silent when the flagellum is in its rest (straight) position, movement away from which causes excitation of those units on the side towards which it is bent. A few, small units behave like pure position units, but the majority have a marked phasic component which, in most cases, is velocity sensitive. In the latter, movement towards the rest position produces phasic inhibition (Taylor, 1967a, b). Similar tonic units with a large phasic component have been recorded from the antennular proprioceptors of *Panulirus* (Laverack, 1964) but, in addition, Laverack recorded

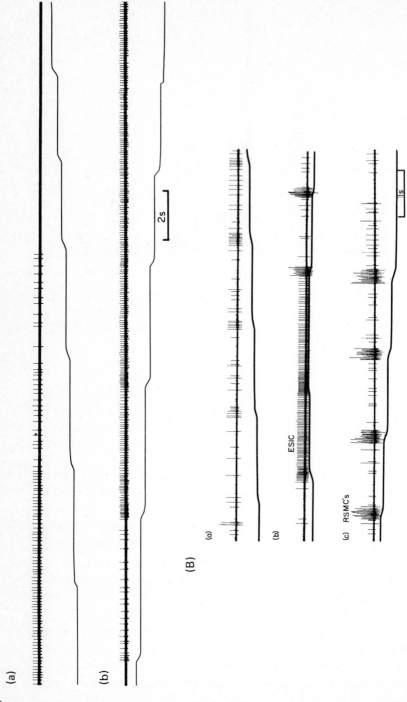

Fig. 6.14 (A) An asymmetric RSPC (relaxation sensitive position cell) from the PD organ of *Homarus gammarus*. a and b are continuous. (B) An ESIC (Elongation sensitive intermediate cell) and several RSMC's (relaxation sensitive movement cells) from the PD organ of *Homarus gammarus*. The traces are continuous. Lower traces in (A) and (B) indicate elongation (upwards) and relaxation (downwards) of the receptor strand. (From Mill and Lowe, 1972.)

other units which became more active as the rest position was approached.

The position units of the J2—J3 organ of *Panulirus* are interesting in that those of receptor strand 1 respond exclusively to position of segment 3; those of receptor strand 2 exclusively to flagellum position (Hartman and Austin, 1972).

The response pattern of those units with a phasic component is typically asymmetric as far as the tonic level a few seconds after movement has ceased is concerned. Thus ESPCs show a higher firing frequency for a given length of the strand if that length is achieved by elongation rather than by relaxation; RSPCs show the reverse (Fig. 6.9c) (Wiersma, 1959; Taylor, 1967a, b; Mill and Lowe, 1972). This 'hysteresis' also occurs in the almost pure position units of MCO2 (Clarac, 1968b). Possibly, in the long term, the adapted frequency becomes the same irrespective of the direction of the initial stimulus and indeed Taylor (1967b) reported that adaptation to a background discharge is complete in 20—30 s. Furthermore, the background discharge disappears below 12°C; response to movement below 10°C (Taylor, 1967b).

6.5.3 Intermediate units

Intermediate units were described by Wiersma and Boettiger (1959) as rapidly adapting phasic units over part or all of the movement arc which, in addition, fired tonically when they were fully elongated (ESICs) or fully relaxed (RSICs). Bush (1965b) referred to units of this type as movement units since he observed that the tonic discharge was at 50 Hz and suggested that this may be a result of 'mains feed-back' vibrations in the recording system, which only become manifest when such units are in the most sensitive part of their range. However, in other cases the tonic discharge is well below 50 Hz (Wiersma, 1959). These units may also show some background discharge throughout their range (Mill and Lowe, 1972) (Fig. 6.14b). Wiersma (1959) found intermediate units relatively scarce, with less in *Carcinus* than in *Palinurus*, and he suggested the possibility that they represent abnormal sense cells.

Various other units have been described as 'intermediate' in the literature (Taylor, 1967b; Clarac, 1968b; Mill and Lowe, 1972) but they have all been treated here as position units. Perhaps those with the best claim to be considered as intermediate are those tonic units with a very marked phasic component. However, since a range in the magnitude of the phasic component exists, it is impossible to draw any meaningful dividing line. An alternative of course, would be to consider all tonic units with a phasic component as intermediate.

6.6   Proportions of elongation and relaxation sensitive units

The principal types of movement and position units, as well as intermediate units, occur in most or all of the pereiopod joint chordotonal organs. In CB and PD, ESCs and RSCs occur in approximately equal numbers (Wiersma and Boettiger, 1959; Wiersma, 1959; Bush, 1965b; Mill and Lowe, 1972) (Table 6.4).

However, in CP1, CP2, MC1, MC2 and IM1 of *Carcinus* the response to relaxation is considerably greater than that to elongation, both as regards movement and position units (Wiersma, 1959; Bush, 1965a; Clarac, 1968b). The difference is least in CP1, but is progressively more pronounced in MC1 and IM1, Clarac not recording any ESPCs from the latter. The extreme situation is reached in CP2 and MC2. Wiersma (1959) found that ESCs in CP2 are rare, while Bush (1965a) did not find any in either CP2 or MC2.

Table 6.4

		Afferents	
Organ	Elongated by	Elongation sensitive	Relaxation sensitive
PD	Dactylopodite closing	+++	+++
CP1	Propodite bending	++	+++
CP2	Propodite stretching	θ[6], θ[7]	+++
MC1	Carpopodite flexion	+	+++
MC2	Carpopodite flexion	θ	+++
IM1	Meropodite reduction	θ[3], +[4]	+++
IMa	Meropodite reduction[1] and production	+++	++
MCO2	Carpopodite flexion[2]	+++	++[5], +[4]
MCO1	Carpopodite extension	+++	+(+)
BI	Ischiopodite retraction	θ[3]	+++
CB	Basipodite extension	+++	+++
TC	Coxopodite?	?	?

[1,2]N.B. Homology of IMa and MCO2; [3]*Astacus;* [4]*Carcinus;* [5]*Cancer;* [6]a few reported in CP2 of *Carcinus;* [7]chela of *Pagurus;* [8]*Homarus* (from various sources). The crosses (+) indicate only the *relative* proportions of afferents.

Wiersma pointed out that the directional differences are somewhat less marked in younger crabs.

IM1 in the pereiopods of *Astacus* and the 3rd maxillipeds of *Homarus* is apparently lacking in ESCs (Clarac, 1970; Wales *et al.*, 1970) and this is also the case with BI of *Astacus* (Clarac, 1970) and CP2 in the chela of *Pagurus* (Field, 1974). CP2 in the 3rd maxillipeds of *Homarus* does not conform to this pattern. It shows the complete opposite to its counterpart in the pereiopods with virtually no RSCs (Wales *et al.*, 1970). IMa (*Astacus*) also shows the reverse situation with rather more elongation than relaxation units (Clarac, 1970).

In MCO1 of *Cancer* (Hwang, 1961) and *Astacus* (Clarac, 1970) most of the units are ESCs. In MCO2 of *Cancer* Cohen (1963a) found a slight preponderance of ESCs. RSMCs are present, but RSPCs are apparently lacking. In *Carcinus* Clarac (1968b) found very few RSCs and, in contrast, these were RSPCs not RSMCs. The homology of IMa and MCO2 has been discussed in Section 6.3 and it is interesting to note that ESCs are most abundant in both of these organs.

In the antennal J2–J3 organ of *Panulirus* receptor strand 1 contains more ESCs than RSCs; receptor strand 2 approximately equal numbers of both (Hartman and Austin, 1972). All of the units in the antennal flagellar chordotonal organ of *Petrocheirus* are relaxation (or flexion) sensitive (Taylor, 1967b).

6.7   Transduction

It has been established that the dendrites of most pereiopod joint chordotonal organs are paired, each pair sharing a single scolopidium (Chapter 9), and that the members of a pair can, in most cases, be distinguished ultrastructurally as 'ciliary' and 'paraciliary'. In MC2 and CP2 the paraciliary dendrite is poorly developed or absent (Whitear, 1960, 1962; Mill and Knapp, unpublished observations), while in CB, both dendrites are ciliary (Whitear, 1962). Dendritic pairing also occurs in the antennal flagellar chordotonal organ of *Petrocheirus*, although in this case the dendrites are not embedded in a receptor strand (Taylor, 1967a).

On the basis of the lack of ESCs in MC2 and CP2, Whitear (1962) suggested that the ciliary dendrite responds to relaxation of the receptor strand, the paraciliary to elongation. According to this theory CB should not have any ESCs, but Bush (1965b) found similar numbers of ESCs and RSCs in this organ. Bush proposed a mechanical arrangement involving collagen fibres which would lead to bending of the scolopidium in one direction during strand elongation and in the opposite direction during strand relaxation. He suggested that, if the bending were approximately equal in both directions then a pair of ciliary dendrites would suffice but that, if greater bending occurred in one direction, a paraciliary dendrite might be necessary to provide additional sensitivity to bending in the opposite direction. Mendelson (1963) had earlier proposed a similar theory but this required that the distal end of each scolopidium should lie at a marked angle to the longitudinal axis of the strand, whereas they remain parallel (Whitear, 1962; Mill and Lowe, 1973). However, he was able to show that twisting of the receptor strand and the presence of the accessory strands did not affect the nature of the sensory response. Taylor (1967a) also suggested that dendritic flexion, by causing stretch on one side and compression on the opposite side of the dendrite, is the adequate stimulus in the antennal flagellar chordotonal organ. However, his application of this hypothesis to the joint chordotonal organs necessitates that the dendrites cross an interface between two types of connective tissue, which does not occur (Mill and Lowe, 1971, 1973; Lowe, Mill and Knapp, 1973). An alternative theory which Bush (1965b) proposed involved ephaptic inhibition.

Although dendritic flexion cannot entirely be ruled out as the adequate stimulus for the joint chordotonal organs it seems, at present, the less likely. The principal alternative is dendritic stretch. It has been suggested that the distal ends of ESCs are stretched by strand elongation, while those of RSCs are stretched by strand relaxation (Wiersma and Boettiger, 1959; Wyse and Maynard, 1965; Hartman and Boettiger, 1967; Mill and Lowe, 1971, 1973; Lowe *et al.*, 1973), and it is convenient at this stage to look separately at the differences between ESMCs and RSMCs, between ESPCs and RSPCs and between movement and position sensitive cells.

6.7.1   Movement sensitive cells

In the PD organ of *Carcinus* and *Cancer* the ESMCs insert into the anterior (or occasionally posterior) surface of the receptor strand and are especially numerous at the proximal end of the movement group, whereas the RSMCs insert into the dorsal surface and are more abundant at the distal end of this group (Wiersma and Boettiger, 1959; Hartman and Boettiger, 1967). Hartman and Boettiger also made the significant observation that movement cells paired within the same scolopidium respond to the *same* direction of movement, albeit with different sensitivities (Fig. 6.15). Furthermore they noted that the RSMCs are inserted into a collagen–rich area of the strand, while the ESMCs (and position sensitive cells) insert into a very different material which can be stripped away, leaving the uninjured RSMCs behind. ESMC and RSMC scolopidia have several marked differences (Mill and Lowe, 1973). In ESMC, but not in RSMC, scolopidia the enveloping cells contain scolopale-like material. The ESMC scolopidia have a canal cell surrounding the distal region of

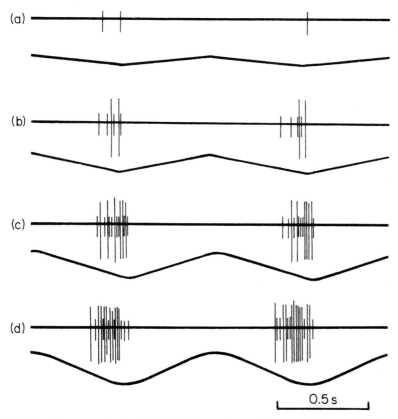

Fig. 6.15   Responses of paired ESMCs to increasing strand elongation (lower trace downwards). Movement is (a) 50 μ s^{-1}; (b) 150 μ s^{-1}; (c) 200 μ s^{-1}; (d) 300 μ s^{-1}. Spikes retouched. (After Hartman and Boettiger, 1967.)

the enveloping cells and most of the outer surface of this cell and of the enveloping cells not surrounded by it, is juxtaposed by strand cells. On the other hand, RSMC scolopidia do not possess a canal cell and the enveloping cells have little contact with the strand cells. Rather they are surrounded by longitudinally oriented collagen fibres which form hemi-desmosomes where they abut against the enveloping cells (Fig. 6.16).

In both ESMC and RSMC scolopidia the dendrites are firmly fixed by desmosomes to the scolopale cell and their terminal regions fit fairly closely into the tube. Mill and Lowe (1971, 1973) proposed that the arrangement of the ESMC scolopidia ensures that they are firmly attached to the cellular matrix of the strand and hence any elongation of the strand will stretch the dendrites. However, in RSMC scolopidia the lack of firm attachment to the cellular matrix will prevent stretch of the dendrites during strand elongation; rather this may serve to store energy in the collagen fibres. When the strand is relaxed this stored energy is released and a force exerted against the enveloping cells via the hemi-desmosomes. Since the enveloping cells taper distally there will be more hemi-desmosomes, and hence a greater force exerted, in the scolopale region and this will force the scolopale region away from the tube region, thereby stretching the dendrites. The RSMC dendrites are generally longer than those of ESMCs and this may be correlated with the greater sensitivity necessitated by the extremely small movements to which the RSMC dendrites would be subjected if the above theory holds (Mill and Lowe, 1973). Differences between the dendrites of a pair could be concerned with differences in sensitivity (Hartman and Boettiger, 1967; Mill and Lowe, 1973).

6.7.2   Position sensitive cells

The situation with regard to the position cells is less clear. They all apparently enter the collagen-rich area of the strand (Hartman and Boettiger, 1967; Boettiger and Hartman, 1968) but at the ultra-structural level the scolopidia appear to be intermediate between ESMC and RSMC scolopidia. Thus the canal cell is absent but there is some scolopale-like material in the enveloping cells and the latter are juxtaposed by a mixture of strand cells and collagen (Lowe *et al.*, 1973). The most distal scolopidia of the group lie in a region of large haemocoelic lacunae (Lowe *et al.*, 1973). Unfortunately there is no

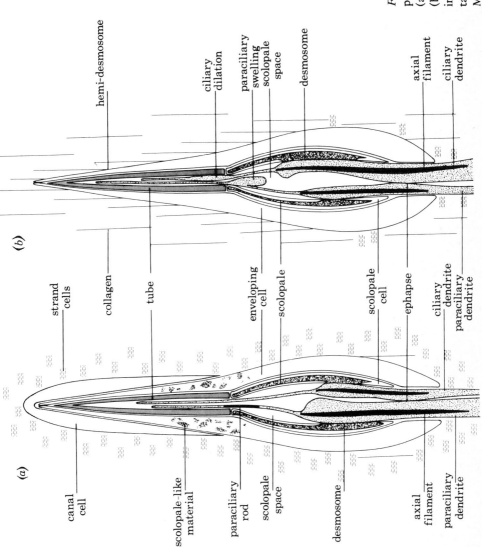

Fig. 6.16 Diagrammatic representation of the structure of (a) an ESMC scolopidium and (b) a RSMC scolopidium. Both in longitudinal section. Horizontal hatching, strand cells. (From Mill and Lowe, 1973).

(a)

canal cell
scolopale-like material
paraciliary rod
scolopale space
desmosome
axial filament
paraciliary dendrite

strand cells
collagen
tube
enveloping cell
scolopale
scolopale cell
ephapse
ciliary dendrite
paraciliary dendrite

(b)

hemi-desmosome
ciliary dilation
paraciliary swelling
scolopale space
desmosome
axial filament
ciliary dendrite

clear physiological or structural evidence as to which cells are ESPCs and which are RSPCs. Perhaps the explanation lies in the dendrites being held at different 'levels' with respect to their maximum sensitivity; ESPCs approaching their maximum sensitivity when the strand is elongated, RSPCs approaching theirs when it is relaxed (Lowe *et al.*, 1973).

6.7.3   Movement and position sensitivity

Mendelson (1963) suggested that the tubes of movement scolopidia may be more loosely associated with the surrounding matrix than those of the position scolopidia; those of intermediate scolopidia having an intermediate relationship (Bush, 1965b). (The Pacinian corpuscle has a phasic-tonic generator potential, the purely phasic output being a function of the structures surrounding the dendritic terminations (Loewenstein and Mendelson, 1965; Loewenstein and Skalek, 1966)). However, there is no indication of any such differences and, if anything, the most distal position cells look to be the least firmly attached. An explanation in purely mechanical terms could involve the relationship between the dendrites and their tube. It could be that the phasic response is achieved by the dendrites slipping rapidly back in their tube, while the tonic ones are more firmly fixed and indeed the tubes of the latter do generally appear to contain some extracellular material (Lowe *et al.*, 1973).

Alternatively the explanation may lie partly or wholly in physiological differences, Nakajima (1964) and Nakajima and Onodera (1969) suggesting that the differences between the phasic and tonic crustacean muscle receptor organs are due to differences in the electrically excitable membrane at their spike initiating loci. Mendelson (1966) raised the possibility that in movement units the generator site may be close to the spike initiating zone, which would thus be affected by the phasic component; whereas in position units it may be further away, when the capacity of the intervening membrane would attenuate this component.

6.7.4   The site of spike initiation

In the sensory cells of the PD organ the site of spike initiation is in the dendrites in the scolopale region; i.e. at a considerable distance distal to the soma (Mendelson, 1963, 1966; Hartman and Boettiger, 1967).

6.8   Resistance reflexes and the role of proprioception

The involvement of the joint chordotonal organs in 'resistance reflexes', i.e. excitation of the antagonist and, where a specific inhibitor is present, inhibition of the synergist, is now well established (Eckert, 1959; Bush, 1962, 1965c; Murayama, 1965; Muramoto, 1965; Evoy and Cohen, 1969).

Each limb muscle is innervated by at least one excitatory axon and the pattern for the distal segments is the same in the Astacura, Palinura, Brachyura and Anomura (Fig. 6.17a). The axons can be divided into 'slow' and 'fast' on the basis of their effect on the muscles (Wiersma, 1941; Wiersma and Ripley, 1952) or into 'tonic', which fire most of the time and are concerned with normal movements and posture, and 'phasic', which fire rhythmically as a result of some specific stimulus and are involved in vigorous activity (Atwood, 1973). The dactylopodite opener and propodite stretcher muscles are unusual in sharing an excitatory axon and require selective inhibition to operate independently, Each muscle also receives either one or two inhibitory motor axons but the innervation pattern is different in each of the above groups (Fig. 6.17b). A few muscles have a 'specific inhibitor' or share an inhibitor with another muscle ('shared inhibitor'). Also in each group there is a 'common inhibitor' which innervates several muscles. This latter is somewhat different in the Astacura in that it only innervates three of the seven most distal limb muscles, compared with five in the Brachyura, six in the Palinura and all seven in the Anomura (Wiersma, 1941; Wiersma and Ripley, 1952).

In the Brachyura Bush (1962), working on *Carcinus*, demonstrated that *passive* closing elicits a reflex response in the opener/stretcher motor axon, while passive opening produces a response in the opener inhibitor, in the 'slow' closer and, with rapid opening, in the 'fast' closer. Similarly, passive bending of the propodite produces a reflex response in the opener/stretcher motor axon, while passive stretch of the propodite elicits a response in the stretcher inhibitor and the 'slow' bender, with faster movements also exciting the 'fast' bender. Bush (1965c) also described resistance reflexes associated with passive movements of the carpopodite, meropodite and coxopodite and he was able to demonstrate that the joint chordotonal organs are responsible for these reflexes; PD eliciting the opening and closing reflexes of the dactylopodite, CP1 and CP2 the stretch and bending reflexes respectively of the propodite (Bush, 1962, 1965c).

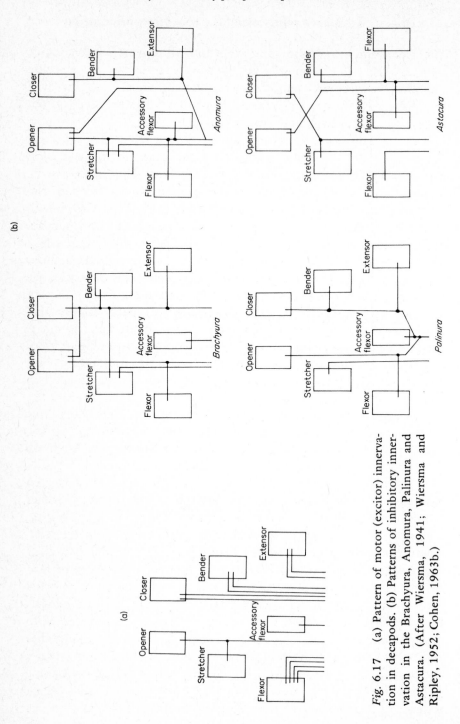

Fig. 6.17   (a) Pattern of motor (excitor) innervation in decapods. (b) Patterns of inhibitory innervation in the Brachyura, Anomura, Palinura and Astacura. (After Wiersma, 1941; Wiersma and Ripley, 1952; Cohen, 1963b.)

Similar results for the distal joints of brachyurans were obtained by Spirito (1970) in the chela of *Uca* and Spirito, Evoy and Barnes (1972) in the walking legs of *Cardisoma*, except that they also recorded reflex responses in the stretcher inhibitor during passive closing of the dactylopodite and in the opener inhibitor during bending of the propodite (Fig. 6.18). Conversely, they did not record any activity in the 'fast' closer or bender, presumably because they did not use high speeds of movement.

In *Carcinus* the reflex response frequency increases with velocity of movement and, in the opener/stretcher motor axon, it also increases with displacement at constant speeds up to $100°$ s^{-1}, (Bush, 1962, 1965c). Furthermore, in *Cardisoma* at least, the reflexes increase in strength when the imposed movement is towards the extreme joint positions. In the chela of *Uca* and the pereiopods of *Cardisoma* there is close coupling between the appropriate excitor and inhibitor during dactylopodite closing and propodite bending, but little or none during the opposite movements (Fig. 6.19). In *Uca*, the two inhibitors each have a different phase relationship with the opener/stretcher motor axon, but in each case the coupling is optimal for a reduction in height of the muscle ejp's and thus of muscle tension. Furthermore, the stretcher muscle functions at a higher frequency range than the opener. These mechanisms are probably both concerned with the functional separation of the opener and stretcher muscles (Spirito, 1970; Spirito *et al.*, 1972).

In anomurans similar reflexes to those in the Brachyura occur at the two most distal joints (Bush, 1963; Field, 1974). However, in the Astacura, in which group an inhibitor is shared by the stretcher and closer muscles, there seems to be greater variability. In the chela of *Astacus*, Eckert (1959) observed a reflex discharge in the opener inhibitor on passive opening and a discharge in the opener/stretcher motor axon on passive closing. This was confirmed in the chela of *Procambarus* by Marayama (1965) who also recorded strong responses in the slow closer and weak responses in the fast closer on passive opening, and a discharge in the closer inhibitor on passive closing. However, he noted that these reflexes were often centrally inhibited and those associated with closing were inconsistent. Furthermore Wilson and Davis (1965) found (also in the chela of *Procambarus*) that the opener excitor has a background activity under normal conditions and that passive opening may produce reflex activity in the opener excitor as well as in the opener inhibitor,

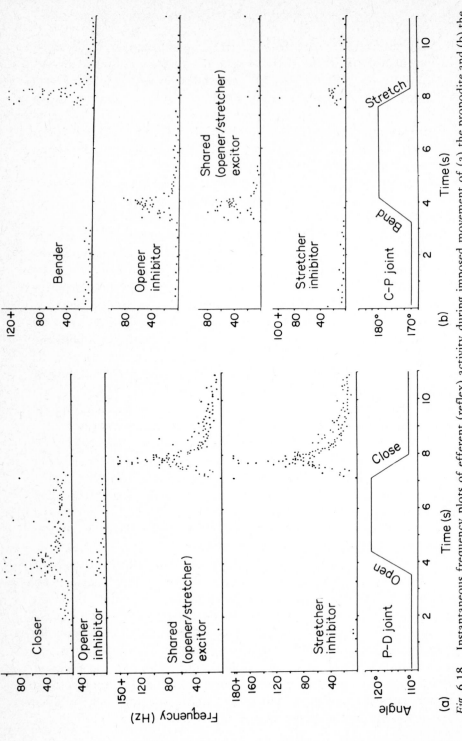

Fig. 6.18 Instantaneous frequency plots of efferent (reflex) activity during imposed movement of (a) the propodite and (b) the dactylopodite of *Cardisoma guanhumi* (Brachyura). (From Spirito *et al.*, 1972.)

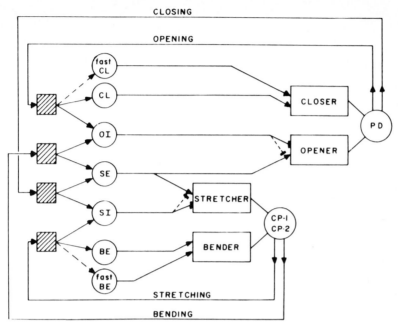

Fig. 6.19 Model of a control system for resistance reflexes in the Brachyura. The shaded squares represent 'control units' in the central nervous system. (→) excitatory connections; (⊣) inhibitory connections. CL—closer; OI—opener inhibitor; SE—shared excitor; SI—stretcher inhibitor and BE—bender motor axons. The fast closer and fast bender axons are probably only activated during rapid movements. (From Spirito *et al.*, 1972.)

although the latter gives the earlier and larger response. In the legs of *Astacus* and *Homarus* Bush (1963) found that the reflexes produced by passive movements of the dactylopodite and propodite are generally weaker than in brachyurans with the opener/stretcher motor axon showing little if any response to closing and the fast axons predominating over the 'slow' ones. Also a strong response occurs in the stretcher closer inhibitor axon to passive stretch, but only a weak one to passive closing. In *Astacus* the reflex response frequency is virtually independent of movement velocity above 45° s⁻¹ (Vedel, Angaut-Petit and Clarac, 1975).

During normal walking, resistance reflexes are apparently absent. In *Cardiosoma* Barnes, Spirito and Evoy (1972) have shown that the locomotory motor output pattern differs from that elicited by passive movements. Thus the coupling during closing and bending alluded to above is not present. However, opening and closing movements imposed during walking do produce resistance reflexes, with a tendency for them to be stronger when the imposed movement opposes the direction produced by the locomotory pattern. Hence Barnes *et al.* suggest that the resistance reflexes compensate for any changes in loading which may occur during normal locomotion on a rough substrate or on an incline. This is similar to the idea of an 'efference copy' system (von Holst, 1954).

The mero-carpopodite joint is particularly important for support and there are two receptor systems concerned with movement at this joint, the MC organs and the myochordotonal organs. The myochordotonal organs are primarily stimulated by AFM activity, augmented by changes in passive tension in this muscle brought about by movements of the carpopodite. Thus they are under efferent control with the AFM serving a sensory role. On the other hand the MC organs are more sensitive to joint movement than tension changes in the AFM (Cohen, 1963a; Evoy and Cohen, 1969; Clarac and Vedel, 1971). These two systems have been studied separately by cutting the AFM tendon between the proximal and distal heads. The MC organs alone can then be stimulated by passive joint movements, the myochordotonal organs alone by pulling and releasing the part of the tendon attached to the proximal head.

In *Cancer*, if the carpopodite is kept stationary in its normal rest position there is backround activity in at least one axon to the extensor, flexor and AF muscles; the tonic level to the flexor muscle and the AFM increasing with maintained extension. Stimulation of the MC organs produces resistance reflexes which are strongest at the extremes of the movement arc. Thus carpopodite flexion produces a burst in the extensor excitors and depression of activity in the flexor and AF excitors; while carpopodite extension inhibits activity in the extensor muscle excitors and elicits a burst in apparently all of the flexor excitors and in the AF excitor. However, no peripheral inhibition has been observed (Fig. 6.20).

As far as the flexor and extensor muscles are concerned the effects of myochordotonal organ stimulation are similar to, but generally weaker than, those of the MC organs. However, the reflex response to the AFM depends on the length of the muscle with respect to that in the normal resting position. Elongation of the distal head (extension) from the rest position increases the excitatory activity, but so also does relaxation (flexion) from the rest position. Conversely, return to the rest position from either the elongated

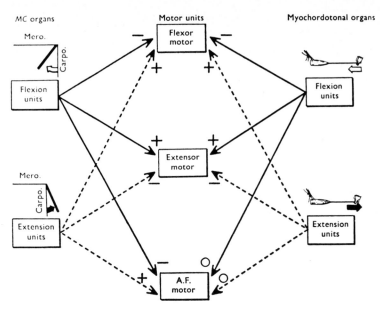

Fig. 6.20   Summary diagram of the probable proprioceptor connections with motor units in the thoracic ganglion of *Cancer*, which control the muscles of the mero-carpopodite joint. (———) extension; (— — — —) flexion; (+) excitatory connection; (−) inhibitory connection; O—excitatory if movement is away from 'rest', inhibitory or no effect if movement is towards 'rest'. (From Evoy and Cohen, 1969).

(extended) or relaxed (flexed) positions reduces or suppresses the excitatory activity (Fig. 6.20) (Evoy and Cohen, 1969, 1971; Cohen and Evoy, 1971).

Similar responses are found in the motor axons to the corresponding muscles of the other legs but the reflexes are weaker, especially contralaterally (Evoy and Cohen, 1969). Evoy and Cohen suggest that the primary resting position of the joint can be reset by altering the tonic output to the AFM. In the intact animal, while stimulation of the AFM excitor causes contraction of both heads, the dominant effect of the distal head actually causes *elongation* of the proximal head. Thus stimulation of the AFM excitor and passive extension of the carpopodite will both produce a similar effect on the myochordotonal organs (Cohen, 1965). When both receptor systems are intact the resistance reflex response tends to dominate all three muscles, except that flexion from the rest position produces a variable effect on the AFM (Evoy and Cohen, 1969).

The effect of the MC resistance reflexes is to drive the carpopodite towards its rest position, the final adjustments being carried out by the myochordotonal system, especially those reflexes concerning the AFM. Indeed, after a period of walking the carpopodite extensor and flexor muscles show oscillatory activity which declines to a low level background discharge, producing a gradual adjustment of the carpopodite into its rest position (Evoy and Cohen, 1969, 1971; Cohen and Evoy, 1971).

Cohen (1965) demonstrated that removal of MC1 usually produces extension of the carpopodite, whereas MC2 removal either has no effect or causes slight flexion. Removal of the myochordotonal organs from one leg has little or no effect (Barth, 1934; Cohen, 1965), but when removed from several legs the animal becomes generally less responsive to external stimuli (Cohen, 1965). However, there is no basic effect on walking and no readjustment of posture (Cohen and Evoy, 1971; Evoy and Cohen, 1971). Extreme flexion of the carpopodite occurs if the proximal head of the AFM is kept elongated in several legs. Conversely, carpopodite extension is elicited if the proximal heads are kept relaxed (Cohen and Evoy, 1971; Evoy and Cohen, 1971).

Fixing the mero-carpopodite joint in the extended position (in *Carcinus*) causes an exaggerated elevation of the basipodite if the leg is leading, but reduces this movement if the leg is trailing. These effects are reversed if the MC organs are removed and the myochordotonal organs left in their relaxed positions. Fixing the joint in the flexed position reduces elevation of the basipodite if the leg is leading, but has no apparent effect on it if the leg is trailing (Clarac and Coulmance, 1971).

These experiments involving proprioceptor ablation and joint fixation still present a somewhat confusing picture, probably as a result of the almost certain dual function of the MC (and possibly myochordotonal) organs — in locomotion and posture.

At the mero-carpopodite joint in *Astacus* the AFM excitor and one of the flexor excitors continue to fire when the carpopodite is stationary. Passive extension increases the level of activity, passive flexion inhibits it. Apart from occasional bursts the extensor excitors remain silent in the absence of movement and only one shows a reflex response to flexion. As in *Cancer* the MC organs and the myochordotonal organ (only MCO1 is present) both produce resistance reflexes (Fig. 6.21), those caused by the latter being

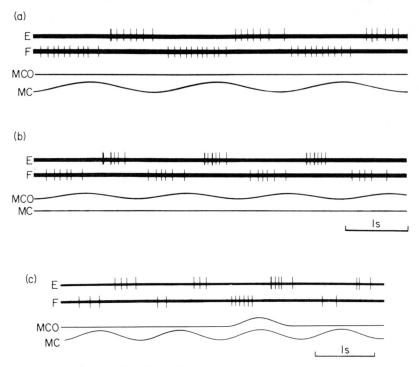

Fig. 6.21   Reflex modulation of excitor motoneuron activity of extensor (E) and flexor (F) muscles by (a) stimulation of the MC organs alone; (b) stimulation of the myochordotonal organ (MCO1) alone and (c) the effect of a single elongation and relaxation of the proximal head of the accessory flexor muscle (myochordotonal organ stimulation) superimposed on regular stimulation of the MC organs. (From Vedel *et al.*, 1975.)

somewhat weaker. However, unlike in *Cancer*, the reflexes involving the AFM are similar to those involving the flexor muscle. Elongation of the proximal head always increases the response in the accessory flexor excitor and vice versa (Vedel *et al.*, 1975). This could be a result of the absence of MCO2 in *Astacus*, since MCO1 contains mainly units responding to elongation of the proximal head of the AFM; while MCO2 contains mainly units responsive to relaxation of the proximal head; thereby depriving the system of flexion sensitive units (Section 6.6).

Light stimulation of the cephalic or abdominal appendages ('alerting stimuli') superimposed on rhythmic, passive movements of the carpopodite, reduces carpopodite extension by increasing the excitatory response to the flexor muscle (increases activity in the tonic excitor and initiates activity in two of the other excitors) and the AFM caused by carpopodite extension (Fig. 6.22a, b); and produces a response in the common inhibitor. Stronger stimulation increases these responses and, in addition, blocks the excitatory response to the extensor muscle elicited by carpopodite flexion (Fig. 6.22b); thus producing a general flexion. Stimulation of a contra-lateral leg has a similar effect on the flexor excitors and on the inhibitory discharge, but enhances excitation of the extensor muscle (Fig. 6.22c) (Angaut-Petit, Clarac and Vedel, 1974; Vedel *et al.*, 1975). Prosser (1935) noted that flexion and extension produce reflexes in homolateral limbs. Posture is also controlled by the cuticular stress detector, CSD2 (Chapter 7). The tonic flexor unit and the AFM excitor are closely linked physiologically, as might be expected with two muscles having a similar qualitative effect, although their degree of coupling is variable (Vedel *et al.*, 1975).

In *Cancer* and *Astacus* only central inhibition is involved in the resistance reflexes elicited by the MC and myochordotonal organs, even though a specific flexor inhibitor is present in the Astacura. However, in *Astacus* the common inhibitor is activated as a response to 'alerting stimuli'.

Evoy and Cohen (1971) suggest that there are two different motor scores available to the efferent pathways of the legs and that they are in a state of balance. Thus, during locomotion, the antagonistic muscles are coordinated by strong central reciprocal inhibition between the motoneuron pools but, as the excitation of the walking oscillator decreases, the effectiveness of the proprioceptive input increases.

Little is known of the basal joints. Evoy and Fourtner (1963) found that, if a leg is held maximally elevated, but with the distal joints free to move, the normal motor rhythm to the leg ceases, but is resumed if the animal walks against a load.

Burke (1954) suggested that the PD receptor of *Carcinus*, apart from signalling the rate and extent of dactylopodite movement, also served as a vibration receptor and the vibration response which he observed is undoubtedly a response by the most sensitive pure movement units. Muramoto (1965) found that the overall discharge frequency of the PD organ in the claw of *Procambarus* increases with the velocity of dactylopdite movement, but she only used stimu-lation frequencies up to 80 Hz. Horch (1971) has shown that the leg myochordotonal organs (not specifically MCO1 as indicated by

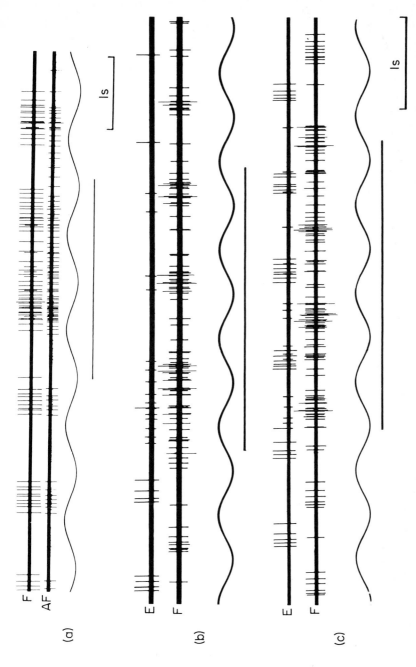

Fig. 6.22 Reflex modulation of (E) extensor, (F) flexor and (AF) accessory flexor muscles by stimulation of (a) head, (b) abdomen and (c) a contralateral leg. Horizontal line indicates period of stimulation. (From Vedel *et al.*, 1975.)

Horch) of *Ocypode* are sensitive to high frequency vibrations and air-borne sound. This is of particular interest since *Ocypode* can produce sounds by stridulation (Horch and Salmon, 1969). Hwang (1961) found that MCO1 in *Cancer* only responded to vibrations up to about 150 Hz.

It has been established in *Procambarus* that the rhythmic central command to the swimmerets can continue in the absence of sensory feedback (lkeda and Wiersma, 1964). Nevertheless, proprioceptive feedback does occur. In *Homarus*, if a swimmeret is passively retracted and held in this position, a resistance reflex is initiated and maintained, presumably by the coxal proprioceptors (Section 6.2.1). Thus the excitors of the main return-stroke (protractor) muscle show increased activity; the inhibitor decreased activity. This effect is opposed during movement by a negative feedback reflex initiated by sensory setae on the rami. Simultaneously, the proprioceptors elicit a positive feedback reflex to the main powerstroke (retractor) muscle (increased activity of the excitors and decreased activity of the inhibitor) but this is reinforced by stimulation of the sensory setae. The overall effect of these intrasegmental reflexes could be to amplify the central motor command by strengthening the power stroke and increasing the linkage between power and return strokes (Davis, 1969a, 1974). However, these responses are to passive movements of the swimmerets, so it may be that the reflexes only operate to correct errors in loading on individual swimmerets, as with limb movements during normal locomotion (p. 285). The reflex discharges normally occur whenever the swimmeret is retracted beyond its mid position, the intensity at any given position being greater when that position is approached by retraction. The responses are maximal when the swimmeret is in its fully retracted position (Davis, 1969b).

Although there is a posterior—anterior decrease in the excitability of the central oscillator (Ikeda and Wiersma, 1964) the power produced by individual swimmerets is similar. This smoothing out could be effected by the intersegmental reflexes, which only operate in a posterior—anterior direction (Davis, 1968b, 1974).

References

Alexander, C. G. (1969) Structure and properties of mechanoreceptors in the pereiopods of *Ligia oceanica* Linn. (Crustacea, Isopoda). *Comparative Biochemistry and Physiology*, **29**, 1197—1205.

Alexandrowicz, J. S. (1958) Further observations on proprioceptors in Crustacea and a hypothesis about their function. *Journal of the Marine Biological Association, U.K.,* 37, 379–396.

Alexandrowicz, J. S. (1967) Receptor organs in the coxal region of *Palinurus vulgaris. Journal of the Marine Biological Association, U.K.,* 47, 415–432.

Alexandrowicz, J. S. (1972) The comparative anatomy of leg proprioceptors in some decapod Crustacea. *Journal of the Marine Biological Association, U.K.,* 52, 605–634.

Alexandrowicz, J. S. and Whitear, M. (1957) Receptor elements in the coxal region of decapod Crustacea. *Journal of the Marine Biological Association, U.K.,* 36, 603–628.

Angaut-Petit, D., Clarac, F. and Vedel, J. P. (1974) Excitatory and inhibitory innervation of a crustacean muscle associated with a sensory organ. *Brain Research,* 70, 148–152.

Barnes, W. J. P., Spirito, C. P. and Evoy, W. H. (1972) Nervous control of walking in the crab, *Cardisoma guanhumi.* II. Role of resistance reflexes in walking. *Zeitschrift für vergleichende Physiologie,* 76, 16–31.

Barth, G. (1934) Untersuchungen über Myochordotonalorgane bei dekapoden Crustaceen. *Zeitschrift für wissenschaftliche Zoologie,* 145, 576–624.

Boettiger, E. G. and Hartman, H. B. (1968) Excitation of the receptor cells of the crustacean PD organ. *Symposium of Neurobiology of Invertebrates,* pp. 381–390. Hungarian Academy of Sciences.

Burke, W. (1954) An organ for proprioception and vibration sense in *Carcinus maenas. Journal of Experimental Biology,* 31, 127–137.

Bush, B. M. H. (1962) Proprioceptive reflexes in the legs of *Carcinus maenas* (L.) *Journal of Experimental Biology,* 39, 89–105.

Bush, B. M. H. (1963) A comparative study of certain limb reflexes in decapod crustaceans. *Comparative Biochemistry and Physiology,* 10, 273–290.

Bush, B. M. H. (1965a) Proprioception by chordotonal organs in the mero-carpopodite and carpo-propodite joints of *Carcinus maenas* legs. *Comparative Biochemistry and Physiology,* 14, 185–199.

Bush, B. M. H. (1965b) Proprioception by the coxo-basal chordotonal organ, CB, in legs of the crab, *Carcinus maenas. Journal of Experimental Biology,* 42, 285–297.

Bush, B. M. H. (1965c) Leg reflexes from chordotonal organs in the crab, *Carcinus maenas. Comparative Biochemistry and Physiology,* 15, 567–587.

Clarac, F. (1968a) Proprioceptor anatomy of the ischio-meropodite region in legs of the crab, *Carcinus mediterraneus. Zeitschrift für vergleichende Physiologie,* 61, 203–223.

Clarac, F. (1968b) Proprioception by the ischio-meropodite region in legs of the crab *Carcinus mediterraneus. Zeitschrift für vergleichende Physiologie,* 61, 224–245.

Clarac, F. (1970) Fonctions proprioceptives au niveau de la région basi-ischio-méropodite chez *Astacus leptodactylus. Zeitschrift für vergleichende Physiologie,* 68, 1–24.

Clarac, F. and Coulmance, M. (1971) La marche latérale du crabe (*Carcinus*). Coordination des mouvements articulaires et régulation proprioceptive. *Zeitschrift für vergleichende Physiologie*, 73, 408–438.

Clarac, F. and Masson, C. (1969) Anatomie comparée des propriocepteurs de la région basi-ischio-méropodite chez certains Crustacés décapodes. *Zeitschrift für vergleichende Physiologie*, 65, 242–273.

Clarac, F. and Vedel, J. P. (1971) Étude des relations fonctionelles entre le muscle fléchisseur accessoire et les organes sensoriels chordotonaux et myochordotonaux des appendices locomoteurs de la langouste *Palinurus vulgaris*. *Zeitschrift für vergleichende Physiologie*, 72, 386–410.

Cohen, M. J. (1963a) The crustacean myochordotonal organ as a proprioceptive system. *Comparative Biochemistry and Physiology*, 8, 223–243.

Cohen, M. J. (1963b) Muscle fibres and efferent nerves in a crustacean receptor muscle. *Quarterly Journal of Microscopical Science*, 104, 551–559.

Cohen, M. J. (1965) The dual role of sensory systems: detection and setting central excitability. *Cold Spring Harbour Symposia of Quantitative Biology*, 30, 587–599.

Davis, W. J. (1968a) The neuromuscular basis of swimmeret beating in the lobster. *Journal of Experimental Zoology*, 168, 363–378.

Davis, W. J. (1968b) Quantitative analysis of swimmeret beating in the lobster. *Journal of Experimental Biology*, 48, 643–662.

Davis, W. J. (1969a) Reflex organization in the swimmeret system of the lobster. I. Intrasegmental reflexes. *Journal of Experimental Biology*, 51, 547–563.

Davis, W. J. (1968b) Reflex organization in the swimmeret system of the lobster. II. Reflex dynamics. *Journal of Experimental Biology*, 51, 565–573.

Davis, W. J. (1974) Neuronal organization and ontogeny in the lobster swimmeret system. In: *Control of Posture and Locomotion*. Stein, R. B., Pearson, K. B., Smith, R. S. and Redford, J. B. (eds) pp. 437–455. Plenum Press, New York.

Eckert, B. (1959) Über das Zussamenwirken des erregenden und des hemmenden Neurons des M. abductor der Krebsschere beim Ablauf von Reflexen des myotätishen Typus. *Zeitschrift für vergleichende Physiologie*, 41, 500–526.

Evoy, W. H. and Cohen, M. J. (1969) Sensory and motor interaction in the locomotor reflexes of crabs. *Journal of Experimental Biology*, 51, 151–169.

Evoy, W. H. and Cohen, M. J. (1971) Central and peripheral control of arthropod movements. *Advances in Comparative Physiology and Biochemistry*, 4, 225–266.

Field, L. H. (1974) Sensory and reflex physiology underlying cheliped flexion behaviour in hermit crabs. *Journal of Comparative Physiology*, 92, 397–414.

Gordon, S. E. (1973) A chordotonal organ in the leg of *Cancer pagurus* Linn.: some ultrastructural observations. *Ph.D. Thesis*. University of Leeds, U.K.

Hartman, H. B. and Austin, W. D. (1972) Proprioceptor organs in the antennae of Decapoda Crustacea. I. Physiology of a chordotonal organ spanning two joints in the spiny lobster *Panulirus interruptus* (Randall) *Journal of Comparative Physiology,* **81**, 187–202

Hartman, H. B. and Boettiger, E. G. (1967) The functional organization of the propus-dactylus organ in *Cancer irroratus* Say. *Comparative Biochemistry and Physiology,* **22**, 651–663.

von Holst, E. (1954) Relations between the central nervous system and the peripheral organs. *British Journal of Animal Behaviour,* **2**, 89–94.

Hwang, J. C. C. (1961) The function of a second sensory cell group in the accessory-flexor proprioceptive system of crab limbs. *American Zoologist,* **1**, 453.

Horch, K. (1971) An organ for hearing and vibration sense in the ghost crab *Ocypode. Zeitschrift für vergleichende Physiologie,* **73**, 1–21.

Horch, K. W. and Salmon, M. (1969) Production, perception and reception of acoustic stimuli by semiterrestrial crabs (Genus *Ocypode* and *Uca,* Familiy Ocypodidae). *Forma et Functio,* **1**, 1–25.

Ikeda, K. and Wiersma, C. A. G. (1964) Autogenic rhythmicity in the abdominal ganglia of the crayfish: The control of swimmeret movements. *Comparative Biochemistry and Physiology,* **12**, 107–115.

Laverack, M. S. (1964) The antennular sense organs of *Panulirus argus. Comparative Biochemistry and Physiology,* **13**, 301–321.

Loewenstein, W. R. and Mendelson, M. (1965) Components of receptor adaptation in a Paccinian corpuscle. *Journal of Physiology, London,* **177**, 377–392.

Loewenstein, W. R. and Skalek, R. (1966) Mechanical transmission in a Paccinian corpuscle. An analysis and a theory. *Journal of Physiology, London,* **182**, 346–378.

Lowe, D. A. and Mill, P. J. (1972) The relationship between the PD proprioceptor, the propodite-dactylopodite joint and the dactylopodite flexor muscle in the walking legs of *Cancer pagurus. Marine Behaviour and Physiology,* **1**, 157–170.

Lowe, D. A., Mill, P. J. and Knapp, M. F. (1973) The fine structure of the PD proprioceptor of *Cancer pagurus.* II. The position sensitive cells. *Proceedings of the Royal Society, Series* B, **184**, 199–205.

Mendelson, M. (1963) Some factors in the activation of crab movement receptors. *Journal of Experimental Biology,* **40**, 157–169.

Mendelson, M. (1966) The site of impulse initiation in bipolar receptor neurons of *Callinectes sapidus* L. *Journal of Experimental Biology,* **45**, 411–420.

Mill, P. J. and Lowe, D. A. (1971) Transduction processes of movement and position sensitive cells in a crustacean limb proprioceptor. *Nature,* **229**, 206–208.

Mill, P. J. and Lowe, D. A. (1972) An analysis of the types of sensory unit present in the PD proprioceptor of decapod crustaceans. *Journal of Experimental Biology*, 56 509–525.

Mill, P. J. and Lowe, D. A. (1973) The fine structure of the PD proprioceptor of *Cancer pagurus*. I. The receptor strand and the movement sensitive cells. *Proceedings of the Royal Society, Series B*, 184, 179–197.

Muramoto, A. (1965) Proprioceptive reflex of the PD organ of *Procambarus clarkii* by passive movement and vibration stimulus. *Journal of the Faculty of Science, Hokkaido University. Series 6*, 15, 522–534.

Murayama, K. (1965) Proprioceptive reflex responses of the efferent axons to passive and active movements in the cheliped of the crayfish. *Journal of the Faculty of Science, Hokkaido University. Series 6*, 15, 510–521.

Nakajima, S. (1964) Adaptations in stretch receptor neurons of crayfish. *Science*, 146, 1168–1170.

Nakajima, S. and Onodera, K. (1969) Adaptations of the generator potential in the crayfish stretch receptors under constant length and constant tension. *Journal of Physiology, London*, 200, 187–204.

Prosser, C. L. (1935).

Sandeman, D. C. (1963) Proprioceptor organs in the antennules of *Squilla mantis*. *Nature*, 201, 402–403.

Sigvardt, K. A. (1974) Sensory-motor interactions in antennal reflexes of the american lobster. *Ph.D. Thesis*, University of Iowa, U.S.A.

Spirito, C. P. (1970) Reflex control of the opener and stretcher muscles in the cheliped of the fiddler crab, *Uca pugnax*. *Zeitschrift für vergleichende Physiologie*, 68, 211–228.

Spirito, C. P., Evoy, W. H. and Barnes, W. J. P. (1972) Nervous control of walking in the crab, *Cardisoma guanhumi*. I. Characteristics of resistance reflexes. *Zeitschrift für vergleichende Physiologie*, 76, 1–15.

Taylor, R. C. (1967a) The anatomy and adequate stimulation of a chordotonal organ in the antennae of a hermit crab. *Comparative Biochemistry and Physiology*, 20, 709–717.

Taylor, R. C. (1967b) Functional properties of the chordotonal organ in the antennal flagellum of a hermit crab. *Comparative Biochemistry and Physiology*, 20, 719–729.

Vedel, J. P., Angaut-Petit, D. and Clarac, F. (1975) Reflex modulation of motoneurone activity in the leg of the crayfish *Astacus leptodactylus*. *Journal of Experimental Biology*. 63, 551–567.

Wales, W., Clarac, F., Dando, M. R. and Laverack, M. S. (1970) Innervation of the receptors present at the various joints of the pereiopods and third maxilliped of *Homarus gammarus* (L.) and other macruran decapods (Crustacea). *Zeitschrift für vergleichende Physiologie*, 68, 345–384.

Wetzel, A. (1934) Chordotonalorgane bei Krebstienen (*Caprella dentata*). *Zoologischer Anzeiger*, **105**, 125–132.

Whitear, M. (1960) Chordotonal organs in Crustacea. *Nature*, **187**, 522–523.

Whitear, M. (1962) The fine structure of crustacean proprioceptors. I. The chordotonal organs in the legs of the shore crab, Carcinus maenas. *Philosophical Transactions of the Royal Society, London*, B **245**, 291–325.

Wiersma, C. A. G. (1941) The inhibitory nerve supply of the leg muscles of different decapod crustaceans. *Journal of Comparative Neurology*, **74**, 63–79.

Wiersma, C. A. G. (1959) Movement receptors in decapod Crustacea. *Journal of the Marine Biological Association, U.K.*, **38**, 143–152.

Wiersma, C. A. G. and Boettiger, E. (1959) Unidirectional movement fibres from a proprioceptive organ of the crab Carcinus maenas. *Journal of Experimental Biology*, **36**, 102–112.

Wiersma, C. A. G., van der Mark, F. and Fiore, L. (1970) On the firing patterns of the 'movement' receptors of the elastic organs of the crab, Carcinus. *Comparative Biochemistry and Physiology*, **34**, 833–840.

Wiersma, C. A. G. and Ripley, S. H. (1952) Innervation patterns of crustacean limbs. *Physiologia Comparata et Oecologia*, **2**, 391–405.

Wiersma, C. A. G. and Ripley, S. H. (1954) Further functional differences between fast and slow contraction in certain crustacean muscles. *Physiologia Comparata et Oecologia*, **3**, 327–336.

Wilson, D. M. and Davis, W. J. (1965) Nerve impulse patterns and reflex control in the crayfish claw motor system. *Journal of Experimental Biology*, **43**, 193–210.

Wood, J. (1974) Activity recorded from the propodite dactylopodite organ of Pachygrapsus crassipes at rest and at constant speeds of movement. *Comparative Biochemistry and Physiology*, **50A**.

Wyse, G. A. and Maynard, D. M. (1965) Joint receptors in the antennule of Panulirus argus Latreille. *Journal of Experimental Biology*, **42**, 521–535.

7 Crustacean cuticular stress detectors

F. CLARAC

7.1 Introduction

The crustacean Cuticular stress detectors (CSDs) are two sensory structures located in the basi-ischiopodite region of the walking legs, and were first described in detail by Wales, Clarac and Laverack (1971). One of these organs referred to as CSD2 by Wales *et al.* (1971) was first seen in the proximal ischiopodite of the walking legs of *Astacus leptodactylus*. (Clarac and Masson, 1969). The other (CSD1 of Wales *et al.*) was initially described by Wales, Clarac, Dando and Laverack (1970) in the basipodite of the third maxilliped and pereiopods of *Homarus gammarus*. They suggest that, considering its position, this receptor could be used to control autotomy.

The most striking feature of these two organs is their close association with an external area of soft cuticle. Even though their discovery is very recent, the main anatomical characteristics of the CSDs are known, as are their sensory responses to different stimuli. However, we have yet to determine the CSD function in behaviour where the walking legs are involved. The close linkage of these organs with the breakage plane implicates them in the process of autotomy and suggests the possibility of its control by CSD1 in particular.

7.2 Anatomical aspects of the CSDs

It is necessary to describe the region where the CSDs are situated before considering their functional organization.

7.2.1 External description of the basi-ischiopodite region

In all the pereiopods of the crab, the basipodite and the ischiopodite have fused to form the basi-ischiopodite. The preformed breakage plane corresponds to a complete basi-ischiopodite division; in addition there are other externally visible lines. From a point between the insertion of the levator muscle tendons a 'furrow' runs in the cuticle of the basipodite. It appears to be the same structure as that described by Paul (1915) as a 'slanting joint' (Fig. 7.1).

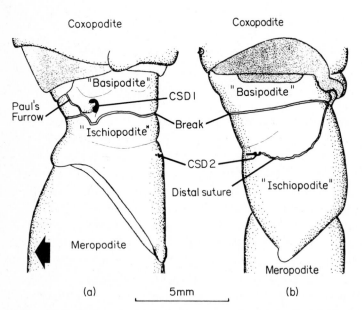

Fig. 7.1 External anatomy of the basi-ischiopodite region of the 2nd right pereïopod of *Carcinus maenas*. Although the basipodite and ischiopodite have fused to become a single entity, the portions of the basi-ischiopodite which approximate to the basipodite and ischiopodite have been indicated separately. (a) the anterior surface; (b) the ventral surface of the limb. The arrow indicates the dorsal surface. Break—preformed breakage plane; CSD1—soft cuticle associated with CSD1; CSD2—thin cuticle associated with CSD2. The articular membrane of the limb joints is indicated by fine hatching (From Wales, Clarac and Laverack, 1971.)

The position of CSD1 is located externally by a discrete area of easily deformable cuticle, which is connected to the fracture plane by a line of less heavily calcified cuticle. In the proximal, ventral ischiopodite region there is a 'suture' which runs from the dorsal basipodite onto a discrete area of soft cuticle, where CSD2 is inserted.

In the Astacidea, a greater variety of form is found. Taking *Homarus* for example, the walking legs have, except for the chelae, a movable basi-ischiopodite joint and an incomplete breakage plane. This is associated with a poor ability to autotomize. In general the basi-ischiopodite joint of the astacideans can be compared with the ischio-meropodite joint of brachyurans, both in general structure and in the degree of movement which they permit (Fig. 7.2). The walking leg of *Homarus* is not as heavily calcified as that of *Carcinus* and some of the cuticular features are less distinct. The soft cuticle associated with CSD1 is larger in area but is still relatively disposed to the levator muscle tendons. There is also a structure analogous to Paul's furrow, but it terminates at the soft cuticle of CSD1 rather than the breakage plane. It appears that the diameter of the soft cuticle associated with CSD1 is largest in the Scyllaridae, Galatheidae and Paguridae and smallest in the more heavily calcified Brachyura. Similarly the thin cuticle associated with CSD2 decreases in size from the Nephropsidae, Scyllaridae and Anomura to the Brachyura, whereas the development of the distal suture is apparently dependant on the degree of calcification and is equally well developed in *Palinurus, Galathea* and *Carcinus*. The importance of the area of soft cuticle associated with CSD1 and CSD2 seems not to be linked with the ability to autotomize a limb.

7.2.2 The breakage plane and the coxo-basipodite musculature

The musculature of the C-B joint of *Carcinus* walking leg consists of two antagonistic muscle groups, the levators and the depressors. There are two levators muscles; the anterior levator muscle (AL) is considerably larger than the other and traverses both the thorax-coxopodite and coxo-basipodite joints. Its tendon consists of a block of cuticle fused to the basipodite at the preformed fracture plane. The posterior levator muscle (PL) is disposed entirely in the coxopodite. Its tendon is peculiar and was carefully studied by McVean (1973). The main blade for muscle attachment is approximately perpendicular to the muscle fibres. It tends to rotate but is prevented from so doing by its close apposition to the tendon of the anterior levator muscle.

There are three depressor muscles; two small ones originating in the coxopodite and a larger, main one, the origin of which is almost entirely located in the thorax. The position of the main tendon on the basipodite is midway between the condyles.

The musculature of the C-B joint in the other decapod groups is similar to that in the Bachyura with regard to the different tendons, but the anterior levator muscle does not have a block of cuticle interposed between the blade of the tendon and its attachment to the basipodite in the Astacidea at least.

7.2.3   Receptor anatomy

(See Wales *et al.*, 1971). CSD1 and CSD2 have certain features in common concerning their structure and their position. They are orientated perpendicularly to the chordotonal organs of the limb joint. They do not cross a limb joint nor are they attached to a muscle or a tendon, rather they are associated with a region of soft external cuticle. Thus they are clearly not part of the series of joint chordotonal organs of the limbs (Chapter 6).

In *Carcinus*, CSD1 lies dorsally in the basipodite proximal and parallel to the preformed breakage plane. Its strand inserts anteriorly to the soft cuticle and is posteriorly attached to a flexible peg. It is innervated by more than 40 bipolar neurons (Fig. 7.3). The majority of the sensory cells (up to 60 μm diameter) lies in the nerve at its junction with the strand; some smaller cells are distributed along the strand. The distal end of the CB chordotonal organ lies close to the CSD1 strand. CSD1 in other species exhibits some minor anatomical differences which are described by Wales *et al.* (1971).

In *Carcinus*, CSD2 is deeply embedded in dense connective tissue. It is a very small structure, with a strand length of about 150 μm for an animal measuring 50 mm across the carapace. Because of this, CSD2 was first studied in *Astacus leptodactylus*, where it is located ventrally in the ischiopodite quite close to the basi-ischiopodite joint. CSD2 is composed of two strands (Fig. 7.4). The main one a broad dorso-ventrally flattened sheet originates on the ventro-posterior wall of the ischiopodite. From this posterior attachment, which is on both soft and calcified cuticle, the strand runs ventro-anteriorly. It

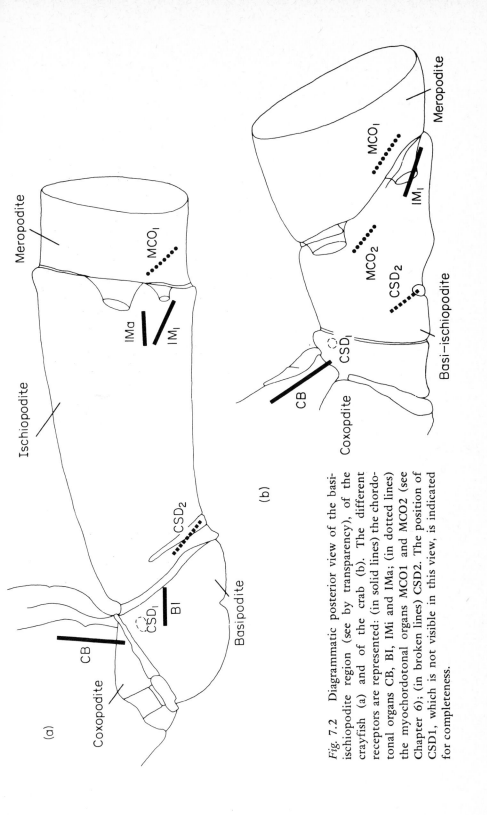

Fig. 7.2 Diagrammatic posterior view of the basi-ischiopodite region (see by transparency), of the crayfish (a) and of the crab (b). The different receptors are represented: (in solid lines) the chordotonal organs CB, BI, IMi and IMa; (in dotted lines) the myochordotonal organs MCO1 and MCO2 (see Chapter 6); (in broken lines) CSD2. The position of CSD1, which is not visible in this view, is indicated for completeness.

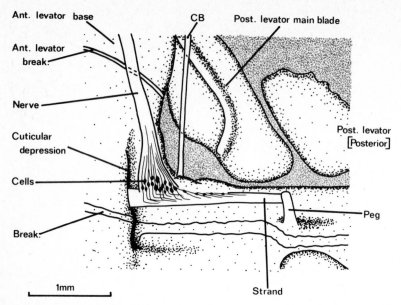

Fig. 7.3   The general disposition of CSD1 in the right 2nd pereïopod of *Carcinus maenas* as it appears when all other soft tissue is removed from the limb. The top of the figure corresponds to proximal, the left side to anterior, in the walking leg. The receptor strand, which lies close to the breakage plane, (Break) is attached posteriorly to a peg and anteriorly to an area of soft cuticle situated in a cuticular depression. The chordotonal organ of the coxo-basipodite joint (CB) is attached to a small protrusion close to CSD1. The block of cuticle at the base of the anterior levator muscle tendon (Ant. levator base) is connected to the 'basipodite' at a preformed breakage plane (Ant. levator break.). The tendon of the posterior levator muscle is normally in two parts; the main blade, which is orientated perpendicular to the plane of the figure, and a smaller posterior portion (Post. levator [Posterior]) (From Wales, Clarac and Laverack, 1971).

appears to converge, forming a spiral half turn. Its anterior insertion on the ventral area of soft cuticle is almost separated from the more posterior portion by a peninsula of more heavily calcified cuticle. The majority of the bipolar cells (up to 60) lie close to the posterior attachment of the strand. The accessory strand is situated distally. Its bipolar sensory cells are distributed over a great length of it and their dendritic processes run towards the anterior insertion, which is in the opposite direction to the CSD1 dendritic arrangement. Electron microscope studies have revealed (i) that the support is entirely cellular and (ii) that the dendritic endings are embedded in

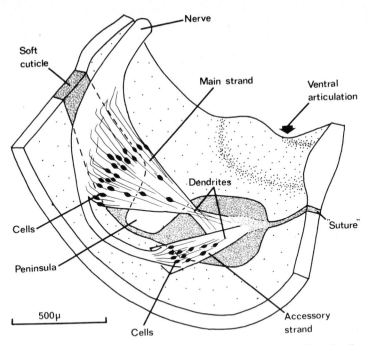

Fig. 7.4   A stereodiagram of CSD2 in the 2nd right pereïopod of *Astacus leptodactylus* as it would appear with all other soft tissue removed. CSD2 is located ventrally in the proximal ischiopodite, close to the ventral articulation of the basi-ischiopodite joint. The receptor is situated on soft cuticle which is narrowed anteriorly to resemble a 'suture'. The area of soft cuticle is almost divided into two parts by a peninsula of more heavily calcified cuticle. The top of the illustration corresponds to proximal; the left side to posterior in the limb (From Wales, Clarac and Laverack, 1971.)

scolopidial structures (Moulins and Clarac, 1972). The receptor nerve, which is short, passes dorsally to join the main leg nerve just distal to the breakage plane.

7.3   Responses of the CSD receptors to different modes of stimulation

If a *Carcinus* walking leg is amputated at the ischio-meropodite joint to remove all of the distal proprioceptive organs, it is then easy to detect in the coxopodite the activity of the CSD receptors by recording from small bundles of the main leg nerve. When the receptor is not mechanically stimulated, a number of units are always active. Pressure applied to the basi-ischiopodite produces a dramatic response in CSD1 and CSD2. and the degree of receptor activity is dependant on the magnitude of the force applied (Clarac, Wales and Laverack, 1971).

7.3.1   Response to water- and substrate-borne vibration

It is possible that the CSD's are vibration sensitive, as suggested by the ultrastructure of the strand (see Chapter 9). Tapping the experimental bench produces a good response, but in the experimental condition the ischio-and basipodite were applied to the substrate, whereas in the standing animal substrate borne vibrations would have to be conducted by four or five leg joints to reach the receptors. Horch and Salmon (1969) have shown that *Ocypode* can detect sound waves by means of receptors in the pereiopods. In a later publication (Horch 1971) it was suggested that the receptor responsible was the myochordotonal organ (Chapter 6) which is anatomically modified in this species. It would, nevertheless be interesting to determine whether or not the CSDs, which are close to the myochordotonal organs, are also involved.

7.3.2   Stimulation of the CSD soft cuticle area

Clarac, *et al.* (1971) used a probe to deform the discrete area of soft cuticle where the CSDs are inserted. When pressure is applied to the soft cuticle of CSD1, both phasic and tonic unit responds (Fig. 7.5). On release, a short burst of phasic activity occurs whilst the tonic activity ceases. In CSD1, the majority of units are 'ON' units, that is they respond to the application of pressure to the external surface of the soft cuticle. Fig. 7.5b shows two tonic 'ON' units, the larger of which is active in the resting condition, whereas the smaller ceases to be active in the absence of stimulus. In Fig. 7.5c-d, the phasic-tonic unit responds spontaneously. At the onset of the stimulus, the frequency discharge is initially high, but rapid adaptation occurs. Removal of the stimulus inhibits the discharge of the units which become active again after about one second.

The area of CSD2 soft cuticle is smaller in the crab but the same sort of stimulation was, nevertheless, applied. Considering either the

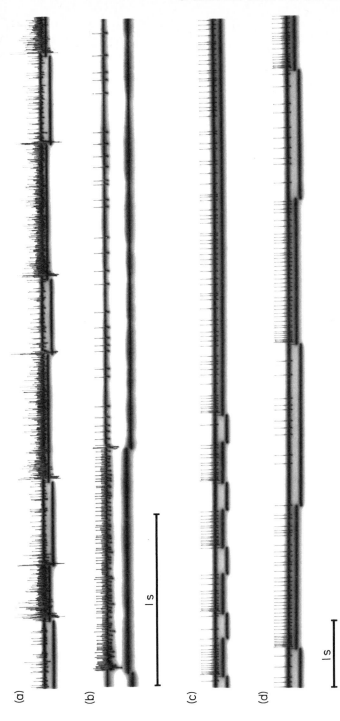

Fig. 7.5   The activity from CSD1 in the walking leg of *Carcinus maenas* in response to the application of constant pressure. The force was applied to the area of soft cuticle by a blunt probe. Upward movement of the lower trace indicates application of the stimulus. (a) Activity in a large bundle of axons. The majority of the units are tonic or phasic-tonic, but some units respond phasically to application and removal of the stimulus. (b) A recording from a small bundle of axons. The smaller, tonic unit is active only during the period of stimulation but the larger tonic unit continues to fire at a reduced frequency on removal of pressure. (c, d) This unit responds phasic-tonically to stimulation with a burst but rapidly adapts to a constant frequency with maintained pressure. On removal of the stimulus, the activity ceases and then returns at a reduced frequency. (From Clarac, Wales and Laverack, 1971.)

general response, or the types of units detected, the response is basically similar to that of CSD1. However, less phasic activity was observed and the 'OFF' response was reduced.

7.3.3   Responses to increased tension in the coxo-basipodite joint muscles

The CSD receptors respond to tension in both of the muscle groups of the coxo-basipodite joint, although the effect of the depressor muscle on CSD1 was smaller than that of the anterior levator muscle. This is particularly interesting because of the primary importance of this latter muscle in walking leg autotomy (Wood and Wood, 1932).

Maintained tension in the muscles evokes a high level of tonic activity with little adaptation in CSD1. The tension in the muscle tendon was not measured but it was insufficient to produce autotomy.

7.3.4   Responses to joint movement

Clarac, *et al.* (1971) restricted their investigation to the effect of passive joint movements. There is a good response from CSD1 and CSD2 both *Homarus* and *Carcinus*. The responses are very similar in the two receptors. Fig. 7.6a shows the CSD2 response to protraction of the ischiopodite (only a small afferent bundle was recorded from). The activity occurs over most of the range of imposed movement. Fig. 6b-c, showing the CSD2 response of *Carcinus* to ischiopodite manipulation, illustrates some of the unit types. The response of most units is unidirectional, as is the chordotonal joint receptors (Wiersma, 1959). Fig. 6C presents a single unit which responds to production for most of the range of movement. Two opposite direction units are recorded in Fig. 6D. The threshold is relatively high for this sort of stimulation and the ischiopodite had to be moved several degrees away from the rest position to elicit these responses.

7.4   Autotomy and CSD1 function

7.4.1   The autotomy mechanism

The ability to break off a limb is a property possessed by several Arthropod groups (Pieron, 1907, 1908; Bliss, 1960; Bauer, 1972).

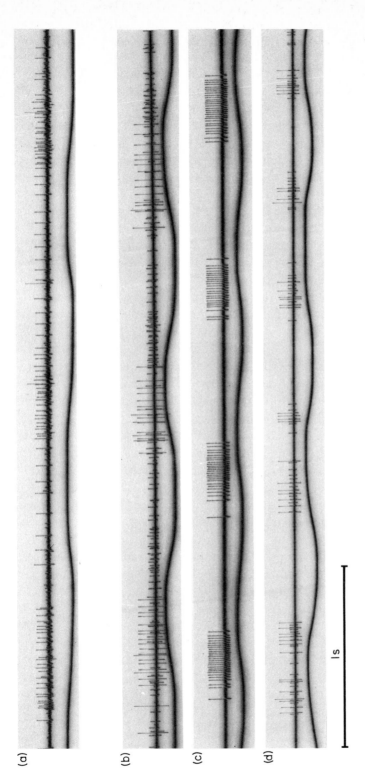

Fig. 7.6 (a) The response in a small number of units of CSD2 in the walking leg of *Homarus gammarus* to passive movement at the basi-ischiopodite joint. Downward movement of the lower trace indicates protraction of the ischiopodite. (b-d) The activity from small bundles of sensory axons from CSD2 in the walking leg of *Carcinus maenas* to passive movement at the ischio-meropodite joint. Upward movement of the lower trace indicates production of the joint. (b) Generally the units which respond to production and reduction are different; (c) a single unit active only during production; (d) two units which respond to both production and reduction of the joint. (From Clarac, Wales and Laverack, 1971.)

True autotomy involves the breaking of appendages along a preformed fracture plane as a result of muscular contraction in the coxo-basipodite region (Wood and Wood, 1932). Leg injury is the normal stimulus for autotomy in most cases, though in certain crabs we must consider it as a defensive tactic (Robinson, Abele and Robinson, 1970). CSD1 is anatomically close to the preformed breakage plane and the soft cuticle of CSD1 appears to be structurally associated with it.

The papers of Fredericq (1882, 1883, 1891) and Wood and Wood (1932) are concerned with identifying which of the muscles is instrumental in the production of the cuticular break, but it is apparent from the work of Demoor (1891) and McVean (1970) that more than one muscle is involved. CSD1 is responsive to change in tension as produced by activity in the muscles of the coxo-basipodite joint, and this suggests that CSD1 must be involved in the breaking phenomenon.

Autotomy can be divided into different phases. The first one is the injury response stage in the different coxo-basipodite muscles, the second consists of the fracture of the cuticle at the breakage plane and the last stage is the rupture of the soft internal structures. It results in minimal tissue damage because only the leg nerve and circulatory pathways cross the fracture plane and so good regeneration is permitted (Bliss, 1960; Gomez, 1964). Autotomy must be preceded by strong elevation of an injured limb. Thus the anterior levator (AL) muscle was considered to play a fundamental role in production of the fracture due to its strength and to the insertion of its tendon in direct association with the fracture plane (Fredericq, 1891; Paul, 1915; Wood and Wood, 1932; Clarac *et al.*, 1971; McVean, 1973, 1974). In the literature, contradictory hypotheses have been proposed concerning the role of the smaller, posterior levator (PL) muscle during autotomy. Some authors consider that it is not involved in the phenomenon, others that it may play an important role. McVean (1970), studying the mechanical forces required for autotomy, states that in *Carcinus maenas* the PL muscle, due to its perpendicular position, initiates autotomy by fracturing the AL muscle tendon at its insertion onto the basi-ischiopodite and thus re-directing the forces exerted by the muscle. The fact that stimulation of CSD1 evokes an excitatory reflex response of the PL motoneurons (Clarac and Wales, 1970) is, in agreement with this hypothesis.

As tension increases in the anterior levator muscle the CSD1 activity increases, and it continues to increase until the second phase of the breakage occurs at which time the receptor activity decreases dramatically (Clarac *et al.* 1971). The degree of calcification and thickness of cuticle seem important in autotomy. For example, Paul (1915) proposed that the furrow in the basi-ischiopodite of Carcinus is an important structure. Even though the mechanism that he presented was wrong, as demonstrated by Wood and Wood (1932), this furrow exists in all species of large decapod *Crustacea* and, therefore, must be involved in some aspect of the phenomenon.

7.4.2 Electrophysiology of the autotomy

To elucidate some questions concerning the autotomy mechanism, electrophysiological studies have been made. They are, however somewhat contradictory (Clarac and Wales, 1970; McVean, 1974; Moffett, 1973, 1975). We must consider the different behavioural acts involving the levator/depressor muscles, i.e. 'different routine limb elevation or depression' (Moffett, 1975) and autotomy.

As Bush (1965) states, at the coxo-basipodite joint, resistance reflexes are elicited by the CB joint chordotonal organ. PL is recruited simultaneously with AL in response to depression of the basipodite. In *Carcinus maenas*, two motor units have been found to innervate the PL. (Clarac and Wales, 1970); while McVean (1970) described three AL motoneurons, two of which are involved in the resistance reflex. Moffett, working on the land crab *Cardisoma guanhumi* recorded two to four units sensitive to basipodite movement. However, cross sections of the AL nerve of the *Cardisoma* reveal the presence of nine units. Thus less than half of them are involved in this reflex response. In walking, PL and AL appear to be synergists and their activities are antagonistic to the depressor muscles (Clarac and Wales, 1970; Moffett, 1975; Table 7.1). Considering autotomy, the more accurate electrophysiological work is that of Moffett, (1973, 1975) because she recorded simultaneously from all three muscles of the coxo-basipodite joint. Three different results were obtained:

(i) When the leg is crushed, the AL tendon being cut to allow study of the first phase of autotomy only and to eliminate the reflexes elicited by CSD1, a large burst of activity is recorded from AL.

Table 7.1

Muscles	Autotomy	Posture and walking
Anterior LEVATOR	Autotomizer muscles (*Fredericq*, 1883, 1891) Muscle activity recorded by *Moffett* (1974). It elicits autotomy.	*Bush* (1965) *Moffett* (1974) } Resistance reflexes
	In the two phenomenon different motoneurones are involved (*Moffett*, 1974)	
Posterior LEVATOR	Stimulated by CSDI (*Clarac* & *Wales*, 1970) Initiates autotomy (*McVean*, 1970, 1974) Inhibited during real autotomy (*Moffett*, 1974)	Ablation experiments confirm that it is not necessary (*Wood* and *Wood*, 1932) In Resistance reflex synergistic with AL (*Moffett*, 1974) Muscle activity recorded during walking (*Clarac* & *Wales*, 1970)
DEPRESSORS	Inhibited during real autotomy (*Moffett*, 1974)	*Bush* (1965) *Moffett* (1974) } In Resistance reflexes antagonistic with the levator muscles

In contrast, activity ceases in the PL nerve and in the depressor nerve. In the AL nerve, the units elicited are not the same as the units activated during routine limb elevation (Fig. 7.7a).

(ii) When autotomy is elicited by pulling on the AL tendon, CSD1 is activated (Clarac *et al.*, 1971) as well as motor units in both levator nerves. Moffett (1975) states that in the case where it is the PL tendon only is stretched, the same AL units that respond during resistance reflexes, (and not the large one) are elicited. When the real fracture occurs (second and third phases), the AL response has a longer latency and a shorter duration than the PL one (Fig. 7.7d). On-going activity in the depressor is, in contrast, inhibited.

(iii) During natural autotomy produced by injury to the limb, only the AL 'injury response' contraction, which produces the fracture, occurs and it is followed by PL activity. Moffett (1975)

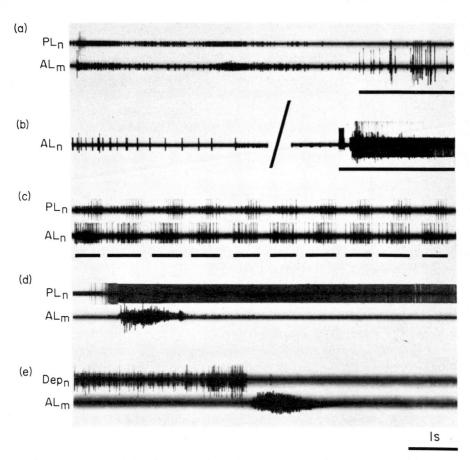

Fig. 7.7 The activity from AL (anterior levator) PL (posterior levator) Dep (depressor) motor nerves (n) or muscles (m) in the walking leg of *Cardisoma guanhumi* in response to injury and artificially produced fracture. (a) ongoing activity in the PL nerve was inhibited during AL muscle response to injury (indicated by horizontal bar). (b) AL nerve units activated by forced depression are shown before the break in the record; those units activated by injury are shown after the break. There is a three second interval between the two sections. (c) activity is evoked simultaneously from both levator nerves by pulling on the AL tendon (indicated by horizontal bars). (d) shows the response of PL_n and AL_m to limb fracture, elicited by pulling the AL tendon. (e) the ongoing activity in Dep n ceased when AL m responded to limb fracture with a burst of activity (From Moffett, 1975.)

states that in this case, PL contraction is not necessary for the success of the fracture (Fig. 7.8). McVean's theory is in absolute contradiction to this last idea (McVean, 1974). He suggests that autotomy is initiated by a stress in the limb; this stimulates CSD1, PL excitatory activity is reflexly increased and initiates the fracture, which is followed by firing of a large AL motor burst unit. This last hypothesis, which cannot as yet, be disregarded, seems nevertheless, inadequate, since section of the CSD1 afferent nerve does not prevent the autotomy. Also the PL muscle may be ablated without suppressing the autotomy reflex.

A question comes to mind concerning the effective role of CSD1. CSD1 stimulation elicits a clear response of the posterior levator, but seems also to modify the AL response. This reflex is easily obtained with *Carcinus maenas, Palinurus vulgaris* and *Homarus gammarus*, but is much more labile in *Astacus leptodactylus* (Clarac *et al.*, 1971). According to McVean's hypothesis the CSD1—PL reflex activation indicates CSD1 as the possible initiator of the autotomy phenomenon. However, since Moffett (1975) shows in her records that PL is centrally inhibited during the time which preceeds the autotomy, it seems necessary to consider only the second hypothesis proposed by Clarac and Wales (1970). Thus, CSD1 might serve as a brake to increase the motor output to the PL muscle when stress in the cuticle distal to the AL muscle signals the danger of accidental autotomy. CSD1 regulating the motor output of the PL muscle could assume a greater share of the levatory load when unitential fracture is imminent. It would be important for such a reflex to be suppressed when the autotomy motor pattern is activated. The injury input that activates AL muscle contraction may or may not directly suppress the PL response (Fig. 7.9). The scheme presented in Fig. 7.9 is not, of course, a definitive one; it must be considered that, if CSD1 can play a role in controlling levator and depressor tension and movement, the real autotomy mechanism is centrally organized and under command fibre influence.

Autotomy seems to be a process elicited at a certain level of injury. Regulation is possible through CSD1 afferents onto the motor command of the limb muscles. If the stimulation reaches too high a level, the autotomy central network is directly stimulated inducing the breakage of the limb.

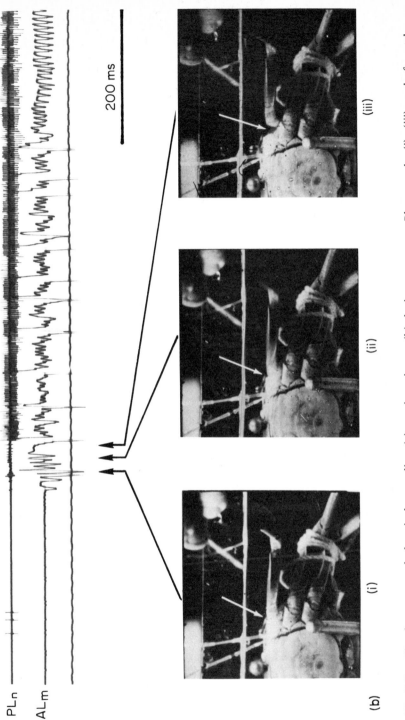

Fig. 7.8 Simultaneous and electrical recording (a) motion picture (b) during autotomy. Photographs (i)–(iii) made from three consecutive frames of a film showing autotomy (64 frames/s.). Black arrows indicate correspondence of frames to film monitor (third trace). White arrows point to fracture plane. The coxopodite does not move between frames (i) and (ii), but the distal portion of the limb is displaced; fracture may have started by (ii). In (iii) fracture is unquestionably complete; this frame coincides with the beginning of the high-frequency PL burst. PL activity coinciding with (b) is atypical, and may be muscle activity picked up from AL. AL_m, anterior levator muscle; PL_n, posterior levator nerve (From Moffett, 1975.)

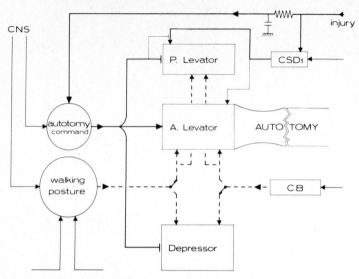

Fig. 7.9   General scheme to illustrate the possible coxo-basipodite joint activity during posture, walking and autotomy. CSD1 is sensitive to injury (arrow with broad black line). The other arrow to CSD1 summarizes the other influences which can affect it (joint movements, increasing tension, etc.). The arrow to CB indicates its sensitivity to basipodite movement. The CB joint chordotonal organ can regulate levator and depressor activity during walking and posture patterns. Autotomy seems to be elicited at a certain level of injury (schematized by a threshold system) and depends on CNS influence. Following Moffetts results (Moffett, 1975), the AL command is synchronized with PL and Depressor muscle inhibition.

7.5   CSD activity and regulation of muscle activity in the walking legs

The CSDs (CSD1 and CSD2) regulate other joints and segments of the same appendage as well as the coxo-basipodite one. Two joints in particular seem to be activated by CSD stimulation, the mero-carpopodite and the pro-dactylopodite i.e. those joints at which movement occurs in the same plane as at the coxo-basipodite joint.

Correlations between the mero-carpopodite and the coxo-basipodite joints seem very numerous in sideways walking of the crab. Clarac and Coulmance (1971) have shown that the influence of the mero-carpopodite joint on basipodite elevation varies depending on the direction of walking. Vedel, Clarac and Bush, (1975) have demonstrated the direct influence of CB chordotonal afference on the accessory flexor (AF) muscle excitatory motoneuron. Also a flexor (F) motoneuron is excited when CB is released and the tonic extensor motoneuron is excited when CB is stretched. Angaut-Petit, Clarac and Vedel (1974) and Vedel, Angaut-Petit and Clarac (1975) have described the direct influence of CSD2 on the AF and F motoneuron discharge. A pressure applied to the CSD2 soft cuticle in *Astacus leptodactylus* evokes a response in both tonic motoneurons of F and AF. The frequency of the F discharge is always higher than that of the AF one (Fig. 7.10c, d). The F phasic motoneuron does not seem to be elicited by this sort of stimulus. In the extensor nerve, the tonic unit is not noticeably modified. It appears that the AF excitatory motoneuron discharge is somewhat coupled with the discharge of one tonic F motoneuron. The percentage of coupled impulses is approximately the same before, during and after CSD2 stimulation (Angaut-Petit and Clarac, 1976). Stimulation of CSD1 also seems to modify the AF activity (see Fig. 7.10b), increasing the discharge of its tonic excitatory motoneuron. As far as the pro-dactylopodite joint is concerned, CSD1 and CSD2 appear to influence both opener and closer activity (Clarac, unpublished observations).

A problem arises concerning the type of stimulation applied. If the specificity of the stimulus cannot be described, a problem occurs concerning the real modification of the receptor when the soft area is pressed. It could be that it relaxes the strand and then, when the pressure is off, the strand returns to its initial length. The responses in the different motoneuron described above are obtained when the soft area is pressed.

In conclusion, it appears that the coxo-basipodite, mero-carpopodite and pro-dactylopodite joints are under the control of the CSDs. This may explain the main role to these receptors. Thus, they may modulate the posture of the entire leg depending on the weight supported by the animal. Because of their response to distortion of the exoskeleton, they can regulate the movement of the different joints, considering the mechanical resistance to muscle tension. It is proposed that they regulate the posture of the different legs depending on the forces exerted by each of them. Experiments on contralateral reflexes would be a useful test of this suggestion.

The CSDs can also be used by the animal, along with other receptors, to detect when the leg is or is not in contact with the

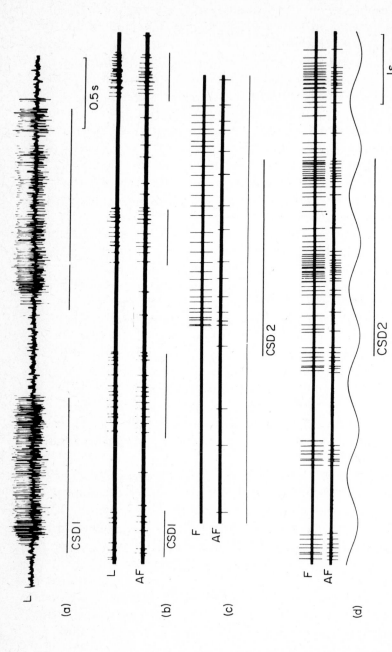

Fig. 7.10   Modulation of the levator, flexor and accessory flexor activities induced by cuticular stress detector stimulation (CSD1 and CSD2) in the crab *Carcinus maenas* (a, b) and in the crayfish *Astacus leptodactylus* (c, d). The stimulation is applied to the external soft cuticle area of the receptors. Duration is represented by a horizontal bar. (a) Electromyogram response of the levator muscle (L). (b) Levator (L) and accessory flexor (AF) motor nerve responses (it has not been possible in these experiments to distinguish AL from PL). (c, d) Flexor (F) and accessory flexor (AF) tonic motoneuron responses. In (c), the mero-carpopodite joint is held stationary; in (d), the carpopodite is moved passively (potassium 1)

ground. In contrast with the joint chordotonal organs, they are not receptors of specific joints; rather they record joint movement and changes in stress, and regulate the entire leg posture. The possible sensitivity of the CSDs to vibration of the substrate, as is suggested by their ultrastructure (Moulins and Clarac, 1972), may also be involved in the regulation of leg muscle activity. This general role in response to muscle activity is analogous to the CSD activity in autotomy. These receptors regulate elevator/depressor muscles in normal 'routine' activity. By their action, they can coordinate the first autotomy phase mechansim more to prevent it than to induce the following phases.

The CSD receptors seem novel in their anatomy and in their function. Their study is yet very incomplete, lacking data on chronic animal preparations and determination of the nature of the specific stimulus to which they respond.

Acknowledgements

I would like to thank Dr. J. P. Vedel very much for his help during the program of this research, and Dr. W. Wales for his detailed criticism of a draft of this manuscript.

References

Angaut-Petit, D., Clarac, F. and Vedel, J. P. (1974) Excitatory and inhibitory innervation of a Crustacean muscle associated with a sensory organ. *Brain Research*, **70**, 148–152.

Angaut-Petit, D., Clarac, F. (1976) A study of a temporal relationship between two excitatory motor discharges in the crayfish. *Brain Research*, **104**, 166–170.

Bauer, K. H. (1972) Funktions-mechanisms der Autotomie bei spinnen (Aranae) und seine morphologischen Voranssetzungen. *Zeitschrift Morphologie der Tiere*, **72**, 173–202.

Bliss, D. E. (1960) Autotomy and regeneration. In : *Physiology of Crustacea*. Waterman, T. H. (ed.) Vol. 1, Ch. 17, 561–589.

Bush, B. M. H. (1965) Leg reflexes from chordotonal organs in the crab *Carcinus maenas*. *Comparative Biochemistry and Physiology*, **15**, 567–587.

Clarac, F. and Coulmance, M. (1971) La marche latérale chez le crabe *Carcinus*. *Zeitschrift für vergleichende Physiologie*, **73**, 408–438.

Clarac, F. and Masson, C. (1969) Anatomie comparée des propriocepteurs de la région basi-ischio-méropodite chez certains crustancés décapodes. *Zeitschrift für vergleichende Physiologie*, **65**, 242–273.

Clarac, F. and Wales, W. (1970) Contrôle sensoriel des muscles élévateurs au cours de la marche et de l'autotomie chez certains crustacés décapodes. *Comptes rendus de l'Académie des Sciences, série D*, **271**, 2163–2166.

Clarac, F., Wales, W. and Laverack, M. S. (1971) Stress detection at the autotomy plane in decapod Crustacea. II the function of the receptors associated with the cuticle of the basi-ischiopodite. *Zeitschrift für vergleichende Physiologie*, **73**, 383–407.

Demoor, J. (1891) Etude des manifestations motrices des Crustacés au point de vue des fonctions nerveuses. *Archives de Zoologie expérimentale et générale*, **2**, 191–227.

Fredericq, L. (1882) Amputation des pattes par mouvement réflexe chez le crabe. *Archives de Biologie*, **3**, 235–240.

Fredericq, L. (1883) Sur l'autotomie ou mutilation par voie réflexe comme moyen de défense chez les animaux. *Archives de Zoologie expérimentale et générale, 2 ème série*, **1**, 413–426.

Fredericq, L. (1891) Nouvelles recherches sur l'autotomie chez le crabe. *Archives de Biologie*, **12**, 169–197.

Gomez, R. (1964) Autotomy and regeneration in the crab *Paratelphusa hydrodromous*. *Journal of Animal Morphology and Physiology*, **11**, 97–104.

Horch, K. W. (1971) An organ for hearing and vibration sense in the Ghost crab Ocypode. *Zeitschrift für vergleichende Physiologie*, **73**, 1–21.

Horch, K. W. and Salmon, M. (1969) Production, perception and reception of acoustic stimuli by semiterrestrial crabs (genus Ocypode and Uca, family Ocypodidae). *Forma et functio*, **1**, 1–25.

McVean, A. R. (1970) Neuromuscular coordination in *Carcinus maenas*. Ph.D. Thesis, University College of North Wales, U.K.

McVean, A. R. (1973) Autotomy in *Carcinus maenas*. *Journal of Zoology, London*, **169**, 349–364.

McVean, A. R. (1974) The nervous control of autotomy in *Carcinus maenas*. *Journal of Experimental Biology*, **60**, 423–436.

Moffett, S. B. (1973) The motor pattern of the basi-ischial muscles in walking and autotomy in the land Crab, *Cardisoma guanhumi*. Ph.D. Thesis, University of Miami, U.S.A.

Moffett, S. B. (1975) Motor patterns and structural interactions of basi-ischiopodite levator muscles in routine limb elevation and production of autotomy in the land crab *Cardisoma guanhumi*. *Journal of Comparative Physiology*, **96**, 285–305.

Moulins, M. and Clarac, F. (1972) Ultrastructure d'un organe chordotonal associé à la cuticule dans les appendices de l'écrevisse *Comptes rendues de l'Académie des Sciences*, **274**, 2189–2192.

Paul, J. H. (1915) A comparative study of the reflex of autotomy in decapod Crustacea. *Edinburgh Proceedings of the Royal Society*, **35**, 232–262.

Pieron, H. (1907) Autotomie protectrice et autotomie évasive. *Comptes rendus de l'Académie des Sciences*, **144**, 1379—1381.

Pieron, H. (1908) Le problème de l'autotomie. *Bulletin scientifique de France et de Belgique*, **42**, 185—246.

Robinson, M. H., Abele, L. G. and Robinson, B. (1970) Attack autotomy : a defense against predators. *Science*, **169**, 300—301.

Vedel, J. P., Clarac, F. and Bush, B. M. H. (1975) Coordination motrice proximo-distale au niveau des appendices locomoteurs de la Laupouste. *Comptes rendus de l'Academie des Sciences Paris*, **281**, 723—726

Vedel, J. P., Augaut-Petit, D. and Clarac, F. (1975) Reflex modulation of motoneurone activity in the leg of the crayfish *Astacus Leptodactylus*. *Journal of Experimental Biology*, **63**, 551—567.

Wales, W., Clarac, F., Dando, M. R. and Laverack, M. S. (1970) Innervation of the receptors present at the various joints of the pereïopods and third maxilliped of *Homarus gammarus* (L) and other macruran decapods (Crustacea). *Zeitschrift für vergleichende Physiologie*, **68**, 345—384.

Wales, W., Clarac, F. and Laverack, M. S. (1971) Stress detection at the autotomy plane in decapod crustacea. I. Comparative anatomy of the receptors of the basi-ischiopodite region. *Zeitschrift für vergleichende Physiologie*, **73**, 357—382.

Wood, F. D. and Wood, H. E. (1932) Autotomy in decapod Crustacea. *Journal of Experimental Zoology*, **62**, 1—55.

Wiersma, C. A. G. (1959) Movement receptors in decapod crustacea. *Journal of the Marine Biology Association, U.K.*, **38**, 143—152.

8 Limb and wing receptors in insects, chelicerates and myriapods

B. R. WRIGHT

8.1 Introduction

The appendages of insects and arachnids and probably also of myriapods are richly supplied with proprioceptors. These receptors can, conveniently, be separated into two fairly distinct, functional groups. Into one group may be gathered the many kinds of receptor that supply information concerning the movement and angular position of one segment of an appendage relative to its neighbour (e.g. movement of the tibia relative to the femur). Into the other may be gathered those receptors in muscle or cuticle, which are directly sensitive to muscle tension or cuticular stress and which provide information about various changes, the majority of which directly concern the mechanical activity of muscles (e.g. isometric muscle tension). The physiological properties of the receptors in these two groups are sufficiently distinct to warrant separate consideration.

It is now known that proprioceptors in the limbs of insects and arachnids and in the wings of insects play a role in regulating motor output during a number of different behaviour patterns. Interest in the significance of these proprioceptors in the control of behaviour probably arose from early experiments on insects which demonstrated that leg amputation could dramatically change the coordination of the legs in locomotion (von Buddenbrock, 1921; Ten Cate, 1936). With the introduction of electrophysiological techniques in the early thirties interest in insect proprioceptors grew and it was quickly revealed that the plan of proprioceptors in the insect limb was markedly different from that in the vertebrate limb (Pringle, 1938a, 1938b). Recent advances include the development of techniques for recording intracellularly from central neurons (Hoyle and Burrows, 1970, 1973a) and for the specific staining of functionally identified nerve cells (Stretton and Kravitz, 1968; Iles and Mulloney, 1971; Pitman, Tweedle and Cohen, 1972).

There are many reviews of the structure and distribution of proprioceptors in insects (Debaiseaux, 1938; Pringle, 1961; Dethier, 1963; Bullock and Horridge, 1965; Finlayson, 1968; Howse, 1968) and arachnids (Pringle, 1961; Bullock and Horridge, 1965; Finlayson, 1968) and several dealing with the role of sensory feedback and its integration (e.g. Pringle, 1961; Wilson, 1966; Hoyle, 1970; Huber, 1974; Miller, 1974). In this chapter, particular attention will be given to more recent information and especially to advances in our knowledge concerning the mechanisms of integration of proprioceptive information in insects. The merostome *Limulus*, since it is closely related to the arachnids and has been studied intensively, will be included in the scope of this chapter.

8.2   Receptors monitoring joint position and movement

To move its appendages effectively it is of importance for an animal to be provided with information concerning the movements and angular position of the joints. In insects, arachnids and myriapods this information is provided by internal proprioceptors whose dendrites are closely associated with the articular membrane of the joint or are attached to muscle or connective tissue that spans the segments of the joint.

8.2.1   Variety and distribution

Insects and arachnids are ideal animals for the study of proprioceptors. They possess an array of well-defined cuticular structures and numerous proprioceptors within the body, and the structural variety of these organs is enormous. Sensory cell bodies are, with few exceptions, peripherally located and sensory axons are often large enough to separate the activity of single units.

In both classes proprioceptors are common in the limbs and they are profuse at the insect wing base and in the dipteran haltere, although uncommon in the wing itself. In the legs they are found at, or are associated with, the joints and the insect leg may contain connective chordotonal organs, simple multiterminal receptors, hair-plates and hairs, a muscle receptor organ (only found so far in the bee (Markl, 1965)) and campaniform sensilla; the arachnid leg: joint receptors and slit sense organs of various kinds. In addition, special sense organs receptive to sound or to vibration of the substratum also occur in the legs of arachnids and in many insect groups (Finlayson, 1968; Chapter 12). However, myochordotonal organs, typical of the legs of crustaceans (Section 6.2.2), have not been reported in the legs of insects or arachnids.

In the leg of *Limulus* there are articular membrane receptors, a muscle receptor organ and nerve proprioceptors. Very little is known about proprioceptive structures in the myriapod leg, but a muscle receptor organ has been found at some limb joints of the chilopod, *Lithobius* (Rilling, 1960). In other myriapods multiterminal cells that possibly function as 'stretch receptors' (Fuhrmann, 1922) and campaniform-like sensilla (Leydig, 1860; Fuhrmann, 1922) have been described.

Connective chordotonal organs

The term 'chordotonal organ' applies to any arthropod sense organ whose sensory cells are associated with rod-like terminal structures (scolopales) (Chapter 9). They occur in various situations, but the ones which are the concern of this chapter are all integumentary organs that span, or connect two structures. Howse (1968) has proposed that these sense organs be called 'connective' chordotonal organs (Chapter 9).

Connective chordotonal organs are found in the legs of all insects (Debaiseaux, 1936, 1938; Slifer, 1936; Richard, 1950; Nijenhuis and Dresden, 1952; Becht, 1958; Kendall, 1970) and their distribution in the leg has been mapped for several orders (Debaiseaux, 1938). There are usually coxal (Nijenhuis and Dresden, 1952), femoral, tibial (Richard, 1950; Howse, 1965), tarsal and pretarsal organs (Slifer, 1936; Kendall, 1970). Many insect orders have two groups of scolopidia scoloparia sharing a common tendon in the

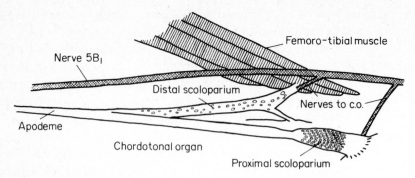

Fig. 8.1   Diagram of a prothoracic or mesothoracic femoral connective chordotonal organ of the locust, *Schistocerca gregeria*, drawn with its apodeme to the right:—The distal scoloparium is attached to the cuticle and the flexor tibiae muscle; The proximal scoloparium is attached only to the cuticle; Branches of nerve $5B_1$ innervate the chordotonal organ (After Burns, 1974).

femur, one being a discrete structure, the other somewhat dispersed (Fig. 8.1). One or two connective organs are present in the tibia, except in Ephemeroptera and Pscocoptera, where they appear to be absent (Debaiseaux, 1938) and there are usually three tarsal-pretarsal organs. Coxal connective organs were not studied by Debaiseaux (1938) and there is little comparative information concerning them. In the cockroach there are three coxo-trochanteral organs and one thoracic-coxal organ (Nijenhuis and Dresden, 1952).

Connective chordotonal organs occur at the wing base of insects (Leydig, 1860; Lehr, 1914; Eggers, 1928; Zaćwilichowski, 1934a, b; Pringle, 1957; Gettrup, 1962; Bullock and Horridge, 1965; Möss, 1967, 1971) and in the halteres of dipterans (Weinland, 1891; Pflugstaedt, 1912; Zaćwilichowski, 1934b; Pringle, 1957). In the wing base they are located near the radial or subcostal veins or arranged within the lumen of the radial vein. Pringle (1957) examined the distribution of connective chordotonal organs at the wing base in different insects and found that ante-alar, radial, medial and cubital organs may be present. Four organs are present in Mecoptera (*Panorpa*) but in other orders various organs drop out in a way that is parallel to the reduction in wing venation (Bullock and Horridge, 1965). The bee is exceptional in retaining three organs. In some orthopterans chordotonal organs appear to be absent from the wing base (Erhardt, 1916; Zaćwilichowski, 1934a), Crickets and locusts have only a single organ (Gettrup, 1962; Möss, 1967, 1971).

Two or more chordotonal organs occur in the basal region of the dipteran haltere (Pflugstaedt, 1912). Often the ante-alar organ and radial organ are enlarged and in certain diptera the terminal processes of the scolopidia are attached to the cuticle over a wide area (Pringle, 1957). The functions of chordotonal organs in the wings and halteres is not well understood. It has been assumed that they are important in flight and, in the case of the haltere, in its gyroscopic function (Pringle, 1957; Section 8.2.3).

Simple multiterminal receptors

These receptors ('joint receptors', 'stretch receptors') generally consist of one or two cells, the dendrites of which are associated with a connective tissue strand.

Simple multiterminal receptors in the legs are known from a number of insect groups (Richard, 1950; Denis, 1958; Barbier, 1961; Guthrie, 1967; Coillot and Boistel, 1968) and in Orthoptera (and probably other orders), they are a feature of most joints (Guthrie, 1967). These cells differ a good deal in detail (Section 8.2.1) but are generally similar to the multiterminal stretch receptor cells of abdominal segments (Section 4.2). In Orthoptera, typically one to five are associated with each leg joint (Guthrie, 1967), but information is insufficient to indicate their abundance and distribution in the legs of insects in general. In the cockroach two multiterminal cells have been found at the trochantero-femoral joint and three at the femoro-tibial joint (Guthrie, 1967). In the locust five cells occur at the femoro-tibial joint (Coillot and Boistel, 1968, 1969) and at least eight have been found in the tarsus (Kendall, 1970) (two occurring in the first tarsal segment, three in the third and at least three in the ariolium). In the termite two cells are found at the coxo-trochanteral joint (Denis, 1958) and one at the tibio-tarsal joint (Richard, 1950). Physiological studies indicate that these receptors are specialized movement and position detectors (Section 8.2.3).

Multiterminal cells are also found at the wing base. In Orthoptera there is a single receptor in the region of the subalar sclerite (Gettrup, 1962, 1963; Pabst, 1965). This receptor produces information concerning the amplitude and velocity of each wing upstroke (Gettrup, 1962, 1963; Pabst, 1965; Möss 1971).

Hair-plates

Hair-plates are an unusual kind of proprioceptor found in insects. They are position receptors (Pringle, 1938b; Wendler, 1964; Pearson, Fourtner and Wong, 1974) and are composed of long, fine hair sensilla (Pringle, 1938b). In the leg they are only found at the thoracic-coxal and coxal-trochanteral joints (Pringle, 1938b; Hoffman, 1964; Lombardo, 1974). The leg of the cockroach, for example, has three hair-plates: an inner coxal, an outer coxal and a trochanteral (Pringle, 1938b). Hair-plates serve to signal and reflexly maintain a particular joint position (Wendler, 1964; Pearson *et al.*, 1974).

Chelicerate joint receptors

In spiders and scorpions internal proprioceptors, important in signalling joint position and movement, have been reported in the walking legs and the pedipalps (Rao, 1964; Laverack, 1966; Rathmayer, 1967; Rathmayer and Koopmann, 1970; Bowerman and Larimer, 1973). These proprioceptors are very closely associated with the articular membrane of the joint and, in spiders at least, they are probably a feature of every joint in the leg. Thus, in the tarantula, each leg and pedipalp joint is equipped with at least two of these organs, there being a total of 135 neurons in 18 groups in the whole leg (Rathmayer and Koopmann, 1970) (Fig. 8.2). The greatest number (five) are found at the ball and socket joint between the coxa and the trochanter. Less is known about the distribution of similar proprioceptors which occur in scorpions and so far, two groups have been reported at the patello-tibial joint of the walking leg (Laverack, 1966) and two at the femoro-patellar joint of the pedipalps (Bowerman and Larimer, 1973).

In *Limulus* articular membrane receptors, muscle receptor organs ('tendon receptor organ') and, possibly, nerve proprioceptors serve as position and movement detectors (Pringle, 1956; Barber, 1956, 1958, 1960; Barber and Segel, 1960; Barber and Hayes, 1964; Hayes and Barber, 1967). The distribution of articular membrane receptors in the legs is very similar to that of multiterminal receptors in insects and proprioceptors within the joints in the legs of arachnids. One or more articular membrane receptors are found at each leg joint, except the tarso-claw joint, a total of nine in each walking leg (Fig. 8.3).

A muscle receptor organ ('tendon receptor organ') is found in the trochanter of each walking leg of *Limulus* (Barber and Hayes, 1964; Hayes and Barber, 1966). This organ is concerned with movements of the coxo-trochanteral and trochantero-femoral joints (Hayes and Barber, 1967; Section 8.2.3).

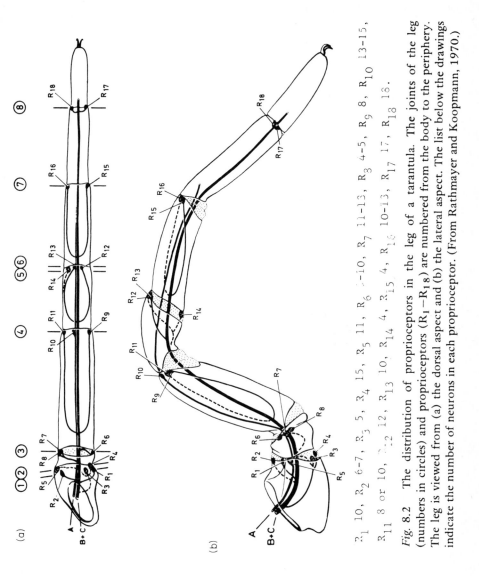

Fig. 8.2 The distribution of proprioceptors in the leg of a tarantula. The joints of the leg (numbers in circles) and proprioceptors ($R_1 - R_{18}$) are numbered from the body to the periphery. The leg is viewed from (a) the dorsal aspect and (b) the lateral aspect. The list below the drawings indicate the number of neurons in each proprioceptor. (From Rathmayer and Koopmann, 1970.)

R_1 10, R_2 6–7, R_3 5, R_4 15, R_5 11, R_6 7–10, R_7 11–13, R_8 4–5, R_9 8, R_{10} 13–15, R_{11} 8 or 10, R_{12} 12, R_{13} 10, R_{14} 4, R_{15} 4, R_{16} 10–13, R_{17} 17, R_{18} 18.

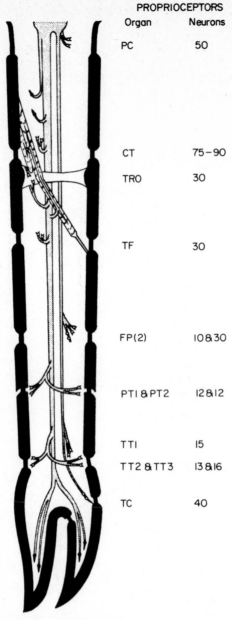

PROPRIOCEPTORS	
Organ	Neurons
PC	50
CT	75–90
TRO	30
TF	30
FP(2)	10&30
PTI&PT2	12&12
TTI	15
TT2&TT3	13&16
TC	40

Fig. 8.3 Diagrammatic representation of the leg of *Limulus* to show the distribution of proprioceptive organs and the number of sense cells which they contain. A single open circle indicates an articular membrane receptor, a double open circle a nerve proprioceptor. Solid circles indicate possible chemoreceptive organs. PC—pleurocoxal; CT—coxo-trochanteral; TRO—muscle receptor organ ('tendon receptor organ'); TF—trochanteral-femoral; FP—femoro-patellar; TT—tibio-tarsal; TC—tarso-claw. (From Hayes and Barber, 1967.)

An unusual type of proprioceptor found in *Limulus* is the nerve proprioceptor. Two are found in the walking leg, one near the femoro-patellar articulation, the other near the tarso-claw articulation (Hayes and Barber, 1967). The neurons of these receptors appear to be bipolar. Preliminary studies (Hayes and Barber, 1967) suggest that these receptors are stimulated by stretching of the nerve in which they are contained, and this is presumed to occur with movement of the joint.

8.2.2 Structure

Connective chordotonal organs
The structure of connective chordotonal organs is known for many insect orders (Chapter 9), the femoral and tibial organs of Orthoptera having received most attention (Usherwood, Runion and Campbell, 1968; Young, 1970; Burns, 1974; Moran, Rowley and Varela, 1975).

The number of scolopidia in leg connective organs varies considerably. Femoral organs contain from seven in a Thysanuran to over 300 in some grasshoppers (Debaiseaux, 1938); tibial organs contain comparatively few, with the exception of this organ in *Apis* which has 60 (Debaiseaux, 1938); while tarsal-pretarsal, coxal and trochanteral organs generally contain few. Thus, in the locust tarsus, for example, each organ contains no more than three scolopidia (Kendall, 1970). Evidence indicates that a scolopidium can often contain more than one sensory cell. In the cockroach tibio-tarsal organ for example (Young, 1970) many scolopidia contain two sensory cells and in the locust femoral chordotonal organ all scolopidia of the proximal scoloparium contain two neurons (Moran *et al.*, 1975).

The femoral organ is generally the largest chordotonal organ in the insect leg. In Orthoptera it is particularly well developed (Debaiseaux, 1938), having two distinct sensory regions. These are attached by a common apodeme to the end of the tibia but may be regarded structurally as separate scoloparia (Slifer, 1935; Debaiseaux, 1938). These scoloparia exhibit marked structural differences, which have been examined in some detail in the locust (Burns, 1974) and the grasshopper (Moran *et al.*, 1975), and some progress has been made in relating structural features to function (Section 8.2.3). The proximal scoloparium has numerous small sensory cells, all of about the same size (Fig. 8.1). There are over 200 in the locust (diameter 10 to 12 μm) and 300 to 400 in the grasshopper. They are arranged in a regular cone-shaped array, somewhat similar to that found in the much more complex subgenual organ (Schwabe, 1906) and Johnston's organ (Eggers, 1928). An ultrastructural analysis of the proximal scoloparium of the femoral organ of the grasshopper shows that each scolopidium contains two sensory cells (Moran *et al.*, 1975).

In contrast, the distal scoloparium contains far fewer cells (about 50 cells in the locust and 150 in the grasshopper). These cells are much larger (diameter 12 to 20 μm) and are scattered throughout the scoloparium. However, their scattered appearance does not infer an absence of structural organization for they appear to be arranged in three groups, each group associated with a different ligament. In addition, the most proximal cells of the scoloparium are the largest, cell size decreasing towards the apodeme. The distal scoloparium also differs from the proximal in being a much flatter structure and it has an attachment to the tibial flexor muscle (Fig. 8.1). These many differences between the two scoloparia of the orthopteran femoral chordotonal organ suggest a high level of functional specialization. Since they insert on a common apodeme they could be expected to have identical mechanical inputs, except that the distal scoloparium is attached to the flexor tibiae muscle. Burns (1974) and Varela, Moran and Rowley (1976) have linked the differences to the phasic and tonic function of the femoral chordotonal organ (Section 8.2.3).

The structure of the femoral organ of the locust metathoracic leg is unusual (Usherwood *et al.*, 1968) in that it contains only 24 cells in a single scoloparium that is located in the distal part of the femur. (In pro-and mesothoracic legs the femoral organ is located proximally in the femur.) The metathoracic scoloparium is probably the homologue of the distal scoloparium of pro-and mesothoracic legs. Probably these differences are in whole or part a consequence of the specialized function of the hind leg for jumping.

The tibio-tarsal connective chordotonal organ of insects conforms to a general pattern (Debaiseaux, 1936, 1938). In most genera the organ inserts on or near the tibio-tarsal intersegmental membrane. The ends of the chordotonal ligaments attach in a 'tent-like' fashion in the cockroach (Young, 1970) (Fig. 8.4) and the termite (Howse, 1965). However, in the bee tibial organ, which contains about 50

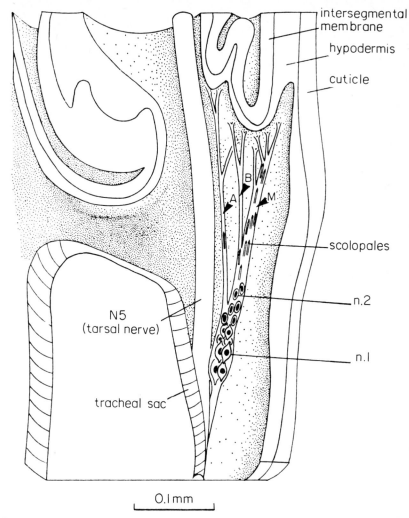

Fig. 8.4   Horizontal section through the distal end of the tibia in the cockroach, *Periplaneta americana* to show the tibio-tarsal connective chordotonal organ. The ligaments of the organ insert on the intersegmental membrane. There are two side branches (A and B) each with one scolopale and the main branch (M) has groups of scolopales scattered along its length n.1: Group 1. scolopales; n.2: Group 2 scolopales (From Young, 1970.)

scolopidia (Lukoschus, 1962), the situation is slightly different, with the scolopidia being arranged in a series of graded length. This attachment of the tibio-tarsal organ contrasts with that of the femoral organ, where attachment is at a single point. Thus there is potential in the tibio-tarsal organ for strands to be stretched differentially by movement of the tarsus.

The structure of the tibio-tarsal connective organ of the cockroach has been studied in some detail (Young, 1970) and it has a surprisingly complex structure. It's 26 cells are associated with 14 scolopales and are arranged in a main branch and two sub-branches (Fig. 8.4). As in the proximal scoloparium of the femoral organ (Burns, 1974), the proximal cells are the largest. Three distinct kinds of scolopidia can be distinguished at the ultrastructural level and the location of each type is consistent, but the functional significance of the different types and their different locations is not yet understood.

Connective chordotonal organs in the tarsus, and in the coxa and trochanter probably occur fairly generally in insects (Debaiseaux, 1938; Nijenhuis and Dresden, 1952), and chordotonal organs are among the many kinds of sense organ found at the wing base and in the dipteran haltere (reviewed by Pringle, 1957; Gettrup, 1962; Möss, 1967, 1971).

Simple multiterminal receptors
Simple multiterminal receptors occur at limb joints and at the wing base (Section 8.2.1). There is a noticeable absence of comparative information concerning these receptors, probably due to the fact that they are very difficult to locate and identify functionally. From the few cases where this has been successfully achieved, it is apparent that they differ a good deal in detail. In the cockroach, for example, a large many-branched multiterminal cell occurs near the antero-ventral condyle of the basal articultation (Guthrie, 1967). Also at this joint are two small cells above the postero-ventral condyle, but these cells have shorter and thicker processes with only two or three branches. Another large cell, found at the trochanteral-femoral articulation of the cockroach has only one or two branches, although these have numerous rami ending in club-shaped terminals (Guthrie, 1967). In the locust the multiterminal cells at the femoro-tibial joint form three distinct units. Two units are postero-lateral (one dorsal (RDPL) and one ventral (RVPL)), while the third is antero-lateral and dorsal. The postero-lateral units each contain two multiterminal cells which are closely apposed, while the antero-lateral unit contains a single cell. All the cells have many dendrites (Coillot and Boistel, 1968).

Hair-plates

Studies on the cervical hair-plate of the bee (Thurm, 1964, 1965) show that each sensillum exhibits the basic structure of a tactile trichoid hair. The long hollow shaft of the sensillum tapers to a tip and is inserted at its base into an annular ring of thin cuticle which is surrounded by a raised, thickened cuticular annulus. Each sensillum is innervated by a single sensory cell.

Chelicerate joint receptors

In each receptor, between 4 and 15 cells form a loose aggregation embedded in a common connective tissue capsule. The receptors are very closely associated with the arthrodial membrane at the joint (Rathmayer and Koopmann, 1970) and their dendrites terminate between the hypodermal cells of the membrane. There is considerable uncertainty concerning the relationship of these proprioceptors to those of other arthropods. They show many similarities to the multiterminal receptors of insects and the 'multiterminal' articular membrane receptors of *Limulus*. However, although in spiders some of these receptors do contain multiterminal cells (Rathmayer and Koopmann, 1970) in both spiders and scorpions have many sensory cells that appear to be bipolar (Rathmayer and Koopmann, 1970; Bowerman and Larimer, 1973). Further ultrastructural studies are required to clarify this.

In *Limulus*, proprioceptors known as articular membrane receptors are found at joints in the walking legs (Barber, 1960; Barber and Segel, 1960; Hayes and Barber, 1967). Each consists of a loose aggregation of large multiterminal cells with long multiply branched dendrites ending among hypodermal cells.

Muscle receptor organs (MROs)

Muscle receptor organs consist of multiterminal cells associated with a special accessory muscle(s), or connective tissue strand, or both. A single MRO (the 'tendon receptor organ') is found in the trochanter of each walking leg of *Limulus* (Barber and Hayes, 1964; Hayes and Barber, 1966, 1967) and it consists of a single tendon and two accessory muscles (Barber and Hayes, 1964; Hayes and Barber, 1967; Fig. 8.5). The tendon is formed of dense connective tissue and is rather inelastic. The accessory muscles have common origins but different insertions, a short accessory muscle spanning the coxo-trochanteral joint, and a longer one spanning this and the tro-

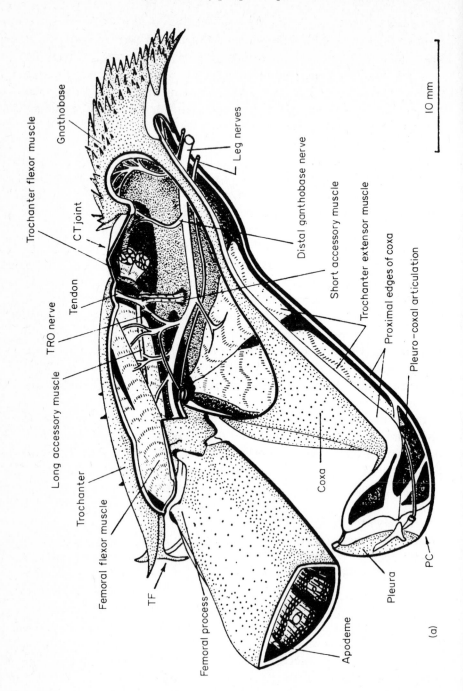

(a)

chanteral-femoral joint. Consequently this MRO can respond to movements of two joints (Section 8.2.3). Extension of the femur stretches only the long accessory muscle, whereas extension of the trochanter stretches both accessory muscles (Fig. 8.5b-d). About 30 multipolar neurons are associated with the tendon of the organ, of which about 20 are associated with the insertion of the short accessory muscle and about 10 with the attachment of the long accessory muscle as it passes over the tendon.

The 'tendon receptor organ' of *Limulus*, thus differs substantially in structure and function from the tendon organs of Crustacea and vertebrates, in which the dendrites lie within the tendon or neighbouring muscle fibres of a leg muscle and serve to monitor the tension developed in this muscle.

A muscle receptor organ similar to that of crustaceans has been found in the legs of a chilopod (Rilling, 1960). In this organ, multiterminal cells with short branched dendrites innervate a specialized receptor muscle of a few muscle fibres.

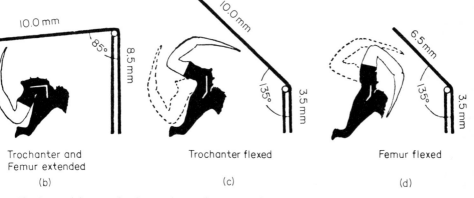

Trochanter and Femur extended	Trochanter flexed	Femur flexed
(b)	(c)	(d)

Fig. 8.5   (a) Proximal portion of a leg of *Limulus* to show the tendon receptor organ, its innervations and relation to nearby structures. The coxo-trochanteral (CT) and trochanteral-femoral (TF) joints are in the extended position. Medial surfaces of the coxa, trochanter and femur are orientated towards top of page. (b, c, d) Relative positions and lengths of tendon receptor organ accessory muscles in different positions of the joints. In the diagrams, the open circles represent the tendon in cross-section and the numbers are the lengths of the muscle segments. The parts of the limb shown in (a) are black, while the more distal parts are white. The locations of the accessory muscles are shown as white lines within the black areas. (b) Both joints extended; (c) Trochanter flexed, femur extended; (d) Both joints flexed. (From Hayes and Barber, 1967.)

In insects, a muscle receptor organ in the legs has been reported in the bee (Markl, 1965). It lies in the coxa and its receptor muscle consists of only four to six fibres and is attached between the coxo-trochanteral joint membrane and the wall of the coxa. A single, presumably multiterminal, neuron innervates the receptor muscle and there are numerous accessory cells (Finlayson, 1968).

8.2.3 Physiology

To understand the role of proprioceptors in behaviour it is necessary to know in some detail what information the different proprioceptors supply to the central nervous system. This information may be phasic, tonic or intermediate (phaso-tonic). Phasic information concerns movement at a joint while tonic relates to angular position.

Connective chordotonal organs

The connective chordotonal organs of the insect leg detect movement and position of the segment upon which their attachment(s) insert. The femoral organ is physiologically and structurally the most complex (Debaiseaux, 1938). In Orthoptera it is particularly well-developed and may function additionally to measure muscle length (Burns, 1974) (Section 8.3.2). Its physiology has been studied in some depth (Usherwood *et al.*, 1968; Burns, 1974; Varela *et al.*, 1976) and it has been shown to produce both phasic and tonic information. In the locust phasic responses are sensitive to displacement of the tibia by as little as 3′ (Usherwood *et al.*, 1968). A large number of phasic units are present and many are unidirectionally responsive to flexion or of extension (Fig. 8.6A). However, the total number and details of directional sensitivity, angular sensitivity and range fractionation of particular units are not known.

Tibial connective organs of insects generally have far fewer cells than femoral organs (Debaiseaux, 1938). The physiological properties of the tibial organ is best known for the tibio-tarsal organ of the cockroach (Young, 1970). All the phasic units of this organ are unidirectionally responsive to elongation of the organ, this being produced by downward and backward deflection of the tarsus.

Responses of the tibio-tarsal organ comprise at least two classes of sensory fibre: 'larger fibres' which arise from Group 1 scolopales and 'smaller fibres' which arise from Group 2 scolopales (Young, 1970) (Section 8.2.2). The 'larger fibres' show unidirectional phasic and

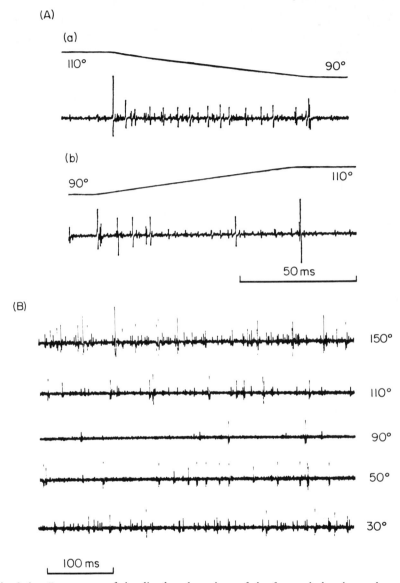

Fig. 8.6   Responses of the distal scoloparium of the femoral chordotonal organ of the locust mesothoracic leg (A) to constant velocity movement of the tibia (a) flexion, (b) extension. A comparison of the recordings reveals that some units are directionally sensitive. (B) with the tibia at rest in different positions. The minimum level of tonic activity is seen in the recording where the femoro-tibial angle is 90°. (Each trace was photographed about 100 s after the tibia was moved to the femoro-tibial angle shown on the right.) ((A) after Burns, 1974; (B) from Burns, 1974).

tonic responses and respond only to extreme deflection of the tarsus; while the 'smaller fibres' are tonic and unidirectional over the full range of deflection of the tarsus. A third type of scolopale (Group 3) is present but has afferent fibres that are too small to record from.

The overall resting spike discharge of a connective chordotonal organ signals joint position. In the locust femoral chordotonal organ, this summed tonic activity has a minimum impulse frequency at a femoro-tibial angle of about 60° for the metathoracic leg (Usherwood *et al.*, 1968) and about 80° for the two scoloparia of the pro-and mesothoracic legs (Fig. 8.6B; Burns, 1974; Varela *et al.*, 1976). These differences are functionally significant since they approximate to the mean angles at which the legs are held during walking (Burns, 1973), and in all legs the total activity/response curve is U-shaped around this angle. However, observations on a small number of large spikes from the response of the distal scoloparium of the mesothoracic organ show that tonic units are active only when the tibia is to one side of the mean leg angle (Burns, 1974). Thus, tonic neurons of the femoral organ become increasingly active as the tibia is moved in one direction away from the mean leg angle and it appears that the CNS of the locust receives detailed information on tibial position from two populations of receptors. Some evidence (Usherwood, Runion and Campbell, 1968) indicates that tonic neurons of the femoral chordotonal organ are recruited with increase in the tonic firing frequency. This would suggest that, as in crustacean chordotonal organs (Section 6.5.2), joint position is signalled by different combinations of neurons as well as by alterations in the firing frequency of neurons.

Examination of the movement of the apodeme of the femoral chordotonal organ may be useful in explaining properties of the tonic neurons. In the locust *Schistocerca gregaria*, examination of the attachment of the metathoracic organ through a 'window' cut from the cuticle of the joint (Wright, unpublished observations) shows that extension of the tibia causes the apodeme to be pulled for all positions of the joint, except for extreme extended positions, when relaxation occurs. Thus, both increasing and decreasing the tension on the apodeme are likely to stimulate tonic scolopidia. However, in the grasshopper *Melanopus bivittatus*, the apodeme of the chordotonal organ attaches over a wide area and both flexion and extension from a femoro-tibial angle of about 60° cause the apodeme to be pulled (Varela *et al.*, personal communication). Thus tonic scolopidia of this organ appear to be stimulated solely by pulling of the apodeme.

The tonic units of the tibio-tarsal connective chordotonal organ of the locust differ from those of the femoral organ in that the minimum overall tonic activity level of the tibio-tarsal organ occurs when the tarsus is fully levated (Young, 1970).

The properties and functions of chordotonal organs in the wing base and in the haltere are very poorly understood. It has been assumed that they are important in flight. However, in the cricket the single chordotonal organ at the wing base responds unspecifically to elevation and depression of the wing, but specifically to stridulatory movements (Huber, 1974). Möss (1971) suggests that this organ is sensitive to vibrations.

Simple multiterminal receptors
Physiological studies have been made on receptors at the tro-chanteral-femur joint and femoro-tibial joint (Guthrie, 1967) of the cockroach and at the femoro-tibial joint of the locust (Coillot and Boistel, 1968, 1969; Coillot, 1974, 1975). At the latter joint there are three multiterminal units (Coillot and Boistel, 1968; Section 8.2.1). Only two of these (RDPL and RDAL) respond to tibial movement or position and they are both unidirectionally responsive to tibial extension (Coillot and Boistel, 1969). The properties of the two cells of RVPL are not understood. They fire only on strong deformation of the overlying arthrodial membrane and they are probably inactive under most conditions of normal loading. The activity of RDPL and RDAL is separable into phaso-tonic and tonic components (Fig. 8.7a). The phaso-tonic component is related to the amplitude and velocity of the movements, the tonic to the stretch level of the receptors (stretch increasing with femoro-tibial angle). A few minutes are required for the receptors to completely adapt to their new tonic level.

The most significant difference between receptor units at the femoro-tibial joint is that they have different threshold angles for excitation (Coillot and Boistel, 1969). At a femoro-tibial angle of 90° only RDPL cells are tonically active, the single receptor of RAL becoming tonically active on moving the tibia to 120° (Fig. 8.7b).

In Orthoptera, and probably in other insects, there is a simple multiterminal receptor ('wing stretch receptor') at each wing base. In the locust, this receptor responds to elevation of the wing and during

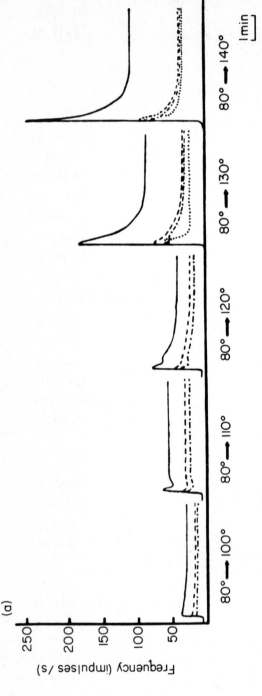

Fig. 8.7   (a) Graphs illustrating the phasic and tonic responses of multipolar cells at the femoro-tibial joint of the locust leg during extensions of the tibia through progressively increasing amplitude. The curve in black shows the combined activity of RDPL and RDAL. The discontinuous lines show the activity of individual cells (. . . . . .) and (– . – . – . –) are for the two cells of RDPL; (.. ...) is for the single cell of RDAL. (From Coillot and Boistel, 1969.)

(b)

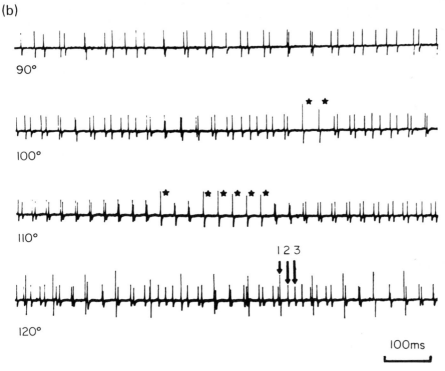

(b) Recordings from the main femoral nerve of the metathoracic leg to show the tonic activity of multiterminal receptors RDPL and RDAL for different angles of the femoro-tibial joint. 1 shows a spike from RDAL; 2 and 3 a spike from each neuron of RDPL. Note that RDAL is tonically active only at 120°. Stars indicate potentials where the spikes of RDPL cells have summated. (From Coillot and Boistel, 1969.)

flight produces a few spikes towards the end of each upstroke (Gettrup, 1962, 1963; Pabst, 1965). If the wing is held elevated the initial phasic response adapts to a tonic level whose frequency depends upon the extent of elevation. Physiological studies show that feedback from this receptor can influence the amplitude of the wing upstroke during flight (Burrows, 1975; Section 8.4.3).

Hair-plates and hairs
Hair-plates are positioned on the limb so that they are deflected by a fold of the cuticle of the joint membrane. Whereas isolated mechanoreceptive trichoid sensilla on the leg are phasically responsive (Runion and Usherwood, 1968; Spencer, 1974) hair-plate

sensilla have a marked tonic response (Pringle, 1938b). This tonic activity of hair-plate sensilla increases with displacement of the hairs by the arthrodial membrane and is related to joint position (Pringle, 1938b). Isolated (tactile) trichoid sensilla are often found in the arthrodial membranes of joints in the insect leg (Pringle, 1938b). However, their very rapid adaptation would allow them to provide information concerning movement only momentarily.

Chelicerate joint receptors

Physiological studies have been made on proprioceptors at the patello-femoral joints (Parry, 1960; Rathmayer, 1967) and the tarsol-metatarsal joint (Mill and Harris, 1976) of the spider leg, and at the patello-tibial joint in the pedipalps of the scorpion (Bowerman and Larimer, 1973). All these receptors monitor joint position and movement, and most contain a sufficiently small number of cells to allow the analysis of individual units. Three types of unit are found: phasic, tonic and intermediate, although in the scorpion pedipalp, purely phasic units are rare. In the spider and the scorpion phasic units are unidirectionally sensitive to flexion or extension, although at the femoro-patellar joint of the spider leg most units respond to extension. Tonic units comprise the majority of the cells in proprioceptors of the spider and scorpion and always show some degree of range fractionation, this being most marked in the scorpion.

The patello-tibial joint of the pedipalp of the scorpion functions as a ball and socket joint and all neurons of proprioceptors within this joint are biaxially sensitive. Thus a particular sensory unit is responsive to either flexion or extension and to either dorsal or ventral bending of the joint. Biaxially sensitive units have been found in other arthropod proprioceptors (Wyse and Maynard, 1964; Wales, Clarac, Dando and Laverack, 1970). However, these proprioceptors also contain units which are unidirectionally sensitive in a single plane of movement.

Muscle receptor organs

The MRO, or 'tendon receptor organ', found in the trochanter of each walking leg of *Limulus* is a position and movement detector for two joints (Section 8.2.2). Phasic units and a variety of phaso-tonic units are present and all the phasic units are unidirectionally responsive to joint extension (i.e. stretching of the accessory muscles). The phaso-tonic units differ in the proportion of phasic and tonic components, some responding primarily to changes in position, others to joint movement. Some of the tonic units show a considerable degree of range fractionation. Flexion at the joint temporarily inhibits the tonic response, which recovers to a lower discharge frequency (Fig. 8.8A) (Hayes and Barber, 1967). An unusually large number of units of this MRO are phaso-tonic.

Both accessory muscles of the *Limulus* MRO receive a motor innervation (Fig. 8.8B) and the neuromuscular properties of the accessory muscles indicate that they develop tension in a slow, graded fashion (Eastwood, 1971). Fountain (1973) has found that four motoneurons are shared between the accessory muscles and the trochanteral depressor muscle. Activation of these shared moto-neurons would normally occur with contraction of the trochanteral depressor muscle and consequently with flexion at the coxal-trochanteral joint. Thus the shared motoneurons could serve to compensate for tension loss in the accessory muscles during flexion. There are similarities of this control system to that of the crustacean abdominal MROs, where receptor muscles each share one moto-neuron with extensor muscles although these shared motoneurons of the crustacean MROs do not have many endings in the extensor muscles and are not the main tension inducing motoneurons for these muscles (Chapter 2). A similar situation may exist in *Limulus* since the accessory muscles also receive the endings of numerous unshared motoneurons (Fountain, 1973). Thus, as in crustaceans, excitation of the shared motoneurons in *Limulus* is unlikely to produce sufficient tension to offset the inhibitory action of flexion on sensory activity of the organ. The slow responses of the accessory muscle may further limit the rapidity of any compensation (Eastwood, 1971).

8.2.4   Mechanical sensitivity and correlation with structure

Connective chordotonal organs
It is often extremely difficult to relate the anatomical features of sense organs to their functional parameters, even when the physi-ology of the sense organ is well known. This is particularly true of connective chordotonal organs. So far, it has been possible to correlate relatively few of the structural features of these organs with

(A)

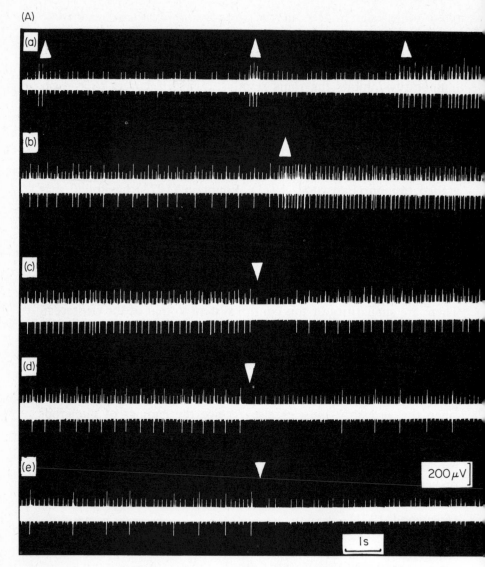

Fig. 8.8   (A) Responses from an intact muscle receptor organ of the leg of *Limulus*. The femur is extended and, initially, the trochanter is flexed. In (a) and (b), the trochanter is extended in 0.25 mm steps (pointers up) (A) large unit gives a phasic response to the first two movements and a tonic response to the third movement. The interval between (a) and (b) is 30 s. (c–e) flexion of the trochanter in 0.25 mm steps temporarily inhibits activity in the large unit. After (e) this unit does not respond to further flexion of the trochanter. The interval between (c and d) is 5 s and between (d and e) is 35 s. (From Hayes and Barber, 1967.)

(B)

(a)

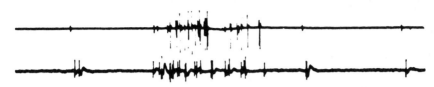

(b)

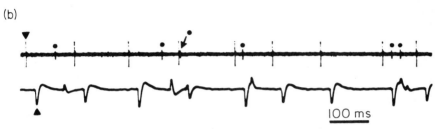

100 ms

Fig. 8.8 (B) Recordings from the leg nerves of *Limulus* to show co-activation of motoneurons of the trochanteral depressor muscle and accessory muscle of the muscle receptor organ. In (a) tactile stimulation of the gill opercula elicits a bursting discharge of units in the depressor motor nerve (upper trace) and the long accessory muscle (lower trace) (recorded extracellularly). (A) one-to-one correspondence between recordings of one unit is clearly visable. (b) One-to-one correspondence between units that are tonically active in the depressor motor nerve (upper trace) and potentials from the long accessory muscle (lower trace) (recorded extracellularly). The large spikes (indicated by a triangle) precedes the downward potential and the small spikes (indicated by a dot) precede small upward potentials (two sizes). (From Fountain, 1973.)

a specific function. In Orthoptera, the tibial and femoral connective chordotonal organs have been studied in some detail and show surprising structural complexity (Section 8.2.2). In these organs, and probably in other leg connective organs of insects, movement sensitive neurons are generally larger than those signalling joint position, are more proximally positioned in the organ and have larger spikes; a situation paralleled in the Crustacea (Chapter 6). Their information thus reaches the CNS quicker although the differences in conduction time between phasic and tonic fibres are unlikely to be of significant importance in behaviour (Chapman and Pankhurst, 1967). In the PD organ of crabs the flexion and extension sensitive movement cells insert into different surfaces of the receptor strand

(Section 6.2.1), but it is not known if a similar arrangement exists in the connective chordotonal organs of the insect leg.

In the pro-and mesothoracic legs of the locust the larger neurons are solely contained in the distal scoloparium (Burns, 1974) and all phasic information is derived from this scoloparium (Varela *et al.*, 1976); while the proximal scoloparium contains only tonic units (Varela *et al.*, 1976). Positional information is produced by the distal scoloparium but the accuracy of this information is considerably degraded by hysteresis. Burns (1974) has suggested that the locust could use the proximal scoloparium as the preferred source of information on tibial position since the distal scoloparium has no connections with neighbouring muscle and thus its responses may show less hysteresis.

An important functional feature of the tibio-tarsal organ in the cockroach leg, and perhaps chordotonal organs in general, can be related to the length differences of the scolopales (Young, 1970). Young suggests that the threshold of the response of different neurons in the tibial organ may be largely determined by scolopale length. Thus, depression of the tarsus to its extreme position (N.B. extension of the chordotonal ligaments corresponds to depression of the tarsus) would cause neurons with shorter scolopales to respond first, those with longer scolopales responding later in the movement. In the tibio-tarsal organ Group 1 scolopidia are the most proximally positioned and consequently have the longest scolopales (Fig. 8.4). These scolopidia respond only to extreme depression of the tarsus and thus length differences of the scolopales can account for their angular sensitivity.

A striking feature of the neurons of almost all scolopidia is some form of ciliary dilation in the dendrite (Chapter 9). This generally occurs in the part of the cilium just proximal to the cone-shaped cap that encloses the nerve terminal. Howse (1968) suggests that for most scolopidia the adequate stimulus may be stretch of the membrane of this dilation by flexion at the scolopale-cap junction and recent work by Moran *et al.*, 1975, supports this view. Moran *et al.* propose that slight bending of the dilation of the cilium can cause an active stroke in the cilium which travels to its base, where the movement distorts the cell membrane near the basal body. Young (1970), however, argues that in the tibio-tarsal connective organ of the cockroach, the adequate stimulus appears to be stretch of the ciliary apparatus.

Hair-plates

Hair-plate sensilla are unidirectionally sensitive to movement. In the hair-plate sensillum a ciliary structure separates the terminal segment of the dendrite from the distal nerve process. This terminal segment is filled with a cap structure (Thurm, 1964). Thurm (1965) suggests that the adequate stimulus in the hair-sensillum is compression of the terminal segment in the region of the terminal body.

Multiterminal cells

Multiterminal receptors of insects and arachnids are unidirectionally responsive. The simple multiterminal receptors of insect appendages whose responses have been examined are stimulated by joint extension if they are in the leg (Guthrie, 1967; Coillot and Boistel, 1968, 1969) and wing elevation, if at the wing base (Gettrup, 1962; Pabst, 1965). The multiterminal cells of the MRO of *Limulus* respond to tension in the accessory muscles (Hayes and Barber, 1967; Eastwood, 1971). Increasing the tension in the accessory muscles, as occurs during joint extension, increases the activity of each sensory neuron in the organ, while loss of tension, as occurs during joint flexion, results in an abrupt loss of activity in all units.

The absence of cilia and accessory structures in multiterminal receptors suggest a direct effect of the mechanical stimulus on the membrane of the dendrites. The ultrastructure of simple multiterminal cells in the legs has not been examined but studies on similar cells in the body of insects show that their sensory dendrites bear numerous fine terminations along their lengths (Chapter 4) which presumably serve to increase the sensitivity of the membrane to mechanical deformation.

8.3   Receptors monitoring muscle activity

In insects, arachnids, and probably also in myriapods, distinct receptors are found in the limbs that respond to cuticular stress. Insects have campaniform sensilla, while arachnids have slit sensilla and, in orthopteran insects, the femoral connective chordotonal organ may also be important.

It is well known that campaniform sensilla and slit sensilla measure cuticular stress (Pringle, 1961). Recent studies on the sensitivity of these organs (Barth, 1972a; Chapman, Duckrow and Moran, 1973; Spinola and Chapman, 1975; Chapman and Duckrow, 1975) and behavioural studies (Pringle, 1940; Pearson, 1972; Pearson and Iles,

1973; Pearson *et al*., 1974) show that campaniform sensilla and probably also slit sensilla, are active under normal loading conditions. In the legs these organs can serve to monitor (a) tension developed in muscles through the cuticular strain resulting from muscle activity; (b) resistances that exist or are imposed to movement at a joint, either by the animal's weight, external forces or both and (c) external forces which move or tend to move a part of the limb from its spatial relationship with other parts of the limb, or with the body. In myriapods, neurons with free nerve endings that supply hardened sutures have been described (Fuhrmann, 1922). These are probably directional strain receptors but their physiological properties have still to be examined. In *Limulus* distinct cuticular sensilla are absent and sensory neurons send processes between hypodermal cells into the lower layers of the cuticle (Pringle, 1956).

8.3.1   Variety and distribution

Campaniform sensilla

Campaniform sensilla are widely distributed on the body and in the appendages of insects (Fig. 1.17). In the legs they are generally associated with the joints and are particularly numerous on the trochanter (Pringle, 1938a; Bullock and Horridge, 1965). The cockroach leg for example has four groups of sensilla on the trochanter, one at the base of the femur, one on the base of the tibia, and one group on each segment of the tarsus. There are a total of about 100 sensilla on each leg, of which about 70 per cent are found on the trochanter (Pringle, 1938a). Campaniform sensilla normally occur in groups, but can be generally distributed as on the locust tarsus (Kendall, 1970) or found singly as at the base of each spine on the metathoracic tibia of the cockroach (Chapman, 1965). The function of the latter is unusual since the sensillum detects movement and position of the spine through changes in cuticular stress at the spine base. Campaniform sensilla are profuse at the wing base (Erhardt, 1916; Zaćwilichowski, 1934a, b; Pringle, 1957; Gettrup, 1965) and on the dipteran haltere (Weinland, 1891; Pflugstaedt, 1912; Pringle, 1948; Pringle, 1957; Uga and Kumabara, 1967; Smith, 1969; Chevalier, 1969). Their arrangement at the wing base is complex and there are at least two groups. Typically, sensilla of a group are arranged in packed parallel rows. A similar situation is found on the haltere. In addition, on the wing, there are scattered campaniform sensilla on the veins (Pringle, 1957).

Slit sensilla

Slit sensilla are numerous on the legs of arachnids (Fig. 1.14). Their topology is well known for web-building spiders (McIndoo, 1911; Vogel, 1923; Barth and Grill, 1970) and hunting spiders (Barth and Libera, 1970), for a scorpion (Barth and Wadepuhl, 1975) and for phalangids (Edgar, 1963).

In spiders slit sensilla occur as isolated single slit sensilla or as compound organs (generally known as lyriform organs). In a lyriform organ several slits lie closely side by side. Intermediate cases also occur (grouped single slit sensilla), where the slits are more distributed and lack the neat parallel arrangement of lyriform organs.

In scorpions only isolated and grouped slit sensilla are found in the legs (Barth and Wadepuhl, 1975), lyriform organs being completely absent. The legs of phalangids (Edgar, 1963) and amblypygids (Barth, unpublished observations) show a condition intermediate between spiders and scorpions, a single lyriform organ being present in the distal part of the trochanter.

Each walking leg of the hunting spider *Cupiennius salei* has over 3000 slit sensilla, of which about half are combined into groups (Barth and Libera, 1970).

In both spiders and scorpions, the slit sensilla are found on all segments of the leg, including the claw and post-tarsus. The vast majority are on the lateral surfaces and close to the joints, where they are orientated roughly parallel to the long axis of the leg (Barth and Grill, 1970; Barth and Libera, 1970; Barth and Wadepuhl, 1975). However, slit sense organs are far less restricted in their distribution than the campaniform sensilla of insects and they occur singly far more frequently.

8.3.2 Structure

Campaniform sensilla

The structure of the campaniform sensillum is well documented (Dethier, 1963). Each sensillum has a single bipolar neuron whose dendrite contains a ciliary process that inserts into a convex cuticular cap. The details of this organ are described in Chapter 1 (Fig. 1.16).

The dendritic ending of the neuron, together with two accessory ('enveloping') cells and an extracellular space, are contained within a cuticular canal (Uga and Kumabara, 1967; Moran and Chapman, 1968; Chevalier, 1969; Moran, Chapman and Ellis, 1969, 1971; Smith, 1969). The cap of the sensillum contains the rubber-like protein resilin (Chevalier, 1969) and is ringed by a border of denser cuticle. In surface view campaniform sensilla are circular or oval, the oval ones being directionally sensitive (Section 8.3.3).

Typically, two accessory cells are present. The 'inner enveloping cell' (accessory supporting cell) surrounds the dendrite, and the 'outer enveloping cell' (enveloping cell) surrounds the inner one. A dense secretion tightly binds the tip of the neural process to the cap membrane and an extensive extracellular space surrounds this dense secretion and the distal part of the neural process. The ultrastructure of campaniform sensilla in the haltere (Smith, 1969; Chevalier, 1969) is almost identical to that of sensilla in the limbs (Moran *et al.*, 1969, 1971).

Slit sensilla
In the slit sensillum (Fig. 1.13) the distal process of one or two sensory cells traverses a fluid filled cavity to be inserted on, or near cuticle within the slit (Barth, 1971). The slit 'proper' consists of two parts. The upper part is a trough-shaped chamber in the exocuticle. This is covered by a thin membrane whose ultrastruture resembles that of the dermal layer of the epicuticle. Viewed from above a flat tear-shaped depression can be seen in the outer membrane. This depression deepens towards the middle to form a cylinder that projects like a finger to the interior of the slit. As in campaniform sensilla, the exocuticle is thickened around the edge of the slit outer membrane.

The upper part of the slit opens out like a bell into the lower which is contained by the exo-and endocuticle. This lower part forms a large fluid-filled cavity.

The slit is normally innervated by one dendrite (Pringle, 1961), but two can be present (Barth, 1971). The dendrite ends directly below the outer membrane, and the other dendrite, if present, ends close to the inner membrane. Both dendrites are composed of three portions that are remarkably distinct in their fine structure (Barth, 1971). The most distal portion contains a modified ciliary process. Two sheath cells surround the dendrites, the inner terminating at the level of the base of the ciliary structure, the outer at the outer membrane of the slit. Studies on a spider indicate that the ultrastructure of different kinds of slit sensilla is essentially the same (Barth, 1971).

Connective chordotonal organs
The femoral connective chordotonal organ of Orthoptera, as well as being a movement and position detector for the tibia may function to measure the length of the muscle that flexes the tibia. In the locust (Burns, 1974) the distal scoloparium of pro-and mesothoracic organs gives rise at its proximal end to two separate ligaments, of which the largest is attached to the flexor tibiae muscle (Fig. 8.1) and physiological studies (Burns, 1974) show that the femoral organ responds to stretching of the flexor tibaie muscle. A more complex situation exists in the jumping leg of the locust (Usherwood *et al.,* 1968). Here several ligaments link the femoral organ to the flexor muscle and there is an attachment to the extensor tibiae muscle. Also the main ligament connecting this organ to the flexor muscle appears to contain dendrites.

8.3.3   Physiology

It is well established that campaniform sensilla and slit sensilla respond to cuticular strain. Both are phaso-tonic receptors and their properties are very similar (Pringle, 1938a, 1955, 1961; Barth, 1972a). In campaniform sensilla the phasic component of the response is dependant upon the speed and amplitude of the change in tension, the tonic upon the level of tension.

Mechanical sensitivities and correlation with structure
The mechanical properties of the slit sensillum and the campaniform sensillum have been intensively studied. Several common features have been found. In both, compression deformation is the effective stimulus (Chapman, 1965; Barth, 1972a; Chapman *et al.,* 1973; Spinola and Chapman, 1975) and in both this causes indentation of the 'cap' membrane. The indentation of the cap necessary to produce a response has been measured in campaniform sensilla and is extremely small, 10−20 nm being the estimated minimum threshold (Chapman *et al.,* 1973; Chapman and Duckrow, 1975), and indentations between 10 and 50 nm the normal range for the sensillum.

Both slit sensilla and campaniform sensilla have pronounced directional properties. Studies on model slits made from plexiglass have shown that deformation of the slit is greatest to stress applied perpendicular to the long axis (Barth, 1972a). These directional properties are conferred by the slit's elongate shape and exaggerated by the exocuticular thickening that surrounds the covering membrane. The campaniform sensillum is also specifically sensitive to compression in the cuticle perpendicular to the long axis and its cap has a similar thickened cuticular ring. The minimum threshold on the cap to punctate stimulation occurs at the junction of the centre of the dome of the cap and the mid-line of the long axis of the sensillum (Chapman *et al.*, 1973).

The frequent occurrence of campaniform sensilla and slit sensilla in groups on the leg is of interest, since members of a group generally show a common orientation (Pringle, 1938a; Barth and Grill, 1970; Barth and Libera, 1970; Barth and Wadepuhl, 1975). Since these sensilla have pronounced directional properties the common orientation indicates that the group detect a particular line of cuticular stress. Considering the importance of campaniform sensilla in behaviour (Pearson *et al.*, 1974) it would be of interest to determine how sensilla of a group differ in their stimulus threshold.

8.4   Integration and behaviour

Proprioceptors are numerous on the appendages of insects and arachnids and the number of neurons in an appendage may be very large. (e.g. the spider *Cupiennius salei* has over 3000 slit sensilla on each walking leg (Barth and Libera, 1970)). Thus the quantity of proprioceptive information reaching the nervous system from an appendage at a particular time (e.g. during walking) may be considerable. Evidence to date generally supports the view that in insects, and probably also in arachnids, central processing of this information occurs exclusively in the neuropil.

Central processing of proprioceptive information is complex as it is influenced by a wide number of factors such as the topography of connections; integrative properties of afferent synapses; intrinsic properties of central neurons and properties of the neural network. Tyrer and Altman (1974) suggest that the complex structure of insect central neurons indicates that considerable integration may occur in the dendritic tree. Details of central processing are given in Chapter 16.

8.4.1 Receptor inputs

Our knowledge of the mechanisms underlying the integration of proprioceptive information from the limbs and wings of insects has greatly advanced in recent years. In particular, intracellular studies on the locust and the cockroach have begun to unravel the neural circuitry of integration in posture and locomotion and to show the part played by interneurons in different neural pathways (Hoyle and Burrows, 1973a, b; Burrows and Horridge, 1974; Pearson *et al.*, 1974; Burrows, 1975). Studies on the locust (Burrows and Horridge, 1974) indicate that proprioceptive information from the femoro-tibial and tibio-tarsal joints of the leg acts through combinations of interneurons to influence motoneurons that control leg muscles. However, this generallization cannot be extended to include joints of the coxa and trochanter since in the cockroach proprioceptors in these segments can be monosynaptically connected to motoneurons that control coxal muscles (Pearson *et al.*, 1974) (Section 8.4.3). Monosynaptic connections have also been found between the simple multipolar receptor at the wing base ('wing stretch receptor') and motoneurons that control flight muscles (Burrows, 1975).

Contralateral effects of proprioceptive input from the limbs and wings in insects appear to be very weak (Burrows, 1973a; Burrows and Horridge, 1974; Burrows, 1975; Fig. 8.9). Anatomical studies show that sensory arborizations and motoneuron dendrites rarely cross the midline in thoracic ganglia (Burrows, 1973b; Burrows and Hoyle, 1973; Tyrer and Altman, 1974; Burrows, 1975) and thus contralateral effects of proprioceptive input are due to interactions between interneurons.

Intersegmental effects of proprioceptive input from the limbs and wings have been found to be of importance in insect flight and insect walking (Pearson and Iles, 1972; Wendler, 1972, 1974; Burrows, 1975), and intersegmental projections of proprioceptive afference have been demonstrated anatomically (Bentlage, 1972; Tyrer and Altman, 1974; Burrows, 1975) by the use of cobalt staining techniques (Iles and Mulloney, 1971; Pitman *et al.*, 1972). Cobalt staining of wing stretch receptors of locusts (Bentlage, 1972; Tyrer and Altman, 1974; Burrows, 1975) reveal that central projections of these receptors are as complex as those of any insect thoracic motoneuron. The fore-wing stretch receptor has projections in all three thoracic ganglia and the hindwing receptor has projections in meso-and metathoracic ganglia (Fig. 8.10). Despite this

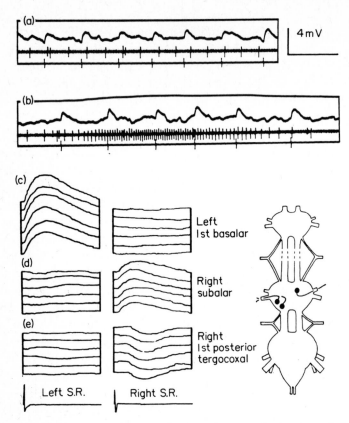

Fig. 8.9   Recordings to show that the forewing stretch receptor of the locust does not synapse onto contralateral mesothoracic motoneurons. (a) The ipsilateral stretch receptor (fourth trace) and contralateral stretch receptor (third trace) spike tonically when the fore wings are held stationary (first trace), but only the ipsilateral stretch receptor evokes EPSPs in the impaled 1st basalar motoneuron (second trace). (b) Elevation of the contralateral wing causes a high frequency of spikes in its own stretch receptor but no EPSPs in the contralateral 1st basalar motoneuron. (c–e) Signal averaged PSPs from the first basalar (c), a right subalar (d) and a right first posterior tergocoxal monoteuron (e) of the same locust. The left-hand column shows excitatory potentials in the 1st basalar linked to the spike of the left forewing stretch receptor. The right hand column shows potentials (excitatory in subalar, inhibitory in tergocoxal) linked to that of the right stretch receptor. There are no contralateral effects. Calibration, horizontal: (a, b) 200 ms; (c–e) 26 ms. (From Burrows, 1975.)

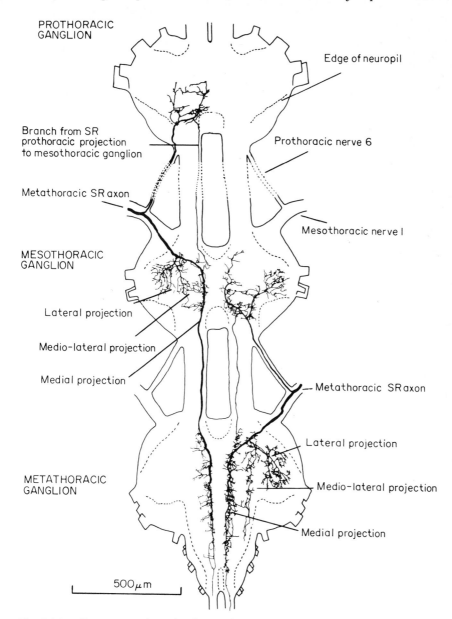

PROTHORACIC
GANGLION

Edge of neuropil

Branch from SR
prothoracic projection
to mesothoracic ganglion

Prothoracic nerve 6

Metathoracic SR axon

Mesothoracic nerve I

MESOTHORACIC
GANGLION

Lateral projection

Medio-lateral projection

Medial projection

Metathoracic SR axon

Lateral projection

METATHORACIC
GANGLION

Medio-lateral projection

Medial projection

500 μm

Fig. 8.10   Summary of projections of the meso- and metathoracic stretch receptors to the pro-, meso- and metathoracic ganglia in three species of locust. Main branching divisions of the projections are indicated. Characteristically, the projection of each receptor in the more anterior ganglion sends a branch ('loop-line') back through the connective to the next posterior ganglion. (After Tyrer and Altman, 1975).

complexity all branches of these receptors are entirely ipsilateral and lie within the dorsal neuropil. Anatomical studies on the wing stretch receptors support the conclusions of physiological studies (Burrows, 1975).

Projections of the wing stretch receptors, like those of other insect neurons (Pitman *et al.*, 1973), show a great deal of intraspecific variation. Major dendritic branches are consistent in position (Tyrer and Altman, 1974; Burrows, 1975) as are some fine branches, but small branches show considerable variation and the ramifications of very fine branches appear to have no consistent pattern. Nevertheless, tufts of fine fibres often occupy a particular region of the neuropil and the projections of the wing stretch receptors have a characteristic overall shape which is distinct from that of any motoneurons or interneurons.

The wing stretch receptors are monosynaptically connected to flight motoneurons (Burrows, 1975) and connectivity may be deduced from their coincidence in the neuropil, although this requires a close comparison of afferent and dendritic fields since the number of possible sites for synaptic contact is enormous. From serial sections of ganglia with stained profiles of both afferent and dendritic fields Tyrer and Altman (1974) have been able to demonstrate several close associations of sensory afference with dendrites of flight motoneurons.

The wing stretch receptors are a suitable system for developmental studies on proprioception and Altman 1975b has studied changes that occur in the hindwing stretch receptor during development in *Locusta* and *Chortoicetes*. The main branches of the receptor first appear in the meso-and metathoracic ganglia in the third instar, finer branches appearing in the fourth instar. Altman (1975a) suggests that a major effect of changes in sensory input from the wing base during the development of *Chortoicetes* is to progressively increase the average wingbeat frequency.

8.4.2   Reflexes

Pringle (1961) suggested that proprioceptive reflexes are of two basic types: 'regulatory', where sense organs provide negative feedback, and 'compensatory', where sense organs produce positive feedback that assists a regulatory process, and argues that all other kinds of adjustment may be produced by action or interaction of these two basic types.

In the limbs and wings of insects, arachnids and myriapods stretch receptor elements similar to those of the vertebrate muscle spindle appear to be absent and 'regulatory reflexes' are likely to involve proprioceptors that lie in parallel with muscle or muscle groups (e.g. connective chordotonal organs, joint receptors of arachnids, multi-terminal receptors of various kinds). 'Compensatory reflexes' are likely to involve force-sensing elements in series with muscle (campaniform sensilla and slit sense organs) (Fig. 8.12b).

In the leg, regulatory reflexes are of major importance in the control of posture. A characteristic regulatory reflex is the 'resistance reflex'. This is generated by forced flexion or forced extension at a joint. In insects, resistance reflexes are a feature of all major joints of the leg (Wendler, 1961, 1964, 1965; Wilson, 1965; Bässler, 1972a, b; Burrows and Horridge, 1974) as they are also in decapod crustaceans (Section 6.8). Intracellular studies on the locust (Burrows and Horridge, 1974) show that forced movement at the femoro-tibial joint or tibio-tarsal joint produces proprioceptive feedback that excites one set of motoneurons and inhibits their antagonists (Fig. 8.11). For example, forced flexion of the tibia excites extensor tibiae motoneurons and inhibits 'slow' flexor tibiae motoneurons (Fig. 8.12a).

At the femoro-tibial joint in the stick insect, Bässler (1972a, b) has demonstrated that resistance reflexes are generated by the femoral connective chordotonal organ. Bässler stimulated this organ inde-pendantly of movement of the tibia and found 'resistance-like' reflex movements of the tibia. Even slow tension changes on the femoral organ (equivalent to 0.002 mm/min movement of the apodeme) are effective and these reflexes can markedly slow the return of the tibia to its original position after passive movement. In the metathoracic leg of the locust. resistance reflexes of the femoral chordotonal organ appear to involve extensor tibiae muscle motoneurons and only 'slow' motoneurons of the flexor tibiae muscle; 'intermediate' and 'fast' motoneurons receiving an excitatory input on flexion of the tibia (Wright, 1976). This excitatory input has a longer latency and is more phasic than that produced when the chordotonal organ is stimulated by extension of the tibia through the same angle. Also, Burrows and Horridge (1974) find that the tension generated in femoral muscles by reflexes involving movement-sensitive proprio-

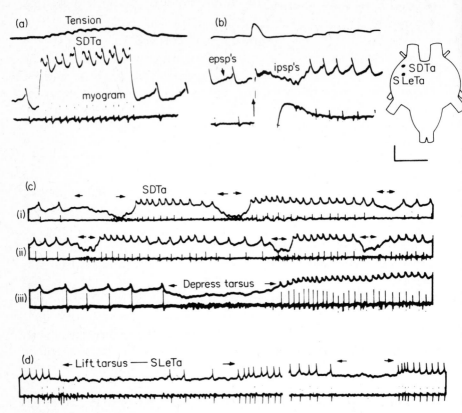

Fig. 8.11   'Resistance reflexes' from the locust metathoracic segment involving the tarsal levator muscle and tarsal depressor muscle. (a) Identification of a 'slow' tarsal depressor motoneuron. Upper trace registers tension developed to cause tarsal depression; middle trace is intracellular from the soma; lower trace is extracellular myogram from the muscle. The slow depressor motoneuron is spontaneously active and a depolarizing pulse accellerates spiking and causes increased tension development. (b) A stimulus was given to the depressor muscle (arrow). The brisk contraction was immediately followed by a train of summating ipsp's which inhibited the tonic discharge. (c) Spontaneous tonic activity in a 'slow' depressor motoneuron (SDTa) is inhibited during tarsal depression (between arrows). Upper traces: intracellular from the motoneuron somata; lower traces myograms. On release from depression, firing recurs, starting at a higher rate than normal and then accommodating. (d) Corresponding discharge in a 'slow' tarsal levator motoneuron (SLeTa) is inhibited by lifting the tarsus (between arrows). The inhibition is the result of a continued train of ipsp's. Calibration: vertical, 15 mv (a, b), 20 mv (c, d); horizontal, 400 ms (a, b), 250 ms (c, d). (From Hoyle and Burrows, 1973b.)

ceptors at the femoro-tibial joint is very low and passive resistance of flexor muscles to stretching appears to be of far greater importance in restoring the leg to its original position than in the stick insect.

At the proximal, leg segments of the cockroach and stick insect resistance reflexes involve hair-plates (Wendler, 1961, 1964; Pearson *et al.*, 1974). These receptors are located on the trochanter in the cockroach, while in the stick insect they are on the coxa. Stimulation of the trochanteral hair-plate in the cockroach monosynaptically excites the slow motoneuron that controls the coxal depressor muscle and inhibits motoneurons controlling coxal levator muscles (Section 8.4.3) (Fig. 8.14a). A similar control system exists for the coxal hair-plates of the stick insect, although the details of the pathway have not been analysed to the same extent (Wendler, 1961, 1964). In the stick insect, and probably also in the cockroach, hair-plate reflexes assist in maintaining the body of the animal at a constant height relative to the substratum.

Interaction of proprioceptors at one joint with motoneurons controlling muscles of another joint has been shown to be important in the control of the posture of the tarsus in the locust (Burrows and Horridge, 1974). Burrows and Horridge found that movement-sensitive proprioceptors at the femoro-tibial joint partly control the posture of the tarsus by influencing the level of activity (or 'set point') of tarsal motoneurons. Extension at the femoro-tibial joint tonically excites slow tarsal depressor motoneurons while inhibiting the slow tarsal levator motoneuron and visa versa. This inter-joint reflex of the locust tends to maintain the tarsus horizontal to the substratum during movements of the leg.

In insects 'compensatory reflexes' have been found involving proprioceptors on the coxa of the cockroach leg and at the femoro-tibial joint of the locust (Fig. 8.12b). The proprioceptors involved on the coxa are campaniform sensilla (Pringle, 1940; Pearson, 1972; Pearson *et al.*, 1974), and the principle proprioceptors involved at the femoro-tibial joint are probably the same (Burrows and Horridge, 1974). The campaniform sensilla on the coxa are stimulated by leg retraction and form part of a positive feedback loop to increase the tension of the coxal muscles (e.g. when the animal is walking up an inclined surface). It is of interest that reflexes generated by load-sensitive receptors can be far stronger than those produced by receptors sensitive to motion. At the femoro-

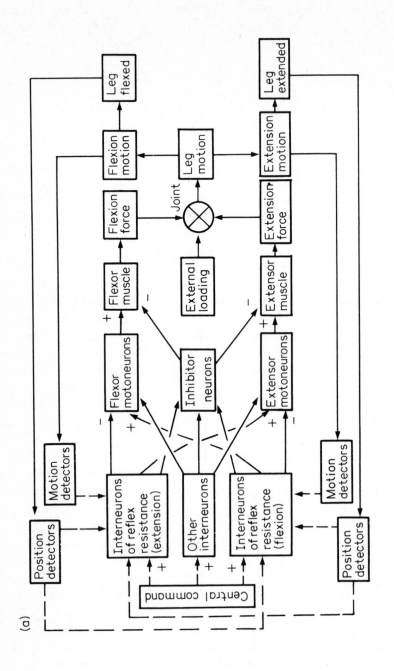

tibial joint effects of load-sensitive receptors are particularly strong on to the 'fast' motoneuron of the extensor tibiae muscle (Burrows and Horridge, 1974) which, in the metathoracic legs, is important in initiating the jump.

8.4.3   Coordination of movement

Neuronal mechanisms underlying the coordinated movements of the limbs and wings of insects have been intensively studied in recent years and significant progress has been made in understanding the role of proprioceptive feedback. In some cases the information is sufficient to construct a wiring diagram (Wendler, 1964, 1974; Pearson *et al.*, 1974; Burrows, 1975). Our knowledge of proprioceptive involvement in arachnid locomotion is far less advanced. In

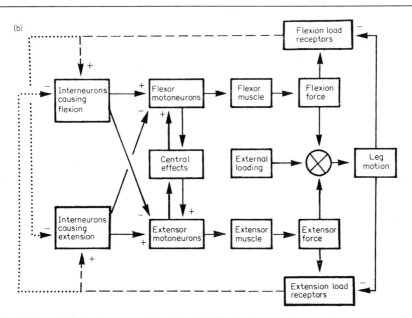

Fig. 8.12   Block diagrams showing feedback loops of proprioceptors at the femoro-tibial joint of the locust metathoracic leg which respond: (a) to movement and position of the joint; (b) when the flexor or extensor muscle of the tibia encounters an external 'load'. Solid lines show observed effects, dashed lines inferred effects. Dotted lines show other possible effects that were shown unlikely by experiment. The sign indicates the excitatory (+) or inhibitory (−) nature of the interaction. Most of the blocks shown are not single, but numerous elements in parallel. (From Burrows and Horridge, 1974.)

this group research has been mainly directed towards analysing leg movements and motor output during walking (Wilson, 1967b; Land, 1972; Bowerman, 1975a) and evidence of proprioceptive feedback comes mainly from experiments involving fixing or amputating appendages (Wilson, 1967b; Bowerman, 1975b). However, some work has been carried out on the role of vision in turning, in jumping spiders (Land, 1972) and on lyriform organs in the kinesthetic orientation of hunting spiders (Barth and Seyfarth, 1971; Seyfarth and Barth, 1972). Little is known of locomotory mechanisms in myriapods. The sequence of leg movements has been analysed (Manton, 1958, 1961; Smolyaninov and Karpovich, 1972) and segmental coordination resulting from sensory feedback (Pringle, 1961) and an angle sense (Banwell, 1965) have been found in the legs of millepedes.

In insects there is now good evidence that the basic pattern of motor output of several kinds of coordinated movements involving the limbs and wings is centrally generated reviewed by (Huber, 1974). In locust flight (Wilson, 1961), cockroach walking (Pearson, 1972), cricket song (Kutsch and Otto, 1972) and probably in cricket courtship (Elsner and Huber, 1969) the neuronal system (or systems) that generates the basic pattern is contained in the thoracic ganglia and its operation is controlled by command interneurons (Chapter 16) that arise in the brain. Proprioceptive feedback appears mainly to modulate the central pattern (Pearson, 1972; Pearson *et al.*, 1974; Burrows, 1975) while other sensory inputs can be important in initiating or terminating the pattern (reviews by Pringle, 1957; Huber, 1974).

Flight

Studies on flight constitute the most advanced analysis of a single kind of insect behaviour to date and several sensory feedback loops have been found. Sensory information can influence flight velocity (Heran, 1959; Heran and Lindauer, 1963; Wendler, 1974), lift (Weis-Fogh, 1956a, b; Gettrup and Wilson, 1964; Gettrup, 1966; Camhi, 1969, 1970a, b), average wing beat frequency (Wilson and Gettrup, 1963; Wilson, 1967a; Gorelkin, 1973), wing beat amplitude (Burrows, 1975) and the phase coordination of wings (Pabst and Scwartzkopff, 1962; Wendler, 1972, 1974).

In Orthoptera, flight is of the neurogenic ('synchronous') type and proprioceptive feedback from the wings is important in maintaining the mechanical coupling between wings (Wendler, 1972, 1974; Burrows, 1975). In contrast, in those insects where flight is of the myogenic ('asynchronous') type (e.g. Coleoptera, Diptera, Hymenoptera) constant phase coordination between wings appears to depend largely upon the mechanical properties of the thorax and its musculature (Pringle, 1965).

In Orthoptera three kinds of proprioceptor may be found at the wing base; a single chordotonal organ; a simple multiterminal receptor and at least two groups of campaniform sensilla (Pringle, 1957; Gettrup, 1962; Möss, 1967). Of these, the simple multiterminal receptor (the 'wing stretch receptor') and the campaniform sensilla (Gettrup, 1965, 1966) have been shown to be involved in flight. The role of the wing stretch receptor has been studied in the locust (Gettrup, 1962, 1963, 1966; Wilson and Gettrup, 1963; Wilson and Wyman, 1965; Gorelkin, 1973; Burrows, 1975). This receptor produces phasic information concerning movement of the wing during the upstroke (Gettrup, 1962, 1963; Pabst, 1965).

For some time, despite intensive study, experiments were unsuccessful in revealing the coordinative role of this receptor in flight. It was found that ablation of the wing stretch receptor in an otherwise intact animal merely lowered the average wing beat frequency (Wilson and Gettrup, 1963; Wilson and Wyman, 1965; Wilson, 1967a, Gorelkin, 1973). No phase reaction could be detected in the flight motor even when the peripheral nerve stump that contained the axon of the receptor was stimulated in the flight rhythm (Wilson and Wyman, 1965). Nerve stimulation simply raised the average wing beat frequency with a period of one or two seconds needed to reach maximum effect. Wilson and Wyman (1965) concluded that the detailed phasic information of the wing stretch receptor appeared not to be used in a coordinating way in controlling flight.

A different approach to studying the role of proprioception in the control of flight has been developed by Wendler (1972, 1974). Using a mechanical armature attached to the wing of a locust, he applied rhythmical forced movements to the wing during flight. Wendler found that the movements of the driven wing affect the phase of the motor output to all wings such as to bring them into a fixed, or preferred, phase with the rhythm of the driven wing. Entrainment is rapid, occurring within two or three wingbeats. However, this proprioceptive control by receptors at the wing base was only evident when the rhythm of the driven wing differed from that of the undriven wings by less than 10–15 per cent. Wendler's results strongly suggest that proprioceptive feedback from the wing could influence the flight motor, and show that proprioceptors at the wing base are of considerable importance in flight.

Recently, Burrows (1975) has shown that the wing stretch receptor of the locust is monosynaptically connected to ipsilateral flight motoneurons. The connections are intersegmental but not contralateral and the pathways are wired for the receptor to provide negative feedback (Fig. 8.13). Thus, elevation of the wing excites the wing stretch receptor. The discharge of this receptor then acts to weakly inhibit wing elevator motoneurons and to excite wing depressors. Stimulation of the wing stretch receptor in a simulated flight pattern causes sub-threshold waves of depolarization in depressor motoneurons. This input when summed with an un-patterned input is able to control the production of spikes. Thus, in flight, the feedback of the wing stretch receptor could influence the time at which flight motoneurons spike and hence affect the amplitude of the upstroke. This feedback could also affect the phase relationship between spikes in motoneurons of antagonistic flight muscles.

Of other receptors at the wing base, the campaniform sensilla of the hind wing of the locust have been shown to be important for forewing twisting in constant lift reactions (Gettrup, 1965, 1966). Intersegmental control is clearly indicated but the specific nature of the pathways are unknown. The functions of chordotonal organs in the wing base are not understood. In the cricket the single chordotonal organ at the wing base does not respond specifically to wing movements and may be a receptor for vibrations or for stridulatory wing movements (Möss, 1971; Huber, 1974).

It is not known if proprioceptors in the legs are important in the control of flight. In flight the legs are held in a characteristic extended posture and certain proprioceptors such as the simple multiterminal receptors at the femoro-tibial joint would be strongly stimulated (Colliot and Boistel, 1969).

Walking

The most common stepping pattern of walking on a smooth surface in adult insects is the alternating tripod gait (Hughes, 1952; Wilson, 1966; Delcomyn, 1971, 1973a, b; Graham, 1972; Burns, 1973; Hughes and Mill, 1974). This gait may vary slightly with increase in the walking speed, and at slow walking speeds (in the cockroach less than 2 steps

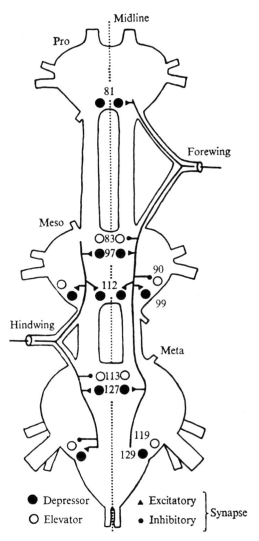

Fig. 8.13 Connections of the left fore-and right hindwing stretch receptors with some flight motoneurons. No contralateral connections were revealed. The diagram shows 10 of the 29 flight motoneurons. Wing depressors: dorsal longditudinal (81 and 112), first basalar (97 and 127) and subalar (99 and 129); wing elevators: tergosternal (83 and 113) and first posterior tergocoxal (90 and 112). (From Burrows, 1975.)

per second) phase relationships between legs often differ from those of an alternating tripod. Pearson (1972) has convincingly demonstrated the existence of a central programme in the thoracic ganglia of the cockroach for generating the motor output of the basic walking pattern.

The walking pattern of arachnids is an alternating tetrapod. Adjacent legs on each side step alternately and corresponding legs on opposite sides also alternate (Wilson, 1967b; Land, 1972). The leg 'oscillators' are more loosely coupled than in insects and there may be great varibility in the timing of individual steps. Nevertheless, the alternating tetrapod appears to be maintained at all speeds (Wilson, 1967b).

The walking patterns of myriapods are quite different to those of insects and arachnids. Legs on the same side move in synchrony and their is a constant phase difference between each leg and the leg in front (Manton, 1961).

In insects it has been known for some time that proprioceptive input from the legs is important in determining the motor output during walking. Experiments involving the amputation of legs (von Buddenbrock, 1921; Ten Cate, 1936; Hughes, 1957) showed that loss of a leg, especially complete removal of the middle leg, markedly alters the intersegmental coordination of remaining legs. Similarily, in arachnids, large changes in leg intersegmental coordination are found after leg amputation (Wilson, 1967b; Bowerman, 1975b). More recent amputation experiments on insects (Delcomyn, 1971) indicate that peripheral receptors strongly influence the timing of leg movements while the cycle of leg protraction and retraction is largely under central control.

Several proprioceptors of the leg have been shown to be important in the control of walking in insects (Wendler, 1961, 1964; Bässler, 1967; Usherwood *et al.*, 1968). Usherwood *et al.*, suggest that the femoral chordotonal organ may play a significant role in walking in the locust. Removal of this organ from the hindleg has the effect of causing the tibia to move through a larger arc than normal during walking. The stepping frequency of the hindleg is reduced and the alternation between hindlegs of the different sides is less regular. There is also a general loss of power in the muscle that extends the tibia.

In contrast, in the stick insect sectioning the apodeme of the femoral chordotonal organ has no noticeable affect on the behaviour of the leg during walking (Bässler, 1967), but if instead the input of the femoral chordotonal organ to the CNS is reversed by attachment of the apodeme of the organ to the tendon of the flexor tibiae muscle there is a marked change in the behaviour of the operated leg (Bässler, 1967). During walking the leg is held up and maintained fully extended. Presumably this posture is due to positive feedback from the chordotonal organ greatly exciting the muscle which extends the tibia. However, if the tarsus of the elevated leg is now touched, the tibia immediately flexes, the leg appears to execute a step and on being lifted clear of the ground returns to its original elevated position. Thus, it appears that in walking the sensory feedback from the tarsus can override the reflex action of the femoral chordotonal organ. Further studies (Bässler, 1972a, b, 1973, 1974) lead Bässler to conclude that during some activity (e.g. searching movements) the resistance-reflex action of the femoral organ may be reversed. Stretching of the apodeme of the organ (this normal occurs on tibial flexion) causes tibial flexion. Releasing the apodeme has no noticeable effect. In contrast, Burrows and Horridge (1974) found that resistance reflexes generated by movement of the femoro-tibial joint of the metathoracic leg of the locust are still present during voluntary movements. However, the tensions generated by proprioceptors at this joint are very low and appear to have an insignificant effect in preventing voluntary movements.

Sensory inputs from the tarsus are of special importance during walking since they provide valuable information concerning the contact of the leg with the substratum. In the stick insect, stimulation of the tarsus can produce leg protraction (Bässler, 1967). In the locust, removal of the tarsus results in the tibia being held free of the ground (Runion and Usherwood, 1968) and if all tarsi are removed, the animal will not walk. Although proprioceptors may be involved, the tarsal hairs are likely to be primarily involved in producing these effects.

Proprioceptive feedback from receptors in the coxa and trochanter can be of major importance in walking. Indeed, only in the case of receptors on these segments is the precise mechanism understood whereby proprioceptive input acts to pattern the motor output of the walking animal (Pringle, 1940; Wendler, 1961, 1964; Pearson, 1972; Pearson and Iles, 1973; Pearson et al., 1974). Studies on the cockroach show that the hair-plate sensilla on the coxa and the campaniform sensilla on the trochanter are important in regulating

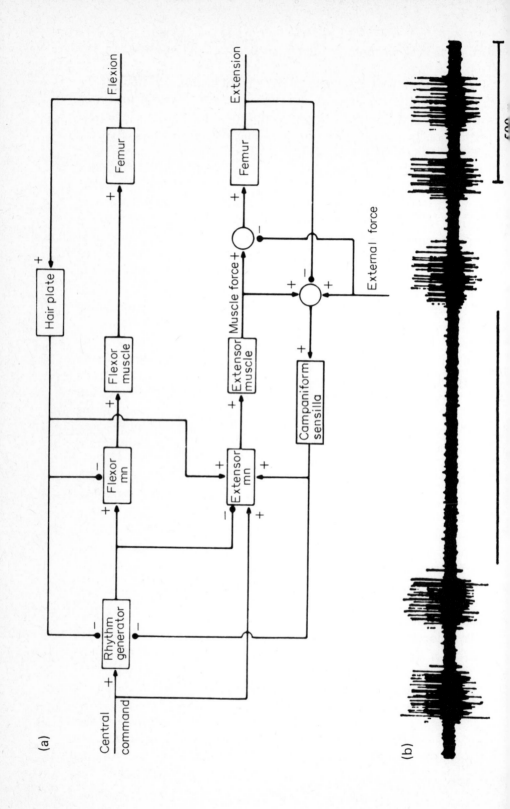

(a)

Central command

Rhythm generator

Hair plate

Flexor mn

Flexor muscle

Femur

Flexion

Extensor mn

Extensor muscle

Muscle force

Femur

Extension

Campaniform sensilla

External force

(b)

500

the motor output of coxal muscles during walking. (Fig. 8.14) (Pearson, 1972; Pearson and Iles, 1973; Pearson *et al.*, 1974). The hair-plates are probably active during leg protraction and removal of these receptors results in a more intense burst in coxal levator motoneurons and, consequently, in a more exaggerated protraction movement of the leg. Removal of the coxal hair-plates of the stick insect has been found to produce similar exaggerated stepping movements (Wendler, 1964). In the cockroach the coxal-hair plates have three distinct actions which influence walking (Pearson *et al.*, 1974) (Fig. 8.14a). First, they elicit a short-latency inhibition of coxal levator motoneurons and this pathway can account for the increased duration of leg protraction when the hair-plates are removed (Wendler, 1964; Pearson *et al.*, 1974). Second, the hair-plates act monosynaptically to excite the 'slow' coxal depressor motoneuron, a reflex which facilitates the initiation of the spike burst in this motoneuron, but doubtless also functions to stabilize the leg during standing (Section 8.4.2). Third, the hair-plates act on the rhythm generator to depress the burst rate of levator moto-neurons.

The campaniform sensilla on the trochanter of the cockroach leg are probably excited during leg retraction and can have at least two effects in the walking insect (Fig. 8.14a). First, they can strongly excite the 'slow' coxal depressor motoneuron and thus serve as a mechanism for load compensation, as they reinforce the activity of this motoneuron at a time when it is producing a burst of spikes. Since forward movement of the animal generally results from the action of this motoneuron on the depressor muscle, the effect of

---

Fig. 8.14   (a) Block diagram to show known and postulated organization of the system controlling rhythmic movements of the femur in the hind legs of a walking cockroach. The sign indicates the excitatory (+) or inhibitory (−) nature of the interaction. The hair-plates are active during flexion of the femur (leg protraction), and the campaniform sensilla are active during extension of the femur (leg retraction). Extensor mn: motoneurons which cause leg extension (e.g. 'slow' coxal depressor motoneuron); flexor mn: motoneurons which cause leg flexion (e.g. coxal levator motoneurons). (b) An example of one of the effects indicated in (a): rhythmic burst activity in coxal levator motoneurons is inhibited by stimulation of the trochanteral campaniform sensilla. Light pressure was applied to the trochanter during the interval indicated by the horizontal bar. ((a) from Pearson *et al.*, 1974; (b) from Pearson and Iles, 1973.)

'load' on the trochanteral campaniform sensilla will decrease. The 'slow' depressor motoneuron will therefore not be continually excited unless forward movement of the limb is impossible. Secondly, the campaniform sensilla can act on the central generating system of the walking pattern to strongly inhibit the production of levator bursts in the same leg (Fig. 8.14b). In an animal under load this reflex would act to decrease the stepping rate since coupling of the rhythm generator systems in adjacent ipsilateral legs ensures that no two adjacent legs step at the same time. Pearson *et al.* (1974) suggest that the inhibitory pathway of the campaniform sensilla may be a mechanism for controlling the movement of individual legs when the cockroach is walking over an uneven surface. Since increased activity in this inhibitory pathway can inactivate the rhythm generator system of the leg the leg can be prevented from, or delayed in, stepping. This inhibition of the rhythm generator system would decrease when an adjacent stepping leg finds solid support and begins to carry some of the animal's weight, since this would reduce the load on adjacent legs. Thus stepping of a cockroach walking on uneven terrain may occur in those legs carrying least load. If so, the stepping pattern of the legs may not be expected to show precise relationships.

Song

Proprioceptive control of song production in insects is generally considered to be small in comparison to that involved in flight and locomotion. Removal, fixing or loading of a leg or wing may lead to a reduction in motor activity, but the effect on the song is very slow to appear (Huber, 1974). However, in insects which stridulate with their wings (such as the cricket) subcostal sensory pads of sensilla trichoidea at the end of the 'file' on the forewing can have an important role in the control of stridulation (Möss, 1971). After laquering these sensory pads in the male cricket, the animal is unable to adopt the normal singing posture and, although stridulatory movements occur, they are soundless. Also, in the male of some species of grasshopper (Huber, 1974) the two hind legs have different stridulatory patterns and a change of stridulatory pattern that normally occurs regularly between the legs disappears when one leg is amputated or fixed. Thus, in this case, peripheral control affects the timing of the motor output.

Courtship

In the male cricket, *Gomphocerippus rufus*, courtship is complex and consists of several neural sub-systems linked in a single programme (Elsner and Huber, 1969). Fixation, amputation and loading of legs does not affect the temporal sequence. Indeed, the whole sequence appears to be mainly centrally controlled, and peripheral feedback from the legs or other parts of the body is only able to influence small details in parts of the courtship pattern.

8.4.4 Connections with higher centres and central effects

In insects, arachnids and myriapods very little is known about mechanisms underlying the integration at higher centres of proprioceptive information from the limbs and wings. In insects, evidence that proprioceptive information from the brain is used at higher centres is demonstrated by the influence of leg proprioceptors in the perception of gravity (Bässler, 1965; Wendler, 1966, 1971) (Chapter 12). Insects do not have statocysts, and gravity is detected through the complex integration of information from several kinds of mechanoreceptor from different parts of the body (Section 12.3). In arachnids, complex behaviour that involves proprioceptors in the legs has been found in hunting spiders (Barth and Seyfarth, 1969; Seyfarth and Barth, 1972). These animals make use of information from lyriform organs in orientation, which are important in food searching as they enable the spider to return to a previous source of food. In myriapods, an angle sense has been found in the legs of a millepede (Banwell, 1965) (Section 14.2.2).

The influence of non-associative learning phenomena in insects and arachnids on proprioceptive feedback from the limbs is poorly understood. Zilber-Gachelin and Paupardin (1974a) have shown that sensitization can increase the size of leg motor reflexes, although effects are transitory and unspecific. In some insects desensitization is known to be important (Zilber-Gachelin and Paupardin, 1974b, c). Physiological studies on the cockroach (Zilber-Gachelin and Paupardin, 1974b) indicate that this phenomenon is solely dependent on proprioceptive feedback from the legs. However, the process is complex, desensitization requiring the stimulation of more than one proprioceptive structure.

Our knowledge of mechanisms underlying learning and memory in arthropods is still at an early stage. In insects the only learning response involving the limbs that has been studied more than superficially is learning of leg position in cockroaches and locusts (Horridge, 1962; Hoyle, 1965, 1970). Such studies emphasize the independance of the sensory input from the learning process, as learning has been shown to be produced by an unspecific stimulus, such as an electric shock, or established in a ganglion completely isolated of sensory input (Hoyle, 1970). However, this feature of learning is more probably related to the general nature of the learning process since the value of learning to the animal would be expected to depend on its association with specific sensory information (Chapter 14).

8.5   Conclusions

Studies on proprioceptors in the limbs and wings of insects, and in the limbs of arachnids have produced a great deal of new information in the last decade. In insects intracellular recordings from central neurons and newly developed neuro-anatomical techniques have lead to a significant advance in our understanding of mechanisms underlying the integration of proprioceptive feedback. In some cases it may now be possible to analyse in detail the physiology and anatomy of several of the neurons (both motor and sensory) involved in a single kind of behaviour (e.g. insect flight (Tyrer and Altman, 1974)). Extensive physiological studies are now required to determine the circuitry underlying proprioceptive reflexes and a quantitative neurophysiological-neuroanatomical approach is required to examine the spatial distribution and other properties of synapses between proprioceptors and central neurons. Indeed, insects are particularly attractive among invertebrates for quantitative studies on mechanisms underlying the integration of proprioceptive information, as their nervous system contains relatively few motoneurons and their behaviour is varied, yet partly stereotyped.

The existence of central programmes which generate several kinds of behaviour involving the limbs and wings of insects appears now to be firmly established (Huber, 1974). The role of proprioceptive feedback in these forms of behaviour has not been neglected, although its importance may have been underestimated. Indeed, Pearson *et al.* (1974) suggest that in locomotory behaviour proprioceptive feedback may be of such importance that an animal walking over an uneven surface may show no precise pattern of leg movements. Thus, in the case of walking, and probably other kinds of behaviour, studies are needed to examine the role of proprioceptive feedback in different environmental situations (e.g. climbing, walking upside-down) as proprioceptors may vary in their relative importance, and even different proprioceptors may be involved in different situations.

The eventual aim of studies on proprioceptors in the limbs and wings, as well as explaining their importance in locomotion and other behaviour, must be to obtain a comprehensive understanding of their structure, variety, distribution and physiology. In arachnids and myriapods there are considerable gaps in our knowledge in all these areas. Without this information it will not be possible to build up an accurate description that shows the full role of limb and wing receptors in behaviour. Some of these gaps represent long-standing problems and perhaps require more immediate attention. Thus, more information is needed about the structure of proprioceptors within the joints of the arachnid leg as it is still not known if the bipolar-like neurons of these receptors are a distinct type, perhaps having similarities to insect chordotonal sensilla. Information is required concerning the specific phasic and tonic properties of the sensory cells of insect chordotonal organs. A systemmatic analysis is needed of the properties of chordotonal organs at the wing base to distinquish vibration receptors from those sensitive to wing movement. These, of course, are just a few of the important problems needing to be answered.

The near future holds great promise for a further significant advance in our knowledge of mechanisms underlying the integration of proprioceptive information from receptors in the limbs and wings. It is also likely that the receptors of the wing base in orthopterans will prove to be ideally suited for developmental studies on proprioceptors.

References

Altman, J. (1975a) Changes in the flight motor pattern during the development of the Australian plague locust, *Chortoicetes terminifera. Journal of Comparative Physiology*, **97**, 127–142.

Altman, J. (1975b) Development of wing sensory projections. (in preparation).

Banwell, F. H. (1965) An angle sense in the orientation of the millepede. *Biological Bulletin*, **128**, 33—50.

Barber, S. B. (1956) Chemoreception and proprioception in *Limulus*. *Journal of Experimental Zoology*, **131**, 51—74.

Barber, S. B. (1958) Properties of *Limulus* articular proprioceptors. *Anatomical Record*, **132**, 409.

Barber, S. B. (1960) Structure and properties of *Limulus* articular receptors. *Journal of Experimental Zoology*, **143**, 283—322.

Barber, S. B. and Segel, M. H. (1960) Structure of *Limulus* articular proprioceptors. *Anatomical Record*, **137**, 336—337.

Barber, S. B. and Hayes, W. F. (1964) A tendon receptor organ in *Limulus*. *Comparative Biochemistry and Physiology*, **11**, 193—198.

Barbier, R. (1961) Contribution à l'étude de l'anatomie sensori-nerveuse des insectes trichoptères. *Annales des sciences naturelles (Zoologies)*, **3**, 173—183.

Barth, F. G. (1971) Der sensoriche Apparat der Spaltsinnesorgane (*Cupiennius salei* Keys., Araneae). *Zeitschrift für Zellforschung und mikroskopische Anatomie*, **112**, 212—246.

Barth, F. G. (1972a) Die Physiologie der Spaltsinnesorgane. I. Modellversuche zur Rolle des cuticularen Spaltes beim Reiztransport. *Journal of Comparative Physiology*, **78**, 315—336.

Barth, F. G. (1972b) Die Physiologie der Spaltsinnesorgane. II. Funktionelle Morphologie eines Mechanoreceptors. *Journal of Comparative Physiology*, **81**, 159—186.

Barth, F. G. and Grill, R. (1970) Versuche zur Bedeutung der Spaltsinnesorgane für das Lokomotionverhalten der Spinnen. Staatsexamensaarbeit (unpublished).

Barth, F. G. and Libera, W. (1970) Ein Atlas der Spaltsinnesorgane von *Cupiennius salei* Keys., Chelicerata (Araneae). *Zeitschrift für Morphologie und Okologie der Tiere*, **68**, 343—369.

Barth, F. G. and Seyfarth, E-A. (1971) Slit sense organs and kinesthetic orientation. *Zeitschrift für vergleichende Physiologie*, **74**, 326—328.

Barth, F. G. and Wadepuhl, M. (1975) Slit sense organs of the scorpion leg (*Androctonus australiss*, L., *Buthidae*). *Journal of Morphology*, **145**, 209—228.

Bässler, U. (1965) Proprioreceptoren am Subcoxal- und Femur-Tibia-Gelenk der Stabheuschrecke *Carausius morosus* und ihre Rolle bei der Wahrnehmung der Schwerkraftrichtung. *Kybernetik*, **2**, 168—193.

Bässler, U. (1967) Zur Regelung der Stellung des Femur-Tibia-Gelenkes bei der Stabheuschrecke *Carausius morosus* in der Ruhe und im Lauf. *Kybernetik*, **4**, 18—26.

Bässler, U. (1972a) Der Regelkreis des Kniesehnenreflexes bei der Stabheuschrecke *Carausius morosus*: Reaktionen auf passive Bewegungen der Tibia. *Kybernetik*, **12**, 8—20.

Bässler, U. (1972b) Der Kniesehnenreflex bei *Carausius morosus*: Ubergangsfunktion und Frequenzgang. *Kybernetik*, **11**, 32—50.

Bässler, U. (1973) Zur Steuerung aktiver Bewegungen des Femur-Tibia-Gelenkes der Stabheuschrecke *Carausius morosus*. *Kybernetik*, **13**, 38—53.

Bässler, U. (1974) Vom femoralen Chordotonalorgan gesteuerte Reaktionen bei der Stabheuschrecke *Carausius morosus*: Messung der von der Tibia erzeugten Kraft im aktiven und inaktiven Tiber. *Kybernetik*, 16, 213–226.

Becht, G. (1958) The influence of DDT and lindane on chordotonal organs in the cockroach. *Nature*, 181, 777–779.

Bentlage, C. (1973) Sensorische innervation der flügel bei der Heuschrecke *Locusta migratoria migratorioiides* mit besonderer Berücksichtigung des Verlaufes der sensorischen Nervenfasern im Zentralnervensystem. Staatsarbeit, Köln.

Bowerman, R. F. (1975a) The control of walking in the scorpion. I. Leg movements during normal walking. *Journal of Comparative Physiology*, 100, 183–196.

Bowerman, R. F. (1975b) The control of walking in the scorpion. II. Coordination modification as a consequence of appendage ablation. *Journal of Comparative Physiology*, 100, 197–209.

Bowerman, R. F. and Larimer, J. (1973) Structure and physiology of the patella-tibia joint receptors in scorpion pedipalps. *Comparative Biochemistry and Physiology*, 46A, 139–151.

Buddenbrock, W. von. (1921) Der Rhythmus der Schreitbewegungen der Stabheuschrecke. *Biol. Zbl.*, 41, 41–48.

Bullock, T. H. and Horridge, G. A. (1965) *Structure and function in the nervous system of invertebrates*. 11, Freeman. San Francisco.

Burns, M. D. (1973) The control of walking in Orthoptera–1. leg movements in normal walking. *Journal of Experimental Biology*, 58, 45–58.

Burns, M. D. (1974) The structure and physiology of the locust femoral chordotonal organ. *Journal of Insect Physiology*, 20, 1319–1339.

Burrows, M. (1973a) Physiological and morphological properties of the metathoracic common inhibitory neuron of the locust. *Journal of Comparative Physiology*, 82, 59–78.

Burrows, M. (1973b) The morphology of an elevator and a depressor motoneuron of the hindwing of a locust. *Journal of Comparative Physiology*, 83, 165–178.

Burrows, M. (1975) Monosynaptic connections between wing stretch receptors and flight motoneurons of the locust. *Journal of Experimental Biology*, 62, 189–219.

Burrows, M. and Horridge, G. A. (1974) The organization of inputs to motoneurons of the locust metathoracic leg. *Philosophical Transactions of the Royal Society Series B.*, 269, 49–94.

Burrows, M. and Hoyle, G. (1973) Neural mechanisms underlying behavior in the locust *Schistocerca gregaria*. III. Topography of limb motoneurons in the metathoracic ganglia *Journal of Neurobiology*, 4, 167–186.

Camhi, J. (1969) Locust wind receptors. III. Contributions to flight initiation and lift control. *Journal of Experimental Biology*, 50, 363–373.

Camhi, J. (1970a) Yaw-correcting postural changes in locusts. *Journal of Experimental Biology*, 52, 519–531.

Camhi, J. (1970b) Sensory control of abdomen posture in flying locusts. *Journal of Experimental Biology*, 52, 533—537.

Chapman, K. M. (1965) Campaniform sensilla on the tactile spines of the legs of the cockroach. *Journal of Experimental Biology*, 42, 191—203.

Chapman, K. M. and Duckrow, R. B. (1975) Compliance and sensitivity of a mechanoreceptor of the insect exoskeleton. *Journal of Comparative Physiology*, 100, p. 251

Chapman, K. M., Duckrow, R. B. and Moran, D. T. (1973) Form and role of deformation in excitation of an insect mechanoreceptor. *Nature*, 244, 453—454.

Chapman, K. M. and Pankhurst, J. H. (1967) Conduction velocities and their temperature coefficients in sensory nerve fibres of cockroach legs. *Journal of Experimental Biology*, 46, 63—84.

Chevalier, R. L. (1969) The fine structure of campaniform sensilla on the halteres of *Drosophila melanogaster*. *Journal of Morphology*, 128, 443—464.

Coillot, J. P. (1974) Analyse du codage d'un mouvement périodique, par des récepteurs à l'étirement d'un insecte. *Journal of Insect Physiology*, 20, 1101—1116.

Coillot, J. P. (1975) La conversion analogique numérique des récepteurs à l'étirement de la patte metathoraque du criquet *Schistocerca gregaria*. *Journal of Insect Physiology*, 21, 423—434.

Coillot, J. P. and Boistel, J. (1968) Localisation and description de récepteurs a l'étirement au niveau de l'articulation tibio-fémorale de la patte sauteuse du criquet, *Schistocerca gregaria*. *Journal of Insect Physiology*, 14, 1661—1667.

Coillot, J. P. and Boistel, J. (1969) Étude de l'activite électrique propagée de récepteurs à l'étirement de la patte métathoracique du criquet, *Schistocerca gregaria*. *Journal of Insect Physiology*, 1449—1470.

Debaiseaux, P. (1936) Organes scolopidiaux des pattes d'insectes. I. Lépidoptères et Trichoptères. *La Cellule*, 44, 271—314.

Debaiseaux, P. (1938) Organes scolopidiaux des pattes d'insectes. II. *La Cellule*, 47, 77—202.

Delcomyn, F. (1971) The effect of limb amputation on locomotion in the cockroach *Periplaneta americana*. *Journal of Experimental Biology*, 54, 453—469.

Delcomyn, F. (1973a) Motor activity during walking in the cockroach *Periplaneta americana*. I. Free walking. *Journal of Experimental Biology*, 59, 629—642.

Delcomyn, F. (1973b) Motor activity during walking in the cockroach *Periplaneta americana*. II. Tethered walking. *Journal of Experimental Biology*, 59, 643—654.

Denis, C. (1958) Contribution à l'étude de l'ontogenèse sensori-nerveuse du termite *Calotermes flavicollis*. Fab. *Insectes sociaux*, 5, 171—188.

Dethier, V. C. (1963) *The Physiology of Insect Senses*. Methuen, London.

Eastwood, A. B. (1971) The comparative physiology of muscles in the walking legs of *Limulus polyphemus*. Ph.D. Thesis, Univ. of Lehigh, U.S.A.

Edgar, A. L. (1963) Proprioception in the legs of phalangids. *Biological Bulletin*, 124, 262–267.

Eggers, F. (1928) Die stiftführenden Sinnesorgane. Morphologie und Physiologie der chordotonalen und der tympanalen Sinnesapparate der Insekten. *Zoologische Bausteine,,* 2, 1–353.

Erhardt, E. (1916) Zur Kenntnis der Innervierung und der Sinnesorgane der Flügel von Insekten. *Zoologische Jahrbücher*, (Anatomie) 39, 293–334.

Elser, N. and Huber, F. (1969) Die Organisation des Werbegesanges der Heuschrecke (*Gomohoceripppus rufus* L.) In Abhängigkeit von zentralen und peripheren Bedingungen. *Zeitschrift für vergleichende Physiologie*, 65, 389–423.

Finlayson, L. H. (1968) Proprioception in the invertebrates. *Symposia of the Zoological Society of London*, 23, 217–249.

Fountain, R. L. (1973) Motor control of accessory muscles in the tendon receptor organ in *Limulus polyphemus*. *Comparative Biochemistry and Physiology*, 44A, 511–517.

Fuhrmann, H. (1922) Beiträge zur Kenntnis der Hautsinnesorgane der Tracheaten. I. Die antennalen Sinnesorgane der Myriapoden. *Zeitschrift für enschaftliche Zoologie*, 119, 1–52.

Gettrup, E. (1962) Thoracic proprioceptors in the flight system of locusts. *Nature*, 193, 498–499.

Gettrup, E. (1963) Phasic stimulation of a thoracic stretch receptor in locusts. *Journal of Experimental Biology*, 40, 323–333.

Gettrup, E. (1965) Sensory mechanisms in locomotion: the campaniform sensilla of the insect wing and their function during flight. *Cold Spring Harbor Symposia on Quantitative Biology*, 30, 615–622.

Gettrup, E. (1966) Sensory regulation of wing twisting in locusts. *Journal of Experimental Biology*, 44, 1–16.

Gettrup, E. and Wilson, D. M. (1964) The lift control reaction of flying locusts. *Journal of Experimental Biology*, 41, 183–190.

Gorelkin, V. S. (1973) The periphoral mechanisms of rhythm stabilization of the wing apparatus activity in the cockroach, *Periplaneta americana*. (In Russian with English Summary). *Zh. Evol. Biokhim. Fiziol.*, 9, 620–622.

Graham, D. (1972) A behavioural analysis of the temporal organisation of walking movements in the 1st instar and adult stick insect (*Carausius morosus*). *Journal of Comparative Physiology*, 81, 23–52.

Guthrie, D. M. (1967) Multipolar stretch receptors and the insect leg reflex. *Journal of Insect Physiology*, 13, 1637–1644.

Hayes, W. F. and Barber, S. B. (1966) Structure and properties of the *Limulus* tendon receptor organ. *American Zoologist*, 6, 74.

Hayes, W. F. and Barber, S. B. (1967) Proprioceptor distribution and properties in *Limulus* walking legs. *Journal of Experimental Zoology*, **165**, 195–210.

Heran, H. (1959) Wahrnehmung und Regelung der Flugeigengeschwindigkeit bei *Apis mellifera* L. *Zeitschrift für vergleichende Physiologie*, **42**, 103–163.

Heran, H. and Lindauer, M. (1963) Windkompensation und Seitenwindkorrektur der Bienen beim Flug über Wasser. *Zeitschrift für vergleichende Physiologie*, **47**, 39–55.

Hoffman, C. (1964) Bau und Vorkommen von proprioceptiven Sinnesorganen bei den Arthropoden. *Ergebnisse Biolie*, **27**, 1–38.

Horridge, G. A. (1962) Learning of leg position by the ventral nerve cord in headless insects. *Proceedings of the Royal Society A.*, **157**, 33–52.

Howse, P. E. (1965) The structure of the subgenual organ and certain other mechanoreceptors of the termite, *Zootermopsis angusticollis* (Hagen). *Proceedings of the Royal Entomological Society of London A*, **40**, 137–146.

Howse, P. E. (1968) The fine structure and functional organization of chordotonal organs. *Symposia of the Zoological Society of London*, **23**, 167–198.

Hoyle, G. (1965) Neurophysiological studies on 'learning' in headless insects In *The Physiology of the Insect Central Nervous System*. Beament J. W. L. and Treherne J. E. (eds), Academic Press, London. pp 203–232.

Hoyle, G. (1970) Cellular mechanisms underlying behavior—neuroethology. In *Advances in Insect Physiology*. Academic Press, London. pp 349–444.

Hoyle, G. and Burrows, M. (1970) Intracellular studies on identified neurons of insects. *Federal Proceedings*, **129**, 1922.

Hoyle, G. and Burrows, M. (1973a) Neural mechanisms underlying behavior in the locust *Schistocerca gregaria*. I. Physiology of identified neurons in the metathoracic ganglion. *Journal of Neurobiology*, **4**, 3–41.

Hoyle, G. and Burrows, M. (1973b) Neural mechanisms underlying behavior in the locust *Schistocerca gregaria*. II. Integrative activity in metathoracic neurons. *Journal of Neurobiology*, **4**, 43–67.

Huber, F. (1974) Neural integration (Central nervous system). In: *Physiology of Insecta*. Vol. IV, pp 3–100. Academic Press: London.

Hughes, G. M. (1952) The coordination of insect movements. I. The walking movements of insects. *Journal of Experimental Biology*, **29**, 267–284.

Hughes, G. M. (1957) The co-ordination of insect movements. II. The effect of limb amputation and the cutting of commissures in the cockroach (*Blatta orientalis*). *Journal of Experimental Biology*, **34**, 306–333.

Hughes, G. M. and Mill, P. J. (1974) Locomotion: Terrestrial. In: *Physiology of Insecta*, Rockstein. (ed), Vol. 3. Academic Press: London.

Iles, J. F. and Mulloney, B. (1971) Procion yellow staining of cockroach motor neurones without use of microelectrodes. *Brain Research*, **30**, 397–400.

Kendall, M. D. (1970) The anatomy of the tarsi of *Schistocerca gregaria* Forskal. *Zeitschrift für Zellforschung und mikroskopische Anatomie*, **109**, 112–137.

Kutsch, W. and Otto, D. (1972) Evidence for spontaneous song production independent of head ganglia in *Gryllus campestris*. *Journal of Comparative Physiology*, **87**, 89–98.

Land, M. F. (1972) Stepping movements made by jumping spiders made during turns mediated by the lateral eyes. *Journal of Experimental Biology*, **57**, 15–40.

Laverack, M. S. (1966) Observations on a proprioceptive system in the legs of the scorpion *Hadurus hirsutus*. *Comparative Biochemistry and Physiology*, **19**, 241–251.

Lehr, R. (1914) Die Sinnesorgane der beiden Flügelpaare von *Dytiscus marginalis*. *Zeitschrift für wissenschaftliche Zoologie*, **110**, 87–150.

Leydig, F. (1860) Über Geruchs- und Gehörorgane der Krebse und Insekten. *Archiv für Anatomie und Physiologie, Lpz.*, **1860**, 265–314.

Lombardo, C. (1974) On the presence of two coxal sense organs in Pterygota insects. *Monit, zool. ital.*, **7**, 243–246.

Lukoschus, F. (1962) Über Bau und Entwicklung des Chordotonal-organs am Tibia-Tarsus-Gelenk der Honigbiene. *Zeitschrift für Bienenforschung*, **6**, 48–52.

Manton, S. M. (1958) The evolution of arthropodan locomotory mechanisms. VI. Habits and evolution of the Lysiopetaloidea (Diplopoda), some principles of leg design in Diplopoda and Chilopoda, and limb structure of Diplopoda. *Journal of the Linneun Society*, (Zoology), **44**, 487–556.

Manton, S. M. (1961) Experimental zoology and problems of arthropod evolution. In: *The Cell and the Organism*. Ramsey J. A. and Wigglesworth, V. B. (eds), The University Press, Cambridge. pp 234–255.

Markl, H. (1965) Ein neuer Propriorezeptor am Coxa-Trochanter-Gelenk der Honigbiene. *Naturwissenschaften*, **52**, 460.

McIndoo, N. E. (1911) The lyriform organs and tactile hairs of araneids. *Proceedings of the Academy of Natural Sciences of Philadelphia*, **63**, 375–418.

Mill, P. J. and Harris, D. (1976). In preparation.

Miller, P. L. (1974) The neural basis of behaviour In: *Insect Neurobiology*. J. E. Treheme (ed.), pp. 359–411 North Holland Publishing Company, Oxford.

Moran, D. T. and Chapman, K. M. (1968) Proprioceptive campaniform sensilla of cockroach tibia: morphological and electrophysiological investigations of large bipolar mechanoreceptor neurons. *Journal of Cell Biology*, **48**, 155–173.

Moran, D. T., Chapman, K. M. and Ellis, R. A. (1969) The fine structure of cockroach campaniform sensilla. *27th Annual Proceedings of the Electron Microscopical Society of America*, Arceneaux, C. J. (ed), Claitor's Baton Rouge, La. 250.

Moran, D. T., Chapman, K. M. and Ellis, R. A. (1971) The fine structure of cockroach campaniform sensilla. *Journal of Cell Biology*, **48**, 155–173.

Moran, D. T., Rowley III, J. C. and Varela, F. G. (1975) Ultrastructure of the grasshopper proximal femoral chordotonal organ. *Cell and Tissue Research*, **161**, 445–457.

Möss, D. (1967) Proprioceptoren im Thorax und Abdomen von Grillen (Orthoptera, Gryllidae). *Zoologischer Anzeiger Suppl.*, **31**.

382 *Structure and function of proprioceptors in the invertebrates*

Möss, D. (1971) Sinnesorgane im Bereich des Flügels der Feldgrille (*Gryllus campestris* L.) und ihre Bedeutung für die Kontrolle der Singbewegung und die Einstellung der Flügellage. *Zeitschrift für vergleichende Physiologie*, 73, 53–83.

Nijenhuis, E. D., and Dresden, D. (1952) A micro-morphological study of the sensory supply of the mesothoracic leg of the American cockroach, *Periplaneta americana*. *Proceedings Koniklijke Nederlandse Akademie van Westenschappen*, 55, 300–310.

Pabst, H. (1965) Elektrophysiologische Untersuchung des Streckrezeptors am Flügelgelenk der Wanderheuschrecke *Locusta migratoria*. *Zeitschrift für Vergleichende Physiologie*, 50, 498–541.

Pabst, H. and Schwartzkopff, J. (1962) Zur Leistung der Flügel-gelenk-Rezeptoren von *Locusta migratoria*. *Zeitschrift für Vergleichende Physiologie*, 45, 396–404.

Parry, D. A. (1960) The small leg-nerve of spiders and a probable mechano-receptor. *Quarterly Journal of Microscopical Science*, 101, 1–18.

Pearson, K. G. (1972) Central programming and reflex control of walking in the cockroach *Periplaneta americana*. *Journal of Experimental Biology*, 56, 173–194.

Pearson, K. G. and Iles, J. F. (1973) Nervous mechanisms underlying intersegmental co-ordination of leg movements during walking in the cockroach. *Journal of Experimental Biology*, 58, 725–744.

Pearson, K. G., Fourtner, C. R. and Wong, R. K. (1974) Nervous control of walking in the cockroach. In: *Control of Posture and Locomotion*. Stein, R. B., Pearson, K. B., Smith, R. S. and Redford, J. B. (eds), Plenum Press, New York.

Pflugstaedt, H. (1912) Die Halteren der Dipteren. *Zeitschrift für wissen-schaftliche Zoologie*, 100, 1–59.

Pitman, R. M., Tweedle, C. D. and Cohen, M. J. (1972) Branching of central neurons: intracellular cobalt injection for light and electron microscopy. *Science*, 176, 412–414.

Pitman, R. M., Tweedle, C. D. and Cohen, M. J. (1973) Dendritic geometry of an insect common inhibitor motoneuron. *Brain Research*, 60, 465–470.

Pringle, J. W. S. (1938a) Proprioception in insects–II. The action of the campaniform sensilla on the legs. *Journal of Experimental Biology*, 15, 114–131.

Pringle, J. W. S. (1938b) Proprioception in insects–III. The function of the hair sensilla at the joints. *Journal of Experimental Biology*, 15, 467–473.

Pringle, J. W. S. (1940) The reflex mechanism of the insect leg. *Journal of Experimental Biology*, 17, 8–17.

Pringle, J. W. S. (1948) The gyroscopic mechanism of the halteres of Diptera. *Philosophical transactions of the Royal Society, Series B.*, 233, 347–384.

Pringle, J. W. S. (1955) The function of the lyriform organs of arachnids. *Journal of Experimental Biology*, 32, 270–278.

Pringle, J. W. S. (1956) Proprioception in *Limulus*. *Journal of Experimental Biology*, 33, 658—667.

Pringle, J. W. S. (1957) *Insect flight*. The University Press, Cambridge.

Pringle, J. W. S. (1961) Proprioception in arthropods. In *The Cell and the Organism*. Ramsay, J. A. and Wigglesworth, V. B. (eds), The University Press, Cambridge. pp. 256—282.

Pringle, J. W. S. (1965) Flight. In: *The physiology of insecta*. Rockstein (ed), Vol. 2. Academic Press, London.

Rao, K. P. (1964) Neurophysiological studies on an arachnid, the scorpion *Heterometrus fulvipes*. *Journal of Animal Morphology and Physiology*, 11, 133—142.

Rathmayer, W. (1967) Elektrophysiologische Untersuchungen an Proprioceptoren im Beim einer Vogelspinne (*Eurypelma hentzi* chamb). *Zeitschrift für vergleichende Physiologie*, 54, 438—454.

Rathmayer, W. and Koopmann, J. (1970) Die Verteilung der Propriorezeptoren im Spinnenbein: Untersuchungen an der Vogelspinne *Dugesiella hentzi* Chamb. *Morphologie und Okologie der Tiere*, 66, 212—223.

Richard, G. (1950) L'innervation et les organes sensoriels de la patte du termite à cou jaune. *Annals des sciences naturelleles (Zoologie)*, 12, 65—83.

Rilling, G. (1960) Zur Anatomie des braunen Steinläufers *Lithobius forficatus* L. (Chilopoda). Skeletmuskelsystem, peripheres Nervensystem und Sinnesorganes des Rumpfes. *Zoologische Jahrbücher (Anatomie)*, 78, 39—128.

Runion, H. I. and Usherwood, P. N. R. (1968) Tarsal receptors and leg reflexes in the locust and grasshopper. *Journal of Experimental Biology*, 49, 421—436.

Schwabe, J. (1906) Beiträge zur Morphologie und Histologie der tympanalen Sinnesapparate der Orthopteren. *Zoologica, Stuttgart*, 50, 1—154.

Seyfarth, E.-A. and Barth, F. G. (1972) Compound slit sense organs on the spider leg: mechanoreceptors involved in kinesthetic orientation. *Journal of Comparative Physiology*, 78, 176—191.

Slifer, E. H. (1935) The morphology and development of the femoral chordotonal organs of *Melanoplus differentialis* (Orthoptera, Acrididae). *Journal of Morphology*, 58, 615—637.

Slifer, E. H. (1936) The scoloparia of *Melanoplus differentialis* (Orthoptera, Acrididae). *Entomological News*, 47, 174—180.

Smolyaninov, V. V. and Karpovich, A. L. (1972) Normal locomotion of the millipede *Julius* sp. (In Russian with English summary)) *Zh. Evol. Biokhim Fiziol.*, 8, 523—529.

Spencer, H. J. (1974) Analysis of the electrophysiological responses of the trochanteral hair receptors of the cockroach. *Journal of Experimental Biology*, 60, 223—240.

Spinola, S. M. and Chapman, K. M. (1975) Proprioceptive indentation of campaniform sensilla of cockroach legs. *Journal of Comparative Physiology* 96, 257—272

Smith, D. S. (1969) The fine structure of the haltere sensilla in the blowfly, *Calliphora erythrocephala* (Meig.), with scanning microscope observations on the haltere surface. *Tissue and Cell,* 1, 443–484.

Stretton, A. O. W. and Kravitz, E. A. (1968) Neuronal geometry: determination with a technique of intracellular dye injection. *Science,* 162, 132–134.

Ten Cate, J. (1936) Beiträge zur Innervation der Lokomotions-bewegung der Heuschrecke (*Locusta viridissima*). *Archives néerlandaises de physiologie,* 21, 562–566.

Thurm, U. (1964) Mechanoreceptors in the cuticle of the honey bee: fine structure and stimulus mechanism. *Science,* 145, 1063–1065.

Thurm, U. (1965) An insect mechanoreceptor. I. Fine structure and adequate stimulus. *Cold Spring Harbor Symposium of Quantitative Biology,* 30, 75–82.

Tyrer, N. M. and Altman, J. S. (1974) Motor and sensory flight neurons in a locust demonstrated using cobalt chloride *Journal of Comparative Neurology,* 157, 117.

Tyrer, N. M. and Altman, J. S. (1975) The central projections of the wing hinge stretch receptors in three species of locust (in preparation).

Uga, S. and Kumabara, M. (1967) The fine structure of the companiform sensillum on the haltere of the fleshfly, *Boettcherisca peregrina. Journal of Electron Microscopy,* 16, 304–312.

Usherwood, P. N. R., Runion, H. I. and Campbell, J. I. (1968) Structure and physiology of a chordotonal organ in the locust leg. *Journal of Experimental Biology,* 48, 305–323.

Varela, F. G., Moran, D. T. and Rowley III, J. C. (1976) A purely tonic ciliated mechanoreceptor-the grasshopper proximal femoral chordotonal organ, (in preparation).

Vogel, H. (1923) Über die Spaltsinnesorgane der Radnetzspinnen. *Jenaische Zeitschrift für Naturwissenschaft,* 59, 171–208.

Wales, W., Clarac, F., Dando, M. R. and Laverack, M. S. (1970) Innervation of the receptors present at the various joints of the pereiopods and third maxilliped of *Homarus gammarus* (L.) and other macruran decapods (crustacea). *Zeitschrift für vergleichende Physiologie,* 68, 345–384.

Weinland, E. (1891) Über die Schwinger (Halteren) der Dipteren. *Zeitschrift für wissenschaftliche Zoologie,* 51, 55–166.

Weis-Fogh, T. (1956a) Biology and physics of locust flight. II. Flight performance of the desert locust (*Schistocerca gregaria*). *Philosophical Transactions of the Royal Society, Series B.,* 239, 459–510.

Weis-Fogh, T. (1956b) Biology and physics of locust flight. IV. Notes on the sensory mechanisms in locust flight. *Philosophical Transactions of the Royal Society, Series B.,* 239, 553–584.

Wendler, G. (1961) Die Regelung der Korperhaltung bei Stabheuschrecken (*Carausius morosus*). *Naturwissenschaften,* 48, 676–677.

Wendler, G. (1964) Laufen und Stehen der Stabheuschrecke *Carausius morosus*: Sinnesborstenfelder in den Beingelenken als Glieder von Regelkreisen. *Zeitschrift für vergleichende Physiologie*, **48**, 198—250.

Wendler, G. (1965) The coordination of walking movements in arthropods. *Symposia of the Society for Experimental Biology*, **20**, 229—249.

Wendler, G. (1971) Gravity orientation in insects: the role of different mechanoreceptors. In *Gravity and the Organism* Gordon, S. A. and Cohen, M. J. (eds), The University of Chicago Press, Chicago and London. pp 195—201.

Wendler, G. (1972) Einfluß erzwungener Flügelbewegungen auf das motorische Flugmuster von Heuschrecken. *Naturwissenschaften*, **59**, 220

Wendler, G. (1974) The influence of proprioceptive feedback on locust flight co-ordination. *Journal of Comparative Physiology*, **88**, 173—200.

Wilson, D. M. (1961) The central nervous control of flight in the a locust. *Journal of Experimental Biology*, **38**, 471—490.

Wilson, D. M. (1965) Proprioceptive leg reflexes of cockroaches. *Journal of Experimental Biology*, **43**, 397—409.

Wilson, D. M. (1966) Insect walking. *Annual Review of Entomology*, **11**, 103—122.

Wilson, D. M. (1967a) An approach to the study of rhythmic behavior. From: *Invertebrate Nervous systems*. Wiersma, C. A. G. (ed), Chicago University Press, Chicago and London, pp 219—229.

Wilson, D. M. (1967b) Stepping patterns in taranula spiders. *Journal of Experimental Biology*, **47**, 133—151.

Wilson, D. M. and Gettrup, E. (1963) A stretch reflex controlling wingbeat frequencies in grasshoppers. *Journal of Experimental Biology*, **40**, 171—185.

Wilson, D. M. and Wyman, R. J. (1965) Motor output patterns during random and rhythmic stimulation of locust thoracic ganglia. *Biophysical Journal*, **5**, 121—143.

Wright, B. R. (1976) Afferent pathways of the femoral chordotonal organ of the locust. (in preparation).

Wyse, G. A. and Maynard, D. M. (1964) Joint receptors in the antennule of *Panulirus argus*. *Journal of Experimental Biology*, **43**, 521—535.

Young, D. (1970) The structure and function of a connective chordotonal organ in the cockroach leg. *Philosophical Transactions of the Royal Society, Series B.*, **256**, 401—442.

Zaćwilichowski, J. (1934a) Über die Innervierung und Sinnesorgane der Flügel von Schabe *Phyllodromia germanica* L. *Bull. Int. Acad. Cracovie* (Acad. pol. Sci.), B, **11**, 89—104.

Zaćwilichowski, J. (1934b) Die Sinnesnervenelemente des Schwingers und dessens Homologie mit dem Flügel der *Tipula paludosa* Mieg. *Bull. int. Acad. Cracovie* (Acad. pol. Sci.), B, **11**, 397—413.

Zilber-Gachelin, N. F. and Paupardin, D. (1974a) Sensitization and dishabituation in the cockroach. Main characteristics and localization of changes in reactivity. *Comparative Biochemistry and Physiology*, **49A**, 441—470.

Zilber-Gachelin, N. F. and Paupardin, D. (1974b) Desensitization by leg contact in the cockroach: localization of the reactivity changes and a study of their possible sensory origin. *Comparative Biochemistry and Physiology*, **49A**, 471–490.

Zilber-Gachelin, N. F. and Paupardin, D. (1974c) Main characteristics of an induced decrease in leg motor reactivity ('desensitization') in the cockroach. *Comparative Biochemistry and Physiology*, **49A**, 491–510.

9 Ultrastructure of chordotonal organs

M. MOULINS

9.1 Introduction

Chordotonal organs (Graber, 1882) are arthropod sensory structures characterized by special sensilla, the *scolopidia*. This definition is a purely morphological one. All of the chordotonal organs investigated physiologically are mechanotransducers, although some of course have nothing to do with proprioceptive mechanisms. However, in this paper we will consider ultrastructural data on all chordotonal organs since (1) in all cases they are morphologically comparable and (2) for many we do not have any functional information.

Chordotonal organs are known only in the insects and crustaceans; they have not yet been found in any other arthropod group (Howse, 1968; Foelix, personal communication). In insects they have been known since the end of the last century (see Graber, 1882; Eggers, 1924; Debauche, 1935; Debaisieux, 1936); their topography, histology and ultrastructure have been studied extensively in that group but little has been done on their physiology. In the crustacea, chordotonal organs have been described more recently (Barth, 1934; Wetzel, 1934). In this group scolopidia are generally much more difficult to see with the light microscope and it was only an ultrastructural study which enabled Whitear (1960, 1962) to show that the joint proprioceptors of the legs are chordotonal organs. Recent physiological work on chordotonal organs has been carried out mainly on these proprioceptors (Chapter 6).

In a chordotonal organ, the scolopidia are associated with a support. As in other arthropod sensilla this can be the cuticle but, contrary to other arthropod sensilla, scolopidia are not associated with any external modification of the cuticle and so can never be identified from the exterior of the animal. The only exception described so far is the crayfish statocyst (Schöne and Steinbrecht, 1968). Also, contrary to other arthropod sensilla, scolopidia have in many cases lost their direct relationship with the cuticle and become internal. The end of this evolutionary trend is the situation in which the scolopidia are embedded in a strand which spans two different regions of the body: it is the condition which is realized in many proprioceptive chordotonal organs.

In the last fifteen years about thirty papers on chordotonal organ ultrastructure have been published. They consider mainly the scolopidia and little has been done on the supports. Also most of these papers concern insect chordotonal organs and more ultra-structural information is needed about crustacean chordotonal organs.

9.2   Ultrastructure of a typical scolopidium

The first ultrastructural study of a scolopidium (Figs. 9.1, 9.2) was carried out by Gray (1960) for the tympanal organ of *Locusta*: this classical work was also the first extensive ultrastructural study of an arthropod sensillum. The work of Whitear (1962) was the first ultrastructural study of a proprioceptive chordotonal organ. The typical structure of a scolopidium has been schematised in Fig. 9.1 and will be briefly described, mainly from the information in the above two fundamental papers.

9.2.1   Structural organization

Like other sensilla in arthropods, a scolopidium is a structural unit consisting of one (or more) sensory cell(s) and two accessory cells (in this case the scolopale cell and the attachment cell). It is also a morphogenetical unit, with all the cells belonging to a scolopidium

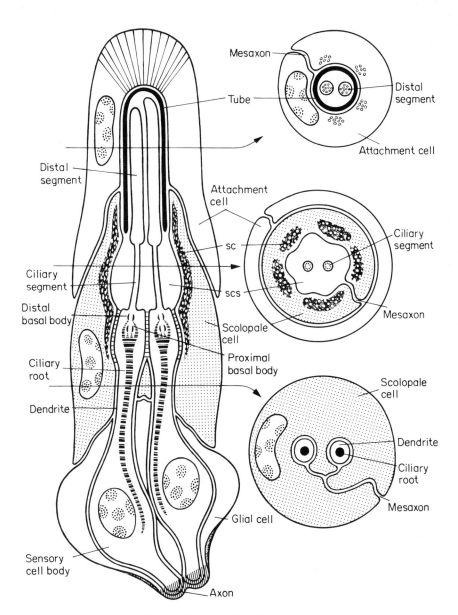

Fig. 9.1   Diagrammatic longitudinal section through a scolopidium (left) with transverse sections at various levels (right) as indicated by the arrows (see Fig. 9.2). *sc*—scolopale; *scs*—scolopale space.

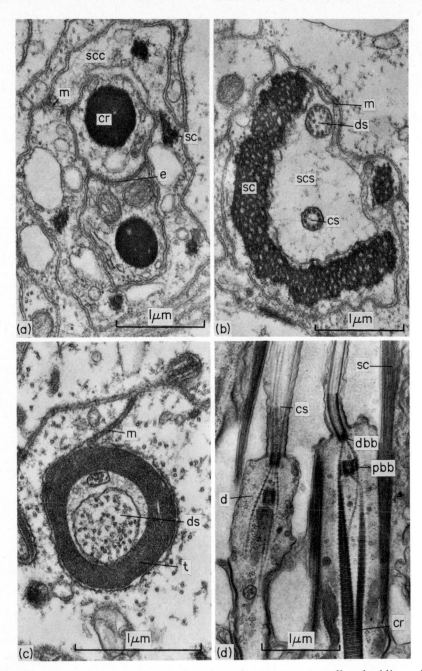

Fig. 9.2 (a, b, c) Transverse sections through a two sensory cell scolopidium of the cuticular stress detector CSD2 of the crayfish *Astacus leptodactylus* (compare with the three transverse sections of Fig. 9.1). (d) longitudinal section of a scolopidium of Johnston's organ of *Chrysopa* (Planipennia) (compare with the longitudinal section of Fig. 9.1). *cr*—ciliary root; *cs*—ciliary segment; *d*—dendrite; *dbb*—distal basal body; *ds*—distal segment; *e*—ephapse; *m*—mesaxon; *pbb*—proximal basal body; *sc*—scolopale; *scc*—scolopale cell; *scs*—scolopale space; *t*—tube. ((d) from Schmidt, 1969.)

derived from one epidermal mother cell in a mitotic sequence which has been studied by Jägers-Röhr (1968).

The scolopidium is constructed by a special arrangement of the accessory cells round the sensory cell. The scolopale cell is wrapped round the distal region of the dendrite and the proximal region of the centriolar derivative. The attachment cell is wrapped round the distal region of the centriolar derivative. As when a glial cell is wrapped round an axon, a mesaxon is formed by the wrapping of the accessory cells round the sensory cell (surface connecting membrane of Gray, 1960). The centriolar derivative lies in an extracellular space, the scolopale space, the walls of which are formed by the scolopale (electron dense intracellular structure) and the tube, or cap (electron dense extracellular structure).

9.2.2   The sensory cell

The sensory cell is a primary sense cell with a peripheral cell body and the axon runs directly to the central nervous system. It is, as always in arthropod sensilla, an uniterminal cell (Finlayson, 1968), in which the dendrite is prolonged by a centriolar derivative. The apical region of the dendrite possesses two aligned basal bodies which are reminiscent of centrioles. Proximally these are associated with a large cross-banded structure, the ciliary root. Distally the dendrite gives rise to the ciliary segment, which has the structure of a stereocilium, i.e. the axoneme is of the '9 + 0' type. There are two types of centriolar derivatives: in one (type 2) the ciliary segment gives rise to a distal segment invaded by microtubules; in the other (type 1) there is no distal segment (Section 9.3.2).

9.2.3   The accessory cells

The *scolopale cell* is the most important of the accessory cells. It secretes the scolopale, a characteristic structure of the scolopidium which is not found in any other arthropod sensillum. The scolopale is cylindrical and consists of electron dense material. It lies close to the scolopale space and, proximally, to the distal region of the dendrite(s). Distally the scolopale is juxtaposed to an apical, extracellular electron-dense structure (tube or cap) which is also secreted by the scolopale cell.

The *attachment cell* (also known as the cap or canal cell) is apposed to the distal region of the scolopale cell and the tube. This accessory cell is morphologically exterior to the scolopale cell and links the apical region of the scolopidium to the support.

The basal region of the scolopidium, and particularly the cell body, is ensheathed by glial cells. These cells do not belong to the scolopidium because of their origin and because one of these cells can be involved with several scolopidia. Among these cells there is sometimes one which distinctively ensheaths the proximal region of the dendrites and this has been described as the enveloping cell or fibrous sheath cell (Gray, 1960).

Table 9.1 sums up the fine structural studies of chordotonal organs. Concerning the scolopidia, it seems that there are two morphological situations which differ mainly in the nature of the centriolar derivative of the sensory cells (type 1 and type 2) and the nature of the apical structure (cap or tube). This has been noted in the table and is illustrated in Fig. 9.3 (Section 9.3.2). The major difference between those chordotonal organs studied seems to lie in the relationships of the scolopidia with the support and the nature of this support. This is also indicated in Table 9.1 and is illustrated in Fig. 9.8 (Section 9.6).

9.3   The sensory cells

The sensory cells involved in chordotonal organs are uniterminal neurons (Finlayson, 1968) (type I of Zawarzin, 1912) and never multiterminal neurons (type II of Zawarzin), like those of the crustacean MROs (Chapter 2) or insect stretch receptors (Chapter 4). The main difference, however, between these two types of sensory cell lies in the fact that the dendrite of the sensory cell of type I always terminates in a distal process which has the structure of a modified cilium. This structure has never been observed in any type II cell (Moulins, 1974), and appears in some ways to be an additional structure of a type I sensory cell. Since it is associated with a centriole-like structure, it can be termed a 'centriolar derivative'.

9.3.1   Number of sensory cells

It is only with ultrastructural studies that it has been possible to determine without question the number of sensory cells in scolopidia. Until now all those scolopidia studied possess one (Fig. 9.9c), two (Fig. 9.2b) or three (Fig. 9.4a) sensory cells (Table 9.1) and each cell bears only one centriolar derivative. Descriptions of a scolopidium sensory cell with three centriolar derivatives (Kinzelbach-Schmitt, 1968; Vande Berg, 1971) are almost certainly incorrect.

Debauche (1935) was the first to describe scolopidia with two sensory cells and referred to these as 'isodynal' when the two cells have peripheral processes of equal size and 'heterodynal' when the two processes are of different size. The term 'monodynal' was reserved for scolopidia with only one sensory cell. Whitear (1962) adopted this terminology, but with a slightly different meaning. Thus she called scolopidia with sensory cells similar in structure 'isodynal' and those with sensory cells dissimilar in structure 'heterodynal'. The main idea supporting this distinction was that the two sensory cells of a scolopidium could have a different function but, up to now, there is no physiological evidence in favour of this hypothesis (Bush, 1965; Hartmann and Boettiger, 1967; Mill and Lowe, 1971, 1973). These terms have now become more difficult to use since with electron microscope studies, scolopidia possessing more than one sensory cell are always heterodynal in so far as there is always some difference of unknown significance between the sensory cells (e.g. Young, 1970).

In any one chordotonal organ the scolopidia have, most of the time, the same number of sensory cells. This remains true when homologous organs are considered in different species. For example, all of the tympanal organs studied in insects have only one sensory cell in each scolopidium (Table 9.1; e.g. Young and Ball, 1974a, b); all the joint chordotonal organs studied in the legs of decapod crustaceans have two sensory cells in each scolopidium (Table 9.1), except for IMa of Astacidae and MCO2 of other decapods (Fig. 9.9c) (which are homologous and possess only one sensory cell per scolopidium); and all of the Johnston's organs studied in insects have scolopidia with three sensory cells (Fig. 9.4a) (Table 9.1), except in Diptera (Schmidt, 1970). However, in some cases, the same organ possesses scolopidia with differing numbers of sensory cells. For example, the tibio-tarsal organ of the cockroach has both one and two sensory cell scolopidia (Young, 1970) and this is true also for the tarsal organ of *Notonecta* (Wiese and Schmidt, 1974) and for the cuticular stress detector, CSD2, of the crayfish (Fig. 9.7) (Moulins and Clarac, 1972). Furthermore there are some indications that CP2

Table 9.1   Fine Structure of Chordotonal Organs

Chordotonal organs	Species	Authors	Number of sensory cells per scolopidium	Centriolar derivative (Fig. 9.3)	Apical structure (Fig. 9.3)	Relationships with the support (Fig. 9.8)
Tympanal organs of insects:	*Locusta migratoria*	Gray and Pumphrey (1958); Gray (1960)	1	Type 1	Cap.	ci
	Feltia subgothica (noctiud moth)	Ghiradella (1971)	1	Type 1	Cap.	cii
	Cyclochila australasiae (Cicada)	Young (1973)	1	Type 1	Cap.	?
	Gryllus bimaculatus	Michel (1974)	1	Type 1	Cap.	ci
Chordotonal organs of insect legs:						
Subgenual organ	*Periplaneta americana*	Howse (1968)	1	Type 1	Cap.	ci
	Gryllus assimilis	Friedman (1972)		Type 1	Cap.	ci[1]
Tibio-tarsal organ	*Periplaneta americana*	Young (1970)	1 or 2	Type 1	Cap.	dii
Tarsal organ	*Notonecta glauca* (Hemiptera)	Wiese and Schmidt (1974)	1 or 2	Type 1	?	ci
Chordotonal organs of crustacean legs[2]						
CB, MC1, MC2, CP1, CP2, PD organs	*Carcinus maenas*	Whitear (1960, 1962)	2	Type 2	Tube	di
PD organ	*Cancer pagurus*	Mill and Lowe (1971, 1973); Lowe et al. (1973)	2	Type 2	Tube	di

Organ	Species	Reference	No.	Cell type	Attachment	Code
CSD2 organ	*Astacus leptodactylus*	Moulins and Clarac (1972)	1 or 2	Type 2	Tube	dii
IMa organ	*Homarus gammarus*	Mill (Personal communication)	1	Type 2	Tube	di
MCO2 organ	*Cancer pagurus*	Gordon (1973)	1	Type 2	Tube	di
Johnston's organ of insects	*Drosophila melanogaster*	Uga and Kuwahara (1965)	2	Type 1	Tube	aii
	Aedes aegyti	Rislev and Schmidt (1967)	2	Type 1	Tube	aii
	Lepisma saccharina (Thysanura)	Kinzelbach-Schmidt (1968)	3[3]	?	?	aii
	Zootermopsis angusticollis	Howse (1968)	3	Type 2	Tube	ai
	Chrysopa[4] (Planipennia)	Schmidt (1969)	3	1 Type 2 / 2 Type 1	Tube	ai
	Oncopsis flavicollis (leaf-hopper)	Howse and Claridge (1970)	3	?	Tube	ai
	Manduca sexta (hornworm moth)	Vande Berg (1971)	3[3]	1 Type 2[5] / 2 Type 1	?	?
	Camponotus vagus (ant.)	Masson and Gabouriaut (1973)	3	1 Type 2 / 2 Type 1	Tube	?
	Speophyes lucidus (Coleoptera)	Corbière-Tichané[6] (1975)	3	1 Type 2 / 2 Type 1	Tube	ai
			2	Type 1	Cap.	b
Statocyst of crustaceans	*Astacus fluviatilis*	Schöne and Steinbrecht (1968)	3	Type 2	Tube ? (chorda)	a[7]

Continued

Table 9.1   Fine Structure of Chordotonal Organs

Chordotonal organs	Species	Authors	Number of sensory cells per scolopidium	Centriolar derivative (Fig. 9.3)	Apical structure (Fig. 9.3)	Relationships with the support (Fig. 9.8)
Mouth-part organs of insects:						
Mandible	*Rhodnius prolixus* *Triatoma infestans* (Hemiptera)	Pinet (1968) Pinet *et al.* (1968)	2	1 Type 1 1 Type 2	Tube	a?
Maxilla	"	"	1	Type 2	Tube	a?
Mandible and Maxillary lacinia	*Speophyes lucidus* (larva) (Coleoptera)	Corbière-Tichané (1971)	2	Type 2	Tube	ai
Maxillary palp, labium and antenna (scape)	"	"	2	Type 1	Cap.	cii
Antenna:						
Pedicel	Ephemeroptera	Schmidt (1974)	2(8)	Type 1	Cap.	b
Central organ	*Chrysopa*	Schmidt (1969)	2	Type 1	Cap.	ci
—	*Zootermopsis angusticollis*	Howse (1968)	2	Type 1	Cap.	?

(1) From author's Fig. 3A.
(2) For terminology see Section 6.3.
(3) The author noticed only one sensory cell with 3 centriolar derivatives.
(4) Schmidt (1970) has confirmed these results for *Athalia*, *Mellinus* (Hymenoptera), *Tenebrio* (Coleoptera), *Pieris* (Lepidoptera) and *Bibio* (Diptera).
(5) From author's pictures.
(6) Connective chordotonal organ of the pedicel of Howse (1968), Zentral organ of Schmidt (1969).
(7) With a cuticular hair on the top.
(8) An inner ring of scolopidia possesses 3 sensory cells.

and MC2 of *Carcinus* possess a mixture of one sensory cell scolopidia and two sensory cell scolopidia. Whitear (1960) suggested that CP2 had one sensory cell scolopidia but in contrast later (Whitear, 1962) stated that CP2 has two sensory cell scolopidia. Recent work indicates that she may have been right in both of her papers (Mill, personal communication).

The total number of sensory cells in the chordotonal organ, however, varies greatly. Thus, for example, there are 3 sensory cells (3 scolopidia) for the tympanal organ of noctuid moths (Ghiradella, 1971); 25 (15 scolopidia) for the tibio-tarsal organ of *Periplaneta* (Young, 1970); 60 (60 scolopidia) for the tympanal organ of *Locusta* (Gray, 1960); 80 (40 scolopidia) for the PD organ of *Cancer* (Mill and Lowe, 1973); 495 (165 scolopidia) for Johnston's organ of *Camponotus* (Masson and Gabouriaut, 1973).

9.3.2 The centriolar derivative: the two types of scolopidium sensory cells

The centriolar derivative of scolopidium sensory cells is associated with accessory structures (scolopale and tube/cap) which are considered to funnel the stimulus, and it is at this level that the stimulus is transduced (Mendelson, 1966). Therefore, it is important to know in detail the ultrastructure of this region and its relationships with the accessory structures.

Centriolar derivative without distal segment
(Type 1 in Table 9.1).
This type has been studied in detail by Young (1973). The centriolar derivative is represented only by the ciliary segment; in other words the typical '9 + 0' axoneme can be observed from the base to the apex (Fig. 9.3a). The axoneme consists of nine longitudinal fibres, regularly spaced on a ring. Each fibre, originates from one fibre of the distal basal body and contains two microtubules, one of which possesses a dense core and bears a pair of 'arms' (Fig. 9.4d). The characteristic central fibres of motile cilia are always lacking except in the tympanal organ of noctuid moths (Ghiradella, 1971). The diameters of the stereocilia and of the axoneme remain constant, except in a subterminal region where a ciliary dilation occurs (Gray, 1960). This dilation is characterized by (a) a temporary increase in the diameter of the cilium, (b) an associated temporary increase in

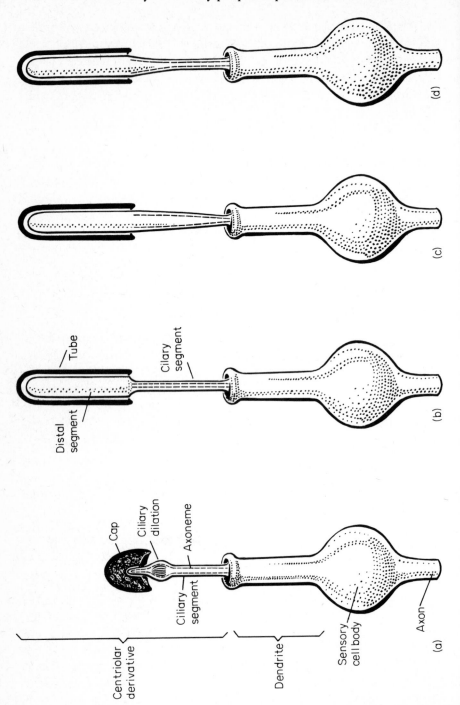

the diameter of the axoneme and (c) the existence of an electron-dense material between the nine fibres of the axoneme. This material seems to be always present and may be organized in a geometric cross pattern deposited on parallel mirotubules (Chu and Axtell, 1972). This can be compared with the apical tubular body of other mechanoreceptor sensory cells (Gaffal and Hansen, 1972) (Figs. 9.4f, 9.10). Distal to the ciliary dilation each fibre of the axoneme consists of two true microtubules, lacking a dense core, and arms. Near the end, as in motile cilia, the axoneme may be disorganized (Fig. 9.5a). This type 1 centriolar derivates has been observed only in insects, but in many organs as varied as tympanal organs, chordotonal organs of the legs, chordotonal organs of the mouth-parts and Johnstons organ (see Table 1). Apart from this last organ, the centriolar derivatives of type 1 are always associated with a special apical structure, the cap (Fig. 9.5a).

Centriolar derivative with a distal segment
(Type 2 in Table 9.1).
It is this type which was briefly described in Section 9.2 and is schematised in Fig. 9.1. The ciliary segment never bears any ciliary dilation containing dense material. The axoneme disappears before the end, usually in the apical region of the scolopale space. At the same time the diameter of the centriolar derivative increases abruptly and a large number of microtubules appear. This gives a distal segment generally longer than the ciliary segment (Fig. 9.3b). The distal segment is ensheathed by a tube (Fig. 9.5b). Again, with the exception of Johnston's organ, (Schmidt 1969; Corbière-Tichané, 1975) there is never any apical tubular body. This type 2 centriolar derivative is known in insects [Johnston's organ (Section 12.3.2), mouth part chordotonal organs (Section 5.3)] and crustaceans [statocyst (Section 12.3.3) chordotonal organs of the legs (Section 6.2.1)] (see Table 9.1).

Whitear (1962) described scolopidia with a 'ciliary cell' and a 'paraciliary cell'. The distinction, confirmed by Mill and Lowe (1973) and Lowe, Mill and Knapp (1973), is based on the structure

---

Fig. 9.3   Sensory cells of scolopidia: Various types of centriolar derivatives (Table 9.1). (a) Centriolar derivative without a distal segment (type 1); (b, c, d) centriolar derivatives with a distal segment (type 2) [(c, d) paraciliary and ciliary cells of Whitear, 1962].

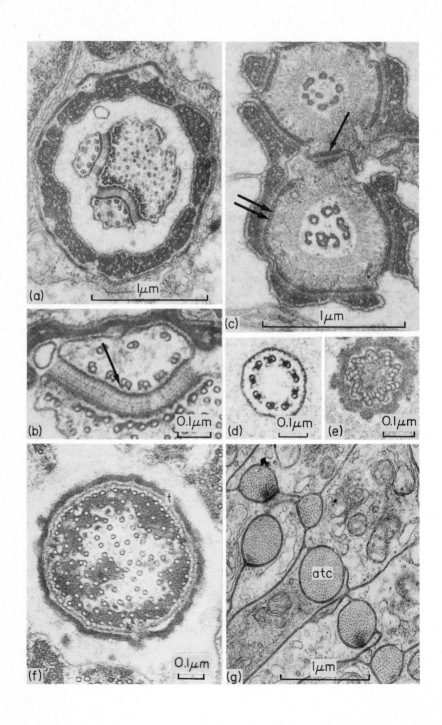

of the centriolar derivatives. In the ciliary cell, the centriolar derivative is represented successively by a ciliary segment, a paraciliary segment and a terminal segment (Fig. 9.3d). In the paraciliary cell, the centriolar derivative is represented only by a paraciliary segment and a terminal segment (and possibly with a very short ciliary segment) (Fig, 9.3c). In both, the paraciliary segment is a cone shaped region where the diameter of the axoneme increases progressively; in other words it is only a modified ciliary structure which can never be compared (as underlined by Whitear) with a ciliary dilation. This modified ciliary structure can be short (as in ciliary cells, Fig. 9.3d) or invade most or all of the ciliary region (as in paraciliary cells, Fig. 9.3c) and intermediate situations can be observed. The important thing is that in both of these cells the centriolar derivative is represented by a ciliary segment (more or less modified) and a distal segment (terminal segment of the above authors) ensheathed in a tube. Thus these cells are morphologically of type 2. Various hypotheses have been made concerning the physiological significance of the morphological differences observed (Section 6.7).

Fig. 9.4   (a–f) Johnston's organ of *Speophyes* (Coleoptera). (a) Transverse section of a scolopidium in the apical region of the scolopale space. The two type 1 centriolar derivatives are associated with the type 2 centriolar derivative (distal segment) by a lamellated intercellular material. (b) Detail of the above intercellular junction. Microtubules facing the junction in the distal segment (of the type 2 centriolar derivative) are connected by a lattice of dense material. Fibres facing the junction in the ciliary segment (of the type 1 centriolar derivative) possess small 'arms' (———→) directed to the membrane. (c) Transverse section of a scolopidium in the apical region of the dendrites. The two dendrites are sectioned between the two basal bodies: at this level the ciliary root (*cr*) is divided into 'fingers'. The scolopale is divided in scolopale rods. A symmetric desmosome (———→) is developed between the two dendrites; an asymmetric desmosome (====⟹) can be observed between the scolopale cell and the dendrites. (d) Transverse section through a ciliary segment. (e) Transverse section through a basal body. (f) Transverse section through the apical, subcuticular, region of a scolopidium. In the tube (*t*) the distal segment of the type 2 centriolar derivative develops a tubular body (electron dense material connecting longitudinal microtubules especially arranged near the cell membrane). (g) Antennal chordotonal organ of Ephemeroptera. Transverse section through attachment cells (*atc*) connecting scolopidia directly to the cuticle (see Fig. 9.8b). These cells are packed with microtubules and so can easily be identified from surrounding epidermal cells. ((a–f) Courtesy of Dr Corbière-Tichané; (g) from Schmidt, 1974.)

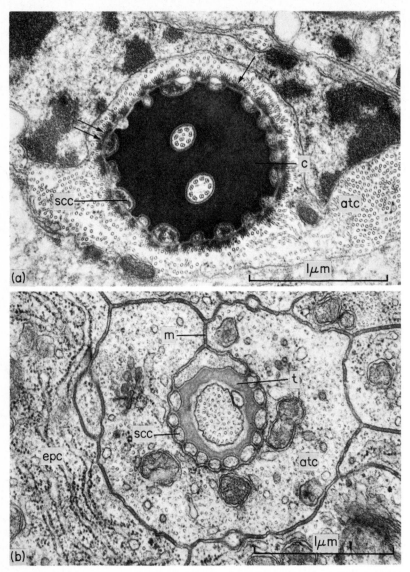

Fig. 9.5 Transverse sections through the apical region of scolopidia. (a) Scolopidium of the antennal chordotonal organ of *Speophyes* (Coleoptera). The two type 1 centriolar derivatives are inserted separately in a cap (*c*). One is sectioned near its apex and the axoneme is disorganized. The scolopale cell (*scc*) has distal prolongations containing scolopale material, lying between the attachment cell (*atc*) and the cap. The attachment cell contains numerous longitudinal microtubules, some of which are associated with hemidesmosomes round the cap (———→) or with the scolopale cell prolongations (═══⇒). This scolopidium is associated with the cuticle by the attachment cell and an epidermal cell (see Fig. 9.8 cii). (b) Scolopidium of the maxilla of *Speophyes*. The

Association of the two types of centriolar derivatives in a scolopidium

We shall see later that application of the stimulus could certainly be applied in different ways for the two types of centriolar derivatives. Therefore, they presumably can not be mixed in the same scolopidium. Until now this seems to be the rule, with the exception of Johnston's organ (see Table 9.1) and perhaps the mandibular chordotonal organs of Hemiptera (Pinet, 1968; Pinet, Bernard and Boistel, 1969). In Johnston's organ scolopidia, two sensory cells are of type 1 and one is of type 2 (Fig. 9.4a) and the apical extracellular structure is a tube attached to the external cuticle. The distal segment of the type 2 sensory cell is the only one which goes to the apex of the tube. Schmidt (1969) and Corbière-Tichané (1975) have clearly shown that the apical region of this distal segment possesses a dense core material which is, as emphasized by Corbière-Tichané, a true tubular body (see Fig. 9.4f). Thurm (1965, 1968) has strongly argued that, in mechanoreceptor sensilla, lateral compression of the cuticle is the active stimulus (Fig. 9.10) and acts on the tubular body. Therefore, this type 2 Johnston's organ sensory cell can be explained in one of two ways: (a) all the type 2 scolopidium sensory cells possess an apical tubular body (never described until now in any other scolopidium?) and the active stimulus is applied to these cells by the same lateral compression at the level of the tubular body as in other mechanoreceptor sensilla; (b) the Johnston's organ scolopidia are composite sensilla with two 'scolopidium' sensory cells and one 'campaniform' sensory cell. This is not too surprising if we remember that mechanoreceptor and chemoreceptor cells can be mixed in the same sensillum, as in the labellar hairs of Diptera.

9.3.3 Centriole-like structures and the ciliary root

The apex of the dendrite possesses two aligned basal bodies (centriole-like structures) (Fig. 9.2d). Distally the distal basal body bears the nine fibres of the axoneme, while proximally the two basal bodies are associated with the ciliary root (Fig. 9.2d).

---

two type 2 centriolar derivatives are ensheathed together by a tube (*t*) (see Fig. 2C). One is sectioned near its apex. The attachment cell (*atc*) does not possess more microtubules than the surrounding epidermal cells (*epc*). This scolopidium is directly associated with the cuticle (see Fig. 3.1A); *m*—mesaxon; *scc*—scolopale cell. (Courtesy of Dr. Corbière-Tichané.)

The basal bodies, as in motile cilia, are formed by a cylinder of nine triplet fibres (Fig. 9.4e). Each fibre is composed of three subfibres which have the structure of a microtubule. The plane through the centres of the three subfibres is inclined at about 40° to the tangent of the basal body, so that the triplets are 'tilted inwards'. The two innermost subfibres of the distal basal body are certainly prolongated by the subfibres of the axoneme. The proximal basal body is never associated with the axoneme. Some satellite structures, such as nine radiating spokes arising from the nine triplets of the distal basal body, have been observed (Young, 1973).

If we refer to what is known of other sensory cells of arthropod sensilla, the presence of two basal bodies is certainly the rule. However, they have only been observed in a limited number of scolopidia (Howse, 1968; Schmidt, 1969, 1974; Howse and Claridge, 1970; Young, 1973; Michel, 1974) and two basal bodies have never been observed in, for example, crustacean chordotonal organs. In some cases (Gray, 1960; Young, 1970) the two basal bodies have been observed, but the distal one is described as a 'ciliary base'. It is difficult to know what the significance of the of two basal bodies could be. If the distal one is considered as an organizer for the development of the centriolar derivative, it is possible that in all arthropod sensilla the proximal basal body becomes functional when the centriolar derivative is lost (Moulins, 1967). This is the situation which apparently occurs at each moult for different types of cuticular sensilla (Moran, 1971; Schmidt and Gnatzy, 1971) and certainly also for cuticular scolopidia (Richard, 1957; Schmidt, 1974). However, as pointed out by Schmidt (1969), it is difficult to believe that this is also true for 'internal' scolopidia. Internal scolopidia also possess two basal bodies (Schmidt, 1969; Young, 1973).

The ciliary root, is a cross-banded structure (Fig. 9.2d), regularly associated with the two basal bodies and it is much more developed in scolopidium sensory cells than in any other sensillum sensory cell. This is the axial filament of light microscopists. The ciliary root runs proximally in the dendrite and can be seen in the cell body and sometimes in the axon (Schmidt, 1969). In most cases, contrary to what appears in other sensillum sensory cells, the ciliary root of scolopidia is a single cylinder of cross-banded material. This cylinder is sometimes divided proximally when it reaches the cell body and always distally where it is associated with the basal bodies. This latter gives concentric finger-like processes (Fig. 9.4c) which penetrate the basal bodies. The ring is sometimes clearly composed of nine processes and it is possible that each process is associated with one fibre of the basal body.

9.3.4   Relationships between sensory cells in a scolopidium

Physiological interactions between sensory cells in a scolopidium have sometimes been postulated (e.g. Bush, 1965; Howse, 1968). Thus it is interesting to know if there is some morphological evidence for such interactions and this has been studied particularly by Corbière-Tichané (1969, 1971 and 1975).

In a scolopidium, sensory cells can come into contact with each other at the level of the dendrites or at the level of the centriolar derivatives. In their distal region the dendrites are not separately ensheathed by the scolopale cell and the resulting undifferentiated contact between sensory cells has been described as an 'ephapse' by Whitear (1962) and Mill and Lowe (1971) (Fig. 9.2a). We do not know the meaning of these ephapses which can, for example, also be observed between small sensory axons in arthropod nerves. In the apical region of the dendrites a differentiated intercellular junction reminiscent of the vertebrate desmosome (Kelly, 1967) is generally developed. It is a belt desmosome (zonual adherens) and is developed centrally between the sensory cells and laterally between the scolopale cell and the sensory cells (Fig. 9.4c). The same junction can be observed in scolopidia with only one sensory cell (Ghiradella, 1971). In other words this junction does not concern especially the possibility of interactions between sensory cells and must be compared with apical zonula adherens known in many epithelia where they are generally considered as structural support (Satir and Gilula, 1973) (Section 9.6). In the scolopale space specialized junctions between centriolar derivatives have also been observed. One such type of junction concerns the scolopidia of Johnston's organ of *Speophyes* (Corbière-Tichané, 1975) and can be identified in every scolopidium (Fig. 9.4a, b), where it associates the type 2 with the two type 1 centriolar derivatives. In the large intercellular space a lamellated dense material can be observed. The two cell membranes are internally underlined by dense material and the microtubules of the type 2 centriolar derivative are embedded in an electron-dense matrice. In this organ the two type 1 centriolar derivatives are not inserted into a cap: possibly the junction provides these centriolar derivatives with a mechanical anchorage. The other junction described in the scolopale space (Corbière-Tichané, 1969) is a gap junction and is found in the scolopidia of the maxilla. Gap junctions are known to provide an electronic coupling between cells and this may be important for an interpretation of the physiology.

9.4   The scolopale cell

9.4.1   The scolopale

The scolopale is an *intracellular* structure which cannot be compared with any other sheath protecting a centriolar derivative in other arthropod sensilla (Fig. 9.10). These latter sheaths are always extracellular structures and must be compared with the apical extracellular structure (tube or cap) of scolopidia. Unfortunately they have been described as a 'scolopale' in many papers. The scolopale is composed of an electron dense material deposited round longitudinally oriented microtubules Mill and Lowe (1973) (Fig. 9.4a). This material must perhaps be compared with the electron-dense latice material which is associated with microtubules in various accessory cells (Masson and Gabouriaut, 1973). The scolopale material constitutes a cylinder round the distal part of the dendrite(s) and the proximal part of the centriolar derivative(s). This cylinder is often irregular (Fig. 9.2b) and mostly fenestrated, particularly at the level of the scolopale space. This can give the scolopale rods described by Gray (1960). In this case the scolopale space is in communication, between the scolopale rods, with lateral extracellular spaces hollowed out in the scolopale cell and constituting part of the 'labyrinth' (Section 9.4.3).

9.4.2   Relationships of the scolopale cell with the dendrites and with the apical structures

The scolopale cell and the scolopale are in relation proximally with the dendrites and distally with the apical extracellular structure.

As seen above, in the apical region of the dendrites a belt desmosome can be developed between the dendrites and the

Table 9.2   Terminology of scolopidium accessory structures

This paper	Gray (1960) Young (1970)	Whitear (1962) Mill and Lowe (1973)	Howse (1968) Howse and Claridge (1970)	Schmidt (1969)	Corbière-Tichané (1971)
Attachment cell	Attachment cell	—(1)	Attachment cell	Hülzelle or Kappenzelle	Cellule de fixation
Scolopale cell	Scolopale cell	Scolopale cell	Scolopale Sheath cell	Stiftzelle	Cellule scolopale
Scolopale	Scolopale	Scolopale	Scolopales	Wandrippen	Cylindre scolopale
Tube	—	Tube		} Stift	Gaine cuticulaire
Cap	Cap		} Cap		Chapeau
Scolopale space	Extracellular space	Scolopale space	—	Extrazellularraum	—

(1) The canal cell described by the authors in some scolopidia is certainly an attachment cell.

scolopale cell (Fig, 9.4c). This junction is characterized by (a) inter-cellular dense material, (b) intracellular dense material against the two cell membranes and (c) tonofilaments in the dendrites, associated with the intracellular material and arranged as in vertebrate desmosomes (Kelly, 1967). The filaments give a ring of dense material round the basal bodies and the ciliary root. The ring sometimes appears lamellated ('lamellar structure of the collar' of Ghiradella, 1971; Plates 3 and 5 in Vande Berg, 1971). In the crustacean PD organ the desmosomes between the scolopale cell and the sensory cells are often associated with projections of the scolopale (Mill and Lowe, 1973). The junction seems to connect the axial structures (basal bodies and ciliary root) to the scolopale (Howse, 1968). It has been observed in many scolopidia and is more elaborate in scolopidia with centriolar derivative of type 1. In all cases this junction certainly provides good fixation of the apical region of the dendrite (Mill and Lowe, 1973).

Distally the scolopale cell gives rise to prolongations round the apical extracellular structure (Fig. 9.5). When this is a cap, i.e. the centriolar derivatives are of type 1, the scolopale penetrates into the prolongations and an hemidesmosomal system is developed (Fig. 9.5a). When the apical structure is a tube, i.e. the centriolar derivatives are of type 2, there is no such arrangement (Fig. 9.5b). In other words, if we assume some mechanical significance of the above differences, the dendrites, the scolopale and the apical structure are much more in mechanical solidarity in scolopidia with type 1 centriolar derivatives than in scolopidia with type 2 centriolar deriva-tives. In the former the apex of the centriolar derivative is inserted in a cap and longitudinal stretching of the scolopidium, which seems to be the adequate stimulus, results in a direct (without inertia) stretching of the centriolar derivative. In the second case, the centriolar derivative (distal segment) is independent of the tube and longitudinal stretching may not directly stretch the centriolar derivative. This could be achieved indirectly by the extracellular material which is in the tube ('glue' of Whitear, 1962) and the scolopale space, the visco-elasticity of which gives some inertia to the system.

9.4.3 The scolopale cell as a glandular cell

The above considerations raise the question of the origin of extracellular material. There is accumulating evidence that the scolopale cell is, in most cases, very 'vacuolated' and this has been briefly mentioned by many authors or can be easily seen on published pictures (Uga and Kuwabara, 1965; Howse, 1968; Young, 1970; Corbière-Tichané, 1971, 1975; Vande Berg, 1971; Friedman, 1972; Chu and Axtell, 1972; Michel, 1974; Wiese and Schmidt, 1974). The vacuoles constitute a complex system of extracellular cavities, a 'labyrinth', which communicates with the scolopale space. Similar labyrinths are known in secretory cells and it is possible to suggest that it is the scolopale cell which secretes the material contained in the scolopale space and in the tube. The same situation is known in chemoreceptor sensilla, where an equivalent accessory cell, the trichogen cell (Fig. 9.10), possesses a similar labyrinth (Moulins, 1971; Moulins and Noirot, 1972). Generally, in electron micrographs the material cannot be visualized after osmium fixation and is slightly electron-dense after aldehyde fixations. The situation is the same for the chemoreceptors (Moulins, 1971), where there is some histochemical indications suggesting that this material could be an acid muccopolysaccharid. If so, this is not contrary to the above hypothesis concerning the function in scolopidia. This material could also play the role of an ion barrier and so must be considered with the hypothesis of Thurm (1970) concerning the initiation of the generator potential in sensilla (Thurm, 1970; Barth, 1971).

9.5   The other accessory structures

9.5.1   The apical extracellular structure

In all scolopidia the apical region of the centriolar derivative is enclosed in an extracellular structure: the cap (Gray, 1960) or the tube (Whitear, 1962). Until now these two names have sometimes been used indiscriminately (Howse, 1968). However, it seems useful to keep a distinction for two main reasons:

(a) The structures are morphologically very different. The tube is an elongated canal-like structure closed at the apex and is composed of an homogenous electron-dense material resembling the external cuticle. The centriolar derivatives are ensheathed together by the tube and are not firmly attached to the tube wall (Figs. 9.2c, 9.5b). The cap is generally short and often made of a spongy, amorphous electron-dense material. The centriolar derivatives are ensheathed separately by the cap and seem to be firmly attached to the cap material (Fig. 9.5a). Sometimes the apical region of the centriolar derivative goes through the cap and is terminated on the cap by a bulbous region (Schmidt, 1969; Corbière-Tichané, 1971 and 1975).

(b) The centriolar derivatives of type 2 are always associated with a tube, never with a cap, whereas the centriolar derivatives of type 1 are always associated with a cap, never with a tube. (N.B. The only exception is Johnston's organ, which is discussed in Section 9.3.3, Fig. 9.4). In other words these two types of apical structure are characteristic of the two types of sensory cell and so it is possible to talk of two types of scolopidia.

Graber (1882) was certainly the first to note a distinction between two types of scolopidia and refers to 'amphinematic' scolopidia which are drawn into a distal thread and 'mononematic' scolopidia which lack this structure. When scolopidia are in direct relationship with the cuticle they are always amphinematic, i.e. they always possess a tube. When scolopidia are 'internal' (without any direct relationship with the cuticle) they can be mononematic or amphinematic. For example, Whitear (1962), referring to the original definition of Graber, talks of amphinematic scolopidia for the joint chordotonal organs of crustacean legs, where the scolopidia clearly do not enter any relationship with the cuticle. Unfortunately there is now a common tendency to call only those scolopidia which are associated with the cuticle amphinematic and refer to all the others as mononematic.

9.5.2   The attachment cell

This accessory cell generally envelops the apical extracellular structure of the scolopidium. Its main role seems to be to provide the apical anchorage of the scolopidium to the support. We will see that numerous variations occur and so it is not surprising that the development and the structure of this cell vary greatly from one chordotonal organ to another.

In strand chordotonal organs it is difficult to identify the attachment cell. Whitear (1962) has shown that, in the joint chordotonal organs of *Carcinus* legs the scolopale cell extends distally and completely ensheaths the tube. Nevertheless a canal cell, perhaps equivalent to the attachment cell, can be observed round the apical region of the tube in MC1, CP1 and PD organs. Mill and Lowe (1971, 1973) and Lowe *et al.* (1973) have made a detailed study of the PD organ of *Cancer* and shown that, in this chordotonal organ, only the elongation sensitive scolopidia possess a canal cell.

In chordotonal organs without a strand, or when the strand is entirely cellular (Young, 1970; Moulins, unpublished observations) an attachment cell can be identified. This cell is generally differentiated as a 'tendon' cell, i.e. the cytoplasm is rich in longitudinal microtubules (Figs. 9.4g, 9.5a) connected to apical and basal hemidesmosomes. This must be compared with the differentiation of the epidermal tendon cells of muscle (Laï-Fook, 1967; Caveney, 1969; Smith, Järlfors and Russell, 1969) and of the supporting connective tissue structure of Arthropods (Moulins, 1968). This cell can also be compared with the epidermal cell which takes part in the cuticular anchorage of the connective tissue strand of stretch receptors in Insects (Moulins, 1974).

9.6   The supports

The term 'chordotonal organ' is used to refer to two morphologically different types of organ:

(a) Chordotonal organs in which the scolopidia are only topographically associated, without lateral connections between them. In this case the scolopidia can be packed in small independent groups (Scoloparium of Schmidt, 1969; 'complexe récepteur' of Masson and Gabouriaut, 1973) and no strand material can be identified (Fig. 9.6). The scolopidia of these organs are more or less directly associated with the integument by their apex. A basal anchorage of the scolopidia on some internal or external structure has been described sometimes and is thought to be the rule.

(b) Chordotonal organs in which the scolopidia are embedded in a common strand (connective chordotonal organs of Howse, 1968). This strand usually spans two different regions of the body and the stimulus is applied to the scolopidia via tension, stretching or relaxation of the strand (Figs. 9.7, 9.9b).

9.6.1   Chordotonal organs without a strand

As can be seen from Table 9.1, the only one which has been studied in Crustacea, is the statocyst of the crayfish (Schöne and Steinbrecht, 1968). On the contrary in insects, most of the chordotonal organs studied are without a strand.

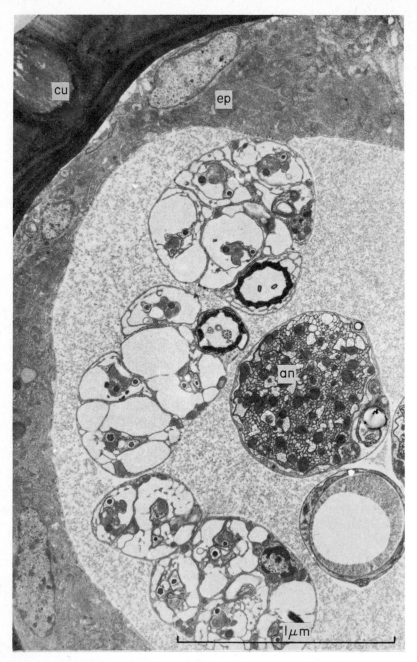

Fig. 9.6 Transverse section through a chordotonal organ without a strand: Johnston's organ of *Speophyes* (Coleoptera). In the antennal cavity scolopidia are associated in small groups but no strand material can be observed in these groups. Most of the scolopidia are sectioned in the distal region of the dendrites: at this level the scolopale cells are particularly 'vacuolated'. *an*—antennal nerve; *cu*—external cuticle; *ep*—epidermis. (Courtesy of Dr Corbière-Tichané.)

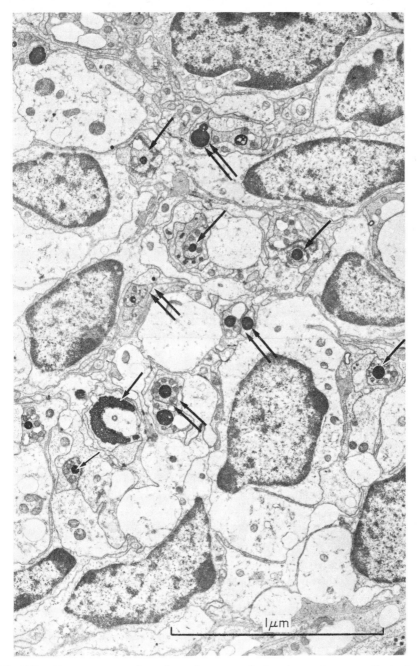

Fig. 9.7 Transverse section through a strand chordotonal organ: Cuticular stress detector CSD2 of the crayfish *Astacus leptodactylus*. The scolopidia are embedded in a purely cellular strand without connective tissue. In this organ there are scolopidia with one sensory cell (⟶) and others with two sensory cells (⟹).

Apical anchorage of scolopidia.
The scolopidia can be attached to the integument in three different ways (Fig. 9.6):

(a) The attachment cell is an epidermal cell and the extracellular apical structure of the scolopidium is inserted in the cuticle (Fig. 9.8a). The conditions realized are the same as for all other arthropod mechanoreceptor sensilla (Fig. 9.10) and it has been described for Johnston's organ and for some mouthpart chordotonal organs of insects (Table 9.1). The apical structure is always a tube, never a cap. The tube remains closed at its apex and can sometimes penetrate a long way into the cuticle (Howse and Claridge, 1970). The attachment cell is not modified (Fig. 9.5b). In the crayfish statocyst, the tube is replaced by a long 'chorda' inserted at the base of the external cuticular hair (Fig. 9.9a). In some chordotonal organs, like Johnston's organ of Diptera, the anchorage of the tube is on an invaginated part of cuticle (Uga and Kuwabara, 1965; Risler and Schmidt, 1967) (Fig. 9.8aii).

As in other arthropod sensilla, in these scolopidia the tube is lost at ecdysis (Richard, 1957; Schmidt, 1974). This supports the hypothesis that, as often claimed for other sensilla, the tube is a cuticular structure.

(b) The attachment cell is still an epidermal cell, but the apical structure is not inserted in the cuticle. (Fig. 9.8b). So far this situation has only been described by Schmidt (1974) and Corbière-Tichané (1975) for antennal chordotonal organs. Because of its position the attachment cell is a tendon cell connecting the apical extracellular structure (a cap) to the cuticle. This cell becomes differentiated and like the epidermal tendon cell of muscle contains numerous microtubules (Fig. 9.4g) associated with basal and apical hemidesmosomes.

(c) The attachment cell becomes a subepidermal cell (Fig. 9.8c) and is connected at its apex to the base of an epidermal cell. It is the hypodermal cell of Gray (1960), the microtubular shaft of Ghiradella (1971), the secondary attachment cell of Young and Ball (1974a), the 'akzessorische Zelle' of Schmidt (1969) and Michel (1974) and the 'cellule de fixation 2' of Corbière-Tichané (1971). This situation has been described in tympanal organs,

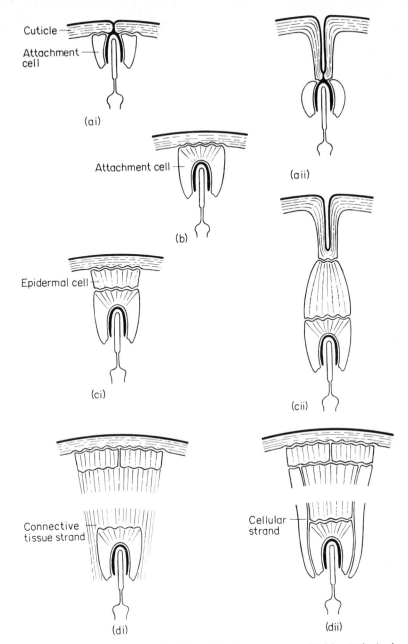

Fig. 9.8   Relationships of scolopidia with the tegument (Table 9.1). (a, b, c) chordotonal organs without a strand. (a) apical structure (tube) directly associated with the external cuticle without (i) or with (ii) internal invagination. (b) apical structure associated with the cuticle by the attachment cell (c) apical structure associated with the cuticle by the attachment cell and an epidermal cell without (i) or with (ii) internal invagination; (d) strand chordotonal organ with a connective tissue strand (i) or with a cellular strand (ii).

subgenual organs, tarsal organs and antennal organs of insects (Table 9.1). The 'tendon' is composed of the two cells with microtubules and hemidesmosomes. The apical extracellular structure is always a cap. In this case also the scolopidia can be in a more internal position by the development of an integumentary invagination (Ghiradella, 1971; Corbière-Tichané, 1971; Fig. 9.8cii).

Basal anchorage of scolopidia

Very little is known about the basal anchorage of scolopidia. It seems to occur individually with each sensillum being anchored to the tegument or to some internal structure. In the crayfish statocyst (Schöne and Steinbrecht, 1968), in Johnston's organ of ants (Masson and Gabouriaut, 1973) and in the antennal chordotonal organ of *Speophyes* (Corbière-Tichané, 1971), the scolopidia are associated laterally with the integument by specialized junctions between the scolopale cell and the epidermal cells. As for apical anchorage, a hemidesmosal system associated with microtubules is present (see Fig. 3a in Corbière-Tichané, 1971). In the Johnston's organ of *Chrysopa* (Schmidt, 1969) the same situation occurs but it is, a glial cell enveloping the cell body which takes part in the anchorage on the integument. In many organs, the basal anchorage occurs much more proximally, perhaps involving the sensory nerve itself.

9.6.2   Strand chordotonal organs

All of the chordotonal organs studied in Crustacea, with the exception of the crayfish statocyst, are strand organs. In Insecta the only one on which an ultrastructural study has been carried out is the tibio-tarsal organ of *Periplaneta* (Young, 1970). It is possible to distinguish between connective tissue strands and cellular strands.

Connective tissue strands
(Fig. 9.8di)
These have been studied by Whitear (1962) and Mill and Lowe (1971, 1973). The scolopidia are embedded in an extracellular matrix which is more or less rich in cellular components (strand cells) (Fig. 9.9c). In this matrix longitudinal cross-striated fibres of collagen can be observed. The strand cells are branching cells and

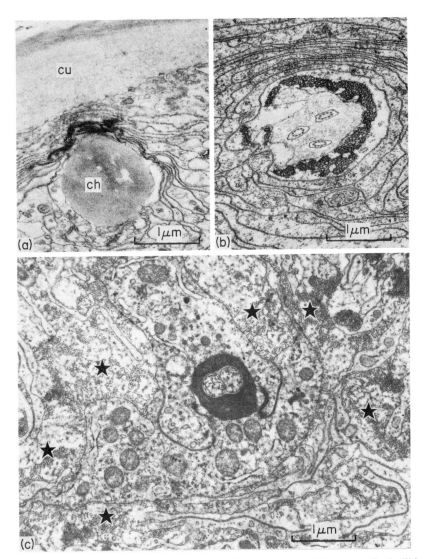

Fig. 9.9 (a, b) Scolopidium of the crayfish statocyst (*Astacus fluviatilis*). (a) transverse section through the chorda ch—an apical, extracellular structure equivalent to the tube and connected to the cuticle cu—at the base of an external hair. (b) transverse proximal section through the scolopale region of the scolopidium. In the scolopale space three ciliary segments can be observed. (c) transverse section through a strand chordotonal organ: MCO2 of *Cancer pagurus*. A scolopidium is embedded in a strand where large extracellular spaces are invaded by connective tissue (*); compare with Plates 4 and 5. ((a, b) Courtesy of Dr Steinbrecht; (c) Courtesy of Drs. Mill and Lowe.)

may be fibroblasts (Mill, personal communication). Certainly there is evidence that the cells of the collagenous strand of MCO2 (Section 6.2.2) are fibroblasts (Gordon, 1973). The situation is very different from that seen in the connective tissue strands of insect stretch receptors, where no fibroblast has been identified (Moulins, 1974). There is an outer layer of amorphous connective tissue without fibres and without strand cells. Taylor (1967a, b) has suggested that stretching of the strand can result in the bending of dendrites if the scolopidia lie between two different layers of connective tissue, but Mill and Lowe (1973) have shown that this does not occur in the PD organ of *Cancer*, i.e. there is never a scolopidium in the interface between the outer layer and the central region of the strand (Section 6.7). It is important to consider the structure of the strand in relation to the nature of the physiological response of the scolopidia. Thus, elongation sensitive scolopidia are embedded in a part of the strand rich in cellular components (and poor in extracellular connective tissue), while relaxation sensitive scolopidia are embedded in a part of the strand rich in extracellular connective tissue (and poor in cellular components) (Mill and Lowe, 1973). This is confirmed indirectly by Young (1970) who has shown that, in a purely cellular strand, it was not possible to identify any scolopidium responding to relaxation.

Cellular strands (Fig. 9.8dii).   These have been studied in insects by Young (1970) and in crustaceans by Moulins and Clarac (1972). In both cases the scolopidia are embedded in a purely cellular tissue (Fig. 9.7). The surrounding cells are sometimes difficult to differentiate from the glial cells, enveloping cell bodies and axons of the sensory cells (Young, 1970). The only extracellular connective tissue which can be observed is a thin outer layer of the nerve perineurium. Surrounding cells are not modified, except in the anchorage region onto the integument where they also become tendon cells with longitudinal microtubules and hemi-desmosomes. In the cuticular stress detector CSD2 of the crayfish, the situation is more complex as in some scolopidia the tube is connected to the cuticle.

9.7 Conclusions

Ultrastructural studies of chordotonal organs are of interest (a) in terms of the comparative morphology of sensory structures and (b) to complement physiological data.

Comparative Morphology

There is no question about the nature of scolopidia which are sensilla, i.e. morphogenetical units derived from an epidermal mother cell. In all sensilla, accessory structures associated with the sensory cell can be compared and this has been done in Fig. 9.10 for the main types of mechanotransducer sensilla. It is possible to consider that the scolopale cell and the attachment cell are homologous respectively with the trichogen and tormogen cells of other sensilla (For another interpretation, see Schmidt, 1973). The originality of the scolopidium is represented by the presence of the scolopale and the absence of a tubular body. This has certainly something to do with the function of the system, and particularly the way in which the stimulus is applied to the dendrites. What remains to be understood is how scolopidia can migrate from an original epidermal position and become internal, sometimes included within a strand.

Relationship with physiology

As far as understanding physiological results are concerned ultra-structural studies can help to answer two questions:

(i) What is the mechanism through which the stimulus is applied to the sensory cell? There is now accumulating evidence (see Mill and Lowe, 1973) that longitudinal stretching is the adequate stimulus. This is indicated by the axial symmetry of the accessory structures round the centriolar derivative and by the fact that scolopidia are certainly always associated with two different regions of the body. Concerning the way in which nature of the stimulus is applied to the centriolar derivative, ultrastructural studies suggest that the situation is not the same in the two types of scolopidia. In type 1 this can be realized by compression of the apical region of the centriolar derivative (Thurm, 1968), if the cap possesses some elasticity, or by stretching of the ciliary dilation (see Howse, 1968), which contains a special tubular body. In type 2 such mechanisms cannot be assumed and stimulus application is certainly realized by relative movements of the tube and the centriolar derivative.

(ii) What are the mechanisms through which the different responses of scolopidia are obtained (i.e. position or movement responses and stretch or relaxation responses)? Qualitative physiological differences in the responses of scolopidia can be attributed, as in

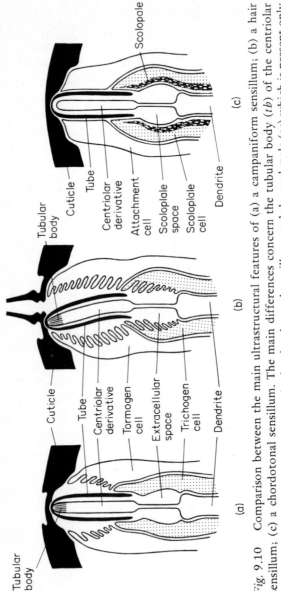

Fig. 9.10   Comparison between the main ultrastructural features of (a) a campaniform sensillum; (b) a hair sensillum; (c) a chordotonal sensillum. The main differences concern the tubular body (*tb*) of the centriolar derivative (*cd*), which does not exist in the chordotonal sensillum, and the scolopale (*sc*) which is present only in the chordotonal sensillum. The scolopale cell (*scc*) and the attachment cell (*atc*) of the chordotonal sensillum are equivalent to the trichogen (*trc*) and tormogen (*toc*) cells respectively of other sensilla. The scolopale space (*scs*) of the chordotonal sensillum is equivalent to the extracellular space (*es*), in which lies the ciliary segment in other sensilla. *cu*–cuticle; *d*–dendrite; *t*–tube. (cp. Fig. 1.25.)

all sensory systems, to the electrical properties of the excitable membrane and to the mechanical properties of the system which applies the stimulus to the membrane. The first point requires electrophysiological techniques. However, concerning the second, ultrastructural studies have shown, in type 2 scolopidia, a direct relationship between the nature of the support (strand) and the nature of the response. Thus relaxation scolopidia are known only in strand regions rich in connective tissue (Mill and Lowe, 1971, 1973). Movement (phasic) and position (tonic) responses, can perhaps be understood in terms of the possibility of relative movement of the centriolar derivative in the tube. For a pure position cell it is possible to imagine that no such movement occurs, the centriolar derivative remaining stretched all the time during stimulus application; whereas for a pure movement cell the centriolar derivative slips back in the tube during movement (Mill and Lowe, 1973) (Section 6.7). Intermediate situations are also easy to imagine. In this respect it will be interesting to know more about the material which is in the tube and whether it can play a role in terms of friction between the tube and the centriolar derivative.

Acknowledgements

I should like to express my thanks to Dr Corbière-Tichané, Dr Mill and Dr Steinbrecht for permission to reproduce unpublished electron microscope pictures. My grateful thanks are due also to Professor Schmidt who allowed me to read his Habilitationsschrift. I am most grateful to Mr Marcel Jean for the drawings and to Mrs Andrée Becherini for photographical and technical assistance in the preparation of the manuscript.

References

Barth, F. G. (1971) Der sensorische Apparat der Spaltsinnesorgane (*Cupiennius salei* Keys., Araneae). *Zeitschrift für Zellforschung und mikroskopische Anatomie*, **112**, 212–246.

Barth, G. (1934) Untersuchungen über Myochordotonal organe bei dekapoden Crustaceen. *Zeitschrift für Wissenschaftliche Zoologie*, **145**, 576–624.

Bush, B. M. H. (1965) Proprioception by the coxo-basal chordotonal organ, CB, in the legs of the crab, *Carcinus maenas*. *Journal of Experimental Biology*, **42**, 285–297.

Caveney, S. (1969) Muscle attachment related to cuticle architecture in Apterygota. *Journal of Cell Science*, 4, 451−559.

Chu, I. W. and Axtell, R. C. (1972) Fine structure of the terminal organ of the house fly larva, *Musca domestica* L. *Zeitschrift für Zellforschung und mikroskopische Anatomie*, 127, 287−305.

Corbière-Tichané, G. (1969) Différents types de jonctions intercellulaires dans le système nerveux sensoriel du *Speophyes lucidulus* Delar. (Coléoptères Bathysciines Cavernicoles). *Journal de Microscopie*, 8, 1003−1016.

Corbière-Tichané, G. (1971) Ultrastructure des organes chordotonaux des pièces céphaliques chez la larve du *Speophyes lucidulus* Delar. (Coléoptère Cavernicole de la sous-famille des Bathysciinae). *Zeitschrift für Zellforschung und mikroskopische Anatomie*, 117, 275−302.

Corbière-Tichané, G. (1975) L'organe de Johnston chez l'imago du *Speophyes lucidulus* Delar. (Coléoptère cavernicole de la sous-famille des Bathysciinae). Ultrastructure. *Journal de Microscopie*, 22, 55−68.

Debaisieux, P. (1936) Organes scolopidiaux des pattes d'Insectes. I. Lépidoptères et Trichoptères. *La Cellule*, 44, 271−314.

Debauche, H. (1935) Recherches sur les organes sensoriels antennaires de *Hydropsyche longipennis*. *La Cellule*, 44, 43−83.

Eggers, F. (1924) Zur Kenntnis der antennalen stiftführenden Sinnesorgane der Insekten. *Zeitschrift für Morphologie und Okologie der Tiere*, 2, 259−349.

Finlayson, L. H. (1968) Proprioceptors in Invertebrates. *Symposia of the Zoological Society of London*, 23, 217−249.

Friedman, M. H. (1972) A light and electron microscopic study of sensory organs and associated structures in the foreleg tibia of the cricket, *Gryllus assimilis*. *Journal of Morphology*, 138, 263−328.

Gaffal, K. P. and Hansen, K. (1972) Mechanorezeptive Strukturen der antennalen haarsensillen der Baumwollwanze *Dysdercus intermedius* Dist. *Zeitschrift für Zellforschung und mikroskopische Anatomie*, 132, 79−94.

Ghiradella, H. (1971) Fine structure of the noctuid moths ear. I. The transducer area and connections to the tympanic membrane in *Feltia subgothica* Harworth. *Journal of Morphology*, 134, 21−46.

Gordon, S. E. (1973) A chordotonal organ in the leg of *Cancer pagurus* Linn.: some ultrastructural observations. *Ph.D. Thesis*. University of Leeds, U.K.

Graber, V. (1882) Die chordotonalen Sinnesorganes und das Gehör der Insekten. *Archiv für mikroskopische Anatomie und Entwicklungsmechanik*, 20, 506−640.

Gray, E. G. (1960) The fine structure of the insect ear. *Philosophical Transactions of the Royal Society, Series (B)*, 243, 75−94.

Gray, E. G. and Pumphrey, R. J. (1958) Ultrastructure of insect ear. *Nature*, 181, 618.

Hartmann, H. B. and Boettiger, E. G. (1967) The functional organization of propus-dactylus organ in *Cancer irroratus* Say. *Comparative Biochemistry and Physiology*, **22**, 651—663.

Howse, P. E. (1968) The fine structure and functional organization of chordotonal organs. *Symposia of the Zoological Society of London*, **23**, 167—198.

Howse, P. E. and Claridge, M. F. (1970) The fine structure of Johnston's organ of the leaf-hopper *Oncopsis flavicollis*. *Journal of Insect Physiology*, **16**, 1665—1675.

Jägers-Röhr, E. (1968) Untersuchungen zur Morphologie und Entwicklung der scolopidial Organe bei der Stabheuschrecke *Carausius morosus* Br. *Biologisches Zentralblatt*, **87**, 393—409.

Kelly, D. E. (1967) Fine structure of desmosomes, hemidesmosomes and adepidermal globular layer in developing newt epidermis. *Journal of Cell Biology*, **28**, 51—72.

Kinzelbach-Schmitt, B. (1968) Zur Kenntnis der antennal chordotonalorgane der Thysanuren (Thysanura, Insecta). *Zeitschrift für Naturforschung*, **23b**, 289—291.

Laï-Fook, J. (1967) The structure of developing muscle insertions in Insects. *Journal of Morphology*, **123**, 503—528.

Lowe, D. A., Mill, P. J. and Knapp, M. F. (1973) The fine structure of the PD proprioceptor of *Cancer pagurus*. II. The position sensitive cells. *Proceedings of the Royal Society, Series (B)*, **184**, 199—205.

Masson, C. and Gabouriaut, D. (1973) Ultrastructure de l'organe de Johnston de la fourmi *Camponotus vagus* Scop. *Cell and Tissue Research*, **140**, 39—76.

Mendelson, M. (1966) The site of impulse initiation in bipolar receptor neurons of *Callinectes sapidus* L. *Journal of Experimental Biology*, **45**, 411—420.

Michel, K. (1974) Das Tympanalorgan von *Gryllus bimaculatus* Degeer (Saltatoria, Gryllidae). *Zeitschrift für Morphologie und Okologie der Tiere*, **77**, 285—315.

Mill, P. J. and Lowe, D. A. (1971) Transduction processes of movement and position sensitive cells in a crustacean limb proprioceptor. *Nature, London*, **229**, 206—208.

Mill, P. J. and Lowe, D. A. (1973) The fine structure of the PD proprioceptor of *Cancer pagurus*. I. The receptor strand and the movement sensitive cells. *Proceedings of the Royal Society, Series (B)*, **184**, 179—197.

Moran, D. T. (1971) Loss of the sensory process of an insect receptor at ecdysis. *Nature*, **234**, 476—477.

Moulins, M. (1967) Les cellules sensorielles de l'organe hypopharyngien de *Blabera craniifer* Burm. (Insecta, Dictyoptera). Etude du segment ciliaire et des structures associées. *Comptes-rendus hebdomadaires des séances de l'Académie des Sciences, Paris*, **265**, 44—47.

Moulins, M. (1968) Etude ultrastructurale d'une formation de soutien épidermo-conjonctive inédite chez les Insectes. *Zeitschrift für Zellforschung und mikroskopische Anatomie*, **91**, 112—134.

Moulins, M. (1971) Ultrastructure et physiologie des organes épipharyngiens et hypopharyngiens (Chimiorécepteurs cibariaux) de *Blabera craniifer* Burm. (Insecte, Dictyoptère). *Zeitschrift für vergleichende Physiologie*, **73**, 139—166.

Moulins, M. (1974) Récepteurs de tension de la région de la bouche chez *Blaberus craniifer* Burm. (Dictyoptera, Blaberidae). *International Journal of Insect Morphology and Embryology*, **3**, 171—192.

Moulins, M. and Clarac, F. (1972) Ultrastructure d'un organe chordotonal associé à la cuticule dans les appendices de l'Ecrevisse. *Comptes-rendus hebdomadaires des séances de l'Académie des Sciences, Paris*, **274**, 2189—2192.

Moulins, M. and Noirot, Ch. (1972) Morphological features bearing on transduction and periphereal integration in insect gustatory organs. *In* Schneider D. (ed): *Olfaction and taste IV*, Wissenschaftliche Verlagsgesellschaft MBH, Stuttgart.

Pinet, J. M. (1968) Données ultrastructurales sur l'innervation sensorielle des stylets maxillaires de *Rhodnius prolixus* (Heteroptera Reduviidae). *Comptes-rendus hebdomadaires des séances de l'Académie des Sciences, Paris*, **267**, 634—637.

Pinet, J. M., Bernard, J. and Boistel, J. (1969) Etude électrophysiologique des récepteurs des stylets chez une punaise hématophage: *Triatoma infestans*. *Comptes-rendus des séances de la Société de Biologie*, **163**, 1939—1946.

Richard, G. (1957) L'ontogénèse des organes chordotonaux antennaires de Calotermes flavicollis (Fab.). *Insectes sociaux*, **4**, 106—111.

Risler, H. and Schmidt, K. (1967) Der Feinbau der Scolopidien im Johnstonschen Organ von *Aëdes aegypti* L. *Zeitschrift für Naturforschung*, **22b**, 759—762.

Satir, P. and Gilula, N. B. (1973) The fine structure of membranes and intercellular communication in Insects. *Annual Review of Entomology*, **18**, 143—166.

Schmidt, K. (1969) Der Feinbau der stiftführenden Sinnesorgane im Pedicellus der Florfliege *Chrysopa* Leach (Chrysopidae, Plannipennia). *Zeitschrift für Zellforschung und mikroskopische Anatomie*, **99**, 357—388.

Schmidt, K. (1970) Vergleichend morphologische Untersuchungen über den Feinbau der Ciliarstrukturen in den Scolopidien des Johnstonschen Organs holometaboler Insekten. *Verhandlungen der Deutschen zoologischen Gesellschaft*, **64**, 88—92.

Schmidt, K. (1973) Vergleichende morphologische Untersuchungen am Mechanorezeptoren der Insekten. *Verhandlungen der Deutschen Zoologischen Gesellschaft*, **66**, 15—25.

Schmidt, K. (1974) Die Mechanorezeptoren im Pedicellus der Eintagsfliegen (Insecta, Ephemeroptera). *Zeitschrift für Morphologie und Ökologie der Tiere*, 78, 193–220.

Schmidt, K. and Gnatzy, W. (1971) Die Feinstruktur der Sinneshaare auf den Cerci von *Gryllus bimaculatus* Deg. (Saltatoria, Gryllidae). II. Die Häutung der Faden- und Keulenhaare. *Zeitschrift für Zellforschung und mikroskopische Anatomie*, 122, 210–226.

Schöne, H. and Steinbrecht, R. A. (1968) Fine structure of statocyst receptor of *Astacus fluviatilis*. *Nature*, 220, 184–186.

Smith, D. S., Järlfors, U. and Russell, F. E. (1969) The fine structure of muscle attachments in a spider (*Latrodectus mactans* Fab.). *Tissue and Cell*, 1, 673–688.

Taylor, R. C. (1967a) The anatomy and adequate stimulation of a chordotonal organ in the antennae of a hermit crab. *Comparative Biochemistry and Physiology*, 20, 709–717.

Taylor, R. C. (1967b) Functional properties of the chordotonal organ in the antennal flagellum of a hermit crab. *Comparative Biochemistry and Physiology*, 20, 719–729.

Thurm, U. (1965) An Insect mechanoreceptor. Part. I: Fine structure and adequate stimulus. *Cold Spring Harbor Symposium on Quantitative Biology*, 30, 75–82.

Thurm, U. (1968) Steps in transducer process of mechanoreceptors. *Symposium of the Zoological Society of London*, 23, 199–216.

Thurm, U. (1970) Untersuchungen zur funktionellen Organization sensorischer Zellverbände. *Verhandlungsbericht der Deutschen Zoologischen Gesellschaft*, 64, 79–88.

Uga, S. and Kuwabara, M. (1965) On the fine structure of the chordotonal sensillum in antenna of *Drosophila melanogaster*. *Journal of Electron Microscopy*, 14, 173–181.

Vande Berg, J. S. (1971) Fine structural studies of Johnston's organ in the tobacco hornworm moth, *Manduca sexta* (Johannson). *Journal of Morphology*, 133, 439–456.

Wetzel, A. (1934) Chordotonalorgane bei Krebstieren (*Caprella dentata*). *Zoologisher Anzeiger*, 105, 125–132.

Whitear, M. (1960) Chordotonal organs in Crustacea. *Nature*, 187, 522–523.

Whitear, M. (1962) The fine structure of crustacean proprioceptors. I. The chordotonal organs in the legs of the shore crab, *Carcinus maenas*. *Philosophical Transactions of the Royal Society, Series (B)*, 245, 291–325.

Wiese, K. and Schmidt, K. (1974) Mechanorezeptoren im Insektentarsus. Die Konstruktion des Scolopidialorgans bei *Notonecta* (Hemiptera, Heteroptera). *Zeitschrift für Morphologie der Tiere*, 79, 47–64.

Young, D. (1970) The structure and function of a connective chordotonal organ in the cockroach leg. *Philosophical Transactions of the Royal Society, Series (B)*, **256**, 401–428.

Young, D. (1973) Fine structure of the sensory cilium of an insect auditory receptor. *Journal of Neurocytology*, **2**, 47–58.

Young, D. and Ball, E. (1974a) Structure and development of the auditory system in the prothoracic leg of the cricket *Teleogryllus commodus* (Walker) I. Adult structure. *Zeitschrift für Zellforschung und mikroskopische Anatomie*, **147**, 293–312.

Young, D. and Ball, E. (1974b) Structure and development of the tracheal organ in the mesothoracic leg of the cricket *Teleogryllus commodus* (Walker). *Zeitschrift für Zellforschung und mikroskopische Anatomie*, **147**, 325–334.

Zawarzin, A. (1912) Histologische Studien über Insekten. II. Des sensible Nervensystem der Aeschnalarven. *Zeitschrift für wissenschaftliche Zoologie*, **100**, 245–286.

10 Arthropod apodeme tension receptors

D. L. MACMILLAN

10.1 Introduction and definitions

In common practice the terms *tension* and *stress* are used synonymously. In physiological usage, however, a *stress* receptor is any receptor which detects changes in force/unit area, irrespective of whether these are produced by an external force acting upon the animal or by the animal's own musculature. The term *tension* receptor is usually reserved for that subset of stress receptors which respond specifically to changes in muscle tension. This traditional division is not entirely satisfactory but will be adhered to here to avoid the necessity of tendering further definitions in a subject already burdened with a plethora.

The problem of determining whether a particular receptor or receptor organ is a generalized stress detector, a tension detector or both, arises in all phyla but is particularly noticeable in the arthropods because of the presence of a hard exoskeleton. There are numerous stress receptors embedded in the exoskeleton which can be shown to respond to forces applied to the animal (Sections 1.4, 5.3, 5.4, 8.3; Chapter 7), but because the arthropod's muscles also attach internally to the same exoskeletal structure, many of these receptors are also potential tension receptors.

Some confusion also arises in determining whether a receptor is specifically, or even primarily, measuring changes in stress because this appears to be monitored by measurement of the concomitant change in strain (change in length/resting length). Thus, a strain receptor may be said to be monitoring macro-strain changes while a stress receptor monitors micro-strain changes. Because receptors do not necessarily fall exclusively into one category or the other it can prove very difficult assessing the functional sensitivity of a particular strain receptor. Some indication of the primary role is often given by the mechanical arrangement of the receptor components in relation to the surrounding skeletal and muscular elements. However, one must be extremely careful not to overlook the possibility of a secondary stress detecting role in an organ which appears to be structurally arranged as a strain detector, or vice versa.

The discussion in this chapter will be limited to those stress receptors which are found on the tendons of arthropod muscles. One of the most interesting features of these receptors is that on structure alone their nature and position virtually precludes the problems of modal definition discussed above and this supposition (confirmed by experimental data) makes them, to date, the only clear-cut examples of tension receptors in the arthropods.

10.2   Structure and distribution

It seems probable that crustacean tension receptors have been described in a variety of situations without actually being recognized as such.

Methylene blue staining techniques have been the primary tool used to study the gross morphology of the crustacean peripheral nervous system (Bullock and Horridge, 1965; Wales, Clarac, Dando and Laverack, 1970). One of the striking results of methylene blue staining, and it has been remarked on by most exponents of the technique, is the vast number of axons, dendrites and nerve cell bodies which are apparent in the connective tissue on the inner surface of the integument and on its internal expansions (e.g. Alexandrowicz, 1933, 1957). These bipolar and multipolar sensory cells have not usually been described as being associated in any particular aggregation, their axons running into integumentary nerves without first collecting into specific bundles. Occasionally, however, as in the case of the detailed descriptions of the chordotonal organs at the mero-carpopodite and carpo-propopodite joints of *Carcinus*, sensory nerves of this general type have been described in association with specific joint and apodeme areas (Whitear, 1962).

In some cases the axons of cells on the muscle apodemes collect together into nerves which only carry fibres from the apodemes and which have therefore been called apodeme sensory nerves (Fig. 10.1a).

In the Brachyura, the cells described in this situation are exclusively bipolar and two different types have been described. Proximally, there are several (up to five) large (50 μm diameter) bipolar cell bodies situated close to the apodeme, but not usually among the muscle fibres (Fig. 10.1b). The long, unbranched dendrites of these proximal sensory cells run down onto the surface of the apodemes where they may be traced for some distance before they no longer stain clearly because of associated dense connective tissue. In the region of these cells the apodeme sensory nerve is not closely associated with the apodeme and the axon from the cells run individually to join it. More distally, the apodeme sensory nerve is closely apposed to the apodeme. Here cell bodies, which are generally smaller than the proximal ones, are found along its length, associated with it. Their dendrites cross the apodeme in the region of the muscle fibre attachments (Fig. 10.1c) (Macmillan and Dando, 1972).

A survey of the distribution of apodeme tension receptors has not been carried out. They occur on most of the peripheral leg muscles in a variety of brachyurans examined. Methylene blue staining of the apodemes in the legs of anomurans, astacideans and palinurans reveals fine fibres and occasional bipolar cells but thus far it has not been possible to determine where, or even whether, the axons from these cells run together to form specific apodeme sensory nerves (Unpublished results in collaboration with L. H. Field and F. Clarac). It is clear, however, that the anatomical situation in these groups must be different from that in the brachyurans where the nerves are large and obvious. Descriptions of cells in the apodeme region of some of the muscles in *Squilla* suggest possible tension receptor aggregations but this also remains to be investigated (Alexandrowicz, 1957).

The only other clear example of apodeme sensory receptors is in *Limulus* (Xiphosura) where they are found in association with the flexor muscles responsible for tailspine movement. As in the Brachyura, the apodeme sensory nerves are also composed of axons from two different cell types. Multipolar cells up to 110 μm in size

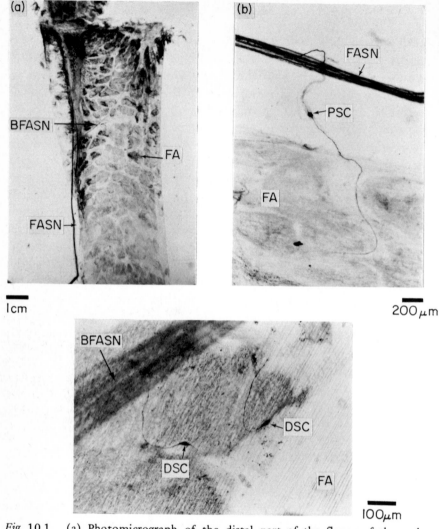

Fig. 10.1   (a) Photomicrograph of the distal part of the flexor of the crab *Cancer magister* where it is attached to the carpopodite at the mero-carpopodite joint. The medial edge of the apodeme is to the left and the distal end, which attaches to the carpopodite, to the top. The muscle fibres of the flexor muscle were gently removed to show the flexor apodeme (FA) and the flexor apodeme sensory nerve (FASN) running along the edge of the apodeme. Branches of the apodeme nerve can be seen running onto the surface of the muscle in the region where the muscle fibres insert.

(b) An example of a large proximal sensory cell (PSC) the axon of which joins the flexor apodeme sensory nerve (FASN) prior to its close association with the edge of the flexor apodeme (FA) and sends its dendrite onto the apodeme.

(c) Examples of the smaller, more numerous distal sensory cells (DSC) which are situated on the apodeme among the muscle fibre insertions. All prepartions are stained with methylene blue. (After Macmillan and Dando, 1972.)

are found around the point where the muscle fibres insert into the apodemes, or within the body of the muscle itself. The dendrites of these cells show extensive arborization around the muscle insertions, where they appear to enter the connective tissue layer. There are also smaller (20 μm) cells, which are thought to be less numerous, although this impression may be given because they stain less readily. These cells are bipolar and are found embedded in the connective tissue around the shafts of the apodemes (Eagles and Hartman, 1975). Similar receptors embedded in the connective tissue of a tendon in the legs of *Limulus* have also been described but these appear to be involved in detecting the movement of the trochanter (Barber and Hayes, 1964).

It seems unlikely that apodeme tension receptors are limited to these two groups or to the limited situations described so far. It is therefore probable that there will be further descriptions of tension receptors of the type already known. It is also probable that tension afference is of general functional importance in other arthropod motor systems so that receptors which are functionally analogous but structurally different may also occur.

10.3   Physiology

In both *Cancer* and *Limulus* it is possible to record from intact apodeme sensory nerves while manipulating the homonymous muscle in a variety of ways. It is also possible, by teasing the nerves, to record the acitivity of single units.

In *Cancer*, recordings from an apodeme sensory nerve during isometric contraction of the muscle which it innervates (by stimulation of the motor nerve) shows that as tension increases there is an increase in both the number of units responding and also in the frequency of the response of some individual units (Fig. 10.2a). Because they respond to muscle tension, these units also respond to imposed movements of the joints if these produce a concomitant change in muscle tension but not if there is no tension change (Fig. 10.2b). Tonic units which increase their rate of firing with increase in tension and phasotonic units which respond to tension onset or increment with a phasic component and then adapt slowly are both found, but units which respond phasically to falling tension have not been described (Macmillan and Dando, 1972).

In *Limulus* there are units which respond either to stretching or to relaxing of the muscles in the absence of motor nerve stimulation but it is not yet clear whether these units are responding to muscle tension or muscle length changes, i.e. whether they also respond to an isometric contraction and whether their response to passive movement is due to residual muscle tension. These units appear to be phasic only and the rate of discharge is increased by increasing the velocity of the movement. There are also phasotonic units which are

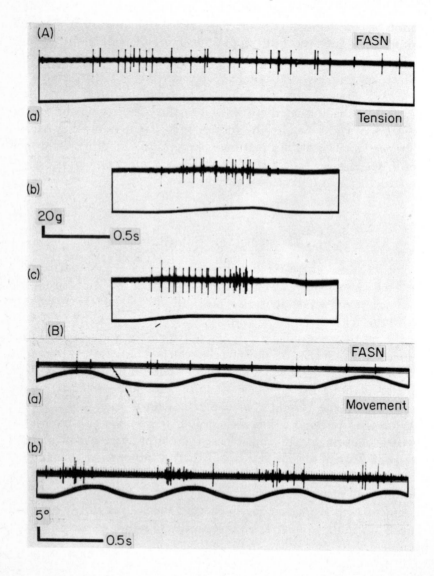

silent during movements imposed on a resting muscle but active during muscular contraction. These units also respond to imposed movements in the presence of muscle tension (Eagles and Hartman, 1975).

In neither *Limulus* nor *Cancer* was a detailed analysis of single fibre responses made, and units responding in other ways may well be present. The important findings to date are that in both cases muscle tension appears to be the adequate stimulus and further that the apodeme receptor is capable of providing continuous information on the phasic and tonic state of the tension in the muscle which it innervates.

10.4   Relationship between tension receptors and chordotonal organs

The relationship between tension receptors and chordotonal organs occurs on two levels: anatomical and functional.

10.4.1   Anatomical

All the apodeme sensory nerves studied in brachyuran limbs have been found in close association with another nerve almost to the point where they run onto the muscle apodeme (Fig. 10.3). This nerve is always the same for each muscle; either the motor nerve to the same muscle or the nerve of the chordotonal organ associated

---

Fig. 10.2   (A) Examples of the response elicited in the flexor apodeme sensory nerve of *Cancer magister* to a series of isometric contractions of the flexor muscle produced by stimulating the flexor motor nerve. The strength of the contractions (lower trace in each pair of records) was graded by increasing the frequency of the stimulation (a) 20; (b) 25; (c) 30 pulses per s. Both the number of units responding and the response frequency of individual units, were increased by increasing tension.

(B) Activity recorded in the flexor apodeme sensory nerve of *Cancer magister* (upper trace in each pair of records) in response to an imposed sinusoidal movement of the carpopodite. The movement (lower trace in each pair of records) was not effective in eliciting an apodeme nerve response if the flexor muscle was completely slack. (a) When the resting tension of the muscle was augmented by continuous stimulation of the flexor motor nerve at 55 pulses per s. (b) the tension on the muscle was increased by lengthening the muscle (upward deflection of movement trace) so that the apodeme sensory nerve responded.

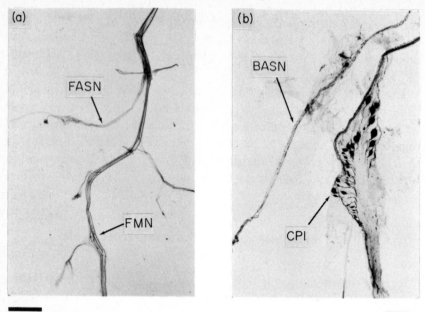

25 mm

200 μm

Fig. 10.3   Examples from *Cancer magister* legs of the close associations between apodeme sensory nerves and the nerves to nearby structures.

(a) The flexor apodeme sensory nerve (FASN) is closely associated with the flexor motor nerve (FMN) to the point where the apodeme nerve runs onto the apodeme and the motor nerve branches to supply the flexor muscle.

(b) The bender apodeme sensory nerve (BASN) is closely associated with the nerve to the CP1 chordotonal organ to the point where the apodeme nerve runs onto the apodeme and the chordotonal organ nerve enters the receptor strand.

with that muscle and joint. In *Limulus* the sensory nerve runs in a bundle with the corresponding motor nerve.

The association between the apodeme sensory nerves and the other nerves gives rise to several problems in interpretation, particularly of previous results concerning chordotonal organ reflexes. For example, in the case of the mero-carpopodite joint of *Cancer magister* where recordings have been made from both intact and sectioned motor nerves to study the reflex motor activity elicited by chordotonal organs the question arises as to whether the motor nerve alone or the motor nerve plus the apodeme sensory nerve was severed (Evoy and Cohen, 1969). This could be expected to have some bearing on the experimental conclusions. Similarly, in the case of chordotonal organ ablation and its effects upon motor patterning and movement (Evoy and Cohen, 1971) the question of whether some of the tension receptor afference was destroyed along with the chordotonal input makes interpretation of the results very difficult.

The effect on the reflexes may well prove to be of minor importance, in which event, the results, together with the conclusions based upon them would remain unaltered. Nevertheless, the knowledge of the close and sustained anatomical association between apodeme receptor nerves and these other nerves demands a careful reassessment of all previous work dealing with resistance reflex physiology as well as caution when planning future experiments.

10.4.2   Functional

It is quite clear from simultaneous recordings from chordotonal organs and tension receptors that in some situations their responses to the same movement are quite different and that they are coding different sensory information (Fig. 10.4). There are, however, still questions concerning the division of stress and strain detecting functions between the two types of organ.

Most chordotonal organs are attached to a variable degree to a muscle apodeme. In his study of the structure and physiology of the chordotonal organ at the pro-dactylopodite joint in the walking legs of *Carcinus* Burke (1954) reported responses from what he assumed to be the receptor itself as a result of isometric contractions of the closer muscle, to the tendon of which the receptor is attached. Clarac and Vedel (1971) demonstrated that, in some cases, tension changes in muscles can modify the response to movement of the chordotonal organs with which they are associated. Thus the chordotonal organ response to a given movement applied to a passive limb may be significantly different from its response to the same movement produced by its own muscles. This finding raised the possibility that chordotonal organs might respond to isometric muscle contraction. Macmillan and Dando (1972) tested this hypothesis for three chordotonal organs, the proximal ends of which are inserted on muscle tendons and found that at least under some limited conditions it was possible for the muscles to contract isometrically without the chordotonal organ responding. In other cases they appear to respond. However, some of the phasic units in chordotonal

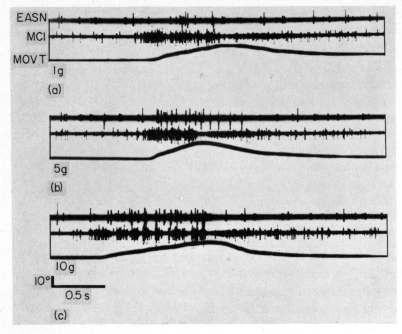

Fig. 10.4   Example of the way in which a chordotonal organ (MC1) and the tension receptor associated with the carpopodite extensor muscle (EASN) of *Cancer magister* can code different sensory information elicited by the same movement. The upper and middle traces in each record show, respectively, the response in the extensor apodeme sensory nerve and the response of MC1 to an isotonic contraction of the extensor muscle. The bottom trace shows the movement of the carpopodite (extension, upwards). The extensor motor nerve was stimulated at 30 pulses per s and the differential response in the two types of receptor demonstrated by allowing the muscle to shorten against different loads of (a) 1 gm; (b) 5 gm; (c) 10 gm. (After Macmillan and Dando, 1972.)

organs are sensitive to very small amplitude vibrations (Chapter 6) and it is exceedingly difficult to ensure that all responses resulting from such vibrations are excluded, and indeed such responses may be of special significance to the animal (Chapter 6). Clarac and Vedel (1975) repeated this type of experiment in the walking legs of *Palinurus* but were unable to eliminate a small residual response to isometric contraction no matter how firmly they fixed the preparation. They argue that the response is so slight that it is probably experimental artifact and unlikely to be of great physiological significance. Nevertheless, they show a residual response to isometric contraction which has 'on' and 'off' phasic components and a low frequency maintained tonic discharge, so the matter still requires further resolution.

One possibility in this context is that a spectrum of degrees of separation of the stress detecting function from the chordotonal organ may be found. In some forms the tension and movement functions may be combined, albeit less efficiently, in the chordotonal organ whereas in other forms there may be complete separation of the two modalities into separate organs. Although this possibility is only speculative at this time it bears serious consideration because of the residual isometric tension response in chordotonal organs and the lack of obvious tension receptor nerves in the Palinura (Clarac and Vedel, 1975); in contrast to the lack of a residual isometric response and presence of obvious tension receptors in the Brachyura (Macmillan and Dando, 1972). This hypothesis also agrees with considerations concerning the internalization and subsequent elaboration of receptors (Alexandrowicz, 1972) (Chapter 1).

10.5   Analogies with vertebrates

As more functional elements corresponding with those of the vertebrate limb motor control system are found in such a phylogenetically different group as the arthropods, it becomes increasingly interesting to speculate on the possibility of analogous functional interactions at the systems level between the various elements: central motor command, movement afference, tension afference and so on. A particularly good example is the situation at the mero-carpopodite joint of the crab (Fig. 10.5) which presents a useful analog of the typical situation in the limb musculature of the vertebrate because each of the elements is separately amenable to electrical recording and stimulation in partially dissected animals. Thus the 'kinesthetic' afferents from the joint capsule of vertebrates are represented by chordotonal organs giving phasic and tonic information on limb position and movement (Chapter 6). The muscle spindle function is represented in a very limited fashion by the myochordotonal organ complex which is in parallel with the working muscles of the leg and has its own receptor muscle and efferent nerve supply. The muscle specific tension receptor provides an analog for the Golgi tendon organ (Dando and Macmillan, 1973). Although this kind of systems analogy might appear to be too speculative, it will be apparent from its use in examining the functional role of tension

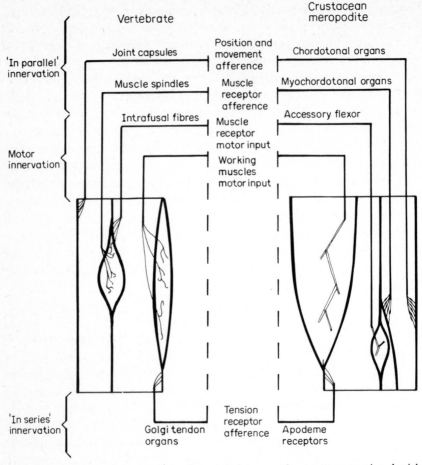

Fig. 10.5   More and more of the functional types of receptors associated with the typical vertebrate muscle and joint system have been discovered in the crustacea. In some cases the functional parallel is quite striking and the two systems may be usefully compared in order to pose questions concerning the roles of the different elements. This figure compares the typical sensory-motor system of the vertebrate limb with that found at the meropodite-carpopodite joint of many crustaceans.

receptors (Section 10.6) that it has already proved a useful guide and raised some interesting questions for the study of Golgi tendon organ function.

One interesting finding in *Limulus* (Eagles and Hartman, 1975) is that due to the mechanical arrangement of muscle fibres and apodemes the distribution of tension about the apodeme may change in different positions and with it the tension receptor response. This suggests a complex tension receptor response which may depend on the direction of apodeme movement and the combination of motor units firing. Although this was not found in a true functional situation in *Cancer*, Macmillan and Dando (1972) demonstrated that a similar system could operate there. This deserves further investigation because of the evidence for a complex vertebrate tendon organ response. Here there is evidence that the mechanical arrangement of the receptor cells along the tendon together with the motor innervation of the muscle fibres that insert into different parts of the tendon can produce a differential Golgi tendon organ response which is dependent on the motor combination producing the muscle contraction (Stuart, Mosher, Gerlach and Reinking, 1972).

10.6   Functional role of tension receptors

It has been known for some time that Golgi tendon organs produce autogenic inhibition, i.e. the receptor input inhibits the motoneurons which activate the homonymous muscle (Granit, 1950). Initial experiments to test for the same effect in *Cancer* produced equivocal results (Macmillan and Dando, 1972; Dando and Macmillan, 1973) which will be discussed below, although Clarac and Dando (1973) found an interesting and related result in *Cancer pagurus*. They found that the resistance reflex responses to the carpopodite flexor and extensor muscles produced by imposed sinusoidal movement of the mero-carpopodite joint (Section 6.8) could be inhibited by stimulation of the tension receptors; the inhibition being restricted to the motoneurons of the homonymous muscle in each case. While this is clearly not the same as the autogenic inhibition described in the vertebrates, demonstration of a negative feedback loop from sense organs in series with a muscle suggests similarities in role.

The investigation of the relationship between the inputs and reflexes from the various receptors is being carried further and while the picture is far from clear several interesting possibilities are emerging. Estimations of central delay times for the resistance reflex to the flexor muscle at the mero-carpopodite joint of *Cancer pagurus* together with conduction velocity measurements in both chordotonal organ and tension receptor nerves indicate that tendon organ afference does not interact with the early part of the resistance reflex. This suggests that the role of the tendon organ reflex in the passive animal is to damp the resistance reflex thus effectively preventing oscillatory interactions between flexor and extensor reflexes. When the animal is active, the situation is less clear and in at least some experimental situations it is possible to produce positive feedback onto the homonymous muscle which can be eliminated by sectioning the tension receptor nerve (Macmillan and Laverack, 1975).

It is suggested that this result could be explained by a central switching of the tension afference during certain active movements and there is some evidence that central switching is involved in turning resistance reflexes off during centrally generated motor activity (Spirito, Evoy and Barnes, 1972; Barnes, Spirito and Evoy, 1972) (Chapter 6). The finding that the different receptor inputs may be organized very differently according to the dynamic state of the animal or its chosen activity makes the analysis of their role considerably more difficult. This type of re-balancing of afferent input by central commands is also likely to be important in the adjustment of the output of the system in response to changing environmental requirements. For example, in the case of tension receptors one might expect the input to be involved in the changes in motor patterning seen when a system is loaded (Macmillan, 1974; Macmillan, Wales and Laverack, 1975; Pearson, 1972; Wales, Macmillan and Laverack, 1975a, b).

Acknowledgements

Author and previously unpublished work supported by S. R. C. Grant B/RG/05857.

References

Alexandrowicz, J. S. (1933) Innervation des branchies de *Squilla mantis. Archives de Zoologie expérimentae et génerale*, **75**, 21—34.

Alexandrowicz, J. S. (1957) Notes on the nervous system in the Stomatopoda. V. The various types of sensory nerve cells. *Publicazioni della Stazione Zoologica di Napoli*, **29**, 213—225.

Alexandrowicz, J. S. (1972) The comparative anatomy of leg proprioceptors in some decapod crustacea. *Journal of the Marine Biological Association, U.K.*, **52**, 605—634.

Barber, S. B. and Hayes, W. F. (1964) A tendon receptor organ in *Limulus. Comparative Biochemistry and Physiology*, **11**, 193—198.

Barnes, W. J. P., Spirito, C. P. and Evoy, W. H. (1972) Nervous control of walking in the crab *Cardisoma guanhumi*. II. Role of resistance reflexes in walking. *Zeitschrift für vergleichende Physiologie*, **76**, 16—31.

Bullock, T. H. and Horridge, G. A. (1965) *Structure and Function in the Nervous Systems of Invertebrates*. Freeman, San Francisco.

Burke, W. (1954) An organ of proprioception and vibration sense in *Carcinus maenas* L. *Journal of Experimental Biology*, **31**, 89—105.

Clarac, F. and Dando, M. R. (1973) Tension receptor reflexes in the walking legs of the crab *Cancer pagurus*. *Nature*, **243**, 94—95.

Clarac, F. and Vedel, J. P. (1971) Etude des relations fonctionelles entre le muscle fléshisseur accessoire et les organes sensoriels chordotonaux et myochordotonaux des appendices locomoteurs de la langouste *Palinurus vulgaris*. *Zeitschrift für vergleichende Physiologie*, **72**, 386—410.

Clarac, F. and Vedel, J. P. (1975) Proprioception by chordotonal and myochordotonal organs in the walking legs of the rock lobster *Palinurus vulgaris*. *Marine Behaviour and Physiology*, In press.

Dando, M. R. and Macmillan, D. L. (1973) Tendon organs and tendon organ reflexes in decapod Crustacea. *Journal of Physiology, London*, **234**, 52—53P.

Eagles, D. A. and Hartman, H. B. (1975) Tension receptors associated with the tailspine muscles of the horseshoe crab, *Limulus polyphemus*. *Journal of Comparative Physiology*, **101**, 289—307.

Evoy, W. H. and Cohen, M. J. (1969) Sensory and motor interaction in the locomotor reflexes of crabs. *Journal of Experimental Biology*, **51**, 151—169.

Evoy, W. H. and Cohen M. J. (1971) Central and peripheral control of arthropod movements. *Advances in Comparative Physiology and Biochemistry*, **4**, 225—266.

Granit, R. (1950) Reflex self-regulation of the muscle contraction and autogenic inhibition. *Journal of Neurophysiology*, **13**, 351—372.

Macmillan, D. L. (1975) A physiological analysis of walking in the American lobster (*Homarus americanus*). *Philosophical Transactions of the Royal Society, Series B.*, **270**, 1—59.

Macmillan, D. L. and Dando, M. R. (1972) Tension receptors on the apodemes of muscles in the walking legs of the crab, *Cancer magister*. *Marine Behaviour and Physiology*, **1**, 185—208.

Macmillan, D. L. and Laverack, M. S. (1975) An extracellular study of resistance and tension receptor reflexes in the walking legs of the crab *Cancer pagurus*. *Marine Behaviour and Physiology*. In preparation.

Macmillan, D. L., Wales, W. and Laverack, M. S. (1975) Mandibular movements and their control in *Homarus gammarus*. II. The effects of load. *Journal of Comparative Physiology*. (In press).

Pearson, K. G. (1972) Central programming and reflex control of walking in the cockroach. *Journal of Experimental Biology*, **56**, 173—193.

Spirito, C. P., Evoy, W. H. and Barnes, W. J. P. (1972) Nervous control of walking in the crab, *Cardisoma guanhumi*. I. Characteristics of resistance reflexes. *Zeitschrift für vergleichende Physiologie*, **76**, 1—15.

Stuart, D. G., Mosher, C. G., Gerlach, R. L. and Reinking, R. M. (1972) Mechanical arrangement and transducing properties of Golgi tendon organs. *Experimental Brain Research*, **14**, 274—292.

Wales, W., Clarac, F., Dando, M. R. and Laverack, M. S. (1970) Innervation of the receptors present at the various joints of the pereiopods and third maxilliped of *Homarus gammarus* (L.) and other Macruran Decapods (Crustacea). *Zeitschrift für vergleichende Physiologie*, **68**, 345—384.

Wales, W., Macmillan, D. L. and Laverack, M. S. (1975a) Mandibular movements and their control in *Homarus gammarus*. I. Morphology. *Journal of Comparative Physiology*. (In press.)

Wales, W., Macmillan, D. L. and Laverack, M. S. (1975b) Mandibular movements and their control in *Homarus gammarus*. II. The normal cycle. *Journal of Comparative Physiology*. (In press.)

Whitear, M. (1962) The fine structure of crustacean proprioceptors. I. The chordotonal organs in the legs of the shore crab, *Carcinus maenas*. *Philosophical Transactions of the Royal Society Series B.*, **245**, 291—325.

11 The structure and function of proprioceptors in soft-bodied invertebrates

D. A. DORSETT

11.1   Introduction

In considering the role of proprioception in the sensory physiology of soft-bodied invertebrates, two points must be borne in mind. The first is that in passing from the coelenterates to the annelids and molluscs, we are looking at some of the earliest stages in the evolution and organization of the nervous system and must ask ourselves at what stage does a true proprioceptive sense arise. Secondly, the structural basis of the body plan in these phyla is quite different from those which possess an internal or an external skeleton, so that precise monitoring of the relationship of some part of the body relative to another may be neither meaningful nor necessary for the execution of the normal biological activities of the organism. The evidence now available suggests that the sequencing of muscular activity necessary for the somewhat limited behavioural repertoire of these simple animals is obtained by integration at the synaptic level within the nervous system, resulting in programmed behaviour which can proceed largely independently of detailed sensory feedback.

In the coelenterates, the shape of the body is determined by the mesogloea, which forms a semi-solid, elastic substrate against which the muscles of the epidermal and gastrodermal layers may contract.

In anemonies, the visco-elastic properties of the mesogloea imply that each animal in the unstrained condition will have a shape which is maintained without exerting continuous muscular or hydrostatic pressure (Alexander, 1962). In most anemonies this probably corresponds to somewhere near the expanded condition, so that *Metridium*, which may double its unstrained length or contract to a flattened disc over a period of 1−2 hours, may depend entirely on the properties of the mesogloea to restore it to its original shape. This is not exactly the same system as in jellyfish, where the distortions of the bell during swimming are completed in the order of 1−2 seconds, but the elasticity of the mesogloea still provides the restoring force against which the muscle fibres can work (Alexander, 1964). Although there is little evidence to show that hydrozoan polyps work in the same way as the *Anthozoa*, there is no reason to invoke an alternative one.

The oligochaetes and polychaetes have a hydrostatic skeleton, supporting the body by the internal pressure of the coelomic fluid acting against a crossed lattice of collagen fibres in the cuticle (Ruska and Ruska, 1961; Brokelmann and Fischer, 1966; Dorsett and Hyde, 1969). This lattice is able to accomodate the extensions and contractions normally undergone by the body wall, while the individual fibres remain relatively inextensible to longitudinal stress. As in the coelenterates, the body has a fundamental shape which it will take up in the absence of imposed muscular activity. Similar principles probably apply to the body wall musculature and botryoidal tissue system which determines the shape of leeches.

The muscular and connective tissue elements of the mantle and body wall of molluscs also act against the haemocoelic fluid to produce a turgidity against which the muscles contract. In the body wall, anatomically defined muscles are rare, but sheets of fibres run in different directions at various levels which enable it to contract in almost any direction.

In all these groups the absence of any fixed points of reference in the mobile and plastic body wall would make it difficult for a receptor monitoring position, tension or movement of two parts relative to one another to operate with any degree of precision, so that proprioception as recognised in certain situations in the arthropods has little meaning. Bearing these limitations in mind we may now examine the evidence for the occurrence and function of these organs.

11.2   Mechanoreceptors in coelenterates

Coelenterates are metazoans of the tissue grade of construction in which the nervous system is present as one or more unpolarized networks of fibres, with some specialization into through conducting systems. Hydrozoan polyps and medusae and the more specialized anthozoans have retained neuroid or epithelial conducting systems in addition to the nerve net, which are associated with particular aspects of behaviour (Josephson, 1965; Mackie, 1967; McFarlane, 1970, 1974). The sedentary hydrozoan polyps, having no appendages other than tentacles, might be expected to orientate with respect to gravity or light, and in pennate forms to respond to currents on the sea floor, so that they present the maximum surface area across the direction of flow. This latter property would require an orientated response to mechanosensory information.

Many hydrozoans appear to show behavioural responses of this type. The manubrium of hydrozoan medusae can accurately locate a region of the sub-umbrellar surface stimulated by touch (Horridge, 1955) and similar dexterity has been observed in the ephyra larva of scypozoans when capturing barnacle nauplii. Josephson (1961) demonstrated the ability of *Syncoryne* to respond to a vibrating probe which created minute disturbances in the water by bending towards the source. As the sensory threshold increases with distance, it suggests that the polyp responds to 'near field' displacements of the water rather than to a propagated pressure wave. By selective stimulation and surgical techniques Josephson was able to show that the sensitivity resided in the capitate tentacles, the polyp bending towards the tentacle perceiving the greatest intensity of stimulation and therefore nearest the source. Tardent and Schmid (1972) have shown that the mechanoreceptor function lies in the stiff hairs which project from the capitate and filiform tentacles in the closely related *Coryne pintneri*. The filiform tentacles bear up to 17 receptors each, which lie in the ectoderm surrounded by an accessory cell (Fig. 11.1). Its most conspicuous feature is the long inflexible cilium (up to 70 μm long and 0.4 μm in diameter), which is set in a small socket formed at its base by the protruding membrane of the accessory cell. The cilium (called by these authors a stereocilium) originates from a complex ciliary apparatus, consisting of a basal lamellary body surrounded by 9 stiff rods in the apical region of the cell. From the base of each rod a striated rootlet extends into the cytoplasm of the

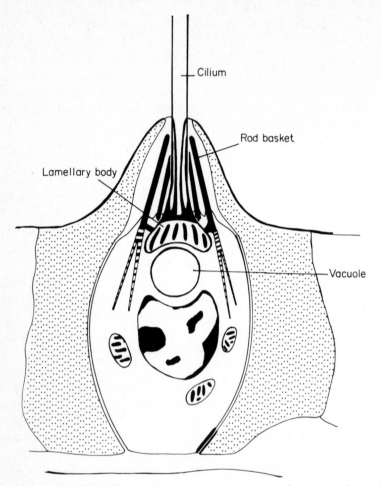

Fig. 11.1   The vibration sensitive mechanoreceptor of *Coryne pintneri*. The receptor is surrounded by an accessory cell which forms the collar. The rod basket surrounding the base of the cilium may represent the outer ring of paired fibrils; the lamellary body a modified basal body. (After Tardent and Schmid, 1972.)

receptor. The lamellary body lies above a large vacuole which is itself distal to the nucleus of the cell.

Horridge (1966, 1969) has described the structure of similar non-motile cilia from the bell margin of medusae and from the 'fingers' of the ctenophore *Leucothea*. The vibration sensitive cilia of these mechanoreceptors also show morphological specialization at the base which can probably be related to the transduction process, but in contrast to those of *Coryne* have a normal 9 + 2 arrangement of fibrils extending the length of the shaft. The mechanoreceptive cilia of *Coryne* may be related to these if one assumes that the basal regions of the outer ring of 9 paired fibrils do not penetrate the shaft but develop into the outer ring of 9 'rods' which surround the base. The lamellary body is probably the equivalent of the 'onion' at the base of the shaft in *Leucothea*. Both Horridge (1966) and Barber (1966) adopt the term kinocilium for these non-motile structures, following the terminology of Dijkgraaf (1963). The term stereocilium is reserved for the shorter hairs which often surround the kinocilium and are filled with fine fibrils which extend a short distance into the cell.

Simple sensory responses such as these require no more than a direct routing to the effector system (usually the longitudinal muscles) which will serve to bend the organism in the appropriate direction. The response is reflexive rather than proprioceptive in that it is not responding to movements generated by or imposed upon the animal itself (Chapter 1). Nevertheless, vibration-sensitive hairs are particularly important in considering the evolution of the statocysts, which are considered in the next section.

11.2.1   Statocysts

Statocysts are sense organs which provide information on the orientation of the body with respect to the gravitational force, and their presence in the coelenterates indicates that they are probably one of the earliest types of proprioceptor to evolve. They are present in the *Scyphozoa* (in the marginal rhopalia or tentaculocysts) but were generally thought to be absent from the hydrozoan polyps and the more primitive *Anthomedusae*, although they occur in the *Leptomedusae*. Recently, Campbell (1972) has described statocysts from the tips of the holdfasts which anchor the polyps of the hydroid *Corymorpha* to the substrate. The statocyst vesicle is a cavity between the two terminal endodermal cells of the holdfast (Fig. 11.2). The lateral walls of the cup-shaped terminal cell are very thin so that only a narrow layer of cytoplasm separates the vesicle from the mesogloea. The statolith is about 5 μm in diameter and is composed of numerous ovoid bodies which are refractile and presumably heavy.

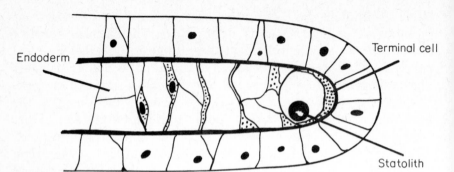

Fig. 11.2   The terminal statocyst in the holfast of *Corymorpha*. The statocyst is contained in a cup-shaped terminal cell. Pressure on the thin lateral walls leads to contraction of the longitudinal fibrils in the underlying musculo-epithelial cells. (After Campbell, 1972.)

When the holdfast is directed vertically downwards, the statolith rests on the distal end of the vesicle, cushioned by a thick layer of cytoplasm from the terminal cell. When rotated horizontally, the statocyst falls against the lateral wall where the cytoplasm is thin. A few seconds after such manipulation the tip of the stolon bends towards the side where the statocyst is resting due to localised contraction of the tissue adjacent to the statocyst. The stimulus appears to be contact of the statolith and the contractile bases of the epidermal cells. The movement may be accompanied by potential changes in these cells but no nervous elements or sensory cilia appear to be involved in the response.

Statocysts are commonly found around the bell margin of the Leptomedusae. In the *Trachylina,* forms such as *Cunina* have up to 60 pendant lithostyles arranged around the bell margin, each containing 2—3 calcareous statoliths secreted by the endoderm. The pendant is surrounded by non-motile cilia arising from the cells around its base (Horridge, 1969), which are believed to be sensory in function. As the lithostyle is free to swing, as the animal is tilted it will strike and deform the surrounding cilia, which may be stimulated in this way. In others, such as *Rhopalonema* and *Geryonia*, the pendant lithostyle becomes enclosed in a vesicle formed by the ectoderm and mesogloea, to form the marginal statocyst. The sensory cilia come to cover the pendant so that, as it swings, the cilia of one side will bear against the wall of the vesicle. Each sensory epithelial cell carries one kinocilium which, in *Geryonia,* is surrounded by about 200 stereocilia. The stereocilia are about 2 μm long and 0.1 μm in diameter. In forms such as *Rhopalonema* the number is reduced to two rings around the kinocilium. In the *Scyphozoa* these fuse into a ridged collar. The statocysts of the *Scyphozoa* are borne on the rhopalia which lie marginally around the bell in the inter-radial position. The flattened ectodermal layer in the apical region of the rhopalium is normally without cilia but the taller cells at the base, and the surrounding ectodermal cells, bear long, non-motile cilia (Fig. 11.3) and may give rise to axons at the proximal end of the cell (Horridge, 1969).

Horridge has suggested a scheme which traces the evolution of the statocyst from an association of lithocytes with vibration-sensitive

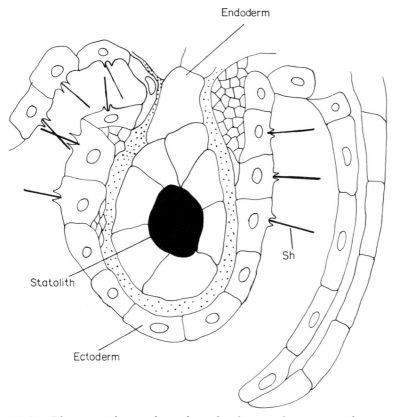

Endoderm

Sh

Statolith

Ectoderm

Fig. 11.3 Diagrammatic section through the pendant tentaculocysts of *Nausithoe*. The statolith is secreted by the surrounding lithocytes, and covered by mesogloea and ectoderm. The vibration-sensitive cells (sh) occur around the base of the pendant and are partially protected by a hood on the exumbrellar side. (After Horridge, 1969.)

hairs. The effectiveness of the hairs may be increased by backing them with a layer of endodermal lithocytes (Fig. 11.4) whose density will convert part of the energy of sound pressure waves transmitted by the water into small displacements, which will be detected by the kinocilium. The isolation of this structure into a pendant will allow it to hang vertically and perhaps improve the gravitational sense, while its subsequent enclosure into a statocyst will protect it from random water movements caused by turbulence. The theory does not fully reconcile with the discovery of statocysts in *Corymorpha* which function effectively without ciliary receptors, but they may represent quite different lines of evolution.

11.2.2   Statocysts of ctenophores

The statocyst of ҫtenophores lies in a depression at the aboral pole, the floor of which is lined by tall ciliated epidermal cells. The statolith consists of a number of calcareous grains, each $5-10$ μm in diameter, which adhere to the apices of 4 groups of partially fused 'balancer' cilia (Fig. 11.5). Each group of balancer cilia consists of $1-200$ cilia of the normal $9 + 2$ variety and they occur in the inter-radial position. From the balancers a groove of elongated cells which bear cilia in their central, nucleated region, connects each balancer to two comb rows. A beat initiated in a group of balancer cilia is transmitted along the ciliated groove and the two comb rows with which it connects. Each beat in the comb row is accompanied by a single action potential of about 30 mv (recorded intra-cellularly in the comb plate cells), which is transmitted from cell to cell with a velocity of $4-7$ cm s^{-1} (Horridge, 1966).

11.2.3   Function of the statocysts

From their similar design to analogous structures in other groups whose function has been analysed experimentally, the statocysts of coelenterates and ctenophores should be capable of responding to gravity and also to sudden accelerations such as those produced by waves or broken water. A jellyfish which is swimming upwards and is tilted to one side regains its equilibrium by introducing an assymetric component into the beat. The margin of the uppermost side contracts more and relaxes less completely than that on the lower side, which brings about a compensatory righting reflex. The reflex

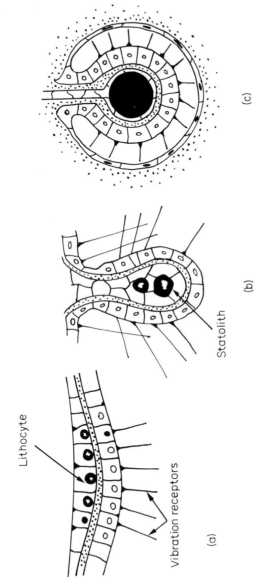

Fig. 11.4   The evolution of statocysts. A. Epithelial vibration receptors backed by a layer of endodermal lithocytes with statoliths. B. Folding into a pendant, such as in *Cunina* allows organ to swing under gravity. C. Enclosure into a statocyst and covering pendant with the receptors enhances gravity responses, as in *Geryonia*. (After Horridge, 1969.)

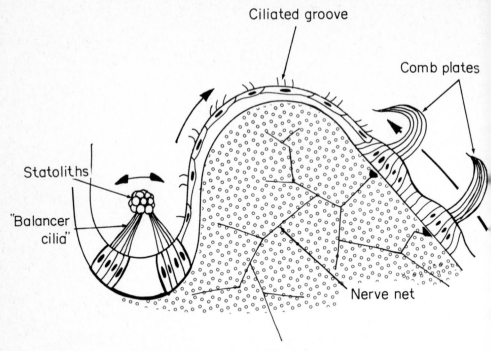

Fig. 11.5   Gravity perception and the righting response of ctenophores. Displacement of the statoliths affects the beating rate of the mechanosensitive 'balancer' cilia and the rate at which impulses are conducted along the cells of the ciliated groove to the comb plate. The beating of the comb plates are also modulated by the synapses with the nerve net. (After Horridge, 1969.)

requires the presence of a single statocyst, which must be located on the uppermost margin of the bell (Fraenkel, 1925; Bozler, 1926; Horridge, 1959). Passano and McCullough (1960) could find no electrical activity in the slow nerve net associated with stimulation of the statocyst in *Cyanea* and *Cassiopea*, and suggest the existence of special pathways to the marginal pacemakers which may mediate this response. In other circumstances the compensatory reflex can be reversed or over-ruled, allowing the jellyfish to swim downwards, or parallel to the surface, for long periods; but the mechanisms which allow them to ignore the input from the statocysts is not known.

Similar problems confront the analysis of the behaviour of ctenophores in an animal swimming downwards, the weight of the statocyst acting on the uppermost balancer increases the rate of initiation of beats in that pacemaker and its associated comb rows, which turn the animal until its oral end is pointing vertically downwards. However, ctenophores also swim upwards or horizontally, which requires the sign of the response to be reversed. The mechanism of the reversal, and indeed that of the balancers and statocyst, is not fully understood.

11.3 Proprioceptors in the Anthozoa

Very little information is available on the structure or function of receptors in the anemonies. The relatively simple behaviour such as elongation, bending and mouth opening are controlled by neuroid conducting systems in the ectoderm or endoderm (McFarlane, 1970, 1974), the elastic properties of the mesogloea being responsible for the fundamental shape of the animal at rest.

A possible proprioceptor has recently been described from the body wall of *Cerianthus*, embedded in the soma of a musculo-epithelial cell in the body wall ectoderm (Peteya, 1973). It is found either at the interface of the muscle layer and nerve plexus or a short distance within the muscle lamellae. Those in the latter position have their receptor axis orientated parallel to the longitudinal axis of the animals body, while those bordering the nerve plexus have no regular orientation. The apical region of the receptor indents the muscle cell membrane and bears a ciliary process 16μm long and 0.3μm in diameter (Fig. 11.6), which has a 9 + 2 arrangement of fibrils in cross section. At its base the cilium is surrounded by a ring of sterocilia 0.2μm in diameter and $3-10 \mu$m long. These are packed with fibrils which extend from the base into the upper part of the cell. The basal region of the cilium is surrounded by a dendritic sheath or collar which interdigitates with the membrane of the muscle cell, to which it appears to be firmly attached by desmosomes. The collar is characterized by a number of densely staining fibre bundles which pass from the basal body region of the cilium to terminate in the desomosomal region of the collar membrane. At their base the fibrils are associated with those descending from the stereocilia and equal them in number, but they divide as they enter the collar. A striated rootlet extends proximally from the basal body of the cilium some way into the cell. The densely staining fibres of the collar strongly resemble the rods that surround the shaft of the cilium of the vibration receptors of *Coryne* and may serve to stiffen the apical region of the receptor, but the transduction mechanism is not

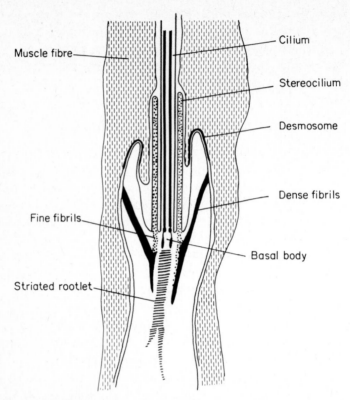

Fig. 11.6   A longitudinal section through the possible proprioceptor in the muscles of *Cerianthus*. The apical region of the receptor is inserted into the muscle fibre to which the receptor cell collar is attached by desmosomes. The shaft of the cilium is surrounded by stereocilia filled with fine fibrils, which are linked to dense fibrils around the basal body A striated rootlet penetrates the upper part of the cell. (After Peteya, 1973.)

understood and speculation is pointless without some experimental evidence. Apart from the fact that its position is ideal for monitoring elongation or tension in the longitudinal musculature, nothing is known of the role of the receptor in the physiology of the neuro-muscular system. Efferent synapses occur on the axon a short distance from the cell body, which may allow some modulation of its sensitivity.

11.4   Proprioception and locomotion in annelids

Although various lines of behavioural and experimental evidence have suggested the existence of proprioceptors capable of registering tension in the muscles and body wall of annelids, neither anatomical nor physiological studies have so far identified receptors giving a tonic response proportional to the intensity of the stimulus.

The locomotory movements of annelids can be broadly classified into; (a) *peristaltic*, in which alternate contraction of the circular and longitudinal muscles are used in burrowing or crawling (*Lumbricus, Arenicola*) or looping (*Hirudo*), (b) *lateral waves* of contraction passing alternately up the longitudinal muscles (fast walking and swimming in *Nereis*), (c) *dorso-ventral waves* of antiphasic contraction in the dorsal and ventral longitudinal musculature (swimming in *Hirudo* and ventilatory movements in *Nereis*). A fourth type, which has no equivalent outside the errant polychaetes, is the parapodial stepping movement made during slow walking.

All these movements are such that appropriately place proprioceptors could provide timing clues so that the completion of one movement provides the signal for initiating the next. In this way coordination is achieved by a series of chain reflexes. The experiments of Friedlander (1895) and Gray and Lissmann (1938) showed that this could not explain the transmission of the locomotory wave in earthworms as it could cross a completely de-afferented section of the nerve cord with the body wall completely removed. They concluded that while tactile receptors or proprioceptors might play a part in maintaining a state of central excitation necessary for locomotion, the mechanism generating the rhythm is endogenous to the central nervous system. This conclusion was generally supported by electrophysiological evidence, which showed that bursts of efferent activity could be recorded from the central stumps of peripheral nerves of a cord in which only a few segments remained intact, but disappeared when the cord was completely isolated (Gray and Lissmann, 1938).

The importance of tension in the longitudinal musculature for maintaining the locomotory peristalsis of the earthworm was shown by experiments which demonstrated that the rhythmic activity of a preparation suspended in air was lost when the body weight was supported in water (Garrey and Moore, 1915; Gray and Lissmann, 1938; Collier, 1939). Gentle tactile stimulation of cuticular receptors in preparations which have come to rest causes extension, by contraction of the circular and relaxation of the longitudinal muscles. If tension is then applied while the longitudinal muscle is relaxing, the rate of relaxation is increased, but if it is applied during longitudinal contraction, the latter is vigorously re-inforced. The range of tensions required to produce this response (2—8 g) is that normally exerted by the worm when crawling. These experiments strongly suggest that inter-segmental reflexes involving some type of tension or stretch receptor may be involved in crawling, although as locomotion from rest normally begins with contraction of the anterior circular muscle, stretch receptors cannot initiate the locomotory rhythm. A model involving the cuticular mechano-receptors and stretch receptors is suggested in Fig. 11.7, but they probably only serve to modulate the endogenous locomotory rhythm. The role of peripheral input is always considered as excitatory, although the transmission of the endogenous wave is speeded up as it crosses a length of cord which has been deafferented.

In a similar way, tactile and proprioceptive information from the suckers is thought to play an important part in coordinating the

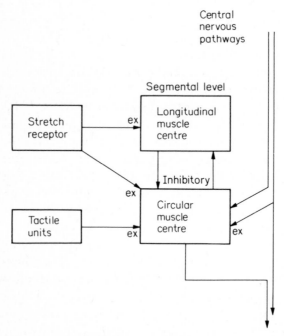

Fig. 11.7   A diagram summarizing the relationship between the sensory and motor systems in one segment of the earthworm. Tactile input excites the circular muscle, causing elongation; while stretch may cause elongation or contraction depending on the motor centre active at the time. Central pathways excite the circular muscle first. ex—excitatory.

sequential activity of the longitudinal and circular muscles during the walking behaviour of the leech (Gray, Lissmann and Pumphrey, 1938). The attachment of the posterior sucker is followed by the detachment of the anterior sucker and elongation of the body, due to contraction of the circular muscles. Upon attachment of the anterior sucker the posterior sucker is detached and the longitudinal muscle contracts. In the absence of tactile stimulation of the ventral surface or suckers, the leech swims.

Leeches swim by dorso-ventral undulations of their extended and flattened body to form a wave that travels posteriorly along the animal. The wave is generated by antiphasic contractions of the dorsal and ventral longitudinal musculature acting against the general tonus of the body wall induced by contraction of the dorso-ventral, and possibly the oblique, muscles (Mann, 1962). The experiments of Uexkull (1905) and Schulter (1933) demonstrated that the alternate sequence of dorsal and ventral contractions can be generated by individual ganglia, the supra-oesophageal and anal ganglia normally serving to modulate the rhythm. After cutting the connectives between two ganglia in the middle region of the body, both anterior and posterior parts continue to swim, but, unlike in the earthworm, they are no longer coordinated. The swimming wave can cross up to three segments which have the body wall completely removed and are only joined by the cord, without losing the coordination of the anterior and posterior sections (Gray, Lissmann and Pumphrey, 1938).

The picture that emerges from these early experiments is that the swimming rhythm of the leech is centrally determined and that the input from possible proprioceptors has little influence on the propagation of the locomotory wave. However, a recent detailed analysis of the dynamics and neuronal mechanisms controlling the swimming activity (Kristan, Stent and Ort, 1974a, b; Ort, Kristan and Stent, 1974) has contributed several new findings. Over a series of cycle periods ranging from 390–1100 ms, the leech maintains one full wave over its body. Whereas, in the intact animal, the intra-segmental phase lag between dorsal and ventral contractions shortens with increase in cycle period; in regions where the body segments have been opened the dorso-ventral phase lag is lengthened. Thus, while a general tonus is not needed to maintain a contractile rhythm matching the swimming movements in the rest of the body, it apparently is necessary to maintain the correct phase relationship of the motoneurons which drive the longitudinal muscle contractions. They suggest that during swimming the motoneurons are driven by a tonic source of excitation, and that the wave period is dependent upon its intensity. The source of this input is not known, but the effect of opening the body suggests that part of it may be peripheral.

In the polychaetes, proprioception has chiefly been considered with respect to the propagation of the parapodial stepping wave in forms such as *Nereis* and *Harmothoë* (Gray, 1939; Horridge, 1963; Dorsett, 1964, 1966). Initially, a centrally conducted wave passes posteriorly down the body, setting the parapodia in the correct relationship for propagation of the slower stepping wave, which then passes from the posterior end forwards. Although the parapodia and body wall possess mechanoreceptors which could form the basis of intersegmental reflexes coordinating stepping, de-afferentation or distortion of the parapodia in a way which would interrupt the normal sensory input over several segments, fails to interrupt progression of the wave or to influence the correct sequence in adjoining segments. This suggests that the wave is centrally propagated and not dependent on sensory feedback. As in the leech, each segmental ganglion has the ability to generate the correct stepping sequence in the parapodia of its own segment, and a short length of three to four segments will maintain a stepping rhythm for periods up to 2 min following a brief tactile stimulus. Peripheral input may, therefore, contribute a background level of excitability required to activate and possibly modulate the output from the segmental locomotory centres.

11.4.1   The epidermal sense organs of annelids

The behavioural evidence suggests that proprioception in annelids may be partitioned between cuticular receptors which signal movement relative to the substrate, and more deep-lying structures which may be sensitive to stretch. It is also possible for the epidermal receptors to monitor stretch if they penetrate or lie under the cuticle, as in earthworms (Knapp and Mill, 1971). The collagen fibres of the cuticle, although inextensible, may pivot about the points where they cross (Fig. 11.8). A receptor penetrating the cuticle will be subjected to transverse or longitudinal pressures as the cuticle distorts during movement, as will a sensory dendrite attached to the cuticle from beneath.

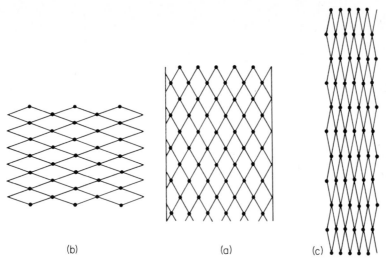

(b)   (a)   (c)

Fig. 11.8   A diagrammatic representation of the cuticular lattice of annelids. (a) When relaxed; (b) With longitudinal muscle contracted; (c) With circular muscle contracted. The individual elements do not change in length but pivot about the fixed points where they cross. Receptors attached to two points on the cuticle could be stretched or distorted during peristaltic locomotion.

The light microscopical studies of earlier workers (Retzius, 1892a, b; Hamaker, 1898; Smallwood, 1926; Dawson, 1928), using methylene blue and silver staining techniques, demonstrated that the sensory cell bodies in the epidermal and sub-epidermal layers of oligochaetes and polychaetes are almost exclusively simple bipolar cells with a small process penetrating the epithelium and terminating in a small bleb or sensory hair. In some cases these are associated in groups which form a small raised papilla on the surface. Dawson (1928) found numerous bipolar or tripolar cells distributed along nerve trunks in the sub-epithelial connective tissue and also large crescent-shaped cells which lie between the dorsal and ventral groups of chaetae. Here they could monitor pro- or retraction of the chaetae in association with the peristaltic waves. Smith (1957) confirmed the predominance of simple epithelial bipolars in *Nereis* and *Perinereis*, together with multipolar cells in a sub-epithelial plexus and he called the latter cells 'association neurons'. He also reported free nerve endings that ramified over the surface of the longitudinal muscles which he thought were proprioceptive as they did not penetrate the muscle sheath. Horridge (1963) considered these were probably slow motor fibres.

Horridge (1963) and Dorsett (1964, 1966) extended the observations on *Harmothoë* and *Nereis* with descriptions of several specialised mechanoreceptors which might serve a proprioceptive function.

11.4.2   Bipolar neurons in the parapodial wall

The ventral wall of the parapodium of *Harmothoë* has four to five large bipolar neurons, each bearing a long peripheral process passing anteriorly, which form a parallel row on the underside (Horridge, 1963). In this position they are so placed that contraction of the underlying anterior or posterior oblique muscles will stretch the long neurite which lies between the muscle fibres and the cuticle. Dorsett (1964) found groups of similarly orientated neurons on the anterior and posterior walls of the neuropodium in *Nereis* (Fig. 11.9b). These sensory cells are 10–15 μm long with a sensory process terminating in a small swelling. The receptors respond to localized prodding of the cuticle with a rapidly adapting burst of spikes, but it is difficult to isolate them physiologically, and their function remains uncertain. Horridge found that the receptors of *Harmothoë* adapted in periods of less than 1 s to a sustained displacement, which indicates they cannot give information about the resting position of the parapodium or coordinate the wave prior to its initiation (Fig. 11.10b).

11.4.3   Bristle receptors

In both *Nereis* and *Harmothoë* each group of chaetae (bristles) are innervated by extremely sensitive mechanoreceptors (Fig. 11.9a, b). In *Harmothoë* the neuropodium has a large bipolar neuron associated with each row of bristles, the long neurite putting out small arborizing side branches to each bristle in the row. These ramify over the surface of the bristles close to their point of exit through the body wall. Another cell is associated with the ventral aciculum. The notopodium has one or two large multipolar neurons which send dendrites to the base of each of the long stiff bristles. The notopodial and neuropodial chaetae of *Nereis* are served by one and two multipolar neurons respectively, which innervate the chaetae in a similar way. The slightest touch to any bristle evokes a rapidly adapting burst of afferent activity in nerve II (Fig. 11.11b), the

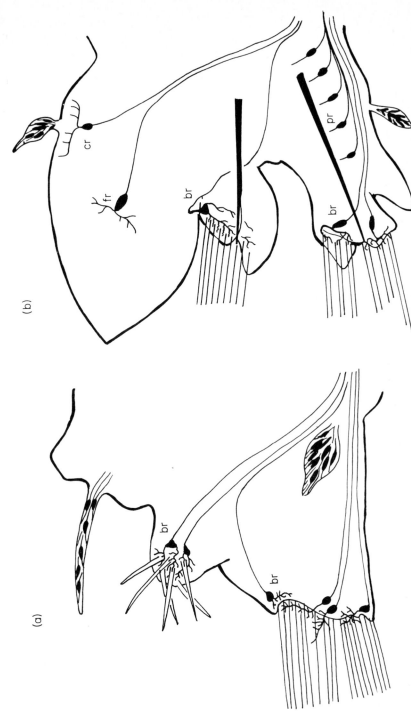

Fig. 11.9   (a) Bristle receptors (br) in the parapodium of the polychaete *Nereis*. br—bristle receptor; cr—cirrus receptor; fr—flap receptor; pr—bipolar receptors in the papapodial wall. ((a) after Horridge, 1963; (b) after Dorsett, 1964.)

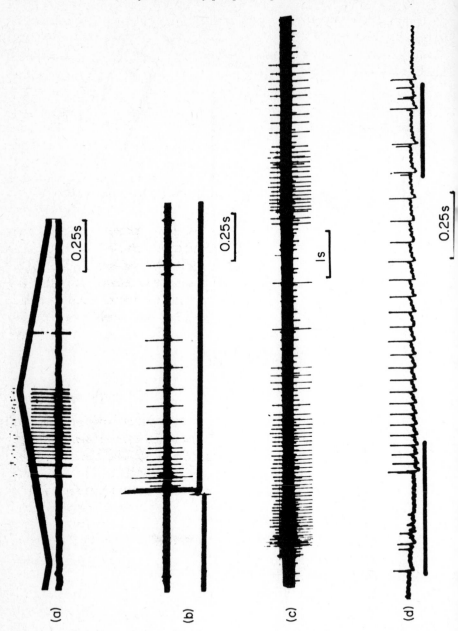

frequency and the number of impulses depending on the intensity of the stimulus and the rate at which it is applied. In all cases the primary transduction appears to be a strain applied to the sensory terminals.

11.4.4   Notopodial flap receptor of Nereis

Centrally placed in the enlarged notopodial flap of *Nereis virens*, a large tripolar mechanoreceptor innervates the line about which the flap flexes as the parapodium makes its stepping movements (Fig. 11.9b). Passive flexion of the flap in a single parapodium evokes a burst of impulses which can be identified with the receptor (Fig. 11.11c), and there is little doubt that the receptor is activated during the normal stepping cycle.

11.4.5   The dorsal cirrus receptor

At the base of the dorsal cirrus is a large bipolar or tripolar neuron with dendritic processes which may be up to 300 μm in length. These ramify among the epidermal cells around the base of the cirrus, sending short branches towards the cuticle (Fig. 11.9b). When the cirrus is moved from side to side, or in an antero-posterior direction, brief bursts of impulses from the receptor are recorded in the parapodial nerve (Fig. 11.11a). Adaptation is rapid and no response is obtained to continuous displacement. Observations on worms living in glass tubes showed that the dorsal cirrus is frequently flicked from side to side, the movement being brought about by a small muscle attached to one side of the base. Because of the mixed nature of the peripheral nerves it is not possible to monitor the sensory response to naturally occurring movements, but these are probably signalled by the receptor. Again it will only signal the movements as they occur, and not the instantaneous position of the cirrus.

Fig. 11.10   Slowly adapting receptors in annelids. (a) From leech body wall. (b) From *Harmothoë*. The smaller unit is the proprioceptor. (c) A unit from nerve IV in *Nereis*, in response to longitudinal muscle stretch. (d) From *Lumbricus*. Obtained by stimulating the integument. b, c and d are from centrally isolated preparations. (a) from Laverack, 1969; (b) from Horridge, 1963; (c) from Dorsett, 1966; (d) from Mill and Knapp, 1967.)

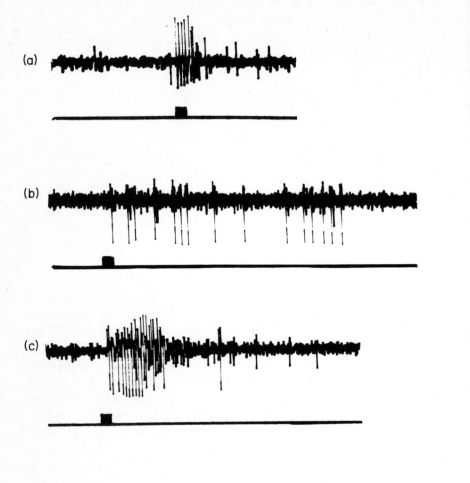

(a)

(b)

(c)

(d)

0.25 s

Fig. 11.11   Responses from the parapodial mechanoreceptors of *Nereis*. (a) Dorsal cirrus receptor. (b) Bristle receptor. (c) Notopodial flap receptor. (d) Epithelial bipolar cells.

11.4.6   Slowly adapting receptors in annelids

Any receptor intended to supply information on the tension in a muscle or on the position of part of the body or appendage relative to a previously set or expected position, would need to adapt relatively slowly over a period of several seconds. Electro-physiological recordings from centrally isolated preparations of the body wall in *Nereis* (Dorsett, 1966), *Lumbricus* (Mill and Knapp, 1967) and the leeches *Hirudo* and *Haemopis* (Laverack, 1969; Smith and Page, 1974) have shown the presence of receptors which respond to an applied stretch with a train of impulses that adapt relatively slowly (Fig. 11.10). In the absence of any anatomically defined structures which can be associated with these responses, it has proved difficult to quantify them or assess their functional significance, and further work is required.

11.4.7   The fine structure of epidermal receptors

As we lack any description of deep-lying sensory structures associated with the musculature, we can only examine here the details of the more peripheral sensory cells which are probably providing the continuous background excitation required by the nervous system to maintain the rhythmic locomotory output. Scanning electron micrographs of the cuticle of *Nereis* and other polychaetes show numerous finger-like hairs or tufts (Fig. 11.12) which project above the cuticle. In sections, these hairs are revealed as the terminal processes of fusiform sensory cells which lie below the epidermis (Dorsett and Hyde, 1969). The cell body tapers into a long distal process 30–40 μm long, which passes between the epidermal cells to terminate in a raised sensory hillock. The distal border of each cell bears a single cilium 7.5 μm long and 0.25 μm in diameter, which passes through the cuticle and emerges invested in a sheath formed from the epicuticle (Fig. 11.13). The cilium contains an outer ring of 9 double filaments and traces of a central pair, which do not stain as deeply as the outer ring. They arise from a basal plate located above the level of the cell border. The basal body is a cylindrical structure which lies below this level and is connected by electron-dense material to the striated ciliary rootlet. The sheath, which is open at the tip, is separated by a narrow space from the outer membrane of the cilium, and may serve to protect it. There are

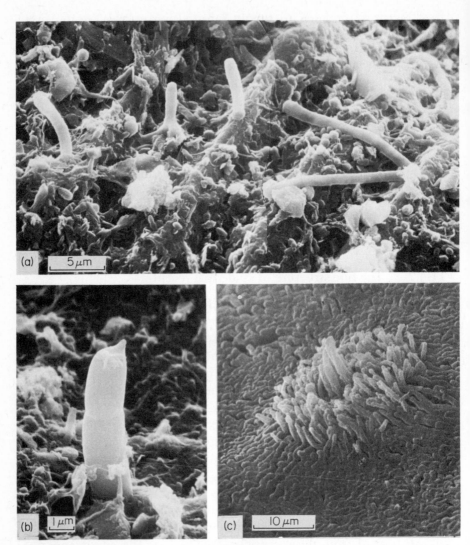

Fig. 11.12   Scanning electron micrographs of sensory hairs on the cuticle of polychaetes. (a) *Aphrodite aculeata* showing three short and two long hairs from the dorsal cuticle (x 4500). (b) A short hair from *Aphrodite* showing an annulated structure (x 10 000). (c) Compound sense organ from the parapodial cuticle of *Nereis*, showing sensory hairs surrounded by cuticular microvilli. (x 2500.) See also Fig. 1.8 which shows a stereoscan electron micrograph of an epidermal sense organ from *Allolobophora*.

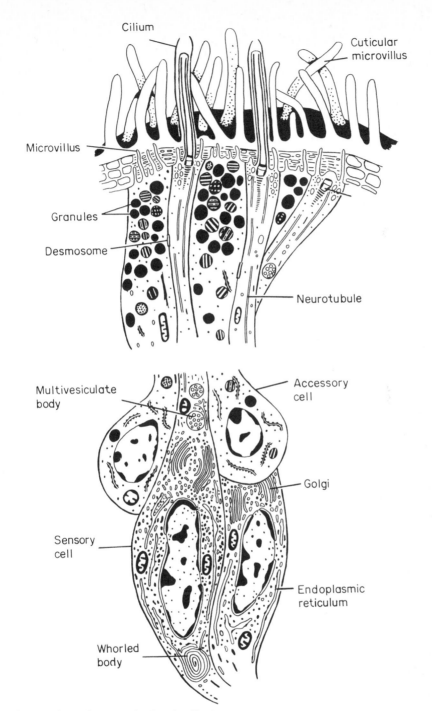

Cilium

Cuticular microvillus

Microvillus

Granules

Desmosome

Neurotubule

Multivesiculate body

Accessory cell

Golgi

Sensory cell

Endoplasmic reticulum

Whorled body

Fig. 11.13   Diagram of a longitudinal section through an epithelial sense organ of *Nereis*. (From Dorsett and Hyde, 1969.)

7—15 ciliated cells in each cluster, separated by a number of supporting cells. At first their arrangement and structure were thought to be consistent with a chemoreceptive function, but as more information has accumulated on the structure of mechano-receptive organelles, this may no longer be valid. Knapp and Mill (1971) found 3 types of cilium-bearing sensory cells in the epidermis of *Lumbricus*. One type of multiciliate cell is only found grouped in small sense organs. The cell bodies lie basally in the epidermis attached to the basal lamina by hemi-desmosomes, and send a long tapering process to the surface (Fig. 11.14b). The distal membrane bears numerous microvilli and 4—18 cilia, in one or two rows. The cilia pass through the cuticle to project several microns above its surface. Each has a standard 9 + 2 arrangements of central fibrils, which arise from the basal plate. The basal structures resemble those of *Nereis* and a striated rootlet passes proximally into the cell from each cilium. The cilia show no preferred orientation with respect to the axis of the central filaments.

Uniciliate sensory cells are also found in small numbers in most of the sense organs. In this type the cilium arises from a depression in the apical membrane and is surrounded by a ring of enlarged apical microvilli which show increased electron density along their inner walls (Fig. 11.14a). These microvilli also contain large numbers of tonofibrils, which emerge from the base and extend proximally to join a ring of electron dense material below the basal body. Both the tonofibrils and this dense cylinder are joined to the basal body by a diffuse layer of less dense material. There is no striated rootlet. The arrangement of large microvilli, tonofibrils and specialized structures at the base of the cilium immediately recalls the stereocilia and structural complex associated with the base of the vibration sensitive hairs of coelenterates (Fig. 11.1, 11.6). Earthworms are also sensitive to vibrations in the soil, which lead to the withdrawal reflexes.

Isolated multiciliate sensory cells are found scattered throughout the epidermis. In this type each cell bears 6—10 cilia, which radiate from the cell horizontally, lying between the epidermis and the cuticle. The cilia have variable arrangement of fibrils, with 9 + 0 and 8 + 1 being the most common. The basal body lacks a striated rootlet. The position of the cilia under the cuticle would seem ideal for monitoring stretch in a radial or longitudinal direction, but experiments show that the stretch reflexes persist in worms that have been superficially anaesthetised, so must depend on more deep lying structures.

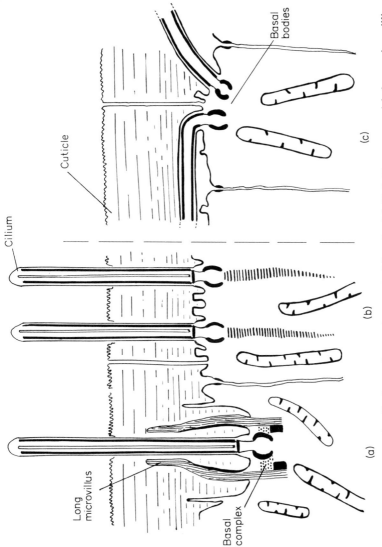

Fig. 11.14   Ciliated cells from the earthworm epidermis. (a) Uniciliate type with long microvilli which resemble stereocilia. (b) Multiciliate type. (c) Multiciliate type where the cilia extend horizontally under the cuticle. (After Knapp and Mill, 1971.) See also Fig. 1.8 which shows a stereoscan electron micrograph of an epidermal sense organ from *Allolobophora*.

11.4.8   The epidermal receptors in leeches

There is little recent information available on the fine structure of epidermal sense organs in leeches. The anterior annulus of each segment usually bears a row of raised papillae, with cells terminating in a small sensory hair (Autrum, 1934; quoted in Grasse, 1959). One would expect these structures generally, to resemble those described in *Nereis* and *Lumbricus*. Damas (1973) has described the ultrastructure of Bayer's organs which are found in the dorsal epithelium. The organs, whose function is not known, each consists of an apical cell, which is surrounded by a ring-shaped muscle cell at its base (Fig. 11.15). When the muscle cell contracts the apical cell is erected and protrudes above the level of the epithelium. The apical cell shows no particular morphological specializations, except for rather elongated microvilli on its distal border. Efferent fibres

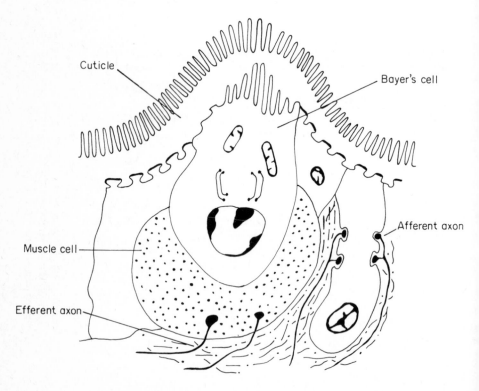

Fig. 11.15   The specialized epidermal and muscle cell of the organ of Bayer, from the epidermis of the leech *Glossiphonia*. The afferent axons are possibly from central mechanoreceptors. (After Damas, 1973.)

synapse with the muscle cell while others, which appear to be afferent, terminate on the lateral walls of the epithelial cells. These may be the peripheral endings of the T, P or N cells described by Nichols and Baylor (1968).

The cell bodies of certain primary mechanoreceptor neurons of *Hirudo* lie within the central ganglion (Nichols and Baylor, 1968; Nichols and Purves, 1970). Seven neurons in each hemi-ganglion respond to mechanical stimulation of the skin, with a rapidly adapting burst of impulses. The cells can be distinguished by their threshold to increasing intensity of stimulation; the 3 T cells responding to gentle touch and water movement, the 2 P cells requiring moderate pressure on the skin to excite them, while the N cells are classified as nociceptive. The receptive fields of the T cells are divided to cover the dorsal, lateral and ventral areas of the skin in each half-segment and extend to the first annulus of the segment on either side. The sensory endings of each cell must divide extensively in the skin, each terminating in some kind of receptor. Only the T cells appear to be capable of providing the continuous background activity required to sustain the excitability of the motoneurons generating the swimming pattern, but they are not active during swimming movements (Kristan; *et al.*, 1974b). Other receptors are probably involved but have not yet been identified.

Slowly adapting receptors which synapse with the large interneuron in the median, or Faivres, nerve, have been identified by Laverack (1969) and Smith and Page (1974). Laverack considered that the primary receptors may be located in the body wall, but the latter authors located them in the connective tissue sheath surrounding the ganglion. The receptors are sufficiently sensitive to be excited by tactile stimulation of the skin of an intact segment. The large interneuron in Faivres nerve makes electrical junctions with the motoneurons of the longitudinal muscles of the body wall and cord sheath (Gardner-Medwin, Jansen and Taxt, 1973, Magni and Pellegrino, 1975) and if the mechanoreceptive endings are sensitive to a stretch stimulus applied to the cord, they could serve an important proprioceptive function in preventing over-extension, by generating tension in the longitudinal musculature.

11.5   Epithelial sensory cells of molluscs

Light microscope studies on the epithelial and sensory cells and organs of molluscs are relatively few (Bullock and Horridge, 1965), one factor being the relative difficulty of obtaining selective staining with methylene blue techniques. Ultrastructural studies have concentrated on specific sensory organs such as the eyes, tentacles and the rhinophores (Storch and Welsch, 1969; Barber and Wright, 1969), with the exception of descriptions of the sensory cells in the non-specialized epithelia of the body wall and mantle of *Nassarius* (Crisp, 1971) and *Lymnaea* (Zylstra, 1972). The several types of primary sensory cells described by these authors consist of a sub-epithelial cell body with a long sensory process ascending to the surface between the epithelial cells (Fig. 11.16). The apical membrane of the receptor carries one or more non-motile cilia. Each has a 9 + 2 arrangement of fibrils and a relatively unspecialized basal body, often with one or more striated rootlets penetrating the upper region of the cell. One cannot be certain of the sensory modality of these receptors from their structural characteristics, but the relative abundance of certain types in areas showing extreme mechano-sensitivity, is perhaps indicative of these having a tactile function.

There are several instances where electrophysiological investigations have indicated the existence of stretch receptors in the body wall (Laverack and Bailey, 1963; Laverack, 1970) or associated with certain muscles (Mellon, 1969; Kater and Rowell, 1973; Janse, 1974), but in only one case have the receptors been identified (Kater and Rowell, 1973) and structural details of the endings are not yet available.

11.5.1 Responses of epithelial mechanoreceptors

Responses to peripherally applied tactile or pressure stimulation generally take the form of a rapidly adapting burst in which the number of spikes depends upon the intensity of the stimulus or the threshold of the unit in question. The receptive fields of individual primary touch sensitive units of *Lymnaea* vary in size; those in the anterior regions of the mantle and body wall are generally small, but more posteriorly and in the foot they are larger and overlap considerably (de Vlieger, 1968; Janse, 1974). Sensory thresholds of individual units are often lower in the centre of their field, suggesting differences in the density or nature of the receptor endings on different branches of the neuron. Touch sensitive units with similar properties have also been reported from the parapodia of *Aplysia* (Anderson, 1967; Hughes, 1971), and also the gill and siphon

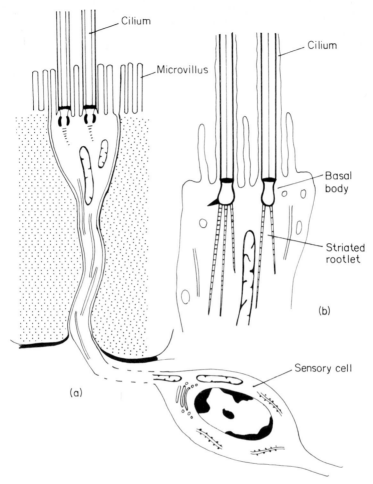

Fig. 11.16   (a) Sensory cell from the unspecialized mantle epithelium of *Nassarius.* The cell body of the receptor lies in a sub-epithelial position. (b) The apical region of a probable mechanosensory cell from *Lymnaea stagnalis.* ((a) after Crisp, 1971; (b) after Zylstra, 1972.)

(Castellucci, Pinsker, Kupferman and Kandel, 1970; Byrne, Castellucci and Kandel, 1974). The somata of the gill mechano-receptor neurons form two clusters of small cells in the right and left halves of the abdominal ganglion and innervate the skin of the siphon, mantle shelf and purple gland through the branchial and siphonal nerves. The mechanoreceptor population is organized in a hierarchical manner according to the field size of individual receptors

(Fig. 11.17). There is considerable overlapping, the smaller fields being overlapped by larger units, which are themselves overlapped by others. Of the 210 cells studied, 190 had receptive fields covering the tip of the siphon, the most sensitive areas not always coinciding with the centre of the receptive field (Byrne *et al.*, 1974). The receptors respond to water jets and tactile stimuli with slowly adapting discharges (Fig. 11.18); either of these stimuli being sufficient to excite the full withdrawal reflex. In fact the function of these units is to mediate withdrawal responses, although it is possible that they might provide information on a moving stimulus, such as a small organism crawling over the surface of the body. Similar touch sensitive units were identified at the base of the siphon in *Buccinum* (Laverack and Bailey, 1963) but the receptive fields were in the region of only 1 mm^2. Two types of receptor adapted rapidly; the third showed tonic activity which was related in some degree to the amount of imposed stretch.

Several authors (Quayle, 1949; Ansell, 1962; Trueman, 1968) have speculated on the role of proprioceptors in bivalve molluscs, particularly with repect to the foot during burrowing movements.

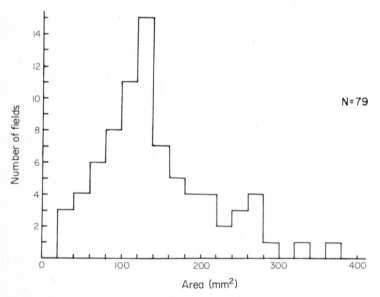

Fig. 11.17 The distribution of sensory units according to the size of their receptive fields on the siphon and mantle of *Aplysia*. (From Byrne, Castellucci and Kandel, 1974.)

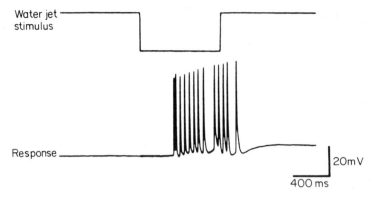

Fig. 11.18   The response of a centrally located receptor cell to a water jet applied to its peripheral sensory field. (From Byrne, Castellucci and Kandel, 1974.)

This aspect has been thoroughly investigated by Olivio (1970), who failed to find any units responding to extension or contraction of the muscles in the highly mobile foot of *Ensis*. All the tactile units present were highly phasic, requiring displacements of high velocity in order to fire them, and these are not provided by the slow contraction of the foot musculature. Digging movements also occurred without the tactile stimulus provided by the sand in which the animal normally burrows, so it appears that the motor sequence of the digging cycle is generated centrally in the nervous system. Trueman (1968) has shown how the digging cycle is modified as the penetration increases or harder layers are encountered, which suggests that the behavioural programme is capable of some modification.

Indirect evidence of the participation of proprioceptors has been obtained by analysis of the neural· mechanism producing the rhythmic swimming movements in the scallop *Pecten* (Mellon, 1969). Experiments on the factors which alter the frequency of fast adductor muscle contractions suggest that, as the muscle is stretched by the elastic pad at the hinge, it excites length or tension sensitive elements serially arranged in the muscle itself. These provide a sensory input, generating the next contraction via the motor neurons. The behaviour can be interpreted as a chain reflex which is terminated by developing tension in the slow adductor, or by adaptation or fatigue in the sensory system.

The feeding mechanism of many gastropods involves cyclical movements in which the buccal mass is alternately pro- and retracted. In conjunction with movements of its own intrinsic musculature, these result in the radula being applied and rasped across the substrate, the cutting stroke coinciding with the retraction of the buccal mass. Activity of this kind is ideally suited to monitoring by proprioceptors and the first indication that these might be involved was provided by Laverack (1970), who found stretch sensitive units which innervated the body wall near the insertion of the buccal mass in *Aplysia dactylomelea*. Recordings from the cerebral nerves showed units sensitive to the velocity of imposed movements, such as would be caused by retraction of the buccal mass during the feeding cycle, but it was not established that this activity did occur (Fig. 11.19). Units having similar properties were also recorded from the second pharyngeal nerve of *Lymnaea*

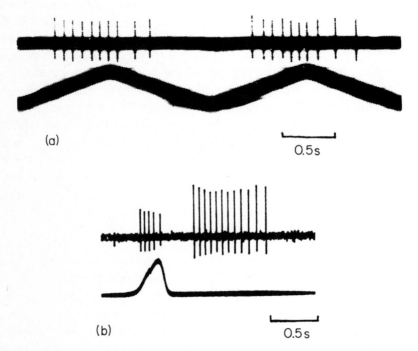

(a)

0.5 s

(b)

0.5 s

Fig. 11.19   (a) Response of a stretch sensitive unit from the body wall in the mouth region of *Aplysia*, recorded in a cerebral nerve. (b) Response in the second pharyngeal nerve of *Lymnaea* to stretching the anterior jugalis muscle. ((a) from Laverack, 1970; (b), from Janse, 1974.)

stagnalis in response to stretching the anterior jugalis muscle (Janse, 1974). Slowly adapting bursts are also recorded from the dorsal buccal nerve in whole animal preparations of *Heliosoma* in response to prodding the posterior end of the buccal mass, and they also mark the transition from protraction to retraction during naturally occurring feeding movements. (Kater and Rowell, 1973). The responses emanate from two clusters of cells embedded in the intrinsic musculature of the buccal mass on either side of the insertion of the oesophagus. There are about 40 cells in each group, each sending an axon into the dorsal buccal nerve, but the details of the stretch sensitive endings have not yet been described.

The mechanoreceptors have a true proprioceptive function in that their discharge is associated with the onset of retraction and the development of tension in the retractor muscles, although it ceases when the buccal mass is fully retracted. The receptors are not an homogenous population; those units with the largest recorded action potentials behave in a phasic manner and discharge in the early part of the retraction, while the smallest units behave tonically and fire throughout the retraction cycle. During feeding the antero-posterior movements of the buccal mass are driven by alternating bursts of impulses in groups of protractor and retractor motoneurons in the buccal ganglia (Fig. 11.20). Mechanical stimulation of the receptors results in an e.p.s.p. volley in the retractor neurons of both buccal ganglia, which correspond 1 : 1 with the receptor discharge recorded in the dorsal buccal nerve. The identical stimulus evokes inhibitory potentials in the corresponding groups of protractor motoneurons. This particular configuration of motoneurons and receptors achieves two ends. Firstly, it ensures the protractor neurons do not fire into the retraction phase. Secondly, the protractors are not inhibited until sufficient tension has been generated in the retractors to ensure a rapid rasping stroke. This effect is enhanced by the positive feedback loop between the receptors and the retractor muscles, which is broken when the rest position is attained and tension relieved by the elasticity of the buccal mass. The feeding behaviour is driven primarily by a neural programme endogenous to the buccal ganglia (Kater, 1974), which can be modulated by sensory feedback which allows for variations in the load imposed upon the system by the characteristics of the substrate upon which is it acting.

So far, this seems to be the best authenticated example of the use of proprioceptors by molluscs, or indeed for any of the invertebrates

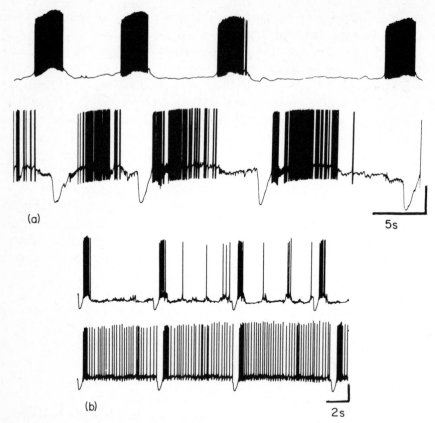

Fig. 11.20   (a) Normal pattern of activity during feeding movements, recorded from a retractor motoneuron (upper) and protractor motoneuron (lower) of *Heliosoma*. (b) Activity of the protractor motoneuron of *Heliosoma* in an intact preparation (upper) and after the dorsobuccal nerves carrying the axons of the stretch receptors have been cut (lower). Note that the discharge continues right up to the strong hyperpolarization in the latter preparation. (From Kater and Rowell, 1973.)

considered in this chapter. The only other behaviour of comparable complexity is the escape swimming response shown by *Tritonia* to predaceous starfish and surf-active compounds (Willows, Dorsett and Hoyle 1973, Dorsett, Willows and Hoyle 1973, Willows and Dorsett in press). The series of alternating flexions of the longitudinal musculature which constitute the swimming movements are generated entirely within the central ganglia, and recordings made from the peripheral stumps of pleural and pedal nerves during the swimming sequences, failed to show any sensory activity which signalled the termination of one flexion or the beginning of the next. Similarly it is not possible to stop the muscular activity by physically restraining the animal so it cannot adopt the flexed postures.

11.6   Conclusions

Insofar as general conclusions can be drawn, it is obvious that proprioceptors in soft-bodied invertebrates have not reached the degree of sophistication and refinement made possible by the possession of an articulated skeleton. In the coelenterates and annelids (with the exception of the leeches), technical difficulties in recording and restraining the preparations have probably been responsible for the shortage of physiological information, and in the molluscs similar considerations apply to the peripheral neuro-muscular relationships. The analysis of the properties of this peripheral component and its interaction with the central nervous system is perhaps the greatest challenge to molluscan neurobiologists at this time. Obviously, abundant opportunities for true proprio-ception occur, yet we have repeatedly seen that the appropriate sense organs are rate sensitive or rapidly adapting, and also that the activity they could supposedly monitor is able to continue after they have been put out of action. This on its own need not disqualify them, for it is their ultimate projection into the central nervous system and the nature of the information they provide which determines their role as proprioceptors, and not whether they can be ignored or over-ridden by centrally organized programmes of motor activity, which may only be receptive to sensory information at some particular phase of its activity cycle.

References

Alexander, R. McN. (1962) Visco-elastic properties of the body wall of sea anemonies. *Journal of Experimental Biology*, **39**, 373–86.

Alexander, R. McN. (1964) Visco-elastic properties of the mesogloea of jellyfish. *Journal of Experimental Biology*, **41**, 363–70.

Anderson, J. A. (1967) Patterns of response of neurons in the cerebral ganglia of *Aplysia californica*. *Experimental Neurology*, **19**, 65–77.

Ansell, A. (1962) Observations on burrowing in the *Veneridae*. *Biological Bulletin, Woods Hole*, **123**, 521–30.

Barber, V. (1966) The morphological polarization of kinocilia in the Octopus statocyst. *Journal of Anatomy, London*, **100**, 685–86.

Barber, V. and Wright, D. E. (1969) The fine structure of the sense organs of the cephalopod *Nautilus. Zeitschrift für Zellforschung,* **102**, 293—12.

Bozler, E. (1926) Weitere Untersuchungen zur Sinnes- und Nervenphysiologie des Medusen. *Zeitschrift für vergleichende Physiologie,* **4**, 797.

Brokelmann, J. and Fischer, A. (1966) Die Cuticulastruktur von *Platynereis dumerlii. Zeitschrift für Zellforschung,* **70**, 131—35.

Bullock, T. H. and Horridge, G. A. (1965) Structure and Function in the nervous systems of Invertebrates. Freeman, London-San Francisco.

Byrne, J., Castellucci, V. and Kandel, E. (1974) Receptive fields and response properties of receptor neurons innervating the skin and mantle shelf in *Aplysia. Journal of Neurophysiology,* **37**, 1041—64.

Campbell, R. D. (1972) Statocyst lacking cilia in the coelenterate *Corymorpha palina. Nature,* **238**, 49—51.

Castellucci, V., Pinsker, H., Kupfermann, I. and Kandel, E. (1970) Neuronal mechanisms of habituation and dishabituation of the gill withdrawal reflex of *Aplysia. Science,* **167**, 1745—46.

Collier, H. O. J. (1939) Central nervous activity in the earthworm I and II. *Journal of Experimental Biology,* **16**, 286—312.

Crisp, M. (1971) Structure and abundance of receptors in the unspecialized epithelium of *Nassarius reticulatus Journal of the Marine Biological Association,* **51**, 865—890.

Damas, D. (1973) Ultrastructure de'l' epithelium tegumentaire de *Glossiphonia complanata* (Hirudinee): Cellules epitheliales et organes sensorielles de Bayer. *Zeitschrift für Zellforschung,* **143**, 355—365.

Dawson, A. B. (1928) Intermuscular cells of the earthworm. *Journal of Comparative Neurology,* **32**, 155—171.

Dijkgraaf, S. (1963) The functional significance of lateral line organs. *Biological Reviews,* **38**, 51—105.

Dorsett, D. A. (1964) The sensory and motor innervation of *Nereis. Proceedings of the Royal Society, Series B,* **159**, 652—667.

Dorsett, D. A. (1966) Overlapping sensory fields of *Nereis*, and their possible role in locomotion. *Proceedings of the Royal Society, Series B,* **164**, 615—23.

Dorsett, D. A. and Hyde, R. (1969) The fine structure of the compound sense organs on the cirri of *Nereis diversicolor. Zeitschrift für Zellforschung,* **97**, 512—527.

Dorsett, D. A., Willows, A. O. D. and Hoyle, G. (1973) Neuronal basis of behaviour in *Tritonia*. IV. Central origin of a fixed action pattern demonstrated in the isolated brain. *Journal of Neurobiology,* **4**, 287—300.

Fraenkel, G. (1925) Der Statische Sinn der Meduseu. *Zeitschrift für Vergleichede Physiologie,* **2**, 658—90.

Friedlander, B. (1895) Beiträge zur Physiologie des Zeutral nerven systems und des Bewegenmechanismus der Regenwürmern. *Pflügers Archiv für gestalten Physiologie,* **58**, 168—207.

Gardner-Medwin, A. R., Jansen, J. and Taxt, T. (1973) The 'giant' axon of the leech. *Acta physiologica scandinavica*, **87**, 30A–31A.

Grasse, P. (1959) Traite de Zoologie V. Annelides. Masson and Cie, Paris.

Gray, J. (1939) The kinetics of animal locomotion. VIII. The locomotion of *Nereis diversicolor*. *Journal of Experimental Biology*, **16**, 9–22.

Gray, J. and Lissmann, H. (1938) Studies in animal locomotion VII. Locomotory reflexes in the earthworm. *Journal of Experimental Biology*, **44**, 93–118.

Gray, J., Lissmann, H. and Pumphrey, R. J. (1938) The mechanism of locomotion in the leech *Hirudo medicinalis*. *Journal of Experimental Biology*, **15**, 408–30.

Garrey, W. E. and Moore, A. R. (1915) Peristalsis and coordination in the earthworm. *American Journal of Physiology*, **39**, 139–148.

Hamaker, J. (1898) The nervous system of *Nereis virens Sars*. *Bulletin of the Museum of Comparative Zoology, Harvard*, **32**, 89–214.

Horridge, G. A. (1955) The nerves and muscles of medusae. II. *Geryonia proboscidalis*. *Journal of Experimental Biology*, **32**, 555–568.

Horridge, G. A. (1963) Proprioceptors bristle receptors, efferent sensory impulses, neurofibrils and number of axons in the parapodial nerve of the polychaete *Harmothoë*. *Proceedings of the Royal Society, Series B*, **157**, 199–222.

Horridge, G. A. (1966) Pathways of coordination in ctenophores. *Symposia of the Zoological Society of London*, **16**, Academic Press, London.

Horridge, G. A. (1969) Statocysts of medusae and evolution of stereocilia. *Tissue and Cell*, **1**, 341–353.

Hughes, G. M. (1971) An electrophysiological study of parapodial innervation patterns in *Aplysia fasciata*. *Journal of Experimental Biology*, **55**, 409–420.

Janse, C. (1974) A neurophysiological study of the peripheral tactile system of the pond snail *Lymnaea stagnalis*. *Netherlands Journal of Zoology*, **24**, 93–161.

Josephson, R. K. (1961) The response of a hydroid to weak water-borne disturbances. *Journal of Experimental Biology*, **38**, 17–27.

Josephson, R. K. (1965) Three parallel conducting systems in the stalk of a hydroid. *Journal of Experimental Biology*, **42**, 139–152.

Kater, S. (1974) Feeding in *Heliosoma trivolvis*. The morphological and physiological bases of a fixed action pattern. *American Zoologist*, **14**, 1017–1035.

Kater, S. and Rowell, H. F. (1973) Integration of sensory and centrally programmed components in generation of cyclical feeding activity of *Heliosoma trivolvis*. *Journal of Neurophysiology*, **36**, 142–155.

Knapp, M. and Mill, P. (1971) The structure of ciliated sensory cells in the epidermis of the earthworm *Lumbricus terrestris*. *Tissue and Cell*, **3**, 623–636.

Kristan, W. B., Stent, G. S. and Ort, C. (1974a) Neuronal control of swimming in the medicinal leech. I. Dynamics of the swimming rhythm. *Journal of Comparative Physiology*, **94**, 97–119.

Kristan, W. B., Stent, G. S. and Ort, C. (1974b) Neuronal control of swimming in the medicinal leech. III. Impluse patterns in the motoneurons. *Journal of Comparative Physiology*, **94**, 155–176.

Laverack, M. S. (1969) Mechanoreceptors, photoreceptors and rapid conduction pathways in the leech *Hirudo medicinalis*. *Journal of Experimental Biology*, **50**, 129–140.

Laverack, M. S. (1970) Responses of a receptor associated with the buccal mass of *Aplysia dactylomelea*. *Comparative Biochemistry and Physiology*, **33**, 471–473.

Laverack, M. S. and Bailey, D. F. (1963) Movement receptors in *Buccinum undatum*. *Comparative Biochemistry and Physiology*, **8**, 289–298.

Mackie, G. (1967) Neuroid conduction and the evolution of conducting tissues. *Quarterly Review of Biology*, **45**, 319–332.

Magni, F. and Pellegrino, M. (1975) Nerve cord shortening induced by activation of the fast conducting system in the leech. *Brain Research*, **90**, 169–174.

Mann, K. H. (1962) *Leeches*. Pergamon Press, New York.

McFarlane, I. D. (1970) Control of preparatory feeding behaviour in the sea anemone *Telia felina*. *Journal of Experimental Biology*, **53**, 211–220.

McFarlane, I. D. (1974) Excitatory and inhibitory control of inherent contractions in the sea anemone, *Calliactis parasitica*. *Journal of Experimental Biology*, **60**, 397–422.

Mellon, de F. H. (1969) The reflex control of rhythmic motor output during swimming in the scallop. *Zeitschrift für vergleichende Physiologie*, **62**, 318–336.

Mill, P. J. and Knapp, M. (1967) Efferent sensory impulses and the innervation of tactile receptors in *Allolobophora longa* and *Lumbricus terrestris*. *Comparative Biochemistry and Physiology*, **23**, 263–276.

Nicholls, J. G. and Baylor D. A. (1968) Specific modalities and receptive fields of sensory neurons in the CNS of the leech. *Journal of Neurophysiology*, **31**, 740–756.

Nicholls, J. G. and Purves, D. (1970) Monosynaptic chemical and electrical connexions between sensory and motor cells in the CNS of the leech. *Journal of Physiology*, **209**, 647–667.

Ort, C., Kristan, W. B. and Stent, G. S. (1974) Neuronal control of swimming in the medicinal leech. II. Identification and connexion of motoneurons. *Journal of Comparative Physiology*, **94**, 121–154.

Olivio, R. F. (1970) Mechanoreceptor function in the razor clam. Sensory aspects of the foot withdrawal reflex. *Comparative Biochemistry and Physiology*, **35**, 761–786.

Passano, L. M. and McCullough, C. B. (1964) Coordinating systems and behaviour in *Hydra*. I. Pacemaker system of the periodic contractions. *Journal of Experimental Biology*, **41**, 643—664.

Peteya, D. J. (1973) A possible proprioceptor in *Cerianthopsis americanus*. (*Cnidaria, Ceriantharia*). *Zeitschrift für Zellforschung*, **144**, 1—10.

Quayle, D. (1949) Movements in *Venerupis* (*Paphia*) *pallustra*. *Proceedings of the Malacological Society*, **28**, 383—390.

Retzius, G. (1892a) Das nervensystem der Lumbricinen. *Biol Unters.* (NF) **3**, 1—16.

Retzius, G. (1892b) Das sensibilen Nervensystem der Polychaeten. *Biologische Untersuchungen*, (NF), **4**, 1—10.

Ruska, C. and Ruska, H. (1961) Die cuticula die Epidermis des Regenwurmes *Lumbricus terrestris*. *Zeitschrift Zellforschung*, **53**, 759—764.

Schulter, E. (1933) Die Bedeutung des Zentralnervensystems von *Hirudo medicinalis* für Lokomotion und Raumorientierung. *Zeitschrift für wissenschaftliche Zoologie*, **143**, 538—593.

Smallwood, W. M. (1926) The peripheral nervous system of the common earthworm *Lumbricus terrestris*. *Journal of Comparative Neurology*, **42**, 35—55.

Smith, J. E. (1957) The nervous anatomy of the body segments of nereid polychaetes. *Philosophical Transactions of the Royal Society, Series B.*, **240**, 135—196.

Smith, P. H. and Page, C. H. (1974) Nerve cord sheath receptors activate the large fibre system in the leech. *Journal of Comparative Physiology*, **90**, 311—320.

Storch, V. and Welsch, V. (1969) Uber aufbau und innervation der Kopfhange der prosobranch schnecken. *Zeitschrift für Zellforschung*, **102**, 419—31.

Tardent P. and Schmid, V. (1972) Ultrastructure of the mechanoreceptors in the polyp *Coryne pintneri*. *Experimental Cell Research*, **72**, 265—275.

Trueman, E. R. (1968) The burrowing activities of bivalves. *Symposium of the Zoological Society of London*, **22**, 167—186.

Uexküll, J. V. (1905) Studien uber den tonus: III. Die Blutgel. *Zeitschrift für Biologie NF*, **28**, 372—402.

Vlieger, T. A. de, (1968) An experimental study of the tactile system of *Lymnaea stagnalis*. *Netherlands Journal of Zoology*, **18**, 105—154.

Willows, A. O. D., Dorsett, D. A. and Hoyle, G. (1973) The neuronal basis of behaviour in *Tritonia*. III. Neuronal mechanism of a fixed action pattern. *Journal of Neurobiology*, **4**, 255—285.

Zylstra, U. (1972) Distribution and ultrastructure of epidermal sensory cells in the F. W. snails *Lymnaea stagnalis* and *Biomphalaria pfeifferi*. *Netherlands Journal of Zoology*, **22**, 283—298.

12 Spatial equilibrium in the arthropods

D. C. SANDEMAN

12.1   Introduction

Orientation in animals has been defined as '. . . a selective process in which environmental stimuli elicit a response sequence that results in a non-random pattern of locomotion, direction of body axis or both' (Adler, 1970, 1971). Within this framework, three types of orientation can be distinguished, the first is that typical of search; the second, a fixed direction of movement; and the third, the establishment of a true position. Spatial equilibrium may be regarded as only a part of the third type of orientation and in this chapter it is considered to be that orientation which depends upon the reception and translation of angular and linear accelerations. The chapter can be said, therefore, to be concerned with how arthropods keep their 'balance', and the emphasis has been placed on those receptor systems which are themselves directly activiated by the forces of linear or angular acceleration. It is possible to derive the direction of the linear acceleration due to gravity by indirect means; the brightest region of the environment, the sky, is normally exactly opposite to the gravitational force and thus a visual input is sufficient for effective 'gravity' orientation; water bugs use a system of sensitive hairs to detect the size of a trapped water bubble, and can thus estimate whether they are moving up or downwards in the water (i.e. in relation to gravity). These are indirect methods of detecting gravity and have been excluded here. On the other hand, gravity reception is not always confined to special receptor organs and examples have been included which show that information about the forces of angular and linear acceleration is gained from proprioceptor systems which monitor the relative positions of the parts of the body.

The chapter has been divided into sections. The first (Section 12.2) deals with some of the general principles of the detection of acceleration and the relationship of the three important parameters involved with spatial equilibrium, namely position, velocity and acceleration. This account is at the intuitive rather than at the mathematical level. Also included are some comments on how the information received by the detecting systems can be coded in the primary sensory nerve axons. The second (Section 12.3) and third (Section 12.4) parts of the chapter deal with specific arthropod receptor systems and organs which are known to respond to linear and angular accelerations respectively. In most cases both behavioural and physiological observations have been made on these systems and, where possible and appropriate, the experimental details of these investigations are included.

At the outset, it must be stressed that, to consider one set of receptors and relate them to orientation behaviour is somewhat artificial because there are probably very few occasions when an animal needs to rely entirely on the input from one single receptor system for its orientation in space. The evidence points instead to a dependence upon multiple cues, although inputs from some receptors may dominate. This principle of overlapping sensitivities of receptor systems is important, and several examples are mentioned later.

Finally, this chapter is not intended to be an exhaustive review of the field of angular and linear acceleration detection in the arthropods. Excellent reviews of the literature in this field have been very recently published and the reader is referred to the work of Markl (1974, 1975) and Schöne (1974).

12.2   General Principles

12.2.1   Position, velocity and acceleration

The type of orientation which concerns us is the establishment and maintenance of a stable position with reference to some constant feature in the animals environment. A change from this position will involve changes in acceleration and velocity and it is important to first explore the relationship between these parameters.

We regard an object to be stationary if its position relative to its immediate environment does not change over a period of time. We could assign a value to this position in relation to some other stationary point in the environment and the value would be a constant. In this situation the velocity of the object is zero and its acceleration is also zero relative to this point in the environment. If the object is now moved at a constant velocity the position changes continually at a constant rate. However, the object does not have to move at a constant velocity; it may begin to move slowly and gradually increase its velocity, so that not only the position but also the velocity of the object is changing. This rate of change of the velocity is the acceleration.

The relationship between these three parameters is shown in Fig. 12.1, where values of position, velocity, and acceleration are plotted against time. In Fig. 12.1a, the position is changed from one value to another at a constant rate and this is represented by a ramp. The velocity is initially zero, is very suddenly raised to a constant value during movement of the object and then returns to zero. The acceleration changes transiently at the beginning and end of the movement and in opposite directions. In Fig. 12.1b, the object

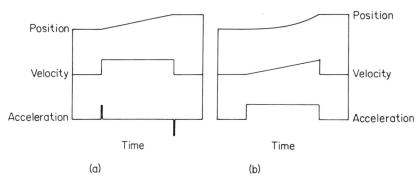

(a)   (b)

Fig. 12.1   The relationship between the position, velocity and acceleration of a moving object. In both diagrams the top line shows the change in the position of the object, the second trace shows its velocity and the third shows its acceleration. The values of these are arbitrary and they change with time, which is displayed from left to right on the abscissae. (a) the object is moved from one fixed position to another at a constant velocity. (b) the object is moved from one fixed position to another under a constant acceleration.

begins to move slowly but increases its *velocity* at a constant rate. The velocity is therefore, a ramp and the acceleration which matches this non-linear change in the position of the object is a step.

Sometimes it is more useful to examine the responses of receptor systems using a dynamic or continually changing stimulus. A sinusoidally oscillating stimulus is convenient and commonly used in the analysis of systems, and for this reason the relationship of the three parameters of position, velocity and acceleration during the sinusoidal oscillation will be given. One of the best examples to consider is a swinging pendulum because many of the equilibrium systems in animals exploit the same principle.

If the position, velocity and acceleration of the bob of a swinging pendulum are traced out we obtain the relationships shown in Fig. 12.2. The upper trace shows the pendulum position. The pendulum is displaced to the side (dotted line) and released from this point. Its maximum rate of change of position occurs as it swings through the centre of its total excursion, after which it begins to slow down,

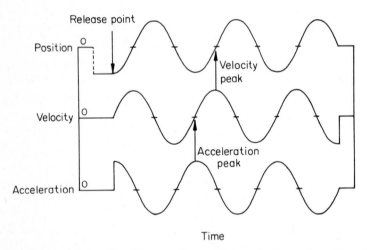

Time

Fig. 12.2   The relationship between the position, velocity and acceleration of an object which is being oscillated sinusoidally between two positions. Arbitrary values for position, velocity and acceleration are plotted against time. The object, a pendulum bob, starts its swing from a position marked by the arrow on the top trace (release point) in which it is displaced from its normal position of rest (i.e. hanging straight down) to the side. Arrows on the second and third traces show how the peaks of velocity and acceleration correspond respectively with periods during which the rate of change of the pendulum's position and velocity are greatest.

reverses direction, and again reaches a peak rate of change of position as it swings back through the centre of its total excursion. The second trace shows the velocity of the pendulum bob. The peak of velocity coincides with that region where the rate of change of position is at its maximum. Similarly, acceleration, shown by the third trace, reaches a peak at the point corresponding with the maximum rate of change of velocity. If the excursion of the pendulum, from its starting position through a swing in one direction and back to its start, is regarded to be a cycle, then it can be said that the peak of velocity leads that of position by 90° and the peak of acceleration leads that of position by 180°. The difference in phase between the three components is extremely useful if we wish to identify which component is represented at some part within the system. The peak discharge frequency of a neuron, for example, may correspond at certain oscillation frequencies with the peak of any one of the three components and thus can lead to a derivation of the component which is effective at the receptor level.

12.2.2   Linear acceleration

Receptor systems used for the detection of both linear and angular accelerations have most probably been derived initially from mechanoreceptors, sensitive to displacement by water. They show a remarkable evolutionary convergence and the device of the 'loaded receptor hair' has been used again and again. Essentially these receptors consist of a secretion of dense material (or in some cases a collection of sand grains) flexibly supported on a base which contains, or is surrounded by, mechanoreceptive hairs. An acceleration imposed on the animal causes the load to shift and thus the hairs beneath it are subjected to a shearing force, which displaces them and activates the receptors below them. It is essentially a heavily damped pendulum and in the simplest case, where the linear acceleration imposed on the animal is due to gravity and is therefore constant, the load on the hairs is displaced until the elastic restoring force of the flexible base is equal to the gravitational force on the statolith. A signal generated by the underlying mechanoreceptors is related to the shearing force and, as the receptor is tilted with respect to gravity, this force will vary as the sine of the tilt angle.

In their normal life animals are often subjected to short duration, linear accelerations in addition to, and perhaps in a different direction from, the constant acceleration due to gravity. The response of the receptors to these transient accelerations can be significantly affected by the elasticity of the structure supporting the load, the elasticity of the mechanoreceptive hairs, and the extent to which the whole system is damped.

Let us suppose that we subject a loaded hair system to a transient linear acceleration, a period of constant velocity and a transient deceleration. In other words we have moved the system from one place to another (Fig. 12.3). The initial acceleration results in the supporting hair being bent in the direction opposite to the direction of travel because of the inertia of the load. If the activation of the receptor neuron of the hair is dependent upon displacement, and the greater the displacement the greater the output signal, then the elasticity of the hairs becomes critically important. If the hairs are very stiff, for example, the load will very rapidly be snapped back to its original position during the period of constant velocity by the high restoring force of the hairs. During the period of constant velocity, there will be no signal from the hair receptor neuron

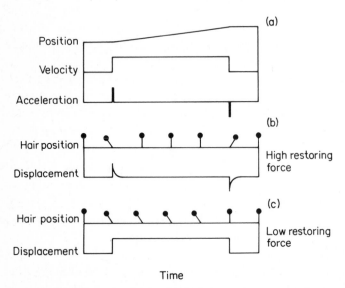

Fig. 12.3   The displacement of a loaded hair in a system in which the restoring force of the hair itself is high (b) or very low (c). The system is moved from one shown in the top three traces (a). The changing positions of the hair and a measure of its displacement are shown.

(assuming that there is no change with respect to gravity). At the end of the movement, there is a deceleration and the load will be again displaced, this time in the same direction as the movement, because it now possesses a momentum imparted to it by its movement (Fig. 12.3b). Therefore, in a system with stiff hairs which have a high restoring force and little damping, the displacement of the hairs corresponds fairly closely with acceleration and deceleration, and no information is given about velocity, nor about the relative change in position of the animal with respect to any other object in the near environment.

A very different situation prevails if the elastic restoring force of the hair is removed. The load is displaced as before by the initial acceleration but it remains in its displaced position until it is restored to its original position by the final deceleration (Fig. 12.3c). The extent of the displacement of the hair corresponds to velocity because, if the system is initially accelerated to a low constant velocity, the displacement of the load will not be as great as if it were accelerated to a higher constant velocity. To work in this way, the system would need to be very critically damped and all real systems have some elasticity. However, it is clear from our example, that by alteration of the elastic restoring force of the hair, either acceleration or velocity, or a combination of the two is measured by the displacement of the load on the hair cells. It is not possible for the system to measure the change in the position of the animal from one point in its environment to another, except in relation to gravity.

12.2.3   Angular acceleration

Accelerometers of the loaded hair type will also respond to angular accelerations provided that these are at right angles to the supporting structure. However, receptor systems have been developed which are primarily sensitive to angular accelerations and which do not respond so well to linear accelerations; again there is a remarkable evolutionary convergence. One of the most effective ways to detect angular accelerations is to enclose the mechanoreceptive hairs within a circular canal. The fluid in such a hollow toroid will move relative to the canal walls, if the canal is rotated so that the angular acceleration is along the circumference of the canal. Angular accelerations applied at right angles to the circumference of the toroid will not produce large displacements of the fluid, nor will

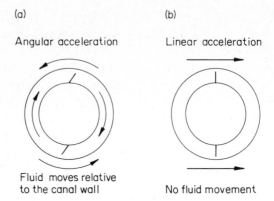

Fig. 12.4   The effect of angular and linear acceleration on the displacement of fluid within a hollow, circular canal. (a) the canal is rotated about its centre and the movement of the fluid relative to the walls of the canal is opposite to the direction of the initial canal movement. A hair with no load and a very small restoring force would be displaced by the relative movement of the fluid as shown by the lines drawn at the top and bottom of the canal. (b) the canal is subjected to a linear acceleration to the right. There is no relative movement of the fluid within the canals and no displacement of the hairs.

linear accelerations in any direction (Fig. 12.4). Thus the canal offers the advantage of providing a transducer for discriminating not only between angular and linear accelerations but also between the different directions of angular accelerations. A measure of the displacement of the fluid can be achieved by introducing into the canal-system mechanoreceptive hairs, preferably offering very little restoring force and at the same time constructed so that they will move with the fluid without impeding its flow. This can be achieved by making the hairs long and very thin.

There are some important physical constraints imposed by viscosity and canal diameter on the flow of fluid in toroidal canals and these should be mentioned here. If three toroidal canals, all filled with the same fluid but having different canal diameters, are subjected to the same angular rotation, the fluid displacement in the canals will be characteristically different. The fluid in the small diameter canal will begin to flow during the initial acceleration and will keep on flowing in the opposite direction to the canal's rotation, until the inertial force of the fluid is overcome by the opposing frictional forces between the walls of the canal and the fluid. The system is thus overdamped and the displacement of the fluid is shown in Fig. 12.5a. The fluid in the large diameter canal would move very rapidly, during the initial acceleration, in a direction opposite to that of the canal, would overshoot the point where its inertia equalled the resistance between the canal walls and the fluid, and would then begin to move more slowly in the same direction as the canal, until it came to a position which it would maintain during the remainder of a constant velocity movement. Thus this system is under-damped (Fig. 12.5b). The fluid in a canal with an intermediate diameter would behave as shown in Fig. 12.5c. The initial acceleration would produce a rapid displacement of the fluid to a point where it would remain during the movement, and when the rotation was stopped it would return, like the others, to its original position. The intermediate-diameter canal system is critically damped and the displacement of the fluid is a measure of the *angular velocity* of the canal (cp. Fig. 12.5c with Fig. 12.1a). Therefore the hydrodynamics of the system, if the fluid viscosity and the canal diameter are critically matched, is capable of performing the conversion of angular acceleration to angular velocity or, in other words, an integration. It must be pointed out that the examples given here greatly simplify the real situation and take no account of frictional forces within the fluid which would prevent the fluid from returning exactly to the same point at the end of a rotational displacement. In addition, the flow of fluid in a cylindrical canal is non-uniform across the canal and this can have significant effects on receptors placed in the canals. Nevertheless, the canal systems in vertebrates and in the crab have been found to be critically damped systems and do effectively convert angular acceleration to angular velocity (Melvill Jones, 1971; Fraser and Sandeman, 1975).

There is a second, ingenious way of measuring angular accelerations and excluding linear accelerations and that is by using the gyroscope principal. A rapidly rotating wheel has inertia in the plane of its angular rotation and an angular displacement out of this plane is resisted and so the gyroscope tends to be stable in a certain position. There are no wheels in nature, but flying insects have rapidly oscillating wings and, more important, the Diptera have retained the hind wings as club shaped structures (halteres) which oscillate rapidly back and forth and behave as gyroscopes. The stability of the fly in flight is not, however, provided by the gyroscopic properties of the halteres themselves, because their mass is so much smaller than the fly. Instead, the torques set up at the

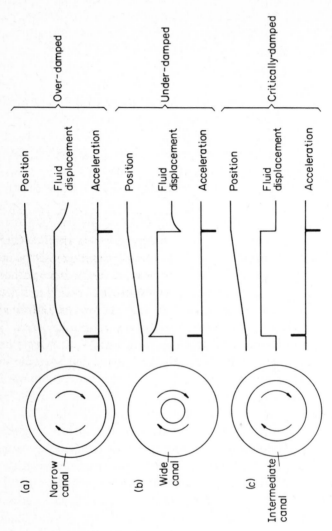

Fig. 12.5   The displacement of fluid in three toroids (a, b and c) containing the same fluid but with different diameter canals. All three are subjected to an angular rotation of the same amplitude with the same constant velocity. The traces alongside each canal represent the change in position of a point on the circumference of the canal (upper traces) the displacement of the fluid relative to the wall of the canal (middle traces), and the angular acceleration at the beginning and end of the movement (lower traces).

base of the halteres during angular rotations of the fly, particularly in the yaw plane, are detected by sense organs placed strategically in this position. The details of this remarkable structure, unique to the insects, are given in Section 12.4.2.

12.2.4 Neuronal coding

In all of the examples of linear and angular acceleration detectors mentioned, the common factor is that the receptors are essentially mechanoreceptors and depend upon the displacement of some associated structure for their activation. The mechanism of transduction from a mechanical displacement to the initiation of nerve impulses in the axons of the primary sensory cells is a subject which will be well covered in other chapters of this book and there is no need to elaborate on it here. The intention here is to show how the nerve can code, in the relative frequency of its discharge, different parameters of the stimulus and how these can be affected both by the mechanical constraints of the receptor structures, which we have mentioned previously and which are well understood, and also by the so-called phasic or tonic properties of the sensory neurons.

Information about intensity is usually conveyed in the nervous system as a modulation of the discharge frequency of the neurons, and the displacement of the hair cells in an acceleration receptor can be similarly coded with an increased frequency of receptor neuron discharge corresponding with larger displacements of the hairs. To take a tonic system first, this could have a basal discharge rate which corresponded with no displacement of the hair. Bending the sensory hair in one direction would increase the discharge rate proportionally, and in the opposite direction would decrease the discharge rate. Directionality is, therefore, measured as a function of the firing frequency in one receptor element. If the system was critically damped and subjected to a change of position, the initial acceleration would produce a displacement of the hairs and this would produce an increased and maintained discharge (Fig. 12.6d). When the object stopped moving, the original position of the hair and firing rate would be resumed. A movement back to the original position would have the opposite effect. In this example the neuron is coding velocity. However, if the elastic restoring forces of the hair are considerably increased the neuron codes a signal which is quite

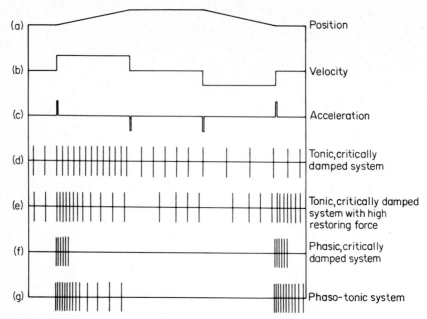

Fig. 12.6   The possible neuronal coding of the displacement of a hair cell in an acceleration receptor. The receptor is moved from one point in space to another at constant velocity, held, and then moved back again. The position, velocity and acceleration are shown by the first three traces (a, b, c) respectively. A tonic and critically damped system (d) with a basal discharge will signal velocity in both directions by increasing or decreasing its firing frequency. The same system in which the hair cells are associated with a structure possessing a high restoring force responds with transient changes more closely related to acceleration (e). A phasic neuron (f) produces a burst of activity associated with acceleration even in a critically damped system, and in only one direction; while a phaso-tonic system (g) reflects both acceleration and velocity.

different. The initial acceleration deflects the hair and a high frequency burst of spikes accompanies the displacement. The hair comes quickly back to its original position and the sensory neuron resumes its original rate of discharge while the system is still moving. When the system stops moving the accompanying deceleration results in a counter movement of the hairs and a transient inhibition of the neuronal discharge. Movement of the system back to its first position results in a repeat of the first two events but in reverse order (Fig. 12.6e). In this case, the output signal is more closely related to the acceleration than to velocity, and it is this former component which is signalled by the neuron. Simply changing the damping and elastic restoring force of the hair system completely alters the component of movement signalled by the neuron.

Similarly, different types of nerve cells can respond differently to the same physical system. If we take a critically damped system again and consider the response of a unidirectional phasic neuron we find that, although the hair is displaced during the period of constant velocity, the phasic neuron responds to the acceleration in only one direction. Again, acceleration is the component which is coded by the neuron (Fig. 12.6f).

Many biological systems show a combination of both the acceleration and velocity components and for them it is no longer meaningful to talk in terms of the actual component of movement which is being conveyed by the neurons (Fig. 12.6g). For very small movements and for oscillatory stimuli over small ranges of frequency some systems do indeed faithfully code either velocity or acceleration. However, it is important not to regard the performance of the receptor solely in terms of these narrow ranges because in natural conditions such limits are often exceeded.

Receptor cells also do not necessarily respond linearly to the displacement of the hair over its entire mechanical range. A hair which may in its initial stages start out as coding velocity, can reach a stage where an increase in velocity no longer produces an increase in the frequency of firing because it has saturated. For a velocity coding system, given a velocity ramp (i.e. an acceleration step) the system could begin to show a pure velocity response but after saturation the frequency is constant and resembles the acceleration.

Another common non-linearity in biological systems is introduced by the process of adaptation. Tonic systems will often show a gradual decline in their discharge frequency after a time, which does not correspond with any change in the displacement of the hair. Because receptors are often subjected to a dynamic stimulus to which they may not adapt, this could be of little importance, but before deciding what information they are really giving to the central nervous system, receptors should be tested under conditions they normally encounter.

The above is an attempt to put in the simplest possible terms the general boundaries imposed by the physical nature of structures which are sensitive to accelerations and by the neurons themselves. What follows are some examples of animal systems in which if can be seen how the above generalizations are exploited to provide information about the direction of gravity or other imposed accelerations.

12.3   Detection of linear accelerations

12.3.1   No specific sense organ

The most significant source of linear acceleration for the spatial orientation of animals is that produced by the earth's gravitational field. This is constant and although a variety of specific gravity detection organs have been evolved, there are many arthropods, particularly insects, which have no specialized gravity organ but which nevertheless show by their behaviour that they can detect gravity (see review by Markl, 1975). Three examples will be considered here: the ant *Formica polyctena*, the stick insect *Carausius morosus* and the mantis shrimp *Squilla mantis*.

The ant *Formica polyctena*, in common with other species of ant, will attempt to escape capture by running in a direction opposite to gravity (Vowles, 1954; Markl, 1971). Ants will make escape runs, on a board which can be tipped over through 180°, even after severe surgical treatment and will detect the direction of gravity in complete darkness. Observations made with the animal and board illuminated with non-directional red light to exclude visual orientation show that a normal ant will turn around and run upwards, if the board is tipped end over end, with a deviation from its previous path of less than 14°. Experiments in which the various body joints of the ant were gently waxed into immovability showed that the input from a single free antenna is enough to allow the animal to orient to gravity. However, ants with both antennae waxed and any one of the joints in the neck, or between the thorax and petiole, or petiole and gaster, or thorax and coxa free, are also able to orient to gravity (Fig. 12.7). There are no statocysts in the antennae of ants but the basal joints do contain complex organs, Johnston's organs, which are well known as gravity receptors. However, gravity sensitivity without the antennae must depend upon the information from the body joints. The relatively large pendulous abdomen of the ant, for example, could on its own act like a sort of statolith, and the deflection of the abdomen due to gravity could give the required information through the activation of hair plates between the gaster and petiole. High speed moving films of ants

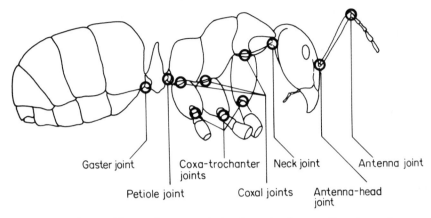

Gaster joint   Coxa-trochanter   Neck joint   Antenna joint
                      joints

    Petiole joint   Coxal joints   Antenna-head
                                           joint

Fig. 12.7   The positions of sensory hair plates in the ant *Formica polyctena* which the insect is able to use to orient to gravity (After Markl, 1971.)

running with different orientations with respect to gravity reveal that the abdomen is in fact displaced by gravity and that the insect tends to correct this displacement with the muscles which control the position of the abdomen in relation to the body. The sensory input for the control of this reflex is from patches of hairs on the gaster joint, and when these are singed off the deflection of the abdomen due to gravity is more pronounced (Markl, 1962, 1963, 1971). The reflex control of appendages which are used for orientation is a common feature of many systems (see p. 505).

Unlike the ant, the stick insect may place more reliance upon the proprioceptive information supplied by its legs. Stick insects are devoid of specialized gravity receptor organs, and yet will also climb upwards in complete darkness. The input from receptors monitoring the joints of the body can be inactivated by waxing the length of the body to a light-weight wooden splint. Such an animal, with its antennae removed (receptors in which come closest to being the primary static receptors (Wendler, 1965)), will still show good gravity orientation. An additional point of interest in the stick insect's gravity response is its dependence upon the linear, rather than the rotational, forces which act on its legs. When the long axis of the stick insect's body is horizontal the centre of gravity of the animal does not lie at the axis of its normal rotation (Fig. 12.8). These two forces, the rotational and the linear, can be separately eliminated by attaching to the animal a collar which shifts the centre

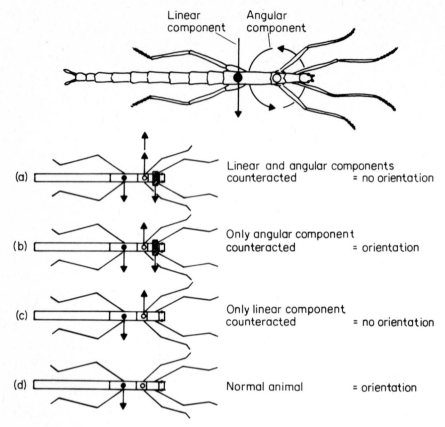

Fig. 12.8   Orientation of the stick insect to gravity. The centre of gravity of the animal (●) does not correspond with the animals axis of rotation (○) and so there are both linear and torsional components acting on the leg proprioceptors. (a) If the animal's centre of gravity is shifted to the centre of rotation by adding a weighted collar, and the weight of the animal plus its collar is counterbalanced, the animal no longer orients to gravity. (b) If only the weight of the collar is counterbalanced, so that the linear but not the torsional component is present, the animal can orient to gravity. (c) With the collar removed and the weight of the animal counterbalanced (i.e. only the torsional component remains) there is no orientation. (d) With no counterbalance, orientation is normal (After Wendler, 1971.)

of gravity, or by counteracting the weight of the animal by attaching it to a counterweight through a pulley system. Tracing the path of the animal's locomotion in the dark clearly shows that the animal cannot orient to gravity when only the torsional component remains, but will orient well whenever the linear component is present, whether on its own or combined with the torsional component (Fig. 12.8; Wendler, 1971). Although a great deal of emphasis has been placed on the limb proprioceptors of the stick insect as the important organs for gravity perception, the hair plates in this animal also play a considerable role (Bässler, 1965).

The use of a general proprioceptive input for gravity perception is not confined to the insects. The mantis shrimp, *Squilla mantis* apparently does not have a specialized static organ and, although these animals do use a dorsal light reaction for their orientation, particularly during their larval stage, as adults they swim and capture prey often in dim light conditions and can orient themselves to gravity when totally blind (Schaller, 1953). It has been suggested that the large eyes are themselves important as static organs, but interference with the eyes of *Squilla*, which among the crustaceans is perhaps the most dependent upon its visual input for prey capture, produces highly traumatic effects. Also, a careful search failed to reveal proprioceptor organs in the eyes of *Squilla*, whereas the antennules do have extensive proprioceptor systems in their joints (Sandeman, 1963). Removal of the swimmerets, or binding them to the abdomen, results in the apparent disorientation of the animal but, as the pleopods are the primary organs of locomotion, it is difficult to draw too far reaching a conclusion about the function of the swimmerets as gravity receptor organs. Certainly, when hanging beneath the animal they could act as part of an overall gravity sensory system.

Most crustaceans with movable eyes, hold them in a fixed relationship to the horizon over a certain range, and displacement of the body is compensated by a change in the eye position. The eye position is therefore a good indication of the animal's perception of a change in its body position relative to gravity. However, in *Squilla* this is a less than satisfactory measure because the animals continually direct their gaze at different objects in their visual field and it is not possible to determine the 'normal' attitude. Blinding the eyes and supporting the animal in different positions in the water failed to give clear eye movements which would indicate an orientation to gravity (Schaller, 1953). So far, research on the gravity orientation response of *Squilla mantis* has not yielded very clear results, perhaps because the behaviour of this remarkable animal is extremely complex. It is quite probable that all of its reactions to surgical interference or to the restraining of its appendages and blinding, are no more than signs either of the animal attempting to escape or of severe operational trauma.

A problem related to gravity perception using an overall proprioceptive and body-joint positional sense is the necessary complexity required at the central nervous level to sort out the incoming information from the sensory receptors and to distinguish those accelerations which are meaningful for effective orientation from those which are the result of the animals own movements. It has been pointed out (Wendler, 1971) that the hair plates between the abdomen and thorax of the bee, known to be important for gravireception (Lindauer and Nedel, 1959) will be maximally stimulated when the animal performs its waggle dance. Similarly, hair plates in the ant will be stimulated during normal walking. The way in which all the information is brought together in those animals which have a 'general' gravity sense is not known but one must assume some kind of abstraction of the relevant information from all inputs. Conceivably, the animal does not depend upon the continual input from the gravireceptors, but 'samples' the input from time to time from different receptors which are not involved for a short time in locomotion. Brief pauses in locomotion are a frequently observed phenomenon in walking insects and perhaps they use these brief pauses to check on their orientation. At a more specific level it is possible to propose systems which will make comparisons between the inputs from strain receptors in the legs of insects and, by substraction, deduce the direction of a constant acceleration due to gravity. There is no evidence yet however for the existence of any such central mechanisms.

12.3.2   Johnston's organ

Johnston's organ in the insects is an elaborate array of receptors which lies in the second joint (pedicel) of the antennae and monitors the movement of the distal joints of the antennae. The movements of the antennae relative to the animal are exploited variously by different insects as a hearing organ, part of a flight control system (see p. 505) and as a gravireceptor. We have already seen (Section 12.3.1) that, even in those insects where a general proprioceptive system is used for gravireception, Johnston's organ can function on its own as a gravireceptor. The gravireceptor properties of Johnston's organs in burrowing beetles have been clearly demonstrated

(Bückmann, 1955), where the overall leg proprioceptive input is perhaps less effective because the animal's whole body is supported for most of the time in the loose sand in which it burrows.

The sensory cells in Johnston's organ are found to be mainly phasic in those insects in which they have been physiologically investigated, and yet the behavioural evidence suggests that they are implicated in the detection of long term changes in linear acceleration. Intuitively one imagines that a tonic receptor would be more appropriate. However, the receptor system of the antennule must operate against a background of continual oscillation caused by the animals own movements, and taken in this context it is easy to appreciate how phasic systems are more appropriate. The best example of this principle is given by the action of Johnston's organs of a fly in flight. In *Calliphora* the pedical of the antenna contains a single campaniform sensillum and many scolopale organs which constitute Johnston's organ (Fig. 12.9; Gewecke, 1967a, b) Chapter 9. The pedicel in flies bears the funiculus which in turn carries the arista. The funiculus is a large flattened, pendulous structure which is derived from the first joint of the flagellum of the basic insect antenna. The segmented arista is the remainder of the flagellum. When the animal is flying, the wind on its head presses the arista and funiculus backwards and laterally and this activates the campaniform sensillum, which is a phasotonic receptor. Also during flight, the movement of the animal's wings results in a continuous vibration of the antenna, which activates the cells of Johnston's organ. These cells are purely phasic and respond to each movement of the flagellum with a single action potential. Within Johnston's organ there is a further distinction between cell types in that some will respond only to lateral deflections and others only to medial deflections of the antenna. Electrodes on the antennal nerve record compound potentials from both groups of cells which correspond respectively with the medio-lateral oscillations of the antennae induced by the movement of the wings (Gewecke and Schlegel, 1970). The compound potentials from the cells sensitive to medial displacement and those sensitive to lateral displacement are of approximately the same amplitude in still air (Fig. 12.10a) but, if the wind on the head of the animal is increased, the funiculus and arista are displaced, their oscillatory excursion is altered and so the amplitude of the phasic potentials of Johnston's organ are modulated. In still air the amplitude of the compound potentials

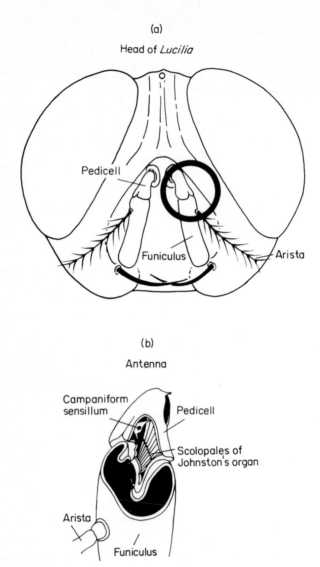

Fig. 12.9   (a) The head of a fly (*Lucilia cuprina*) seen from the front, showing the antennae which are deflected backwards and laterally by wind on the animal's head. Johnston's organ is contained in the pedicellus and monitors the movements of the funiculus which is the enlarged first joint (circled) of the flagellum and bears the feathered arista. (b) The organ (shown in section) consists of a number of scolopale systems which span the pedicellus and insert around the articulation between the funiculus and the pedicellus. In addition there is a single large campaniform sensillum which also monitors the pedicellus-funiculus joint (After Gewecke and Schlegel, 1969.)

from the different receptor groups are almost the same size (Fig. 12.10a) but during acceleration to 1 m s⁻¹ the amplitude of the compound action potentials from one group of receptors is larger than the other, thus signalling the acceleration step (Fig. 12.10b). If the wind speed is maintained at 1 m s⁻¹ a new relationship is established, with one set of compound action potentials being about twice the amplitude of the other (Fig. 12.10c). Deceleration from 1 m s⁻¹ to zero results in large amplitude potentials from both cell groups again. (Fig. 12.10d), and in still air the original pattern is resumed (Fig. 12.10e). Thus no single receptor in Johnston's organ is capable of detecting maintained displacements of the flagellum in relation to the pedicel unless the system is activated first by the oscillations of the wings. Also, the important information is contained in the relative amplitudes of the *different receptor groups* rather than in the amplitude of the compound potential from the whole organ.

The campaniform sensillum also monitors the position of the antennal flagellum and activates an efferent system to bring the flagellum back to its original position, thus ensuring that the antennal receptors remain within their operating range. There is good evidence for the control of flight direction by the antennules so that if they are assymetrically stimulated, the wing stroke angle of the stimulated side decreases thus ensuring that the wind is always directly on the head of the animal. (Schneider, 1953; Gewecke, 1967b).

Another function of the antennal receptors in flies and locusts appears to be the control of flight speed. They have been described as throttle systems, which prevent the flight speed becoming too fast and so uneconomical (Gewecke, 1974). In migrating locusts this is of critical importance. It has been recently shown (Waloff, 1972) that the direction of flight in swarming *Schistocerca gregaria* is oriented, for a substantial proportion of their time, in the direction of the swarm displacement. It is conceivable that what is important for migrating locusts is to stay airborne with the least possible effort, rather than that they should make a determined effort to direct their flight against prevailing winds to reach some goal (Gewecke, 1972).

Johnston's organs are not the only receptors of linear accelerations to be found in flying insects. Locusts, for example, have specialized directionally sensitive hairs on the front of their heads and the output of these receptors has been shown to affect the flight oscillator and also the flight posture of tethered, flying locusts

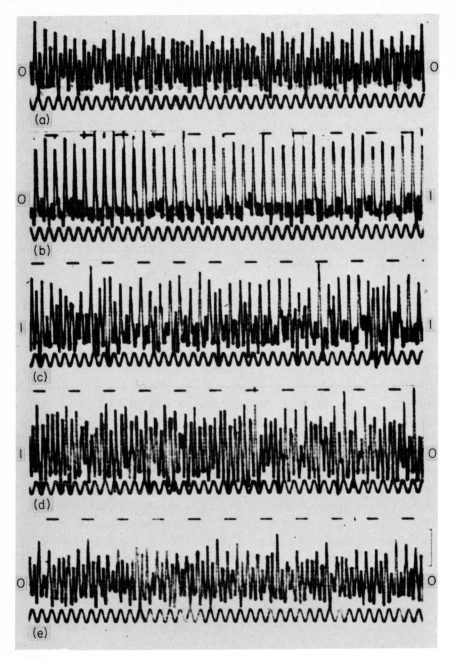

Fig. 12.10   Compound action potentials recorded from the antennal nerve of a fly during tethered flight. Alternate potentials are from the two groups of scolopale cells which respond to medio-lateral or latero-medial displacements of

(Camhi, 1969a, b, c; Camhi and Hinkle, 1973). Similarly, there are a large number of hair receptors on the wings and at the articulation of the wings and thorax of insects, all of which are probably involved in some aspect of the control of the pace and the direction of flight and perhaps also the perception of linear accelerations of short duration. A review of these receptors and their effect on flight, however, needs a chapter on its own. There is apparently no information yet on the action of Johnston's organ in non-flying insects to suggest that the same oscillatory activation of the phasic sense cells could give the animal a method of monitoring gravity.

12.3.3   Statocyst systems

Very few insects have developed true statocysts, in which a weight is suspended or supported on a group of sensory hairs. It is perhaps significant that those that have special static organs are aquatic and the overall sense of the body position or limb proprioception is probably of less value since, whether the animal is swimming or crawling along the bottom, its body is almost weightless in water. The larvae of *Limnophila fuscipennis* have a pair of small sacs in the last abdominal segment with a sensory hair and granules. The granules are thought to be thrown against the sensory hair by movements of the abdomen and the animal can thus gain some indication of gravity. The larvae of *Ptychoptera* have typical vesicular statocysts with movable statoliths in the 10th and 11th segments of the abdomen, but physiological confirmation of the function of these organs is apparently not available. The static organs of *Nepa*, which use a trapped bubble of air and act as differential pressure receptors are well described sensory receptors (Baunacke, 1912; Hamilton, 1931; Thorpe and Crisp, 1947; Markl and Bonke, in preparation).

True statocysts are more common in crustaceans, again perhaps because in water the proprioceptive systems are less useful. The

---

the antennae. The lower trace shows the flight frequency. (a) The fly is in still air, (b) in air being accelerated from 0 to $1 \, m \, s^{-1}$, (c) in air flowing at $1 \, m \, s^{-1}$, (d) in a decelerating air stream and (e) in still air again. Note the different amplitudes of the compound action potentials from the two groups of receptors and how they change for an acceleration step. The records do not show the beginning or end of the acceleration steps. (From Gewecke and Schlegel, 1970.)

decapod statocyst is best known and ranges from being a simple gravity organ in some species to a highly elaborated structure in which separate mechanisms for the detection of linear and angular accelerations are found.

In the Mysids, the statocysts are situated in the base of the uropods in the tail of the animal, and in the decapods they are found in the basal joint of the antennule. In its simplest form the statocyst is an invagination of the exoskeleton which may or may not remain open to the exterior. Where it is closed, the seal is effected by the close apposition of the exoskeleton, and so the statocyst is never a completely internal organ. At ecdysis, the seal is necessarily broken and the whole statocyst is renewed.

Within the lumen of a simple statocyst, such as that of the crayfish, a crescent of sensory hairs surrounds a statolith formed by the accretion of sand grains or calcareous granules (Fig. 12.11). The sensory hairs are each supplied with three sensory cells and are oriented in a particular direction. The articulation of the hair with the floor of the statocyst is arranged so that movement of the hair towards and away from the statolith produces a deformation of the dendritic processes of the nerve cells (Kinzig, 1918; Schöne and Steinbrecht, 1968). The floor of the statocyst is inclined to the horizontal so that when the animal is in its normal position on a horizontal surface, the statolith exerts a shearing force on the supporting sensory hairs. Rotation of the animal around its long axis (roll) will change the shearing force of the statolith upon the hairs by the sine of the angle of tilt. Thus the shear force will be minimal if the statolith is pressing directly down upon the hairs from above and maximal when the floor of the statocyst is vertical. Considerable information about the function of the statolith in crustaceans as a gravity receptor has been derived from changes in the animal's posture and from the attitude of its stalked eyes, following removal or interference with the statolith and its supporting hairs or complete removal of the entire statocyst. For shrimps it has been shown that the receptor cells generate impulses continuously. If the statolith of one side is removed the animal will roll towards the side without a statolith (Fig. 12.12b), and its new posture can be shown to result in a new balance between the input from the two statocysts, because the shearing force produced by the statolith on the intact side, when rolled uppermost, is now much less than before. Removal of both statoliths results in a normal posture

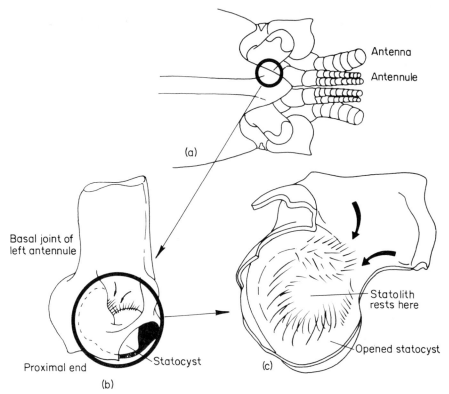

Fig. 12.11 The statocysts of the crayfish (*Cherax destructor*) contained within the basal joints of the antennules (a, b). The opening to the lumen is covered by a screen of hairs (b) and within the lumen (c) a crescent of sensory hairs have their tips directed inwards to enclose the statolith, which is made up of sand particles. (The statolith is not shown here.)

(Fig. 12.12c). Entire ablation of one statocyst will, predictably, result in the continued rolling of the animal toward the damaged side because there is now no position in which the inputs from the two sides can be balanced (Fig. 12.12c; Schöne, 1951).

Deductions drawn from the above types of behavioural observations have been amply verified by physiological studies of the output from the primary receptor cells in the statocysts of lobsters (Cohen, 1953, 1955, 1960; Cohen and Dijkgraaf, 1961). In these animals, several rows of statolith hairs surround a statolith which is made up of a collection of sand grains cemented together and flexibly joined to the statocyst floor. As predicted from the behavioural experiments, a continuous discharge can be recorded

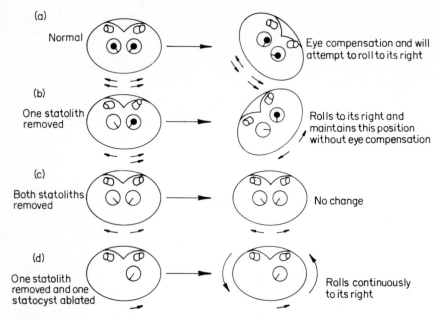

Fig. 12.12   The effect of removal of the statolith or ablation of the statocyst on the postural control of the shrimp. In the diagrams the short arrows show the separate inputs produced by the spontaneous discharge of the hair receptors, and that generated by the load of the statolith on the hairs. (a) If a normal animal is rolled to one side the balanced output of the two statocysts is altered, the eyes compensate for the displacement and, if allowed to, the animal will return to its normal position. (b) With one statolith removed an imbalance is created which is corrected if the animal rolls towards the lithless side. (c) If both statoliths are removed there is no imbalance, but if one statocyst is now ablated (d) the animal rolls continuously towards the injured side because it is unable to achieve a balance between the two sides (After Schöne, 1974.)

from the statolith hair receptors when the animal is held in its normal position. Rotation of the animal about the pitch and roll axes produces changes in the discharge rates of these receptors, but only if the statolith is present. Removal of the statolith does not, however, prevent the continuous discharge of the receptors.

Examination of the electrical responses of a number of single receptor units has shown that there are two types of position receptor. The first has a specific, non-adapting discharge frequency for each maintained position within a certain range about the pitch axis. Rotation of the animal by 180° about the pitch axis results in the frequency of the type I receptors reaching a peak and then declining in frequency. This is understandable considering the change in the shear force produced by the statolith in its different positions, but it means that if the animal were to rely solely on a single position receptor of this kind, there are two positions in space which are not distinguishable, because both positions will produce the same output.

The type II position receptors are also capable of signalling the absolute position of the animal relative to gravity, but they have a dynamic component in their response which provides information on the direction of initial displacement. If the animal is moving from a position at which it discharges at a low frequency, to a position where it discharges at a higher frequency, then the initial acceleration produces a transient excitation. Similarly, if the animal is moved in the opposite direction, the dynamic response is a transient decrease in the frequency of discharge. It is difficult to show that the information given by the nervous response is really used by the animal. The type II unlike the type I, receptors will signal position changes about the roll as well as the pitch axis.

Electrical stimulation of the central stumps of cut statocyst nerves of lobsters results in an increase in the discharge of the oculomotor neurons of the eyes, indicating a link between these receptors and eye movements. A more satisfactory and direct demonstration of the correlation between the direction of statolith movement and the movements of the eyes has recently been provided in the crayfish statocyst, where the hairs surrounding the statolith are polarized so that movement of a hair directly away from the centre of the statolith has the effect of maximally stimulating the receptors attached to it. Careful manipulation of small groups of hairs in an animal from which the statolith has been removed, produces eye movements which show that their direction of movement is determined by the polarization of the individual sensory hairs (Stein, 1975).

The function of the statolith in the Crustacea as a sensor for linear acceleration is therefore well understood. An interesting problem arises, however, when it is considered that the sensitive receptor is lodged, in the decapods, within the basal joints of the antennule, itself a movable appendage. The animal must be able to make allowances for voluntary movements of the antennules. The problem is the same as that experienced by the insects which need to monitor gravity using leg or body proprioceptors which are all activated by its own movements. In the lobster this problem has been studied by observing the position of the eyes while first tilting the whole animal about the pitch axis; secondly, tilting just the antennules, and thirdly, tilting the trunk of the animal while holding the antennules in a horizontal position. The three different situations are designed to activate, in the first case only the statocyst, in the second case, both the statocyst and the joint between the antennule and the animal's body, and in the third case, only the joint between the animal and the antennule (Fig. 12.13).

The result of the experiment is that eye movements are evoked in the lobster in the first and third situations, that is when only the statocyst or only the antennule joint is activated (Fig. 12.13). A search of the antennular joint reveals a proprioceptive system which acts antagonistically to the input from the statolith, provided the animal is the right side up (Schöne and Schöne, 1967). The absence of any eye movements in the second situation, where both the statocyst and the proprioceptive system are activated is, therefore, explained. The central mechanism which achieves the subtraction of these two inputs has not been found. No eye movements occur if the whole animal is tilted about the pitch axis after the statocysts have been removed, so the possible presence of gravity receptors in the trunk of the animal is excluded.

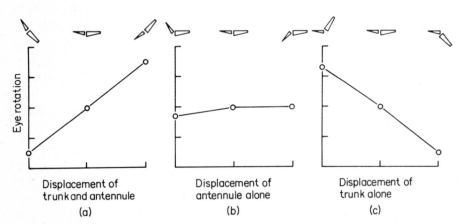

Fig. 12.13 The interaction between the statocyst and the antennule proprioceptor in the production of eye movements in the lobster. The graphs show the eye rotation (ordinate) plotted against the displacement of (a) the trunk and antennule, (b) only antennule and (c) only trunk. No eye rotation occurs when both the statocyst and the antennule base are stimulated (b) (After Schöne and Schöne, 1967.)

The statolith systems in the crayfish have been investigated mainly from the standpoint of static gravity perception. They must also be susceptible to linear accelerations of shorter duration and also to angular accelerations, provided these are directed horizontally in relation to the statocyst floor. The response of statolith systems to yawing of crayfish has apparently not been examined, although one may predict that such rotation would change the shear forces of the statolith on the underlying hairs, particularly as the statocysts in both mysids and decapods lie some distance from the centre of rotation of the turning animal.

12.4 Detection of angular accelerations

12.4.1 No specific sense organ

There is not very much information about the detection of angular accelerations in insects, but blinded bees and ants will compensate for rotations about their dorso-ventral axes and proprioceptors at the bases of the legs may be implicated in this response (Markl, 1975). There are almost certainly other insects which can use the same receptors with which they detect gravity to perceive angular accelerations. *Carausius* for example can make the distinction between linear and torsional forces applied to the proprioceptors of its legs (Wendler, 1971), but it is not known if the animal uses this ability to respond only to angular accelerations.

Antennal receptors are implicated in the detection of angular accelerations in the yaw direction in flies (Schneider, 1953) and in *Apis* (Heran, 1959).

A well known example of the detection of angular acceleration about the roll axis is that of dragonflies (Mittelstaedt, 1950). The large head of the dragonfly is loosely suspended on the prothorax and has a certain inertia. Because the articulation is near the top of the head, and the inertia of the head is larger than the frictional forces of this articulation, a quick roll of the body is not transmitted to the head which keeps its position relative to the horizontal. The misalignment between the head and the body is detected by several hair plates on the neck and the signals from these activate changes in the wing movements which act to realign the body with the head. The reflex alignment of the body also works if the head is turned to the side. Forced movements of the head relative to the body will not reliably produce postural changes of the wings so that the behaviour may be part of the system to ensure that the body follows the direction in which the animal points its head.

When considering the detection of angular accelerations, however, it is necessary to confine ourselves to those examples where a mechanism exists to exclude linear accelerations, at least partially, from the receptor system. There are two good examples of this, the halteres of flies and the circular canal systems of crabs.

12.4.2   Halteres

One of the most remarkable structures developed for the detection of angular accelerations in animals is the haltere of the dipterans. These are small club-shaped structures supported on slender stalks and derived from the hind wings. As sense organs they are probably able to detect more than just angular accelerations during flight, although we assume that this is their primary function (Faust, 1952; Schneider, 1953; Fraenkel and Pringle, 1938; Pringle, 1948).

A haltere in *Calliphora* is about 0.7 mm long, lies just behind the wing and is oscillated through an angle of about 150° in the vertical plane (Fig. 12.14). It has a hinge at its base to allow this movement and a single muscle which moves the haltere against the elasticity of the exoskeleton. In normal animals the oscillation frequency of the halteres is the same as the wings but their direction of movement is in antiphase with the wings. There is some kind of weak coupling between the halteres and wings which helps to keep them beating at the same frequency but this link is broken if the wing frequency is significantly altered by loading the wings, which decreases their wing beat frequency, or by trimming part of the wing off, which increases the wing beat frequency. In such circumstances the frequency of the halteres is not the same as the wings, and it will keep a frequency which is close to its normal value and to its natural mechanical resonance frequency. That the normal value is close to its resonance frequency can be demonstrated by altering the mechanical resonance of the haltere by squeezing the swollen club end. The result is a higher frequency of oscillation which returns to normal when the squeezed club end recovers its normal size, which it will do if the cuticle is not too badly damaged. There is evidence that the change in frequency in the above experiment is not caused by temporarily disabling sensory structures in the tip of the haltere (Sellke, 1936).

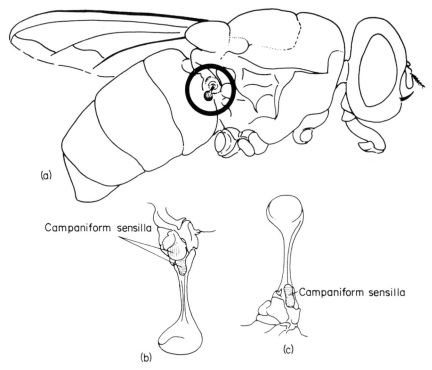

(a)

Companiform sensilla

(b)

Companiform sensilla

(c)

Fig. 12.14   The halteres of the fly *Lucilia cuprina.* (a) The right side wing and the membranous flaps covering the haltere (circled) have been removed to show the haltere lying in a small concavity formed by the abdomen and thorax. (b, c) show enlarged halteres seen from above (b) and below (c), and the positions are shown of rows of campaniform sensilla which are sensitive to the distortion of the cuticle at the base of the haltere.

The shaft of the haltere in *Calliphora* is oval-shaped in cross section and the long axis of the oval lies in the antero-posterior axis of the fly. This means that, while a certain flexibility is allowed in the vertical plane, which is the plane of its primary oscillation, the shaft is quite rigid in the horizontal plane. At the base of the haltere there is a second articulation which allows a small lateral movement of the shaft at right angles to the plane of its primary oscillation. When the halteres are oscillating they behave like a gyroscope, so that if the fly is subjected to an angular acceleration in the yaw plane, because this is at approximately right angles to the primary oscillation plane of the halteres, a torsional force will be set up in the shaft of the haltere which will tend to displace the structures laterally. This displacement is allowed by the secondary articulation.

There are patches of sensilla at the base of the haltere, just where the forces produced by the primary oscillations and imposed torques act on the cuticle, (Fig. 12.14b, c; Pflugstaedt, 1912). These are of the campaniform sensilla type which are known to respond directionally to a compression of the cuticle in which they lie. In *Calliphora* there are also two chordotonal organs, and this type of organ is known from other systems to be activated by extension. Knowing the exact positions of the sensilla at the base of the haltere and the most likely effect of the forces occurring in the haltere during primary oscillation and during imposed yaw, it is possible to deduce that the sensilla lie in the best positions to detect separately the vertical oscillations of the haltere and the strains produced by imposed yaw. There are also sensilla which could detect strains in directions other than those discussed. The above is confirmed by physiological recordings from the haltere nerve, showing that the separate receptor groups are highly selective to the movements of the halteres in different planes (Pringle, 1948). The receptors in the halteres are, like those in Johnston's organ, nearly all phasic, probably for the same reason that tonic receptors would be of less use in monitoring the output of a rapidly oscillating system. The halteres are therefore gyroscopic sense organs. Their mass is too small for them to directly impart stability to the whole fly in the yaw plane, but any movements in the yaw plane act to distort the haltere bases and thus provide information about the direction and extent of imposed yaw. Linear accelerations will not have the same effect as angular accelerations and so the system can distinguish between the two.

It has been suggested that the halteres are essentially organs for stabilizing the fly in the yaw plane and that the system is not able to interpret rolling movements. However, evidence has been produced recently showing that reflex head movements of the fly during rolling are mediated by the halteres (Tracy, 1975).

It is doubtful that flies rely entirely on the halteres for stability in flight. Stability of a haltereless fly can be restored to it by waxing a long thread to the tip of its abdomen (Fraenkel and Pringle, 1938), and flies with their heads waxed immovably to the thorax are unstable (Tracy, 1975). Stability probably depends upon the integration of the inputs at least from the eyes, the antennae, the halteres, and the hair plates which monitor the relative positions of the head, thorax and abdomen.

12.4.3 Semicircular canals

Not all the sensory hairs within the statocysts of some decapods are associated with the statolith. Instead, in some animals there are rows of long, thin hairs known as the thread hairs, which swing free in the fluid in the lumen of the statocyst. The thread hairs in the lobster are arranged around the edge of the statocyst, which is a flattened bowl shaped structure with the large statolith fixed in the centre. Electrical recordings from the axons of these thread hairs shows that they are spontaneously active and that displacement in one direction increases their rate of discharge, while displacement in the opposite direction decreases it. In spite of the hairs lying in a circular canal of sorts, recordings from the system suggest that while the hairs do respond to angular accelerations and not linear accelerations of short duration, they will not respond preferentially to angular acceleration about any specific axis (Cohen, 1960). Thus the system is apparently not so well adapted as the vertebrate labyrinth to this end. Such a refinement has, however, been achieved in the statocysts of the swimming crab *Scylla serrata*, although even here the system consists essentially of two canals, a horizontal and a vertical canal. The vertical canal is oriented at 45° to the transverse axis of the animal and is therefore sensitive to both pitch and roll movements.

The complexities of the crab statocyst were reported very early and the division of the lumen into a horizontal and vertical portion was described by Hensen (1863) and later by Sesar (1927). More detailed descriptions of the types of sensory hairs and the canal system is given by Dijkgraaf (1955, 1956) and Sandeman and Okajima (1972, 1973a, b).

The structure of the crab statocyst is decribed here with the emphasis placed heavily on its sensitivity to rotational acceleration rather than on its static sense, but this does not imply that its static sense is in any way less effective than that already described for the other decapods. The elaboration of the statocyst into canals allows the crab to detect and discriminate angular acceleration in all three axes, which is presumably essential to the animal when swimming freely or when executing the extremely rapid locomotory manoeuvres characteristic of many of the more terrestrial species.

The crab statocyst like all other decapod statocysts is formed from an invagination of the outer exoskeleton but it is then effectively sealed of from the outside by the close apposition of the outer exoskeletal lips of the invagination. Disturbances in the water surrounding the animal therefore cannot possibly directly effect the movement of the fluid enclosed in the statocyst. The elaboration of a spherical vesicle to produce two circular canals is easily achieved by compression, first transversely and then vertically (Sandeman, 1975). Carried to an extreme this results in two circular canals, one lying in the horizontal plane and one in the vertical plane. In *Scylla* the horizontal canal is a complete toroid and the vertical canal, which lies with its circumferential plane at 45° to the transverse axis, is almost joined across its centre by the swelling of the sensory cushion described by previous authors (Fig. 12.15c).

There are four kinds of hairs in the statocysts of crabs. In the extreme lateral part of the horizontal canal is a collection of stiff, large hairs, the group hairs, which lie up against the outer wall of the canal. No nerve supply has ever been shown for them and their function is unknown, although it has been pointed out that they will considerably increase the frictional resistance between the wall of the canal and the fluid in it, thus damping the fluid flow.

Along the posterior arm of the vertical canal there is a patch of hairs which are short and bent at 90° half way along their length. These are called the free hook hairs and they have fine filaments extending from their tips which probably float freely in the fluid of the canal (Fig. 12.15c).

At the bottom of the vertical canal there are two concentric circles of hairs which closely resemble the free hook hairs, except that they are oriented with their hooked ends pointing in towards the centre of the circle and the filaments are bound together with a small statolith which is attached to the floor of the statocyst within the circle of hairs. Like all other crustacean statoliths, when the animal is in its normal horizontal position, the floor of the statocyst beneath the statolith is inclined to the horizontal. The sensory cells of the inner circle of statolith hairs are like those of the free hook hairs but the outer hairs are characterized by having much larger nerve cell bodies at their bases. The significance of this difference is not clear, but it has been suggested that large sudden displacements of the statolith would activate the outer hairs which, with their large cells and perhaps fast conducting axons, could be part of a protective warning system (Sandeman and Okajima, 1973b).

The thread hairs, like those of the lobsters, are long and thin and have a feathered end, with filaments extending laterally from the

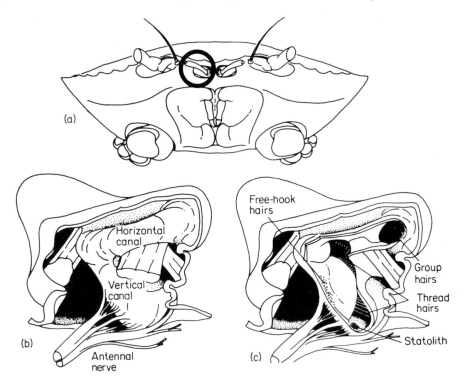

Fig. 12.15   (a) The statocysts of the crab are contained within the basal joints of the antennules (circled). (b, c) show the basal joint of the animals right antennule seen from behind. (b) the back of the basal joint has been removed to show the canals of the statocyst lying at right angles to each other. The nerve supply and some muscles which control the movement of the basal joint are also shown. (c) the canals are opened to show the positions of the receptor hairs, the statolith and the group hairs. Note that the horizontal canal is sealed across its centre to form a toroid. The vertical canal is formed by the dome of the sensory cushion, which carries the single line of thread hairs almost meeting the opposite wall of the statocyst at its centre.

shaft of the hair (Dunn, personal communication) and are arranged in a single row across the sensory cushion of the statocyst. They can be divided into an upper and a lower group of hairs, the upper hairs projecting across the common canal formed by the junction of the vertical and horizontal canals, and the lower group projecting across the lower part of the vertical canal. The lateral filaments of the thread hairs probably act to bind all the hairs of one group into a delicate curtain which, while not impeding the fluid flow in the canal (because the hairs bend so easily at their bases) nevertheless move with the fluid. Thus any displacement of the fluid would be expected to displace the thread hairs similarly.

Electrical recordings from the thread hair and free hook hair nerves show that both are sensitive to water currents. In all cases, units which respond to the hairs being bent in one direction are silent when the hairs are bent in the opposite direction. The receptor neurons of the thread hairs will produce a discharge to hair deflection in the preferred direction which is maintained for some tens of seconds before adapting. No cell has been found to be bidirectional in its sensitivity to deflection of the hair. At the base of each thread hair there are two sensory cells and these lie side by side and are attached to the base of the hair by scolopale organs (Dunn, in preparation). It is not yet known if these two cells have sensitivities to the different directions of hair displacement, but this is strongly suspected on anatomical grounds.

The canal structure of the crab statocyst, the position of the thread hair curtains, and the response of the thread hair neurons to displacement lead to the suggestion that the canal system of the crab functions in the same way as the vertebrate semicircular canals in detecting the angular accelerations about the pitch, roll, and yaw axes and in discriminating angular accelerations from linear accelerations. That the crab statocyst can fulfil these functions has been demonstrated by observing the responses of a primary interneuron which receives its input primarily from the lower group of thread hairs of one statocyst and therefore provides a remarkably convenient opportunity to examine in a single neuron the function of one part of one statocyst (Fraser, 1974a, b; Fraser and Sandeman, 1975). Predictably, the interneuron responds only to angular accelerations about the pitch axis. It is further specialized by responding to movements in the head-up direction only, implying that directional cells of only one kind are activating the interneuron.

The ability of the system to respond to angular accelerations and not to linear accelerations was demonstrated by placing the animal on a swing which could provide short duration linear accelerations by moving back and forth on parallel arms, or angular accelerations, by oscillating about a fixed axis which passed effectively through the centre of the vertical canals. The interneuron responds only to angular accelerations. The discharge frequency of the interneuron at oscillation frequencies ranging from 0.5 Hz to 2 Hz, leads the peak of position by $90°$. Thus, the canal system is apparently behaving as though it were critically damped for this type of stimulus, and the information contained in the interneurons is coded as angular velocity.

The directional sensitivity of the vertical canal system was also demonstrated by observing the responses of this interneuron. As mentioned above, the vertical canal lies at 45° to the true pitch axis of the animal. Thus, if the output of a single vertical canal could be measured, it would be expected to increase as the animal was turned about the yaw axis and set in a position where the plane of the canal lay exactly along the plane of the angular acceleration applied to it. Also, if the animal is positioned so that the angular accelerations are at right angles to the plane of the canal, there should be little or no response. All these predictions were confirmed when the animal was tested on the swing, and show that the conditions for the canal type of angular accelerometer are met by the crab statocyst, at least for sinusoidal oscillations from 0.5 Hz to 2 Hz (Fig. 12.16).

Early workers ascribed a hearing function to the statocyst and, although it is doubtful whether the hairs in the statocyst can detect airborne sound, they are very sensitive to vibrations of the substrate. It is now known that both the thread hairs and the free hook hairs are sensitive to vibrations and to rapid angular accelerations of the bases of the antennules (Fraser, in preparation).

Finally, although the prime function of the thread hairs as angular acceleration detectors has been amply demonstrated, they are also affected by the constant linear acceleration due to gravity. Thread hairs which have been isolated from the statocyst and mounted in saline can be seen to fall over under the effect of gravity if their position is changed (unpublished observations). Similarly, recordings from the sensory interneuron show a significant effect on the dynamic response of the hairs in the vertical position if the animal is oscillated about a mean position out of the horizontal. For example, animals oscillated about a 90° head up or a 90° head down position, receive very little information about angular accelerations from the thread hairs in the vertical canal, presumably because they have been displaced from their normal working range by gravity. It is perhaps no surprise to find that the animal will compensate for imposed pitch movements by opposite rotational movements of the antennule bases, thus keeping them in a more normal position. Whether the input from the thread hairs mediates this response is not known but, like the similar reflex described for the antenna pointing of the fly

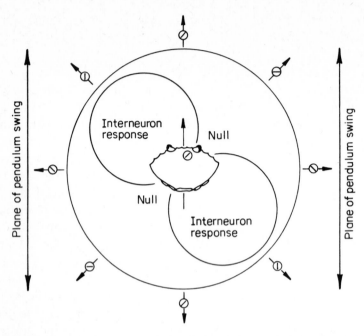

Fig. 12.16 The sensitivity to pitch of the right vertical canal of the crab statocyst. The response of the head-up interneuron is recorded from the intact crab, which is placed in a pendulum swing having its plane of oscillation aligned initially with the longitudinal axis of the animal (shown by the two large lateral arrows in the diagram). The crab is then set in different positions around the yaw axis so that the right vertical canal lies at different angles to the plane of the pendulum, and the output of the interneuron is recorded for a set number of oscillations of a constant amplitude at each position. In the diagram, the circles containing the bar show the orientation of the vertical canal of the right statocyst and the arrows show the direction in which the crab is pointing. The response of the interneuron follows very closely a function of the cosine of the angle which the canal makes with the plane of the pendulum, so that the greatest output of the interneuron corresponds with those points around the yaw axis where the vertical canal lies in the same plane as the pendulum swing. There is no response when it lies across the plane of the pendulum.

and the maintenance of position of the ant's abdomen, it is likely that other proprioceptive systems are carefully monitoring antennule position and a feedback control helps to keep the antennule in the operating range of the receptors. It should be added, however, that compensatory movements of the antennules in *Scylla* are small and the responses from statocyst interneurons when crabs are oscillated about the pitch axis are not noticeably different whether the antennule bases are fixed or free to move.

12.5   Conclusions

Perhaps the most striking thing about linear and angular acceleration detectors is their similarity, even down to fine details of structure. The statolith principle, for example, is exploited widely and must have evolved separately in many animals, probably starting from simple mechanoreceptors, sensitive to displacement of water or air currents, progressing to the enclosed statcoyst (Horridge, 1971). The detection of angular accelerations to the exclusion of linear accelerations seems to have required a higher degree of specialization, but again a single principle, that of the circular fluid-filled canal, has been exploited by quite different groups of animals. In the crustaceans it is even possible to see that the canal system has evolved to different extents; in the lobster as an angular acceleration detector with no apparent directionality and in the crab where directionality is precisely measured.

In all the systems, however, there is a significant overlap, not only with the sensitivities of other receptor systems, but of the capabilities of the different receptor hair types within a single organ. Thus the thread hairs do respond also to gravity, and it will certainly be found that statoliths can be stimulated by angular accelerations in certain planes. The overlap of the sensory systems presents an interesting problem for those interested in the action of the nervous system in producing the appropriate orientation behaviour and we must assume that at least two factors operate here. Firstly, the animal is interested only in the input of its receptors over a certain range and the range is exceeded usually only in situations which demand that the animal make use of some escape behaviour, in which normal orientation behaviours are often overridden. Secondly, there is probably a careful line-labelling of the sensory pathways from the receptors to the second order cells. For example, only the thread hair receptor neurons of one specific group of hairs sensitive to one direction of fluid flow in one canal, are connected to the head-up interneuron. Similar line-labelling will probably be found for the receptors of the halteres and Johnston's organs. In spite of these devices, however, the animal must still be able to interpret the input from the receptors in the light of what it is doing itself, and exclude sensory input irrelevant to its orientation. The realization of even these low level complexities in the nervous system makes the analysis of behaviour at the level of the single cell a challenging affair. Nevertheless it is in the field of orientation behaviour that many of the best opportunities lie for precisely this approach to the nervous system. Orientation behaviour can, in spite of the above qualifications, be reliably evoked by the activation of a limited number of primary receptors, and is often repeatable and obligatory so that, while being potentially more interesting than reflexes, it is still within the realm of a cell by cell analysis.

References

Adler, H. E. (1970) Ontogeny and Phylogeny of orientation. In: *Development and Evolution of Behaviour.* Essays in Memory of T. C. Schneirla. Aronson, L. R., Tobach, E., Lehrman, D. S. and Rosenblatt, J. S. (eds), pp. 303—336. W. H. Freeman, San Francisco.

Adler, H. E. (1971) Orientation in animals and man. In: *Orientation: Sensory Basis. Annals of the New York Academy of Sciences*, **188**, 3—4.

Bässler, U. (1965) Propriozeptoren am Subcoxal-und am Femur-Tibia-Gelenk der Stabheuschrecke *Carausius morosus* und ihre Rolle bei der Wahrnehmung der Schwerkraftrichtung. *Kybernetik,* **2**, 168—193.

Baunacke, W. (1912) Statische Sinnesorgane bei den Nepiden. *Zoologische Jahrbucher (Anatomie)*, **34**, 179—346.

Bückmann, D. (1955) Zur Frage der Funktion der Insekten fühler als Schweresinnesorgane. *Naturwissenschaften*, **42**, 79.

Camhi, J. M. (1969a) Locust wind receptors I. Transducer mechanics and sensory response. *Journal of Experimental Biology*, **50**, 335—348.

Camhi. J. M. (1969b) Locust wind receptors II. Interneurones in the cervical connective. *Journal of Experimental Biology*, **50**, 349—362.

Camhi, J. M. (1969c) Locust wind receptors III. Contribution to flight initiation and lift control. *Journal of Experimental Biology*, **50**, 363—374.

Camhi, J. M. and Hinkle, M. (1973) Response modification by the central flight oscillator of locusts. *Journal of Experimental Biology*, **60**, 477—492.

Cohen, M. J. (1953) Oscillographic analysis of an invertebrate equilibrium organ. *Biological Bulletin, Wood's Hole*, **105**, 363.

Cohen, M. J. (1955) The function of receptors in the statocyst of the lobster *Homarus americanus. Journal of Physiology*, **130**, 9—34.

Cohen, M. J. (1960) The response patterns of single receptors in the crustacean statocyst. *Proceedings of the Royal Society Series B.,* **152**, 30—49.

Cohen, M. J. and Dijkgraaf, S. (1961) Mechanoreception. In: *The Physiology of the Crustacea* Vol II. Waterman, T. H. (ed), Academic Press, New York.

Dijkgraaf, S. (1955) Rotationssinn nach dem Bogengangsprinzip bei Crustaceen. *Experientia*, **11**, 407.

Dijkgraaf, S. (1956) Structure and functions of the statocyst in crabs. *Experientia*, **12**, 394.

Faust, R. (1952) Untersuchungen zum Halterenproblem. *Zoologische Jahrbücher, Abteilung für Allgemeine Zoologie und Physiologie der Tiere,* **63,** 326—366.

Fraenkel, G. and Pringle, J. W. S. (1938) Halteres of flies as gyroscopic organs of equilibrum. *Nature,* **141,** 919.

Fraser, P. J. (1974a) Interneurons in crab connectives (*Carcinus maenas* (L.)): giant fibres. *Journal of Experimental Biology,* **61,** 593—614.

Fraser, P. J. (1974b) Interneurons in crab connectives (*Carcinus maenas* (L.)): directional statocyst fibres. *Journal of Experimental Biology,* 61, 615.

Fraser, P. J. (1975) Integration by crab semicircular canal interneurons. Statocyst input as shown by antennule rotation in *Scylla serrata.* (In preparation).

Fraser, P. J. and Sandeman, D. C. (1975) Effects of angular and linear accelerations on semicircular canal interneurons of the crab *Scylla serrata. Journal of Comparative Physiology,* 96, 205—221.

Gewecke, M. (1967a) Der Bewegungsapparat der Antennen von *Calliphora erythrocephala. Zeitschrift für Morphologie und Ökologie der Tiere,* **59,** 95—133.

Gewecke, M. (1967b) Die Wirkung von Luftstromüng auf die Antennen und das Flugverhalten der Blauen Schmeissfliege (*Calliphora erythrocephala*). *Zeitschrift für Vergleichende Physiologie,* 54, 121—164.

Gewecke, M. (1972) Die Regelung der Fluggeschwindigkeit bei Heuschrecken Flugverhalten der Blauen Schmeissfliege (*Calliphora erythrocephala*). *Zeitschrift für Vergleichende Physiologie,* 54, 121—164.

Gewecke, M. (1972) Die Regelung der Fluggeschwindigkeit bei Heuschrecken und ihre Bedeuting für die Wanderflüge. *Verhandeungsbericht der Deutschen Zoologischen Gesellschaft,* **65,** 247—250.

Gewecke, M. (1974) The antennae of insects as air current sense organs and their relationship to the control of flight. In: *Experimental Analysis of Insect Behaviour.* Ed. L. Barton Browne. Springer-Verlag, Berlin.

Gewecke, M. and Schlegel, P. (1970) Die Schwingungen der Antenne und ihre Bedeutung für die Flugsteuerung bei *Calliphora erythrocephala. Zeitschrift für Vergleichende Physiologie,* 67, 325—362.

Hamilton, M. A. (1931) The morphology of the water scorpion, *Nepa cinerea* L. (Rhynchota-Heteroptera). *Proceedings of the Zoological Society of London.* pp. 1067—1136.

Hensen, V. (1863) Studien über das Gehörorgan der Decapoden. *Zeitschrift für wissenschaftliche Zoologie,* **13,** 319—412.

Heran, H. (1959) Wahrnehmung und Regelung der Flug eigengeschwindigkeit bei *Apis mellifica* L. *Zeitschrift für vergleichende Physiologie,* 42, 103—163.

Horridge, G. A. (1971) Primitive examples of gravity receptors and their evolution. In: *Gravity and the organism.* Gordon, S. A. and Cohen, M. J. (eds), University of Chicago Press, Chicago and London.

Kinzig, H. (1918) Untersuchungen uber den Bau der statocysten einiger dekapoder Crustaceen. *Verhandlung der naturhist—medizinischen Vereins Heidelberg (N.F.),* **14,** 1—90.

Lindauer, M. and Nedel, J. O. (1959) Ein Schweresinnes organ der Honigbeine. *Zeitschrift für vergleichende Physiologie*, 42, 334–364.

Markl, H. (1962) Borstenfelder an den Gelenken als Schweresinnesorgane bei Ameisen und anderen Hymenopteran. *Zeitschrift für Vergleichende Physiologie*, 45, 475–569.

Markl, H. (1963) Die Schweresinnesorgane der Isekten. *Naturwissenschaften*, 50, 559–565.

Markl, H. (1971) Proprioceptive Gravity Perception in Hymenoptera. In: *Gravity and the organism.* Gordon, S. A. and Cohen, M. J. (eds), University of Chicago Press, Chicago.

Markl, H. (1974) Insect Behaviour: Functions and Mechanisms. In: *Physiology of the Insecta.* 2nd Edition. Vol. III. Rockstein, M. (ed), Academic Press, New York.

Markl, H. (1975) The perception of gravity and of angular acceleration in invertebrates. In: *Handbook of Sensory Physiology.* Vol. VI/1, pp. 17–74. Springer Verlag, Berlin.

Melvill Jones, G. (1971) Organization of neural control in the vestibulo-ocular reflex arc. In: *The control of eye movements.* Bachy-Rita, P., Collins, C. C. and Hyde, J. E. (eds), Academic Press, New York.

Mittelstaedt, H. (1950) Physiologie des Gleichgewichtssinnes bei fliegenden Libellen. *Zeitschrift für vergleichende Physiologie*, 32, 422–463.

Pflugstaedt, H. (1912) Die Halteren der Dipteran. *Zeitschrift für wissenschaftliche Zoologie*, 100, 1–59.

Pringle, J. W. S. (1948) The gyroscopic mechanism of the halteres of Diptera. *Philosophical Transactions of the Royal Society Series B.*, 233, 347–384.

Sandeman, D. C. (1963) Proprioceptor organs in the antennules of *Squilla mantis. Nature*, 201, 402–403.

Sandeman, D. C. (1975) Dynamic receptors in the statocysts of crabs. In: *Mechanisms of Spatial Perception and Orientation as Related to Gravity.* Schöne, H. (ed), *Fortschritte der Zoologie*, 23, Gustav Fischer Verlag, Stuttgart.

Sandeman, D. C. and Okajima, A. (1972) Statocyst-induced eye movements in the crab *Scylla serrata* I. The sensory input from the statocysts. *Journal of Experimental Biology*, 57, 187–204.

Sandeman, D. C. and Okajima, A. (1973a) Statocyst induced eye movements in the crab *Scylla serrata* II. The responses of the eye muscles. *Journal of Experimental Biology*, 58, 197–212.

Sandeman, D. C. and Okajima, A. (1973b) Statocyst induced eye movements in the crab *Scylla serrata* III. The anatomical projections of sensory and motor neurones and the responses of the motor neurones. *Journal of Experimental Biology*, 59, 17–38.

Schaller, F. (1953) Verhaltens–und sinnesphysiologische Beobachtungen an *Squilla mantis. Zeitschrift für Tierpsychologie*, 10, 1–12.

Schneider, G. (1953) Die Halteren der Schmeissfliege (*Calliphora*) als Sinnesorgane und als mechanische Flugstabilisatoren. *Zeitschrift für vergleichende Physiologie*, **35**, 416–458.

Schöne, H. (1951) Die statische Gleichgewichtsorientierung dekapoder Crustaceen. *Verhandlungsbericht der Deutschen Zoologischen Gesellschaft*, **16**, 157–162.

Schöne, H. (1974) Spatial Orientation in Animals. In: *Marine Ecology* Vol. II. Kinne, O. E. (ed), John Wiley, London.

Schöne, H. and Schöne, H. (1967) Integrated function of statocyst and antennular proprioceptive organ in the Spiny lobster. *Naturwissenschaften*, **54**,S. 289.

Schöne, H. and Steinbrecht, R. A. (1968) Fine structure of statocyst receptor of *Astacus fluviatilis*. *Nature*, **220**, 184–186.

Sellke, K. (1936) Biologische und morphologische studien an Schädlichen Wiesenschnaken (Tipulidae, Dipt.). *Zeitschrift für wissenschaftliche Zoologie*, **148**, 465–555.

Sesar, M. (1927) Die statocysten der Brachyuren. *Thesis*, Munich.

Stein, A. (1975) Attainment of positional information in the crayfish statocyst. In: *Mechanisms of Spatial Perception and Orientation as Related to Gravity*. Schöne, H. (ed), *Fortschritte der Zoologie* **23**. Gustav Fischer Verlag, Stuttgart.

Thorpe, W. H. and Crisp, D. J. (1947) Studies on plastron respiration. III. The orientation responses of *Aphelocheirus* (Hemiptera, Aphelocheiridae (Naucoridae)) in relation to plastron respiration; together with an account of specialized pressure receptors in aquatic insects. *Journal of Experimental Biology*, **24**, 310–328.

Tracy, D. (1975) Head movement mediated by Halteres in the fly, *Musca domestica* during inertially guided flight. *Experientia*, **31**, 44–45.

Vowles, D. M. (1954) The orientation of ants. II. Orientation to light, gravity and polarized light. *Journal of Experimental Biology*, **31**, 356–375.

Waloff, Z. (1972) Orientation of flying locusts, *Schistocerca gregaria* (Forsk), in migrating swarms. *Bulletin of Entomological Research*, **62**, 1–72.

Wendler, G. (1965) Uber den Anteil der Antennen an der Schwererezeption der Stabheuschrecke *Carausius morosus*. *Zeitschrift für vergleichende Physiologie*, **51**, 60–66.

Wendler, G. (1971) Gravity orientation in insects: The role of different mechanoreceptors. In: *Gravity and the Organism*. Gordon, S. A. and Cohen, M. J. (eds), University of Chicago Press, Chicago.

13 Equilibrium receptor systems in molluscs

B.-U. BUDELMANN

13.1   Introduction

Equilibrium receptor systems are composed basically of two elements: one, a mass whose instantaneous position depends on gravito-inertial forces; the other, a sensory area which is affected by the position of the mass. Equilibrium receptor systems provide an organism with direct information regarding its attitude in space. Combined as a rule with data from other proprioceptive, visual, and tactile systems, this information enables the animal to control its motor activities relative to gravity.

In vertebrates, equilibrium receptor systems are commonly known as the vestibular end organs. In invertebrates, they are known as statocysts (formerly otocysts). Interest in their structure and function has been renewed in recent years, especially in the molluscs (for reviews see: Barber, 1968a; Vinnikov, Gasenko, Titova, Bronstein, Tsirulis, Revznier, Govardovskii, Gribakin, Aronova and Tchekhonadskii, 1971; Wolff, 1973a; Budelmann, 1975; Markl, 1974; Vinnikov, 1974; Wolff, 1975). Statocysts occur throughout the subphylum Conchifera (i.e. the classes: Monoplacophora, Gastropoda, Scaphopoda, Bivalvia and Cephalopoda), with the exception of some sedentary or passively floating species of

Gastropoda and Bivalvia (*Vermetus, Janthinia* and *Pinna nobilis*, cp. Plate, 1924). Statocysts are either absent or at most rudimentary in the subphylum Amphineura (cp. Morton, 1958).

In molluscs, two statocysts are present, each situated near the pedal ganglion. They develop from ectodermal invaginations and range from relatively simple to highly differentiated structures with receptor systems for the detection of linear acceleration (gravity) as well as angular acceleration. Some of the basic concepts concerning linear and angular acceleration are dealt with in Section 12.2.

13.2   Gravity receptor systems

Gravity receptor systems (i.e. linear acceleration receptor systems or statolith organs) are composed of sensory elements and a mass to stimulate them by its mere weight. In molluscs the sensory elements are hair cells, which form part of an epithelium with their hairs (cilia) protruding into a fluid-filled cavity. Attached to the cilia is the stimulating mass. Its specific gravity is greater than that of the medium surrounding it. The position of this system relative to gravity defines the stimulus situation (for details see Section 13.1.2). Central nervous analysis of the sensory output has yielded quantitative information regarding the position of the animal in space.

13.2.1   Morphology

In molluscs, the gravity receptor systems either involve the entire statocyst (in nautiloid cephalopods and in all Conchifera), or only part of it (in octopod and decapod cephalopods). The sensory epithelia are composed of hair cells, which are usually surrounded by supporting cells. The hair cells are considered to be primary receptors, although the possibility of some secondary cells (cp. Klein, 1932) has not been ruled out. The hair cells bear numerous kinocilia and microvilli at their distal end, but no stereocilia. Each kinocilium is morphologically (and physiologically (Section 13.1.2)) polarized by the orientation of its internal 9 x 2 + 2 filaments as well as by the orientation of its basal foot. The number of kinocilia, as well as the number, the size and the polarization pattern of the hair cells differ in the various mollusc groups.

Monoplacophora

In *Neopilina* each of the two statocysts lies between the lateral and pedal nerve cord, near the cerebral ganglion. It is a flattened vesicle (90 x 215 μm in diameter) connected with the pallial groove by a duct. The sense cells are probably situated in the ventral wall only, which is the region from which most of the nerve fibres originate. The nerve joins the second latero-pedal connective and passes out to the lateral nerve cord (Lemche and Wingstrand, 1959).

Gastropoda

The statocysts of gastropods are morphologically well known. For details see: *Lightmicroscopy*—Boll, 1869; Leydig, 1871; Lacaze-Duthiers, 1858; von Ihering, 1876; Solger, 1899; Tschachotin, 1908; Schmidt, 1912; von Buddenbrock, 1915; Pfeil, 1922; Plate 1924; Baecker, 1932; Ulrich, 1942; Dijkgraaf and Hessels, 1969; Detwiler and Alkon, 1973: *Electronmicroscopy*—Quattrini, 1967; Geuze, 1968; Laverack, 1968; Barber and Dilly, 1969; Coggeshall, 1969; Wolff, 1969; Vinnikov *et al.*, 1971; Stahlschmidt and Wolff, 1972; Tsirulis, 1973; Wolff, 1973a; McKee and Wiederhold; 1974.

The statocysts are fluid-filled spheres which sometimes open through a canal, which is thought to represent the embryonic invagination (Schmidt, 1912; Plate, 1924; Geuze, 1968; Dijkgraaf and Hessels, 1969). The statocysts contain either a single statolith or a mass of statoconia; both are freely moveable and smaller in diameter than the cavity of the statocyst (Fig. 13.1a—c). The statoconia (up to 1000; Dijkgraaf and Hessels, 1969; McKee and Wiederhold, 1974) either act as a single mass, or move independently of one another. In *Pomacea* the number as well as the maximum size of the statoconia increases with time; the size being related to the size of the snail rather than to its age (McClary, 1968). Statoliths and statoconia often show a concentric lamellate structure (Fig. 13.1b) and are probably formed by a crystallization process within the fluid of the statocyst (Tschachotin, 1908; Schmidt, 1912; Pfeil, 1922; McClary, 1963). In *Lymnaea* (Geuze, 1968) and *Clione* (Tsirulis, 1973), however, it was clearly demonstrated that the statoconia are formed by the supporting cells.

In pulmonates and opisthobranchs the statocysts vary in diameter from 80 to 200 μm and generally are lined with thirteen large hair cells of up to 60 μm in diameter. The hair cells bear 500 to 700 kinocilia (Fig. 13.1d) (for references see Wolff, 1973a). In some

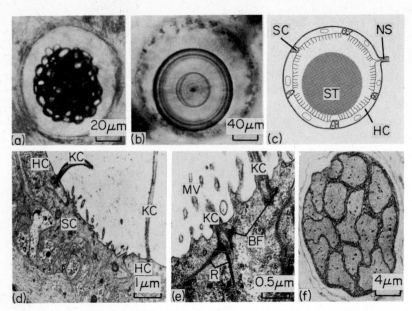

Fig. 13.1   Structure of the gravity receptor systems of gastropods. (a) Photomicrograph of the statocyst of *Hermissenda* (*in vivo* preparation). (b) Section through the statocyst of *Littorina littorea*. (c) Schematic drawing of the statocyst of a pulmonate gastropod. (d) Part of the statocyst wall of *Limax flavus*. (e) Longitudinal section through the base of two kinocilia of a hair cell of the statocyst of *Pomacea paludosa*. (f) Cross section through the statocyst nerve of *Aplysia californica*. BF—basal foot; HC—hair cell; KC—kinocilia; MV—microvilli; NS—Nervus staticus; R—root; SC—supporting cell; ST—statolith (or statoconia). ((a) from Detwiler and Alkon, 1973; (b) courtesy of H.-U. Tüllner; (d) courtesy of H. G. Wolff; (e) from Stahlschmidt and Wolff, 1972; (f) from Coggeshall, 1969.)

prosobranchs, the spheres are larger in diameter (250 to 500 μm) and are lined with several hundred (or up to a few thousand) small hair cells (6 to 9 μm in diameter), which each bear 30 to 40 kinocilia (Stahlschmidt and Wolff, 1972). The hair cells are distributed within the sensory epithelium in a relatively uniform way; only in *Pterotrachea* and some other heteropods are they concentrated into a limited 'macula'-region (Solger, 1899; Tschachotin, 1908; Barber and Dilly, 1969). In several species the kinocilia have been found to be motile (Schmidt, 1912; Geuze, 1968; Laverack, 1968; Dijkgraaf and Hessels, 1969; Wolff, 1969; Detwiler and Alkon, 1973). They are 0.16 to 0.20 μm in diameter and their length varies from 5 to 8 μm (Geuze, 1968; Laverack, 1968; Wolff, 1969), but cilia of up to

20 μm in length have also been described (Stahlschmidt and Wolff, 1972; McKee and Wiederhold, 1974). In pulmonates, the cilia of each hair cell are arranged in more or less random rings with their basal feet (Fig. 13.1e) pointing away from their centre on the cell surface radially (Geuze, 1968; Wolff, 1969, 1973a). However, in some species they may point towards the centre (Wolff, 1975). In the prosobranch *Pterotrachea* basal feet do not occur. Nevertheless, a radial polarization pattern of the kinocilia of each single cell becomes evident through the orientation of the central filaments in the 9 x 2 + 2 arrangement (Barber and Dilly, 1969). However, the available data do not show whether the direction of polarization points away from or towards their centre on the cell surface. By way of contrast, in the prosobranch *Pomacea* the basal feet of the cilia of each single cell and also of those adjacent, all point in the same direction (Fig. 13.1e) (Stahlschmidt and Wolff, 1972).

Since gastropod hair cells are primary receptors, each is represented by a single axon in the statocyst nerve and thus thirteen afferent axons are normally present in opisthobranchs and pulmonates (Coggeshall, 1969; Wolff, 1969; Wood and von Baumgarten, 1972) (Fig. 13.1f). In the prosobranchs *Pomacea* and *Littorina*, the number of axons within the statocyst nerve is apparently greater than the number of hair cells and this is taken as an indication of an efferent innervation. Thus *Pomacea* has 2500—3000 hair cells, but more than 3000 axons (Stahlschmidt and Wolff, 1972) while *Littorina* has about 220 hair cells and 340 axons (Tüllner, personal communication). In pulmonates, a bundle of small axons within, or close to, the statocyst nerve may also be interpreted as efferent (Wolff, 1969, 1975).

In gastropods, the statocyst nerve generally runs more or less directly to the cerebral-ganglion (cp. Wolff, 1973a). In *Hermissenda*, iontophoretic intracellular injection of fluorescent dye has shown that the axons enter the cerebro-pleural ganglion, then partly branch there in the neuropil region and send a branch across to the lateral border of the contralateral cerebro-pleural ganglion (Detwiler and Alkon, 1973).

Scaphopoda

In *Dentalium* the two statocysts are located near the pedal ganglia. They are nearly spherical in form and are lined with an epithelium, probably consisting only of hair cells. The cilia are arranged in rings.

The paired statocyst nerves run directly to the cerebral ganglion (Lacaze-Duthiers, 1858; Plate, 1924).

Bivalvia

The structure of the bivalve statocysts is generally similar to that described for gastropods. Structural details are given by von Ihering (1876); von Buddenbrock (1915); Plate (1924); Barber and Dilly (1969) and Vinnikov *et al.* (1971).

The statocysts are fluid-filled spheres (30 to 500 μm in diameter) which are lined with a sensory epithelium containing hair and supporting cells and contain either a single statolith or a number of statoconia. In some species (often primitive ones) the sphere remains open to the exterior by a canal and the statoconia are thought to be grains of external material (sand, etc.) (cp. von Buddenbrock, 1915; Plate, 1924).

In *Pecten*, differences in structure were found between the left (upper) and right (lower) statocyst. Thus the size of the granular statolith of the left statocyst is larger than that of the statoconial mass of the right one. In addition, the left statocyst has two sorts of hair cells, one with longer and one with shorter cilia; whereas the right statocyst has only one type (von Buddenbrock, 1915). At the ultrastructural level, in the left statocyst hair cells with about 30 kinocilia (polarized in the same direction; Type A) and hair cells with 9–10 kinocilia (polarized radially towards their centre on the cell surface; Type B) have been described; whereas in the right statocyst only those of Type B occur (Barber and Dilly, 1969). In *Pecten jessoensis* only Type B hair cells are present (Vinnikov *et al.*, 1971). The static nerve runs more or less directly to the cerebro-pleural ganglion, although in a variety of species it enters the CNS at the level of the pedal ganglion (cp. Plate, 1924).

Cephalopoda

Within the class of cephalopod molluscs great variability of statocyst structure is evident, ranging from the comparatively simple systems in nautiloids to the highly differentiated systems in octopods and decapods. Within each cephalopod order, however, the structure is quite uniform.

(a) Nautiloidea   The two statocysts of *Nautilus* are embedded in the CNS and are comparable in structure with those of the prosobranch

gastropods. They each consist of an oval sac (2 x 1.5 mm), the cavity of which communicates with the exterior by a canal (Kölliker's canal). The cavity is lined with a sensory epithelium, which is composed of a large number of hair-and supporting cells, and it is filled almost completely with calcareous statoconia (Young, 1965a).

The static nerve enters the magnocellular lobe as several bundles. Some of the fibres turn medially into that lobe and from there apparently run to the posterior suboesophageal cord; others proceed dorsally into the lateral edge of the lobe and enter the central part of the supraoesophageal cord near the cerebral fissure (Young, 1965a).

(b) Octopoda The statocysts of octopods are morphologically the best known of all. For detailed information see: *Lightmicroscopy—* Owsjannikov and Kowalewsky, 1867; Boll, 1869; Hamlyn-Harris, 1903; Ishikawa, 1924; Young, 1960. *Transmission electron microscopy—*Barber, 1965, 1966a, 1966b; Vinnikov, Gasenko, Bronstein, Tsirulis, Ivanov and Pyatkina, 1967; Barber 1968a, b; Vinnikov *et al.*, 1971. *Scanning electron microscopy—*Barber and Boyde, 1968; Budelmann, Barber and West, 1973.

In *Octopus vulgaris*, each of the two statocysts is situated in a cavity within the cartilaginous cranium, directly below the suboesophageal mass of the CNS. The static sac is thin and membraneous. It is about 4 mm in diameter and does not fill the 5—6 mm cavity. Thus it separates endolymph and perilymph (Fig. 13.2). The static sac is attached only at its dorsal end, but is kept firmly in position by numerous strands containing blood vessels (Fig. 13.2). The sac has two main sensory regions: the crista/cupula system, which is primarily for the detection of angular acceleration (Section 13.2), and the macula with a statolith attached to it for the detection of linear acceleration. In addition, a richly innervated posterior sac of unknown function; Kölliker's canal, which is probably involved in endolymph secretion (Amoore, Rodgers and Young, 1959), and a cartilaginous anticrista lobe have been described (Young, 1960). A comparable gross morphology is present in the statocysts of other octopod species (Ishikawa, 1924).

The macula is an oval-shaped plate of cells with an area of about 0.58 mm^2. It is composed of about 5100 hair cells, each of which is surrounded by supporting cells. The hair cells are assumed to be primary receptor cells and each bears up to 200 kinocilia, about 6 (to 10) μm in length and 0.24 μm in diameter (Fig. 13.3a, b). The

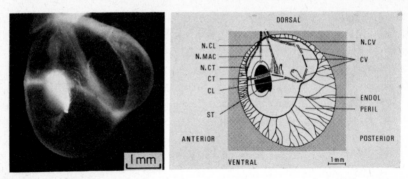

Fig. 13.2   Statocyst of octopod cephalopods. Lateral view of the left statocyst of *Octopus vulgaris*. (a) Photomicrograph showing the statocyst sac. (b) Schematic drawing showing the whole statocyst. CL—Crista longitudinalis; CT—Crista transversalis; CV—Crista verticalis; ENDOL—endolymph; N.CL—N. crista lateralis, N.CT—N. crista transversalis; N.CV—N. crista verticalis; N.MAC—N. maculae; PERIL—perilymph; ST—statolith.

kinocilia are arranged to form an elongated ciliary group (cp. Fig. 13.5a). These groups are arranged in concentric rings. The basal feet of the kinocilia project at right angles to the long axis of the elongated ciliary group, and point radially away from the centre of the macula. The polarization pattern for the whole macula has been determined morphologically (Barber, 1966b, 1968a; Budelmann *et al.*, 1973; Fig. 13.3c—f). Large neurons are located below the hair cell epithelium (Young, 1960; Barber, 1966a; Vinnikov *et al.*, 1967) (Fig. 13.3a) and they number about 2000 and are uni-, bi- and multipolar, as shown by cobalt-chloride staining techniques (Budelmann and Wolff, unpublished observations). In the normal position of the animal each macula is oriented vertically at the medio-frontal sector of the statocyst sac. The macula planes of the right and left statocyst subtend an angle of 90° which is open to the front (Budelmann, 1970a).

The macula nerve is composed of about 9200 fibres which vary in diameter from 18 μm to less than 0.2 μm (Budelmann *et al.*, 1973; Fig. 13.3g). Under the supposition that the hair cells are primary receptors, about 5100 fibres may represent the afference from the hair cells, and about 2000 belong to the large neurons below the epithelium. The remaining 2100 are very likely efferent. Further evidence for efferent innervation can be adduced from lightmicroscopy, histochemistry and electron microscopy (Young, 1960;

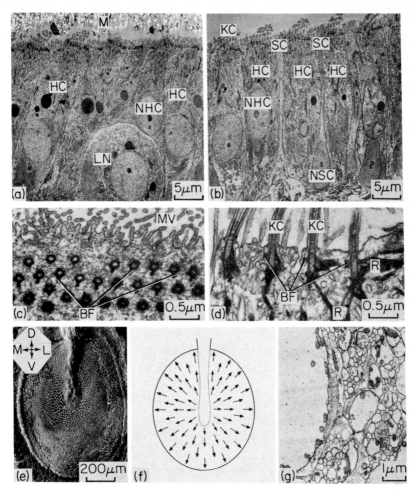

Fig. 13.3   Structure of the gravity receptor system of the statocyst of *Octopus vulgaris* (octopod cephalopod). (a, b) Transverse sections through the macula; (c, d) Orientation of the basal feet of the kinocilia of a single hair cell. (c) Transverse section through some of the basal bodies and (d) longitudinal section through the base of 4 kinocilia. (e) Surface view of the macula with the arrangement of the ciliary groups. The cross-bars indicate dorsal (D), ventral (V), medial (M) and lateral (L). (f) Polarization pattern of the macula. The arrows refer to the direction in which the basal feet point. (g) Transverse section through the macula nerve. BF—basal foot; HC—hair cell; KC—kinocilia; LN—large neuron; M—mucus (attaches the statolith to the ciliary epithelium);. MV—microvilli; NHC—nucleus of the hair cell; NSC—nucleus of the supporting cell; R—root; SC—supporting cell. ((e) from Budelmann *et al.*, 1973.)

Barber, 1966a; Vinnikov *et al.*, 1967). The macular nerve runs straight upwards, passes through the cartilage and enters the CNS; it then divides up into branches in the lateral pedal lobe. Many of these branches end there, others run on to the supraoesophageal lobe (Young, 1971; for further details see also Pfefferkorn, 1915; Thore, 1939).

(c) Decapoda   The morphology of the decapod statocysts has been studied with the lightmicroscope (Owsjannikow and Kowalewsky, 1867; Boll, 1869; Hamlyn-Harris, 1903; Hillig, 1912; Ishikawa, 1924; Klein, 1932), transmission electron microscope (Vinnikov *et al.*, 1967; Barber, 1968b), and with the scanning electron microscope (Boyde and Barber, 1969; Budelmann *et al.*, 1973).

The two statocysts of decapods are situated in a cavity within the cartilaginous cranium, directly below the suboesophageal mass. The cavities are of irregular shape (each about 2 x 4 x 4.5 mm at a maximum in an adult animal). In *Sepia* and *Loligo* eleven to twelve cartilaginous anticrista lobes protrude into the cyst cavity, but the number vary in different decapod species from two to thirteen (Ishikawa, 1924). Kölliker's canal resembles that described in nautiloids and octopods. It does not open to the exterior, but probably ends blindly within the cartilage (Hamlyn-Harris, 1903). The static sac remains in contact with the cartilage. Thus there is no separation into endolymph and perilymph as in octopods. Two receptor systems are present, one for the detection of angular acceleration, the crista/cupula system (Section 13.2); the other for the detection of gravity, consisting of maculae with statolith and statoconial layers respectively (Fig. 13.4a–d).

In contrast to octopods, the gravity receptor system consists of three regions: the Macula statica princeps (MSP), which is attached to the frontal cyst wall, vertically oriented and overlayed by a compact statolith of irregular shape; the Macula neglecta anterior (MNA), which is attached to the medial wall and is also vertically oriented, and the Macula neglecta posterior (MNP), which lies approximately in the horizontal plane. The MNA and MNP are both covered with a statoconial layer (Fig. 13.5c). The three systems are arranged approximately at right angles to each other and thus their topographical orientation is comparable with that of the utricles, saccules and lagenae of vertebrates (Owsjannikow and Kowalewsky, 1867; Hamlyn-Harris, 1903; Budelmann *et al.*, 1973).

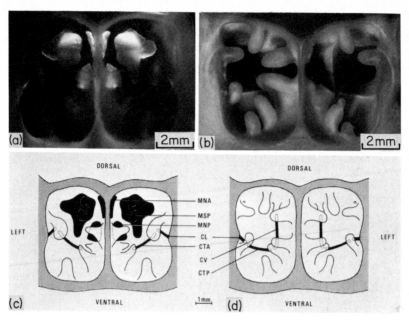

Fig. 13.4 Statocyst of decapod cephalopods. (a, b) Photomicrographs and (c, d) schematic drawings of a transverse section through the statocyst of *Sepia officinalis.* (a, c) Show the forward view; (b, d), the rear view. In the *left* statocyst cavity of (b), five cartilaginous protrusions (anticristae) were removed to expose the outline of the cristae (Sections (a) and (b) are from different individuals. Since the section planes through them do not match perfectly one of the ventral anticrista lobes is seen twice). CL—Crista longitudinalis; CTA—Crista transversalis anterior; CTP—Crista transversalis posterior; CV—Crista verticalis; MNA—Macula neglecta anterior with statoconial layer; MNP—Macula neglecta posterior with statoconial layer; MSP—Macula statica princeps with statolith.

The maculae are composed of hair cells, each surrounded by supporting cells (Fig. 13.5b). The hair cells are considered to be primary receptors and each bears 50 to 150 kinocilia, which are arranged to form an elongated ciliary group. Each kinocilium is about 6 (to 10) μm in length and 0.24 μm in diameter (Fig. 13.5a, b). The form and area of the maculae, as well as the numbers and arrangements of the elongated ciliary groups of the hair cells have recently been described for *Sepia* and *Loligo*. In *Sepia* the MSP (0.47 mm^2) has about 3500 hair cells with their ciliary groups arranged approximately like a figure '3' (Fig. 13.5d). In the MNA (0.33 mm^2) about 2100 hair cells are present, the ciliary groups of which are arranged in semi-circular rings (Fig. 13.5f), and in the MNP

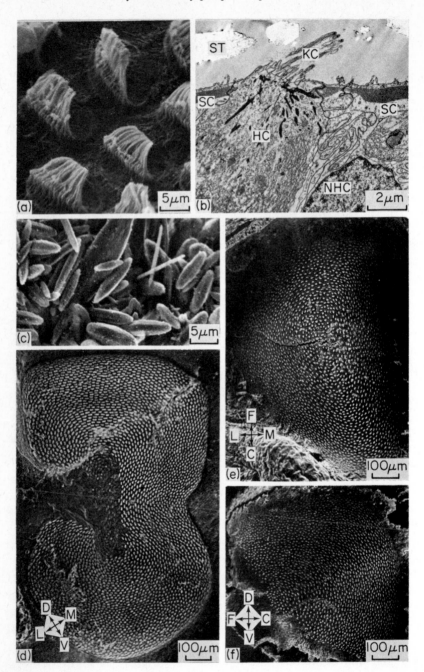

Fig. 13.5   Structure of the gravity receptor systems of the statocyst of *Sepia officinalis* (decapod cephalopod). (a) Ciliary groups of the hair cells of the Macula statica princeps (MSP). (b) Section through the distal end of a hair cell

(0.40 mm^2) about 3100 hair cells are arranged in nearly concentric rings (Fig. 13.5e). In *Loligo* the arrangement of the ciliary groups resembles that of *Sepia*. However, both the area of the maculae and the numbers of hair cells are smaller (MSP: 0.35 mm^2—about 1800 hair cells; MNA: 0.14 mm^2—about 1100 hair cells; MNP: 0.18 mm^2—about 1400 hair cells) (Budelmann *et al.*, 1973).

Comparatively little is known about how the receptor systems connect with the CNS. In *Sepia*, fibres innervating the hair cells of the MNA join as a plexus with the nerve coming from the MSP; whereas fibres from the MNP unite with the nerve coming from the Crista transversalis anterior (Section 13.2.1). Both pass through the cartilage, together with the nerves they joined, and enter the CNS at the level of the lateral pedal lobe. There they branch out (Klein, 1932; Thore, 1939).

(d) Vampyromorpha   The statocysts of the bathypelagic Vampyromorpha seem to be intermediate in structure between the statocysts of octopods and decapods, but more closely related to those of octopods. The static sac does not fill the statocyst cavity; thus endolymph and perilymph are separate. As in octopods, a single gravity receptor system is present as well as an angular acceleration receptor system oriented approximately in three perpendicular planes. The nine anticrista lobes, however, have no direct counterpart in the octopods (Barber, 1968a).

13.2.2   Physiology

As is clear from their morphological organization, to tilt any of these gravity receptor systems results in a local displacement of the statolith (or statoconial layer) relative to the sensory epithelium. This displacement causes a deflexion of all those individual cilia or ciliary groups in contact with the statolith. On the basis of electron microscopy and on neurophysiological and behavioural experiments

---

and supporting cells of the Macula neglecta anterior (MNA). (c) Statoconia of the Macula neglecta anterior (MNA). (d-f) Arrangements of the ciliary groups of the hair cells of the maculae of the left statocyst; (d) Macula statica princeps; (e) Macula neglecta posterior; (f) Macula neglecta anterior. The cross bars indicate dorsal (D), ventral (V), lateral (L), medial (M), frontal (F) and caudal (C). HC—hair cell; KC—kinocilia; NHC—nucleus of another hair cell; SC—supporting cell; ST—statoconia. ((a, d, e, f) from Budelmann *et al.*, 1973.)

it is commonly accepted that the adequate stimulus for gravity and angular acceleration receptors is passive deflexion of the cilia brought about by a shearing motion of the statolith or cupula, i.e. a motion tangential to the surface of the sensory epithelium (for references see Budelmann, 1970c).

By analogy with the results from vertebrate vestibular and lateral-line receptors (for references see Flock, 1971) it is suggested (Barber, 1968a) that in molluscs also, a passive deflexion of the kinocilia of the hair cell in the direction of the basal foot structure (= direction of the outer filaments 5 and 6) causes a maximal excitation (depolarization) of the hair cell, correlated with a maximal increase of the firing rate of the nerve fibre originating from this cell. Conversely there is maximal inhibition (hyperpolarization), correlated with maximal decrease in the firing rate, when deflexion occurs in the opposite direction. The firing rate should vary as a function of the cosine of the angle of the basal foot projection relative to the direction of kinocilia deflexion, i.e. of statolith shear. Consequently, morphological non-polarized hair cells (with their kinocilia polarized radially towards or away from the centre of the cell surface; as in *Pterotrachea* and pulmonates) should respond equally well to stimuli from all directions tangential to the cell surface; whereas in morphologically polarized hair cells (with their kinocilia polarized in the same direction; as in *Pomacea*, octopods and decapods) maximal excitation should be correlated with only one direction. Thus each magnitude and direction of shear from the statolith results in a specific pattern of excited and inhibited hair cells in the sensory epithelium. In *Octopus*, for example, this pattern can be characterized by maximal excitation of those cells which are polarized in the direction of statolith shear (Fig. 13.6). For further information about sensory transduction in hair cells see Flock (1971) and Flock and Lam (1974).

The quantitative analysis of the excitation pattern in regard to the overall amount of excitation (as affected, for instance, by the magnitude of shear; Fig. 13.6a) and/or the angular orientation with respect to epithelium-fixed coordinates (as affected, for instance, by the direction of shear; Fig. 13.6a, b) gives the main information which is necessary to control the general motor activities of the organism relative to gravity. Various neurophysiological studies (intra-and extracellular recordings) and behaviour experiments have been carried out in molluscs to try to gain an understanding of the functional principle of the whole equilibrium receptor system.

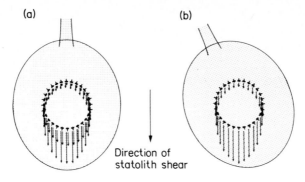

(a)   (b)

Direction of
statolith shear

Fig. 13.6   Model of pattern of excited and inhibited sensory cells of the macula of *Octopus* at two positions with respect to gravity (a and b). Each of the T-shaped symbols represents the group of hair cells which coincides with the orientations of the basal feet of their kinocilia. The arrows indicate stimulation and excitation (inhibition respectively) of these groups, the direction of the arrows referring to the direction of the stimulus as exerted by the statolith. Arrow length refers to the excitation, i.e. the impulse frequency. The two patterns shown in (a) refer to different magnitudes of shear of the statolith. (From Budelmann, 1970c.)

Electrophysiology

Electrophysiological studies on the function of molluscan gravity receptor systems have been carried out mainly in gastropods (Wolff, 1968, 1970a, 1970b; Wood and von Baumgarten, 1972; Alkon, 1973; Alkon and Bak, 1973; Detwiler and Alkon, 1973; Wolff, 1973a, 1973b; Wiederhold, 1974; de Jong and van Wilgenburg, 1975); and there is little data available so far from cephalopods (Budelmann and Wolff, in preparation).

In gastropods, intracellular recordings from hair cells of the opisthobranchs *Aplysia* and *Hermissenda* have shown a resting membrane potential of −40 to −50 mV (Alkon and Bak, 1973; Wiederhold, 1974). The response of the cells (located near or above the equator of the statocyst) varied with their change of position relative to gravity. Thus, when the preparation was tilted (45° upward) so that the statoconia moved away from the recorded hair cell, either no change occured in membrane potential or slight hyperpolarization (−43 mV); on the other hand, when tilted in the opposite direction (45° downward) so that the statoconia moved towards the recorded cell, a depolarization occured, reaching 20 mV at its maximum. Upon return to the starting position the potential returned to its initial value (Wiederhold, 1974).

Extracellular recordings from the static nerve of opisthobranchs and pulmonates show that five to six hair cells are active in the normal position of the preparation (at rest), with a total resting frequency of 15–25 impulses s^{-1}. Any deviation from this position around a horizontal (longitudinal or transverse) axis caused a change in impulse frequency. This reached a maximum at about 180°, i.e. in the upside-down position (Fig. 13.7a). This change has been suggested to be due to different states of adaptation of the hair cells stimulated in the various positions and/or to a (so far unknown) local difference in their number (Wolff, 1970a, 1973b).

During rotation around a longitudinal or transverse horizontal axis single hair cells in the statocysts of pulmonates and *Aplysia* are active only within a limited angular range (120–220°), beyond which the cells are silent. In the various cells this range occurs in different positions but with wide overlap. Maximal frequencies are between 10 and 25 impulses s^{-1} (Fig. 13.7b; Wolff, 1970a).

During clockwise and counter-clockwise rotation around various horizontal axes the changes in activity of individual cells are often very similar (Wolff, 1975). In other instances, considerable differences are displayed both in the position of the maximum (up to 80°) and in its amplitude (up to 6 impulses s^{-1}) (Wolff, 1970a; Wood and von Baumgarten, 1972; Wolff, 1973a; Fig. 13.7c). Nevertheless, the results can be interpreted in terms of a multi-directional sensitivity of the pulmonate and opisthobranch hair cells, and correspond well with the morphological findings (radially arrangement of polarization directions; Section 13.1.1; Wolff, 1975).

The hair cells in pulmonates and *Aplysia* are phasotonic. Upon interruption of rotation at any one position within the active range, the excitation declines very rapidly, reaching a steady response in about 10 seconds. This steady response often turned out to be rather uniform in the different positions (Wolff, 1970a). In *Pleurobranchia*, however, adaptation of the receptors is slower. Moreover, the steady response differs with position (Wood and von Baumgarten, 1972). From these results it was concluded that, in gastropods, the CNS may get the main information about position in space by the spatial orientation of the excitation pattern of the responding sense cells; while the direction of rotation may be detected by the sequence in which different cells become active (Wolff, 1970a; Wood and von Baumgarten, 1972; Wolff, 1973b). Signal transmission through the CNS has been established by cutting the static nerve and

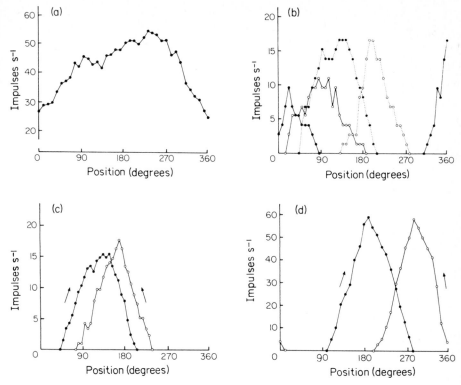

Fig. 13.7 Electrical responses of molluscs gravity receptor cells. (a) Discharge frequencies in the static nerve of *Aplysia limacina* during full circle rotation around the animal's longitudinal axis (rotation velocity 1.67 rpm). (b) Discharge frequencies of four different units in the static nerve of *Arion empiricorum* during full circle rotation around the animal's longitudinal axis. (c) Single unit response from the right static nerve of *Arion empiricorum* during full circle rotation around the animal's longitudinal axis (rotation velocity 1.67 rpm). (d) Response from a few units of the macula nerve of the right statocyst of *Octopus vulgaris* during full circle rotations around the animal's longitidinal axis (rotation velocity 0.44 rpm). In c + d the arrows refer to the direction of tilt (dot, rotation to left; open circle, rotation to right). ((a) after Wolff, 1973b; (b) after Wolff, 1970a; (c) after Wolff, 1973a; (d) from Budelmann and Wolff, in preparation.)

cerebro-pedal connective (de Jong and van Wilgenburg, 1975). The results correspond well with behavioural findings (Dijkgraaf and Hessels, 1969).

It is still uncertain whether the statocyst hair cells in gastropods are spontaneously active or not. Although some cells have been reported to be spontaneously active (Alkon and Bak, 1973), the interpretation of these results is unclear. It is also uncertain whether the silence of pulmonate and *Aplysia* receptors over a limited angular range during horizontal axes rotation appears because of the absence of spontaneous activity or because of inhibitory interaction (Wolff, 1975). In *Pleurobranchea* only a few spontaneously active units have been detected at any one position relative to gravity (Wood and von Baumgarten, 1972).

The physiological evidence for efferent innervation in the gastropod statocyst is also open to interpretation (de Jong and van Wilgenburg, 1975) in that there is some indication, but it has not yet been established unequivocally. Impulses which have been recorded from the two statocyst nerves after either or both were severed from their receptor systems (Wolff, 1970b), may either be efferent, or may represent antidromically propagated excitation (Wolff, 1975). So far, no conclusive influence of efferent excitation on afferent responses of the statocyst receptor cells is evident.

In the cephalopod *Octopus*, recent electrophysiological results from the gravity receptor systems (Budelmann and Wolff, in preparation; Wolff and Budelmann, in preparation) correspond well with the morphological findings. Extracellular multi-fibre recordings from the macula nerve showed that a change of position of the preparation relative to gravity during rotation around various horizontal axes results in a variation of impulse frequency (Fig. 13.7d). The maximum depends on the orientation of the recorded hair cell population within the macula epithelium, i.e. on its direction of polarization. A single hair cell has been found active only within a limited angular range (about 180°), outside of which the cell is silent. In the various cells these ranges of activity occur in different positions but overlap widely. The hair cells adapt rapidly (phaso-tonic in character), and are highly sensitive to vibration. So far, the afferent impulse responses have not been found to be affected by efferent influences. Taken together, the receptor properties of the *Octopus* gravity receptors are basically similar to those known in vertebrates. (But see p. 549—552, for differences in the functional organization of the control system.)

Behavioural experiments
In molluscs, gravity receptor systems are known to control a variety of behavioural reactions: (a) geotactic movements, i.e. locomotion

towards (positive geotaxis), or away from (negative geotaxis) the centre of the earth, (b) reactions for the maintenance of equilibrium, and (c) compensatory reactions, such as righting reflexes, compensatory movements of the eyes (counter-rolling and vertical deviation of the eyes), of the head, and of the eye tentacles. All these reactions are closely related to the direction of the gravito-inertial force, and thus to the function of the gravity receptor systems.

(a) Geotaxis   In gastropods, geotactic behaviour under normal and artificial environmental conditions has been described for several species, and was assumed to be dependent on the statocysts (Baunacke, 1913; Kanda, 1916; for further references see Geuze, 1968). In *Helix*, kept under water, upward movements were observed to be as steep as possible (Jäger, 1932); likewise centrifugation by rotating the animals on a horizontal turntable resulted in a comparable negative geotactic behaviour, i.e. in an orientation and locomotion towards the centre of rotation (Cole, 1926; Fraenkel, 1927a).

The fundamental role of the statocysts on geotactic behaviour is proved, however, when the statocysts are removed. In *Lymnaea*, removal of one statocyst causes the animal to deviate from the perpendicular towards the intact body side during positive geotaxis in air and towards the operated side during negative geotaxis in water. This functional polarity was thought to be based on morphological differences of the receptor systems. *Lymnaea*, with both statocysts removed, however, completely lacks the ability for geotactic orientation (Lever and Geuze, 1965; Geuze, 1968). These results have recently been confirmed for *Pomacea* also. On a vertical creeping plane as well as on a rotating horizontal turn-table animals with both statocysts removed crawl irregularly. After removal of just one statocyst the negative geotaxis remains. A deviation to the intact body side is apparent at first, but disappears with time (Tüllner, 1973; Tüllner and Wolff, in preparation).

When subjected to vertical oscillation, *Helix aspersa* failed to respond differentially to acceleration forces of the magnitude and frequency applied (Mote and Tomlinson, 1965). Neither did an increase in gravito-inertial forces in a centrifuge influence the negative geotaxis of *Lymnaea* (Wolff, 1975); this corresponds well with electrophysiological data (Wolff, personal communication).

Solen ensis is the only bivalve whose geotactic behaviour has been studied. In a rotating vessel filled with wet sand, the animals burrow along the resultant of the centrifugal and gravito-inertial forces (Fraenkel, 1927b).

(b) Maintenance of Equilibrium   In animals with an unstable or meta-stable normal position active movements are required for the maintenance of equilibrium. These movements are known to be partly controlled by gravity receptor systems.

In gastropods, this was first demonstrated for the planktonic *Pterotrachea* (Ilyin, 1900). Removal of both statocysts results in completely disoriented swimming behaviour. Removal of just one statocyst causes the animal to roll towards the intact body side, but this effect is quickly compensated for (within 5 minutes) (Tschachotin, 1908; Friedrich, 1932). In *Aplysia*, bilateral section of the statocyst nerves yields results comparable to the removal of the statocysts; unilateral section, however, has no effect (Dijkgraaf and Hessels, 1969).

In the bivalve *Pecten*, the inequality in structure between the two statocysts seems to be reflected in their physiology as well. Removal of the right (lower) statocyst causes little or no alteration in *Pecten's* swimming behaviour, but removal of the left (upper) one produces major disorientation (von Buddenbrock, 1915).

In octopod cephalopods the role of the statocysts for maintenance of equilibrium is well known (Delage, 1886; von Uexküll, 1895; Fröhlich, 1904; Muskens, 1904; Boycott, 1960; Wells, 1960; Dijkgraaf, 1961). Removal of just one statocyst causes the animal to roll around its longitudinal axis towards the unoperated side; additional removal of the other statocyst results in a complete spatial disorientation during swimming.

In decapod cephalopods, removal of the statolith and statoconial layers unilaterally has little effect; but additional anaesthetization or destruction of the sensory epithelia of the gravity receptor systems of the same side (by injection of Novocain) results in a rolling towards the unoperated side. In animals with the statoliths and statoconial layers removed bilaterally, this rolling becomes evident only after additional anaesthetization or destruction of the sensory epithelia on a single side and has been interpreted in terms of spontaneous activity of the receptor systems (Budelmann, 1975; Budelmann, in preparation).

(c) Compensatory reactions Compensatory righting reflexes of freely swimming gastropods can be clearly correlated with statocyst function. Changed from its normal position, *Aplysia* performs correction movements with its head leading. After unilateral section of the statocyst nerve this compensatory reflex still occurs, but all correction movements disappear when the nerves are cut bilaterally. Additional experiments (section of the cerebro-pedal connective) led to the conclusion that a double reflex pathway originates from each statocyst (Dijkgraaf and Hessels, 1969). A similar reflex pathway was previously described for *Pterotrachea* (Tschachotin, 1908). Compensatory righting reflexes, as well as compensatory movements of the eye tentacles, have also been investigated in pulmonate gastropods; but the conclusion that these reactions are dependent on the statocysts was drawn only by analogy (von Buddenbrock, 1935; Wolff, 1970a).

Compensatory eye movements are the best known reactions which depend on the function of the statocysts. In cephalopods, counter-rolling and vertical displacement of the eyes during rotation of the animal around its transverse (y-axis) and longitudinal (x-axis) body axes have been investigated quantitatively.

In octopod cephalopods (*Octopus*), bilateral removal of the statoliths from the sensory epithelia demonstrated that the compensatory eye movements depend on the function only of the statolith organ of the statocyst. Unilateral removal of the statolith showed that counter-rolling and vertical displacement of the eyes depend on interactions between the afferent input of both statolith organs; furthermore, that a single organ affects the counter-rolling as well as the vertical displacement of both eyes. The effect of each organ on the compensatory reactions was found to be additive, i.e. reactions of unilaterally operated animals were diminished by about 50 per cent in both eyes (Budelmann, 1970a, 1970b, 1970c; Schöne and Budelmann, 1970; Fig. 13.8a–c).

Contrary to Muskens (1904), comparable results were obtained also in *Eledone*. Unilateral and bilateral injection of Novocain into the statocyst cavities produced effects similar to those yielded by statolith removal (Budelmann, unpublished observations).

In octopods, various experiments led to the hypothesis that the compensatory eye movements depend, not on the magnitude, but exclusively on the direction of shear of the statolith on the sensory epithelium. This hypothesis is based on the following facts:

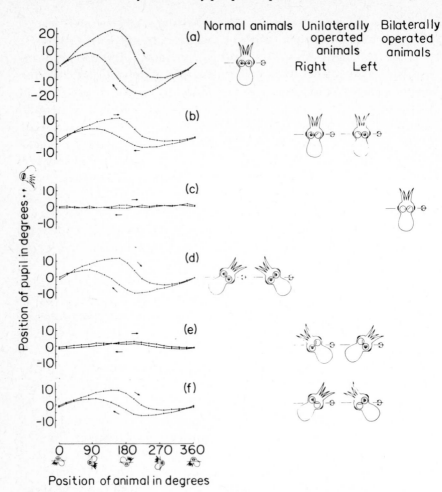

Fig. 13.8   Compensatory counter-rolling of the eyes of *Octopus vulgaris* as a function of body position relative to gravity during rotation around the animal's transverse and diagonal body axes. The degree of counter-rolling (ordinate) is measured as the angular deviation of the pupil slit from an animal-fixed reference line. Each figurine on the right hand side shows an *Octopus* from above, with the position of the macula(e) and statolith(s) relative to the rotation axis. The arrows on the curve refer to the direction of tilt.

(a) normal animals, transverse axis rotation. (b) unilaterally right or left operated animals, transverse axis rotation; (c) bilaterally operated animals, transverse axis rotation; (d) normal animals, diagonal axis rotation with the maculae parallel and perpendicular to the rotation axis; (e, f) unilaterally right or left operated animals, diagonal axis rotation with the macula of the intact gravity receptor system parallel (e) or perpendicular (f) to the rotation axis. (From Budelmann, 1975.)

(i) During rotation of unilaterally operated animals around an axis parallel to the sensory epithelium of the remaining gravity receptor system (i.e. in an experimental condition in which only the magnitude of shear of the statolith—and not its direction—changed during rotation) no compensatory reactions were observed (Fig. 13.8e); (ii) an increase in the magnitude of the gravito-inertial force (up to 1.8 g), and thus an increase in the magnitude of shear, does not influence compensatory eye movements, either in normal animals or in animals with the statolith removed unilaterally; and (iii) during rotation of unilaterally operated animals around their horizontally oriented dorsoventral axis (z-axis), the degree of compensatory counter-rolling of the eyes remains constant independently of the position of the animal relative to gravity, as long as the direction of shear does not alter. However, it changed abruptly when the direction of shear was reversed (Fig. 13.9). Thus, compensatory eye movements could only be obtained as long as the direction of shear of the statolith on the sensory epithelium changed continuously during animal rotation (Budelmann, 1970c).

The compensatory eye movements were consequently described as being a function of the angular orientation of the pattern of all excited sense cells on the macula and its rotatory shift with respect

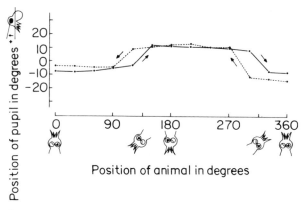

Fig. 13.9 Compensatory counter-rolling of the eyes of a unilaterally left operated *Octopus vulgaris* as a function of body position relative to gravity during rotation around the animal's horizontally oriented dorso-ventral axis. The arrows refer to the direction of tilt. Each figurine at the bottom shows an *Octopus* from above, with the position of the macula(e) and statolith(s). (From Budelmann, 1975.)

to macula-fixed coordinates (Fig. 13.6). Compensatory eye movements have been found independent of the overall amount of excitation, as affected, for instance, by the magnitude of shear (Fig. 13.6a). Thus, these findings reveal a principle of function of a gravity receptor system which differs from that known in crustaceans and vertebrates (Section 12.3). On the other hand, the electrophysiological results are similar to those from crustaceans and vertebrates. Nevertheless, the contradiction indicated by the divergence in the two sets of findings is only apparent. The difference in function between the gravity orientation system in *Octopus* and that in the crustaceans and vertebrates, is due only to differences in the functional organization of the control system, and not to differences in the function of the individual receptor cells (Budelmann, 1970c, 1975).

In decapod cephalopods (*Sepia*), compensatory counter-rolling of the eyes during transverse (y) axis rotation and compensatory head movements during longitudinal (x) axis rotation depend on the function of all six gravity receptor systems of the statocysts. After removal of the statoliths and statoconial layers bilaterally no compensatory reactions could be observed. After unilateral operations, left or right, the reactions were diminished to about 65 per cent (Fig. 13.10). Furthermore, it can be demonstrated that any single system causes compensatory reactions, after removal of the other five, of about 20 per cent of those of normal animals; any two intact systems cause reactions of 35—40 per cent; any three about 65 per cent; any four 80—85 per cent; while removal of just one of the six systems has no effect. Thus, in decapod cephalopods compensatory eye movements were described as a function of the sum of the inputs of the six gravity receptor systems. Since any five of the six systems were sufficient for gravity orientation under the experimental situation, the sixth was thought to have a back-up role, and be in some way redundant (Budelmann, 1975).

In *Sepia*, it is still unclear whether the direction of shear of the statolith, or its magnitude, or both, are responsible for the compensatory reactions. Experiments with normal animals under the influence of increased magnitude of gravito-inertial force (up to 1.8 g) do not show any obvious influences on these reactions. Nevertheless, since these reactions occur even when all of the gravity receptor systems are removed, except those either parallel or perpendicular to the axis of rotation, both the direction as well as

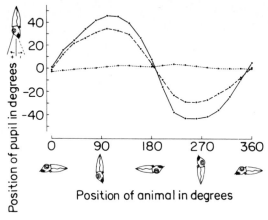

Fig. 13.10   Compensatory counter-rolling of the eyes of *Sepia officinalis* as a function of body position relative to gravity during rotation around the animal's transverse body axis. The curves refer to results from normal animals (● ——— ●), unilaterally left or right operated animals (● – – ●) and bilaterally operated animals (● · · · · · ●). The figurines show a *Sepia* from its left side. (From Budelmann, 1975.)

the magnitude of shear of the statolith are thought to be responsible for the compensatory reactions in the orientation relative to gravity. However, other parameters than the direction and magnitude of shear may also be involved (Budelmann, 1975 and in preparation).

13.3   Angular acceleration receptor systems

In molluscs, angular acceleration receptor systems are basically similar in structure to gravity receptor systems. Sensory elements (hair cells) are part of an epithelium (crista) from which hairs (cilia) protrude into a fluid-filled cyst cavity. Attached to the cilia is a structure (cupula) which is moved passively during angular acceleration of the system, primarily by movement of the fluid relative to the epithelium, thus shearing the cilia (Section 13.2.2). Information regarding angular acceleration is encoded quantitatively in the sensory output.

13.3.1   Morphology

In molluscs, angular acceleration receptor systems are only present in the statocysts of octopod and decapod cephalopods. (For the gross morphology of the statocysts see Section 13.1.1.)

In both cephalopod orders the crista of the angular acceleration receptor system is a ridge of cells lying in three approximately mutually perpendicular planes (transverse, longitudinal and vertical). It is composed of hair and supporting cells, below which large neurons have been described (Klein, 1932; Young, 1960; Barber, 1966a; Vinnikov *et al.*, 1967).

The hair cells are considered to be primary receptors, although the possibility of some secondary cells (in decapods: Klein, 1932) has not been ruled out. The hair cells bear numerous kinocilia and microvilli at their distal end, and no stereocilia (Fig. 13.11d, e). A single hair cell bears up to 200 kinocilia and these are arranged to form an elongated group. Each kinocilium is about 6 (to 10) μm in length, and 0.24 μm in diameter. In the crista, the elongated ciliary groups of the hair cells are aligned in rows along its length (Fig. 13.11a, c). According to the orientation of the basal feet, all kinocilia of a single cell are polarized in the same direction, namely at right angles to the long axis of the ciliary group and, furthermore, at right angles to the course of the crista (Barber, 1966a, 1966b; Vinnikov *et al.*, 1967, 1971).

Octopoda   In octopods, such as *Octopus vulgaris*, the crista has three parts, the Crista transversalis, longitudinalis and verticalis. Each part is further subdivided into three, giving a total of nine sections. These sections alternate between those with a double row of large hair cells on the apex of the ridge, and those with a single row; one double row starts and another ends the crista. Each section has an average length of 0.43 mm with 35 hair cells per row (Fig. 13.11c). In some crista segments, hair cells with different polarization directions have been described (Section 13.2.2). A cupula is attached to each of the nine crista sections. It is a sail-like structure protruding freely into the fluid-filled cyst cavity and reaching a maximum height of 0.4 mm (Young, 1960; Barber, 1968a; Budelmann *et al.*, 1973).

Each of the three crista parts is connected with the CNS by a separate nerve, the Crista transversalis by the anterior crista nerve, the Crista longitudinalis by the middle crista nerve, and the Crista verticalis by the posterior crista nerve (Young, 1960). As recent morphological studies have demonstrated, axons coming from the hair cells of the Crista transversalis (the one adjacent to the sections of the Crista longitudinalis) join the middle crista nerve (Budelmann, unpublished observations). The diameter of the fibres range from 22 μm to

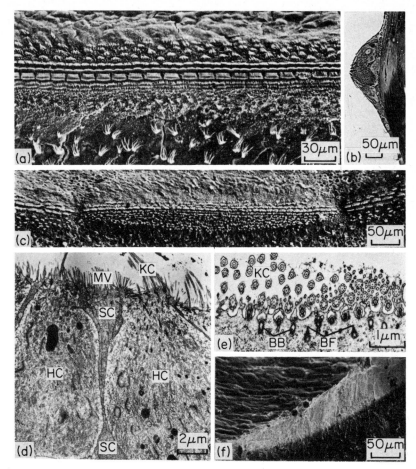

Fig. 13.11   Structure of the angular acceleration receptor systems of cephalopods. (a) Part of the Crista longitudinalis of the left statocyst of *Sepia officinalis*. The cupula is removed. (b) Transverse section through the Crista transversalis anterior of the right statocyst of *Sepia officinalis*. (c) First section (single row, section 4) of the Crista longitudinalis of the left statocyst of *Octopus vulgaris*. The cupula is removed; at both sides of the section parts of double row sections can be seen. (d) Transverse section through the Crista verticalis (double row, section 9) of the right statocyst of *Octopus vulgaris*, showing the two large hair cells at the apex of the crista ridge. (e) Oblique tangential section through the base of the ciliary group of a hair cell of the Crista verticalis of the left statocyst of *Sepia officinalis*. (f) Part of the Crista transversalis posterior of the right statocyst of *Sepia officinalis*. The crista ridge is overlayed by a sail-like cupula. BB—basal body; BF—basal foot; HC—hair cell; KC—kinocilia; MV—microvilli; SC—supporting cell. (a, c, from Budelmann *et al.*, 1973.)

less than 2 μm. The number of fibres (2−22 μm) in the different crista nerves has been determined, using the light microscope, by Young (1965b). However, their number is probably higher since there may be fibres considerably thinner than 2 μm.

The three crista nerves run straight upwards and pass through the cartilage separately. The anterior crista nerve joins the macula nerve intracranially and these, together with the other two crista nerves, enter the CNS at the level of the lateral pedal lobe, where they branch out (Young, 1971; for further details see also Pfefferkorn, 1915; Thore, 1939; Hobbs and Young, 1973).

Decapoda   In decapods such as *Sepia* and *Loligo*, the crista is divided into four sections, the Crista transversalis anterior (CTA), the Crista longitudinalis (CL), the Crista transversalis posterior (CTP), and the Crists verticalis (CV). In each section, the hair cells are arranged in four main rows. The two ventral rows of the CTA, CL and CTP and the two medial rows of the CV are composed of larger hair cells than the other two rows (Fig. 13.11a, b). In *Sepia* the average length of the sections is about 1.60 mm, with about 117 hair cells per row of large cells. In *Loligo* the average section length is shorter (1.19 mm) and the number of hair cells per row of large cells smaller (about 100). For detailed information see Budelmann *et al.* (1973).

A cupula composed of amorphous ground substance and fibrous strands is attached to each of the four sections of the crista and protrudes as a sail-like structure into the fluid-filled cyst cavity (Fig. 13.11f). The height of the cupula increases towards the middle of the crista segment, reaching a maximum height of at least 0.7 mm (Budelmann *et al.*, 1973 and unpublished observations).

Little is known about the exact course of the crista nerves within the statocyst. Two nerves have been described, one apparently containing fibres coming from the Crista transversalis anterior and from the Macula neglecta posterior (Section 13.1.1); the other apparently containing fibres from the other three Crista sections. The two nerves pass through the cartilage and enter the CNS at the level of the pedal lobe (Hamlyn-Harris, 1903; Klein, 1932; Thore, 1939). For details of their course within the CNS see Thore (1939) and Hobbs and Young (1973).

13.3.2   Physiology

According to the morphological organization of the angular acceleration receptor systems, angular acceleration stimuli, applied to the systems by active or passive rotation of the animal, will—because of inertia—result in a movement of the fluid relative to the sensory epithelium. The cupula follows this fluid movement to a certain extent and thereby causes the cilia to deviate (shear) in one or the other direction depending on the direction of angular acceleration. This displacement brings about the change in the receptor potential of the hair cell (Section 13.1.2).

The morphological and topographical arrangements of the crista/cupula systems in cephalopods indicate that the transverse crista systems function best for detection of angular acceleration stimuli during rotation around transverse axes of the animal; the longitudinal crista systems for those around longitudinal axes; and the vertical crista systems for those around vertical axes. This hypothesis has been confirmed electrophysiologically (Wolff and Budelmann, in preparation) as well as in behaviour experiments (Dijkgraaf, 1959, 1961, 1963).

Electrophysiology

In *Octopus*, extracellular recordings from afferent neurons of the three crista systems have shown that each crista system is most sensitive to angular acceleration and deceleration stimuli during rotation around its appropriate axis, i.e. the Crista transversalis to those during rotation around a transverse axis, the Crista longitudinalis to those around a longitudinal axis, and the Crista verticalis to those around a vertical axis. The discharge patterns are generally similar to those in vertebrates. Thus, starting in the normal position, rotation to one side elicits increased activity during acceleration and inhibition during deceleration; rotation to the other side elicits inhibition during acceleration and increased activity during deceleration (Fig. 13.12a, c). These results clearly demonstrate a physiological polarization within the receptor systems (Wolff and Budelmann, in preparation). A unidirectional response has also been mentioned by Maturana and Sperling (1963) for the Crista longitudinalis. However, the interpretation of their results is difficult

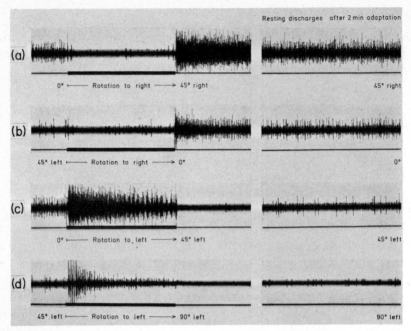

Fig. 13.12 Responses from a few units in the nerve of the right Crista longitudinalis (horizontal angular acceleration receptor system) of *Octopus vulgaris* after intracranial section of all right statocyst nerves. Left—Marker on lower traces indicate the positional changes of 45° with an angular acceleration stimulus at the beginning and a deceleration stimulus at the end of each constant speed rotation of 3.33 rpm. Each horizontal axis rotation was around the animal's longitudinal axis and was carried out after two minutes adaptation to the initial resting position (start position). Right—Resting discharges after two minutes adaptation to the various final resting positions (stop positions). (a) Rotation to right from 0° (normal position) to 45°-right. (b) Rotation to right from 45°-left to 0°. (c) Rotation to left from 0° to 45°-left. (d) Rotation to left from 45°-left to 90°-left (From Budelmann and Wolff, 1973.)

because their recordings were made during rotation around a vertical axis and this axis is not appropriate for the Crista longitudinalis.

In both Cristae longitudinales, ventral deflection of the cupula caused excitation and dorsal deflection inhibition, and in both Cristae verticales, medial cupula deflection caused an excitation and lateral cupula deflection an inhibition. Only in the Cristae transversales were some units found to be excited, and in addition others to be inhibited, during dorsal cupula deflection as well as ventral cupula deflection (Wolff and Budelmann, in preparation). In addition to the sensitivity to angular acceleration and deceleration stimuli, high sensitivity to vibration has been found in all crista systems (Maturana and Sperling, 1963; Wolff and Budelmann, in preparation).

Behavioural experiments
In *Octopus*, Dijkgraaf (1959, 1961) has investigated the role of the Crista verticalis on compensatory reflexes (head and eye movements, including nystagmus and after-nystagmus) on a horizontal turntable. Clear post-rotatory reflexes in response to angular deceleration can still be observed in bilaterally blinded animals after removal of one statocyst and additional transsection of all statocyst nerves of the contralateral side except the one innervating the Crista verticalis. After additional transsection of that nerve the compensatory reflexes are abolished.

Post-rotatory nystagmus has also been investigated in *Sepia* and was related to the function of the crista/cupula systems (Dijkgraaf, 1963). In a more extensive study, Messenger (1970) confirmed these observations also for blinded animals. He also established that post-rotatory nystagmus is almost completely suppressed by visual cues and is abolished by bilateral destruction of the statocyst. In *Sepia*, good reactions were usually found at more than $14.7°$ s^{-1}; at $5.9°$ s^{-1} the reactions were usually much smaller (Collewijn, 1970). These behavioural results correspond well with the electro-physiological findings in *Octopus*: angular acceleration and deceleration stimuli during rotation at $20°$ s^{-1} elicit a clear change in the discharge frequency. Rotation at $4.8°$ s^{-1} elicits only a minimal response. Rotation at $2.67°$ s^{-1} is below threshold (Wolff and Budelmann, in preparation).

Gravity sensitivity of angular acceleration receptor systems
The structural organization of angular acceleration receptor systems does not necessarily exclude their additional function as gravity receptor systems—under the supposition that the cupula has a specific gravity higher (or lower) than its surrounding fluid. (Whether such is the case, however, is still an open question.)

In vertebrates, gravity sensitivity of angular acceleration receptor systems has often been discussed. The interpretation of all experiments to date, however, remains ambiguous. *Octopus* on the other hand has more accessible angular acceleration receptor systems and provides unequivocal results (Budelmann and Wolff, 1973). Extracellular recordings have shown different responses to identical angular acceleration (or deceleration) depending on the position of the preparation with respect to gravity at the time of onset of acceleration (or deceleration) (Fig. 13.12). Furthermore, they have shown different resting discharge rates as a function of angular position with respect to gravity and also changes in the response rate during constant velocity rotation (Fig. 13.12). The interaction of any other system which has been considered responsible for comparable results in vertebrates (Lowenstein, 1972), was excluded. Consequently, the sensitivity to linear acceleration is undoubtedly an additional property of the angular acceleration receptors themselves.

The fundamental role of the angular acceleration receptors as angular accelerometers need not necessarily be affected by the additional sensitivity to linear acceleration (Lowenstein, 1972). However, only behavioural experiments can clarify whether this additional sensitivity actually influences behavioural responses during gravity orientation.

Acknowledgements

The author thanks Dr H. G. Wolff for his critical reading of the manuscript, Dr R. Loftus for improvements to the English text, Miss B. Bey for secretarial assistance and Miss G. Thies for preparing the transmission electron micrographs.

References

Alkon, D. L. (1973) Intersensory interactions in *Hermissenda*. *Journal of General Physiology*, **62**, 185–202.

Alkon, D. L. and Bak, A. (1973) Hair cell generator potentials. *Journal of General Physiology*, **61**, 619–637.

Amoore, J. E., Rodgers, K. and Young, J. Z. (1959) Sodium and potassium in the endolymph and perilymph of the statocyst and in the eye of *Octopus*. *Journal of Experimental Biology*, **36**, 709–714.

Baecker, R. (1932) Die Mikromorphologie von *Helix pomatia* und einigen anderen Stylommatophoren. *Ergebnisse der Anatomie und Entwicklungsgeschichte*, **29**, 449–585.

Barber, V. C. (1965) Preliminary observations on the fine structure of the *Octopus* statocyst. *Journal de Microscopie*, **4**, 547–550.

Barber, V. C. (1966a) The fine structure of the statocyst of *Octopus vulg.* *Zeitschrift für Zellforschung und mikroskopische Anatomie*, **70**, 91–107.

Barber, V. C. (1966b) The morphological polarization of kinocilia in the *Octopus* statocyst. *Journal of Anatomy (London)*, **100**, 685–686.

Barber, V. C. (1968a) The structure of mollusc statocysts, with particular reference to Cephalopods. *Symposium of the Zoological Society of London*, **23**, 37–62.

Barber, V. C. (1968b) The structure of the statocyst of some cephalopods. Ph.D. Thesis, London: University College.

Barber, V. C. and Boyde, A. (1968) Scanning electron microscopic studies of cilia. *Zeitschrift für Zellforschung und mikroskopische Anatomie*, **84**, 269–284.

Barber, V. C. and Dilly, P. N. (1969) Some aspects of the fine structure of the statocysts of the molluscs *Pecten* and *Pterotrachea*. *Zeitschrift für Zellforschung und mikroskopische Anatomie*, **94**, 462–478.

Baunacke, W. (1913) Studien zur Frage nach der Statocystenfunktion (statische Reflexe bei Mollusken). *Biologisches Zentralblatt*, **33**, 427–452.

Boll, F. (1869) Beiträge zur vergleichenden Histologie des Molluscentypus. *Archiv für mikroskopische Anatomie und Entwicklungsmechanik*, **V**, 6, Supplement, pp. 1–108.

Boycott, B. B. (1960) The functioning of the statocysts of *Octopus vulgaris*. *Proceedings of the Royal Society, Series B*, **152**, 78–87.

Boyde, A. and Barber, V. C. (1969) Freeze-drying methods for the scanning-electron-microscopical study of the protozoon *Spirostomum ambiguum* and the statocyst of the cephalopod mollusc *Loligo vulgaris*. *Journal of Cell Science*, **4**, 223–239.

Buddenbrock, W. von (1915) Die Statocyste von *Pecten*, ihre Histologie und Physiologie. *Zoologisches Jahrbuch Abteilung für allgemeine Zoologie und Physiologie*, **35**, 301–356.

Buddenbrock, W. von (1935) Über unsere Kenntnis von der Funktion der Statozysten der Schnecken, mit besonderer Berücksichtigung der kompensatorischen Augenbewegungen. *Biologisches Zentralblatt*, **55**, 528–534.

Budelmann, B.-U. (1970a) Untersuchungen zur Funktion der Statolithenorgane von *Octopus vulgaris*. *Verhandlungen der deutschen zoologischen Gesellschaft*, **64**, 256–261.

Budelmann, B.-U. (1970b) The correlation between statolith action and compensatory eye movements in *Octopus vulgaris*. Proceedings Barany Society, 1st Extraordinary Meeting, Amsterdam–Utrecht 1970, pp. 207–215.

Budelmann, B.-U. (1970c) Die Arbeitsweise der Statolithenorgane von *Octopus vulgaris*. *Zeitschrift für vergleichende Physiologie*, **70**, 278–312.

Budelmann, B.-U. (1975) Gravity receptor function in cephalopods, with particular reference to *Sepia officinalis*. *Fortschritte der Zoologie*, **23**, 84–98.

Budelmann, B.-U. The function of the gravity receptor systems of *Sepia officinalis*. In preparation.

Budelmann, B.-U., Barber, V. C. and West, S. (1973) Scanning electron microscopical studies of the arrangements and numbers of hair cells in the statocysts of *Octopus vulgaris*, *Sepia officinalis* and *Loligo vulgaris*. *Brain Research*, **56**, 25–41.

Budelmann, B.-U. and Wolff, H. G. (1973) Gravity response from angular acceleration receptors in *Octopus vulgaris*. *Journal of Comparative Physiology*, **85**, 283–290.

Budelmann, B.-U. and Wolff, H. G. Electrophysiological study of the gravity receptors of *Octopus vulgaris*. In preparation.

Coggeshall, R. E. (1969) A fine structural analysis of the statocyst in *Aplysia californica*. *Journal of Morphology*, **127**, 113–132.

Cole, W. H. (1926) Geotropism and muscle tension in *Helix*. *Journal of General Physiology*, **8**, 253–263.

Collewijn, H. (1970) Oculomotor reactions in the cuttlefish, *Sepia officinalis*. *Journal of Experimental Biology*, **52**, 369–384.

Delage, Y. (1886) Sur une fonction nouvelle des otocystes chez les invertebres. *Comptes Rendus des Séances de l'Académie des Sciences Paris*, **103**, 798–801.

Detwiler, P. B. and Alkon, D. L. (1973) Hair cell interactions in the statocyst of *Hermissenda*. *Journal of General Physiology*, **62**, 618–642.

Dijkgraaf, S. (1959) Kompensatorische Kopfbewegung bei Aktivdrehung eines Tintenfisches. *Naturwissenschaften*, **46**, 611.

Dijkgraaf, S. (1961) The statocyst of *Octopus vulgaris* as a rotation receptor. *Pubblicazioni della Stazione Zoologica di Napoli*, **32**, 64–87.

Dijkgraaf, S. (1963) Nystagmus and related phenomena in *Sepia officinalis*. *Experientia*, **19**, 29–30.

Dijkgraaf, S. and Hessels, H. G. A. (1969) Über Bau und Funktion der Statocyste bei der Schnecke *Aplysia limacina*. *Zeitschrift für vergleichende Physiologie*, **62**, 38–60.

Flock, A. (1971) Sensory transduction in hair cells. In: *Handbook of Sensory Physiology*, Loewenstein, W. R. (ed.) Vol. 1: *Principles of Receptor Physiology*, pp. 396–441. Springer, Berlin-Heidelberg-New York.

Flock, A. and Lam, D. M. K. (1974) Neurotransmitter synthesis in inner ear and lateral line sense organs. *Nature*, **249**, 142–144.

Fraenkel, G. (1927a) Beiträge zur Geotaxis und Phototaxis von *Littorina*. *Zeitschrift für vergleichende Physiologie*, **5**, 585–597.

Fraenkel, G. (1927b) Die Grabbewegungen der Soleniden. *Zeitschrift für vergleichende Physiologie*, **6**, 167–220.

Friedrich, H. (1932) Studien über die Gleichgewichtserhaltung und Bewegungsphysiologie bei *Pterotrachea*. *Zeitschrift für vergleichende Physiologie*, **16**, 345–361.

Fröhlich, A. (1904) Studien über die Statocysten. I. Versuche an Cephalopoden und einschlägiges aus der menschlichen Pathologie. *Pflügers Archiv für die gesamte Physiologie*, **102**, 415–473.

Geuze, J. J. (1968) Observations on the function and the structure of the statocysts of *Lymnea stagnalis* (L.). *Archive néerlandaises de zoologie*, **18**, 155–204.

Hamlyn-Harris, R. (1903) Die Statocysten der Cephalopoden. *Zoologisches Jahrbuch Abteilung Anatomie und Ontogenie der Tiere*, **18**, 327—358.

Hillig, R. (1912) Das Nervensystem von *Sepia officinalis*. *Zeitschrift für wissenschaftliche Zoologie*, **101**, 736—806.

Hobbs, M. J. and Young, J. Z. (1973) A cephalopod cerebellum. *Brain Research*, **55**, 424—430.

Ihering, H. von (1876) Die Gehörwerkzeuge der Mollusken. Habilitationsschrift, Erlangen.

Ilyin, P. (1900) Das Gehörbläschen als statisches Organ bei Pterotracheidae. *Centralblatt für Physiologie*, **13**, 691—694.

Ishikawa, M. (1924) On the phylogenetic position of the cephalopod genera of Japan based on the structure of statocysts. *Journal of the College of Agriculture Tokyo Imperial University*, **7**, 165—210.

Jäger, H. (1932) Untersuchungen über die geotaktischen Reaktionen verschiedener Evertebraten auf schiefer Ebene. *Zoologisches Jahrbuch Abteilung für allgemeine Zoologie und Physiologie*, **51**, 289—320.

Jong, H. A. A. de and Wilgenburg, H. van (1975) Signal transmission of statocyst information in the central nervous system of the molluscs *Helix* and *Aplysia*. *Fortschritte der Zoologie*, **23**, 51—63.

Kanda, S. (1916) The geotropism of freshwater snails. *Biological Bulletin of the Marine Biological Laboratory* (*Woods Hole*), **30**, 85—97.

Klein, K. (1932) Die Nervenendigungen in der Statocyste von *Sepia*. *Zeitschrift für Zellforschung und mikroskopische Anatomie*, **14**, 481—516.

Lacaze-Duthiers, F.-J. H. (1858) Histoire de l'organization, du développement, des moeurs et des rapports zoologiques du dentale. Librairie de Victor Masson, Paris.

Laverack, M. S. (1968) On superficial receptors. *Symposia of the Zoological Society of London*, **23**, 299—326.

Lemche, H. and Wingstrand, K. G. (1959) The anatomy of *Neopilina galatheae* Lemche, 1957 (Mollusca, Tryblidiacea). In: *Galathea Report*, Bruun, A. F., Greve, S., Spärck, R. and Wolff, T. (eds.), Part 3, pp. 9—72. Danish Science Press, Copenhagen.

Lever, J. and Geuze, J. J. (1965) Some effects of statocyst extirpation in *Lymnea stagnalis*. *Malacologia*, **2**, 275—280.

Leydig, F. (1871) Über das Gehörorgan der Gasteropoden. *Archiv für mikroskopische Anatomie und Entwicklungsmechanik*, **7**, 202—219.

Lowenstein, O. (1972) Physiology of the vestibular receptors. In: *Progress in Brain Research*, Brodal, A. and Pompeiano, O., (eds), Vol. 37: *Basic Aspects of Central Vestibular Mechanisms*, pp. 19—30. Elsevier Publishing Company, Amsterdam-London-New York.

Markl, H. (1974) The perception of gravity and of angular acceleration in invertebrates. In: *Handbook of Sensory Physiology*, Kornhuber, H. H. (ed), Vol. VI/1: *Vestibular System/Basic Mechanisms*, Part B, pp. 17–74. Springer, Berlin-Heidelberg-New York.

McClary, A. (1963) Statolith formation in *Pomacea paludosa* (Say). *Annual Reports of the American Malacological Union*, **30**, 20–21.

McClary, A. (1968) Statoliths of the gastropod *Pomacea paludosa. Transactions of the American Microscopical Society*, **87**, (3), 322–328.

McKee, A. E. and Wiederhold, M. L. (1974) *Aplysia* statocyst receptor cells: fine structure. *Brain Research*, **81**, 310–313.

Maturana, H. M. and Sperling, S. (1963) Unidirectional response to angular acceleration recorded from the middle cristal nerve in the statocyst of *Octopus vulgaris. Nature*, **197**, 815–816.

Messenger, J. B. (1970) Optomotor responses and nystagmus in intact, blinded and statocystless cuttlefish (*Sepia officinalis* L.). *Journal of Experimental Biology*, **53**, 789–796.

Morton, J. E. (1958) *Molluscs*, Ch. 8, pp. 144–161. Hutchinson University Library, London.

Mote, M. I. and Tomlinson, J. T. (1965) The effects of an intermittent stimulus of gravity on the geotactic behaviour of the land snail *Helix aspersa* Müller (Gastropoda: Pulmonata). *The Veliger*, **8**, 3–7.

Muskens, L. J. J. (1904) Über eine eigenthümliche kompensatorische Augenbewegung der Octopoden mit Bemerkungen über deren Zwangsbewegungen. *Archiv für Anatomie und Physiologie, Physiologische Abteilung*, Leipzig, pp. 49–56.

Owsjannikow, P. H. and Kowalewsky, A. (1867) Über das Centralnervensystem und das Gehörorgan der Cephalopoden. *Mémoires de l'Académie imperiale des Sciences de St. Petersbourg*, **11**, 1–13.

Pfefferkorn, A. (1915) Das Nervensystem der Octopoden. *Zeitschrift für wissenschaftliche Zoologie*, **114**, 425–531.

Pfeil, E. (1922) Die Statocyste von *Helix pomatia. Zeitschrift für wissenschaftliche Zoologie*, **119**, 79–113.

Plate, L. (1924) Gleichgewichtserhaltung und Schwerkraftorgane. In: *Allgemeine Zoologie und Abstammungslehre*, Plate, L. (ed.), Teil 2, Ch. IV, pp. 92–131. Gustav Fischer, Jena.

Quattrini, D. (1967) Osservazioni preliminari sulla ultrastruttura delle statocisti dei molluschi gasteropodi polmonati. *Bollettino della Società italiana di Biologia sperimentale*, **43**, 785–787.

Schmidt, W. (1912) Untersuchungen über die Statocysten unserer einheimischen Schnecken. Dissertation, Jena.

Schöne, H. and Budelmann, B.-U. (1970) Function of the gravity receptor of *Octopus vulgaris. Nature*, **226**, 864–865.

Solger, B. (1899) Zur Kenntnis des Gehörorgans von *Pterotrachaea*. *Schriften der Naturforschended Gesellschaft in Danzig*, N. F., Band X, Heft 1, pp. 65–77.

Stahlschmidt, V. and Wolff, H. G. (1972) The fine structure of the statocyst of the prosobranch mollusc *Pomacea paludosa*. *Zeitschrift für Zellforschung und mikroskopische Anatomie*, **133**, 529–537.

Thore, S. (1939) Beiträge zur Kenntnis der vergleichenden Anatomie des zentralen Nervensystems der dibranchiaten Cephalopoden. *Pubblicazioni della Stazione Zoologica di Napoli*, **17**, 313–506.

Tschachotin, S. (1908) Die Statocyste der Heteropoden. *Zeitschrift für wissenschaftliche Zoologie*, **90**, 343–422.

Tsirulis, T. P. (1973) The fine structure of the statocysts of gastropod mollusc *Clione limacina* (Pteropoda). *Zhurnal evolutsionnoi biokhimii i fiziologii*. In press.

Tüllner, H. U. (1973) Die Bedeutung der Statocysten von *Lymnea stagnalis* und *Pomacea paludosa* für die Raumorientierung. Diplomarbeit. Köln.

Tüllner, H. U. and Wolff, H. G. Statocyst controlled orientation in the prosobranch gastropod *Pomacea paludosa*. In preparation.

Uexküll, J. von (1895) Physiologische Untersuchungen an *Eledone moschata*. IV. Zur Analyse der Funktionen des Zentralnervensystems. *Zeitschrift für Biologie*, **31**, 584–609.

Ulrich, J. (1942) Morphologie der Statocyste bei Stylommatophoren im Hinblick auf Körperbau und Lebensweise der Tiere. *Archiv für Molluskenkunde*, **74**, 41–78.

Vinnikov, Y. A. (1974) Evolution of the gravity receptor. *Estratto da Minerva Otorinolaringologica*, **24**, 1–48.

Vinnikov, Y. A., Gasenko, O. G., Bronstein, A. A., Tsirulis, T. P., Ivanov, V. P. and Pyatkina, G. A. (1967) Structural, cytochemical and functional organization of statocysts of cephalopoda. *Symposium on Neurobiology of Invertebrates*, Akadémiai Kiadó, Budapest, pp. 29–48.

Vinnikov, Y. A., Gasenko, O. G., Titova, L. K., Bronstein, A. A., Tsirulis, T. P., Revznier, R. A., Govardovskii, N. A., Gribakin, F. G., Aronova, M. Z. and Tchekhonadskii, N. A. (1971) The balance receptor. The evolution of its structural, cytochemical and functional organization. In: *Problems in Space Biology*, Tchernigovskii, V. N. (ed), Vol. 12. NAUKA, Leningrad.

Wells, M. J. (1960) Proprioception and visual discrimination of orientation in *Octopus*. *Journal of Experimental Biology*, **37**, 489–499.

Wiederhold, M. L. (1974) *Aplysia* statocyst receptor cells: intracellular responses to physiological stimuli. *Brain Research*, **78**, 490–494.

Wolff, H. G. (1968) Elektrische Antworten der Statonerven der Schnecken (*Arion empiricorum* und *Helix pomatia*) auf Drehreizung. *Experientia*, **24**, 848–849.

Wolff, H. G. (1969) Einige Ergebnisse zur Ultrastruktur der Statocysten von *Limax maximus*, *Limax flavus* und *Arion empiricorum* (Pulmonata). *Zeitschrift für Zellforschung und mikroskopische Anatomie*, **100**, 251–270.

Wolff, H. G. (1970a) Statocystenfunktion bei einigen Landpulmonaten (Gastropoda) (Verhaltens-und elektrophysiologische Untersuchungen). *Zeitschrift für vergleichende Physiologie*, **69**, 326–366.

Wolff, H. G. (1970b) Efferente Aktivität in den Statonerven einiger Landpulmonaten (Gastropoda). *Zeitschrift für vergleichende Physiologie*, **70**, 401–409.

Wolff, H. G. (1973a) Statische Orientierung bei Mollusken. *Fortschritte der Zoologie*, **21**, 80–99.

Wolff, H. G. (1973b) Multi-directional sensitivity of statocyst receptor cells of the opisthobranch gastropod *Aplysia limacina*. *Marine Behaviour and Physiology*, **1**, 361–373.

Wolff, H. G. (1975) Statocyst and geotactic behaviour in gastropod molluscs. *Fortschritte der Zoologie*, **23**, 63–84.

Wolff, H. G. and Budelmann, B.-U. Properties of angular acceleration receptors in *Octopus vulgaris*. In preparation.

Wolff, H. G. and Budelmann, B.-U. Electrical responses of the *Octopus* statocyst receptors to vibration. In preparation.

Wood, J. and Baumgarten, J. von (1972) Activity recorded from the statocyst nerve of *Pleurobranchaea californica* during rotation and at different tilts. *Comparative Biochemistry and Physiology*, **43A**, 495–502.

Young, J. Z. (1960) The statocyst of *Octopus vulgaris*. *Proceedings of the Royal Society, Series B*, **152**, 3–29.

Young, J. Z. (1965a) The central nervous system of *Nautilus*. *Philosophical Transactions of the Royal Society, Series B*, **249**, 1–25.

Young, J. Z. (1965b) The diameters of the fibres of the peripheral nerves of *Octopus*. *Proceedings of the Royal Society, Series B*, **162**, 47–79.

Young, J. Z. (1971) *The anatomy of the nervous system of Octopus vulgaris*, pp. 507–530. Clarendon Press, Oxford.

14 *Proprioception and learning*

M. J. WELLS

14.1 Introduction

Other chapters in this book include abundant evidence that proprioceptive information plays a vital part in the control of movements and orientation. The problem to be examined here is whether this class of sensory input is taken into account when invertebrates learn.

Because it is normally impossible to eliminate all the proprioceptors and never quite certain that one has succeeded in eliminating all other sensory cues, it is rarely possible to be certain that an animal is using proprioceptive information when it learns. In vertebrates and among the bulk of the invertebrates the best one can do is to show that they are probably learning to recognize proprioceptive inputs in some instances, and rather unlikely to be doing so in others. One must examine cases where animals learn in circumstances that, *prima facie*, imply that they are taking into account information derived from within their own joints and/or muscles and/or organs of balance and explore these cases rather carefully to see what alternative explanations are possible.

It should be emphasized that the object of this exercise is not to establish whether particular sorts of animal can possibly learn from proprioceptive inputs in *any* circumstances (since that question is unanswerable), but rather whether they normally appear to do so.

Prima facie evidence of proprioceptive learning

Most work on 'learning by invertebrates' has been concerned with changes in behaviour rather than the nature of the stimuli used to bring about these changes. Few experiments have been carried out specifically to investigate the role of proprioception and, in cases where proprioception is held to be involved, it usually appears as a residual hypothesis, the alleged basis of learning when attempts to identify the stimuli concerned have failed. This is clearly unsatisfactory. In most cases we have no complete inventory of the sense organs available. The little we do know about invertebrates in general indicates a widespread dependence on chemical cues, which are notoriously difficult to detect and eliminate from learning situations. We have, moreover, every reason to suspect that electromagnetic senses, as yet unstudied physiologically, may play a part in orientation (e.g. Barnwell and Brown, 1964; Lindauer and Martin, 1972) so that a whole category of evidence, based upon studies of the means by which animals apparently find their way through mazes, or navigate between home and food, is suspect.

These difficulties notwithstanding, it is possible to list several forms of learning that would, on the face of it, appear to necessitate the recognition of inputs arising directly from the movements that animals have made or the postures they have adopted. These include:

(1) The ability to learn to adopt a particular orientation with respect to gravity or to learn to recognize the orientation of things seen or touched relative to gravity, since this implies that the animal is able to take into account its own position in space.

(2) The ability to learn to make turns in a maze or to hold a course in spite of deviations, in the absence of obvious external cues, since these capacities imply that the animal can remember the movements that it has made.

(3) The ability to learn to discriminate between objects of different shapes and sizes by touch, since this implies an ability to monitor and take account of the relative or successive positions of the tactile sense organs in space.

In the review that follows, evidence for these abilities will be considered, first for arthropods and then for the soft-bodied invertebrates. It will be shown that arthropods (like vertebrates) have some, or all of those listed while the soft-bodied invertebrates in general lack them. Instances where worms or molluscs have apparently learned tasks of the above types are uncommon and usually admit of some alternative explanation.

The literature that is potentially relevant to this study is quite large, since it includes all training experiments and behavioural observations that might indicate the use of proprioceptive inputs in learning. Hence the account that follows covers samples only, to illustrate the points that are made. Where possible references are given to review articles, so that readers unfamiliar with the literature on invertebrate learning can view the range from which the samples are selected.

14.2   Arthropods

14.2.1   Learning to orient with respect to gravity

One would expect gravitational forces to be more important to terrestrial animals than their aquatic relatives, since the former have more weight to support. The samples of learned behaviour considered come from web-building spiders and honey-bees, both of which produce structures that are mainly in the vertical plane.

Spiders
Zygiella x-notata builds a vertical web with an open segment along which a single strand runs from the hub of the web to the spider, which waits outside the web with its front legs on the signal thread. It is sensitive to vibration, presumably perceived by lyriform organs close to the tarsal-metatarsal joints (Sections 1.4, 8.3). If a fly lands on the web, *Zygiella* runs down the signal thread to the hub and from there to the prey. The prey is collected and the spider returns directly to the hub and home. Le Guelte (1969) has studied learning in this situation. He reared two groups of spiders. Individuals from one were allowed to build webs on frames; those from the second group were kept in small containers and could not. After 45 days all were allowed to construct webs, which the inexperienced spiders did

quite adequately; web building does not have to be learned. The frames with the webs on were then rotated through 180° and a fly given to lure each spider onto its web. Six experienced spiders required a median of 45 seconds to find their way home, while the six that had built their first web on the day before got home in a median of 3 seconds. This study does not identify the cues that the 'web-experienced' spiders had learned to recognize, but they had clearly learned to orient with respect to something external to the web since rotation upset them so badly.

Experiments on another web builder, *Araneus* (= *Eperia*) *diadema* have been concerned less with learning but give more certain information about the cues that spiders use in orientation. *Araneus* sits in the middle of its web. Blinded spiders can return to the middle after an excursion to the periphery, provided that the web is vertical, but are disoriented if the web is laid flat. Unblinded individuals have difficulty finding their way back to the middle if the vertical web is rotated after luring them to one side (Peters, 1931).

Taken together, the *Zygiella* and *Araneus* experiments constitute strong *prima facie* evidence that spiders can learn to recognize proprioceptive inputs. Since, like insects, spiders have no statocysts, the relevant inputs presumably come from receptors in their joints. Learning by spiders is reviewed by Lahue (1973).

Insects

Honey-bees, *Apis mellifera*, build vertical combs within their hives. A considerable body of literature (von Frisch, 1967) testifies that bees on the comb pay attention to returning foragers and that the latter 'dance' in a fashion that communicates the bearing and distance of the food source to their colleagues. The direction of the food source is indicated by the angle with respect to gravity of the 'waggle runs' that the returning bee makes on the vertical comb. Other bees, following the dance, translate this into sun direction and can set course accordingly. In following the dance leaders, other bees must recognize the stresses set up in their limbs and store this information until they emerge from the hive and set course by the sun, a clear case of proprioceptive learning.

von Frisch's interpretation of his observations on bees has been questioned by Wells and Wenner (1971) who point out that odours may also be very important in determining the recruitment of foragers, and Wells (1973) goes so far as to suggest that 'the entire body of data on recruitment might profitably be reinterpreted in terms of conditional learned relationships between food sources and the odours associated with them'. However, fact remains that bees dance at an angle on the comb; the whole performance seems rather unlikely to be done for the benefit of human observers and on present evidence it is perhaps wiser to adopt Hinde's (1970) view that in all probability both the 'bee language' and olfactory cues contribute to recruitment.

It is, in any case, known (though not apparently to Wells, who does not mention it in his review of bee learning) that bees can be trained to run at particular angles with respect to gravity in order to reach a food source in the dark (Fig. 14.1) (Markl, 1966a, b). Performance is upset by damage to the hair plates (Sections 1.4; 8.3) at the neck, petiole or coxal joints. If all the hair plates are eliminated, the learned orientation with respect to gravity disappears. Markl has discussed his results in relation to bees dancing on their combs, when they show the same small systematic deviations from the 'correct' course dependent upon the angles concerned (Markl, 1966a).

Using similar apparatus, Markl (1964) also succeeded in training ants (*Formica polyctena*) to follow specified angles in relation to gravity in order to reach food. In both series of experiments the disc could be rotated so as to eliminate the possibility that the animals were following odour trails.

There would thus seem to be no doubt whatever that insects and, in all probability, spiders can learn to orient themselves with respect to gravity. Since neither group has statocysts, this must imply the use of proprioceptors signalling muscle tensions, cuticular stresses, or joint angles. Markl's experiments show that the latter is the source of gravitational information in bees (Markl, 1966b).

A number of insects will substitute gravitational for visual information when their orientation is tested under laboratory conditions (Hinde, 1970) and it seems entirely possible that all of them would be shown to be capable of learning on a basis of this class of information if the matter were specifically investigated.

14.2.2   Maze-learning and course control

Arthropods in mazes
Among mammals it has often been held that learning to turn to one side in a maze implies a capacity for proprioceptive learning.

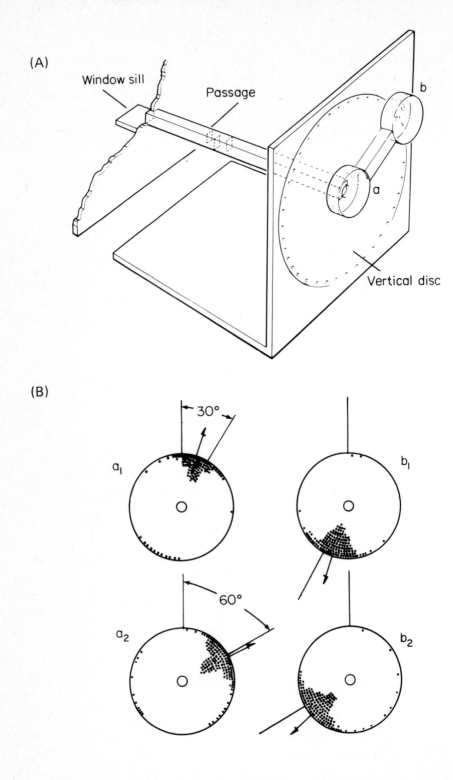

(A)

Window sill

Passage

b

a

Vertical disc

(B)

a₁

30°

b₁

a₂

60°

b₂

Although proof is impossible because the proprioceptors cannot be eliminated, it does appear probable that mammals make use of this class of information in maze learning. In general, they clearly use external cues when these are available and they are relatively slow to learn their way through mazes when the obvious external stimuli are eliminated. This work is reviewed by Munn (1950).

A number of insects and crustaceans have been run in mazes and some at least will learn very complex paths (Fig. 14.2). Tests to determine the cues that they are using suggest that insects, in general, behave rather like mammals. When visual or olfactory cues are available, they use them. Thus Hullo (1948) found that cockroaches, trained to run through a series of 5 successive T mazes, became disoriented when the 5 elements were interchanged, or the floor was covered with cellophane. Neither treatment altered the form of the maze. He found that the animals were unable to run the maze after elimination or immobilization of their antennae. Results like this suggest that the animals are not taking proprioceptive cues into account. But they do not prove that insects ignore proprioceptive cues in all circumstances. Thus in a very thoroughly worked analysis of maze running by ants, Schneirla (1929, 1933, 1941, 1943) showed that while visual cues and chemical trails are certainly important, it is necessary to postulate attention to proprioceptive information as well in order to account for all aspects of the animals' performance. The importance of proprioceptive cues can be shown by cleaning the floor and lengthening any arm of a well-learned complex maze; the ants continue for a while to turn at

---

Fig. 14.1   (A) Apparatus for training bees to run at an angle with respect to gravity. Marked bees alighted on a window sill and ran down a narrow passage to emerge onto a vertical disc. In training their subsequent choice of direction was restricted by a narrow channel (not shown) set always at the same angle, leading from the exit to a feeding place at the edge of the disc. In tests after training the channel was replaced by two inverted petrei dishes (a, b) connected by a corridor at the same angle as the channel used in training. On emerging from the exit the bees could now walk in any direction, and their point of first contact with the periphery of dish a was recorded. Eventually they went down the corridor to food at the centre of dish b. Their move from this to the edge of dish b on the return journey was also recorded. (B) shows the results of tests following training to run at $30°$ (a_1b_1) and $60°$ (a_2b_2) to the vertical. Tests were run in dim red light, invisible to the bees. The disc was rotated to avoid trail-laying. (After Markl, 1966a.)

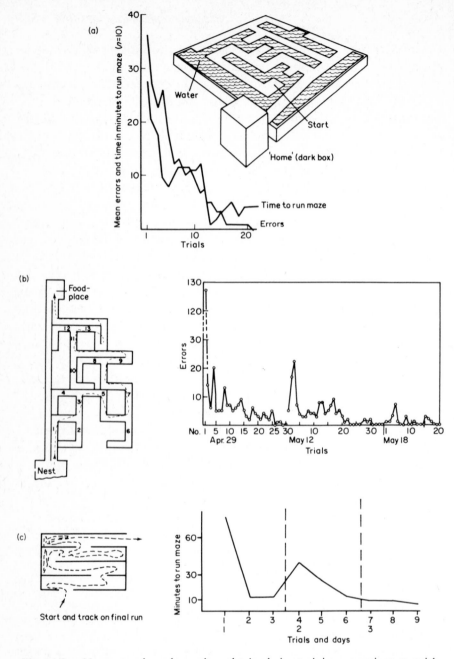

Fig. 14.2   Mazes used and results obtained in training experiments with arthropods. (a) Turner's elevated maze for cockroaches; (b) Schneirla's maze for ants; the dotted path shows a typical track from an ant that has learned; (c) van der Heyde's maze for crabs; the track shown is the route taken at the last trial. (a) After Turner, 1913; (b) After Schneirla, 1933; (c) After van der Heyde (1920).

the old choice point, which suggests that they have learned to carry out a series of movements rather than to make appropriate decisions when they arrive at junctions in the track.

Among the chelicerates, spiders (*Aphonopelma californica*) have been trained to run in a T maze in the absence of obvious directional cues. Henton and Crawford (1966) used three groups of individuals. One group had uniform lighting—'no external cues'—the others had a dark and a light limb to the T, the third illumination by light polarized at right angles in the two limbs. All three learned well. The group with polarized light to guide them performed best; in the other two mean running time and errors fell to the same extent over the 20 training trails, which suggests that proprioceptive cues are quite as important as those derived from light intensity.

Tests with decapod crustaceans suggest that they too may rely on proprioceptive cues, perhaps to a greater extent than do insects. Thus Schöne (1961b) trained spiny lobsters (*Panulirus*) to select the correct escape hatch out of two at the end of a runway leading from the training apparatus to the animals home tank (Fig. 14.3). Errors were reduced to less than 20 per cent in 24 trials if the escape hole was always to the same side. It took *Panulirus* more than 100 trials to reach this standard when the correct alternative was indicated by a light. In another experiment with crayfish (*Pacifastacus*), Gilhousen (1927) found that the animals were still able to negotiate a T maze after removal of both eyes and antennae. Isopods, trained in a T maze, 'exhibit fright and withdrawal movements as soon as they turn in the wrong direction' (Bock, 1942 in Schöne, 1961a).

Accounts of insect maze learning are included in Schnierla (1941, 1943) and Alloway (1973) and in reviews by Markl and Lindauer (1965) and (1974), Schöne (1961a, 1965) and Krasne (1973) have reviewed the subject for crustacea.

Proprioceptors in course control

Insects such as *Carausius* and *Blatella* and millipedes (*Schizophyllum* and *Trigoniulus*) will often run in straight lines in the apparent absence of external cues that they might use to retain their orientation. If forced to deviate by a barrier, or induced to deviate by offering them an external directional cue (gravity, light source), they will revert to their previous bearing when the barrier or stimulus is removed (Mittaelstaedt-Burger, 1972).

This phenomenon has been analysed most thoroughly in the case of millipedes. A millipede will run into a passageway, turn at a bend

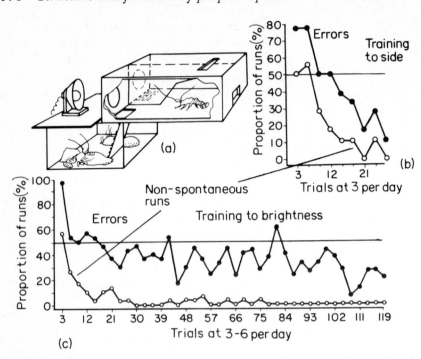

Fig. 14.3   Discrimination learning by the spiny lobster, *Panulirus*, using the apparatus in (a). Training to run to one side (b) is compared with training to run towards the brighter of two lights (c) (After Schöne, 1961b).

in the passage and make a compensatory turn when it emerges at the far end (Barnwell 1965). Memory of the turn made in the passage can be shown to depend on proprioceptive information by altering the bend at different stages during the animal's passing. If the back end of the animal is quickly lined up by moving the passage after the front end has negotiated the turn, the angle of the subsequent compensatory turn is reduced. If the back end is swung to one side while the animal is running straight, a compensatory turn is induced at the end of the passage. In both cases the angle of the compensatory turn is determined by the number of segments that have had to make the turn. (Fig. 14.4.)

If the back end of the animal is lined up with the head at some time after the millipede has begun to emerge and make its compensatory turn, the animal continues along the course that it has adopted until clear of the passage. It then makes a second turn, through an angle proportional to the number of segments prevented

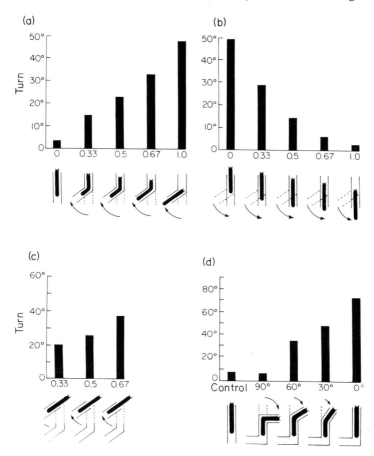

Fig. 14.4   Course control by stored proprioceptive information in millipedes. After passing through a corridor with a bend in it, the animal will make a compensatory turn in the opposite direction. Experiment (a) shows the effect of bending the body of the animal at different stages after the head has passed the bend point. (b) shows the effect of straightening the corridor after the front end has passed. (c) shows what happens when the corridor is lined up after the animal has begun its compensatory turn. The animal continues until it is clear of the corridor and then makes a second turn, proportional to the number of segments that had not made the first compensatory turn. (d) shows the effect of passive initiation of the compensatory turn by bending the front part of the animal. As it emerges it turns to complete the compensatory turn. However, the passively turned part of the animal never made an active turn, and once the whole millipede is clear of the corridor it now makes a further turn (not shown here) proportional to the number of segments that were swung round with the corridor. Ordinate: the angle of the compensatory turn. Abscissa: (a–c) the proportion of the animal that has been swung by bending the corridor; (d) the angle of the passive turn. (After Burger and Mittelstaedt, 1972.)

from making the first compensatory turn on emergence from the passage (Fig. 14.4). If the front part of the body is bent by bending the passageway in the direction of the expected compensatory turn, the animal will again walk clear of the passage, and make a second turn, proportional to the number of segments (in this case at the front end) that were not obliged to made the compensatory turn in the passage.

These results must mean that millipedes store information about the degree and direction of bending of their body segments. The afferent signals (which can hardly be other than proprioceptive) are summed, and the sum is somehow cancelled as the animal makes subsequent turns in the opposite direction (Burger and Mittelstaedt 1972).

14.2.3   Shape and size discrimination

Many arthropods (e.g. crabs and mantids) will manipulate objects that they pick up with their legs, and some of them create quite elaborate structures using their limbs to do so. Spiders make webs and hymenopterans create combs and nests. In the performance of these tasks the animals must measure distances and it would seem inevitable that this involves the perception of joint angles in the limbs.

The matter has been studied rather thoroughly in hermit crabs. These clearly examine and select their homes from among the shells and other objects available to them and there is unequivocal evidence that discrimination is based on the recognition of the movements that they make.

In a very thorough series of experiments with the land hermit crab *Coenobita rugosus*, Kinosita and Okajima (1968) showed that a homeless crab, or a crab seeking a larger shell, always examines concavities in potential homes by feeling into the hole with the chelipeds. It will only attempt to insert its abdomen into holes of suitable depth and diameter. Assessment of depth is normally made by inserting the left cheliped and moving it up and down. If the effective length of the claw is increased by adding a lump of dental wax to the claw tip, the crab underestimates the size of the hole and will not try to move into homes that would in fact be suitable for it (Fig. 14.5a). If the claw tip is filed off, it overestimates (Fig. 14.5b). In the assessment of diameter, both claws are inserted and again the

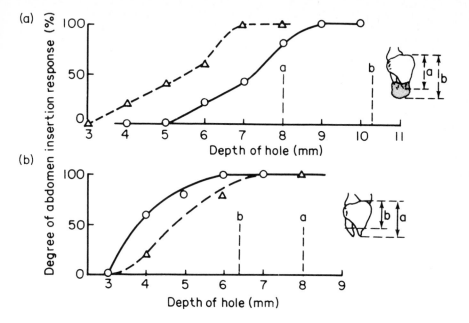

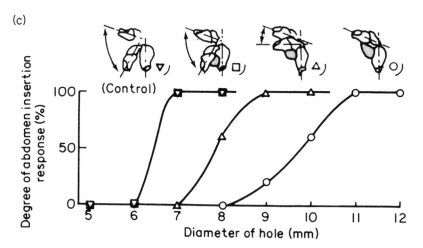

Fig. 14.5   Hole measurement by the hermit crab *Coenobita rugosus*. If dental wax is added to the claw used to feel into the hole, the crab underestimates the depth (a). If the tip of the claw is filed off, the crab overestimates (b) Δ—results before altering claw length, 0—results with the same crabs afterwards. (c) If movement of the carpopodite-meropodite joint is restricted the crab underestimates the diameter of the hole (in this case with its right claw). In each case the ordinate shows the proportion of trials at which there was an attempt to insert the abdomen after examination of the hole. (From Kinosita and Okajima, 1968.)

addition of wax lumps leads to predictable errors of measurement. Loss of one claw does not prevent correct estimation of diameter, but restricting the movement of the remaining claw at the mero-carpopodite joint does so; the more the movement is restricted the greater the underestimate of hole diameter (Fig. 14.5c). Crabs made reliable estimates of hole size in the dark, and the possibility that surface tactile receptors were involved was eliminated by coating the claws in wax. There would seem to be no possible alternative to an explanation of these results based on proprioception. The crab must be measuring hole diameter and depth using receptors, possibly in this case the PD organs (Section 6.2.1) that tell it about the angle through which joints move between contacts with the sides or bottom of the hole.

In these experiments there is no proof that the crabs can *learn* to recognize differences signalled by proprioceptors. Perception of a hole of suitable size may do no more than trigger the abdomen insertion response; there is no indication that a shell once examined, or even inhabited, will be remembered; the crab always re-examines a shell before it risks entry. It might be possible to test the matter by training *Coenobita* to avoid holes of particular shapes or sizes that they would normally enter. But no experiments of this sort have been reported. The only evidence that we have of learning to recognise and handle shells comes from other genera. Thus Drzewina (1910) showed that *Clibanarius* will abandon attempts to enter familiar sealed shells, while continuing to examine new shells that it encounters. Hazlett (1971) reared *Clibanarius* in solitary confinement and showed that individuals lacking tactile experience of sand grains and/or siblings were markedly less competent at entering shells when they encountered these for the first time. In contrast, Reese (1963) working with *Pagurus* and *Calcinus*, found that each species exhibits apparently innate preferences, which are not changed by obliging individuals to live in innappropriate shell types. He claims that larvae reared in isolation examine and move into shells in a wholly 'adult' manner when they first come across them. Clearly there are important innate components to shell-selecting behaviour, and these may well mask a capacity to learn.

A number of other crustaceans manipulate their environment, constructing nests and burrows or decorating themselves with camoflaging materials. The only instance that has been studied from the point of view of possible individual learning is that of *Dromia* which carries sponges, trimmed to shape and held over its back by the last two walking legs. In the absence of sponges, *Dromia* will cut circles of appropriate size out of paper, and individuals have preferred ways of doing this (Dembrowska, 1926). There was no analysis of the cues concerned in this instance and it would be of interest to know whether each crab's characteristic technique is a reflexion of structural variation and whether it would, for example persist in the dark.

There seem to have been no corresponding studies that would suggest proprioceptive learning in the course of tactile examination and manipulation by arachnids or insects. However, one can state unequivocally that at least some arthropods discriminate between objects of differing shapes and sizes on a basis of proprioceptive cues. Since there is some evidence both of individual differences and of changes in performance with experience, it is at least possible that learning based upon proprioceptive cues plays a part in their behaviour.

14.3 Soft-bodied invertebrates

14.3.1 Responses to gravity

A wide range of soft bodied invertebrates will orient with respect to gravity but there appears to be only one case (Sokolov, 1959) of an attempt to train a soft-bodied animal other than *Octopus* to make a response that might depend upon its taking into account the direction of gravity. Sokolov trained *Physa* in the apparatus shown in Fig. 14.6a so that, when a light was switched on, the snail had to move uphill and across the connecting arm to reach the alternative, darker arm of the maze. The initial response required an uphill run from wherever the snail happened to be at the time. Details of a typical experiment are given in Fig. 14.6b. The unconditioned stimulus used in training was a KCl solution, raised from below in the tube with the snail in it. Response to gravity as the orienting stimulus is the simplest explanation of these experiments. But the placing of the lights was asymmetric and *Physa* is known to lay polarized trails which it follows when returning to the surface for air (see p. 584). In the absence of evidence that the snail was not simply retreating along its downgoing path, the learned response to gravity remains unproven.

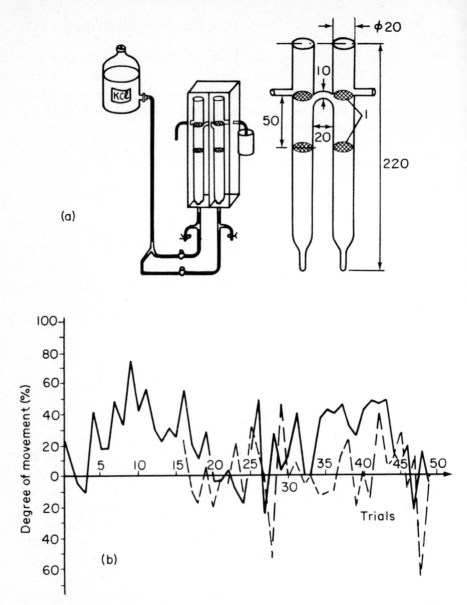

Fig. 14.6   A training experiment with *Physa acuta,*. The animal was placed in one arm of the H-tube (a). The animals movements were limited by wire grids (1). A light was switched on and the level of a 0.2 per cent KCl solution raised from below. The experiment plotted (b) lasted for about two months. After a number of trials light alone induced flight upwards. The ordinate plots the degree of movement as a percentage of the total length of path possible from wherever the snail happened to be at the start of the trial. Dashed lines show responses to a second stimulus (flashing light, or light of a lower intensity) that was not followed by punishment (After Sokolov, 1959).

Recognition of the orientation of objects seen
Octopus vulgaris can recognize differences in the orientation of figures seen, and the discrimination between horizontal and vertical rectangles of the same size has formed the basis of many experiments on the effects of brain lesions on learning (Young 1964). Performance in a series of visual discrimination experiments led Sutherland (1957, 1960) to postulate that the animals classify shapes that they see in terms of their projections on the horizontal and vertical planes. The animals are also able to distinguish between light sources shining through polaroid screens arranged to transmit in the horizontal and vertical planes (Moody 1962). All these experiments suggest that the animal behaves like a higher vertebrate, taking into account its own body position when it assesses the orientation of objects that it sees.

Destruction of the statocysts shows that this is not the case. Where these are eliminated learned discrimination between horizontal and vertical rectangles fails, though the animals remain capable of discrimination between black and white discs. Correct performance can be restored if, and only if, the stimulus orientation is matched to the orientation of the retina at each trial. The same result is obtained when animals are trained to distinguish polarisation plane before and after statocyst removal (Wells and Rowell, 1961). It would appear that information from the statocysts is used to control the orientation of the retina, rather than to inform the central nervous system about the position of the head and sense organs. The animal's visual analysing system operates on the assumption that retinal orientation is fixed relative to gravity.

After statocyst removal, the orientation of the retina is no longer constant with respect to gravity, but it is still consistently related to the posture of the animal. If, for example, the octopus is sitting on the side of its tank the retina will be roughly at right angles to its 'normal' position. If the animal were able to take into account information from other receptors indicating bodily position it would, presumably, eventually learn to make responses dependent upon its position in the tank when shown the test figures. But there are no indications that *Octopus* can do this (Wells, 1960).

We thus have one clear instance where a soft-bodied animal fails to take into account information about its position relative to gravity, and one experiment which indicates that another, less sophisticated, member of the same phylum may just possibly be doing so.

14.3.2 Maze-learning

Molluscs

An account of the few experiments that have been made with gastropods in T and Y mazes is given by Willows (1973). All have been made with pulmonates and all involved punishment for incorrect responses rather than reward for achievement. The most convincing results were obtained by Garth and Mitchell (1926); after 70 trials at 2 per day, the only surviving individual from a group of 10 *Ruminia* made 32 successive correct turns to the right. The floor of the maze was changed regularly to eliminate trails. Although this could be proprioceptive learning, no attempt was made to eliminate the eyes or tentacles and there is no record of precautions taken to ensure absolute uniformity in the maze surroundings. In a more extensive series of experiments, Fischel (1931) appeared to succeed in training *Ampullaria* to turn to one side in a Y maze, but the effects were transient and not repeatable when a T or X maze was used instead; presumably learning based on proprioceptive cues would have been easier in the T (or X) than the Y, which involves a much less extreme turn. Fischel (1931) failed to teach *Lymnaea* to turn to one side on encountering a row of stones and Thompson (1917) failed to teach *Physa* in a vertical T maze.

A possible reason why snails make such poor subjects for maze experiments is discussed in Wells and Buckley (1972). Gastropods leave mucous trails as they move about and at least some species respond to these when they need to return to the surface for air, find other individuals of their own species or catch up with gastropod prey. In at least some instances the track is polarized, so that a second individual can determine the direction taken by its predecessor (Crisp, 1969, Wells and Buckley 1972, Townsend 1974). In this situation, reports of maze learning must be examined warily; to establish dependence upon any other cues, one must be quite sure that the animal has left no trail to guide itself. Work on limpets (Cook, Barnford, Freeman and Teideman, 1969; Cook, 1969) has indicated that the trails may be both long lasting and difficult to eradicate.

The record of erratic success and failure to train snails in T mazes is perhaps not altogether surprising in view of what is now known about their trail-laying habits. The laboratory maze with a removable floor presents a series of problems that the animal would normally solve in the one manner that is forbidden to it, by laying a trail and retracing this on subsequent journeys.

Octopuses do not, so far as we know, lay trails. Their movements about their environment are perhaps too swift even when they are crawling, and they very frequently swim from one rock to another. In all probability their ability to return to a home in the rocks is guided visually.

Boycott (1954) Sanders (1970) and Walker, Longo and Bitterman (1970) all ran octopuses in mazes. Boycott's attempts gave inconsistent results; out of five *O. vulgaris* tested, only one regularly ran around two baffle plates to get a crab. Sanders' Y maze included tactile cues and although there was some improvement in the collective performance of his seven animals, individual scores were erratic. Walker *et al.* (1970) used the maze and obtained the results shown in Fig. 14.7. Their *O. maorum* had to learn to turn to one side in order to get back into the water. There were no obvious cues internal to the maze and, since the external lighting was stated to uniform, this appears to be *prima facie* evidence of response learning on a basis of proprioceptive inputs arising from the animals own movements. Proprioceptive input, would, of course be wholly abnormal since the animal is presumably unused to supporting its weight in air. Perhaps because of this, learning was slow by *Octopus* standards; the initial training reduced errors from about 60 per cent to about 20 per cent in 80 trials, and on reversal from about 100 per cent to about 10 per cent in a further seventy trials (Fig. 14.7).

It is a pity that Walker *et al.* never carried out their original intention of adding visual cues (their apparatus is shown with lights built into the corners of the cross arm of the T) since this would yield a set of results almost directly comparable with Schöne's experiments on *Panulirus* (Schöne, 1961b). Alternatively, training could have continued after blinding the animals, an operation that should not eliminate correct performance by an animal that has learned to recognize proprioceptive cues.

So far as I am aware, only one direct test for response learning has been made with *Octopus*. Wells (1964c) used the apparatus shown in Fig. 14.8 to show that octopuses will detour out of sight of their prey in order to reach it. Running time improves, and abortive entries into the corridor decline, with practice; so that there is some learning in this situation, even though the animals are making a response that is probably a part of their normal behavioural repertoire

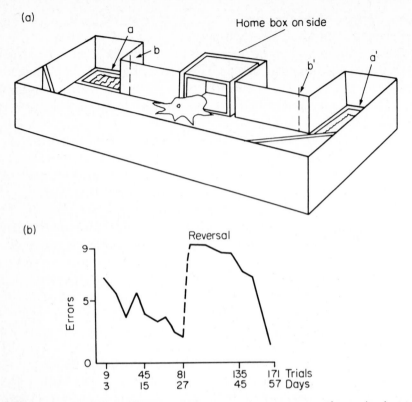

Fig. 14.7   T-maze experiment with *Octopus maorum*. The animal was transferred to the maze in a box that also formed its home in a holding aquarium. The maze (a) was lined with damp paper, but not water filled. The starting box was laid on its side and the octopus had to run to one side of the T to get back to a similar, water filled box (a or a^1). Access to the other side was prevented by a barrier at b or b^1. (b) Results of a training experiment. (After Walker, Longo and Bitterman 1970.)

in the sea. Two octopuses were run in an apparatus similar to that shown in Fig. 14.8, but with a considerably longer corridor. After 30 trials each, always to the same side, each animal was shown the crab on the opposite side of the apparatus, the assumption being that they would carry out the usual series of movements and make an inappropriate turn at the end of the corridor, if they had learned to make a response as a result of the training. Neither of the animals made a mistake. Further experiments in the same series indicate that detouring is guided visually. Blinding in one eye produces systematic errors, while destruction of both statocysts (which include rotation

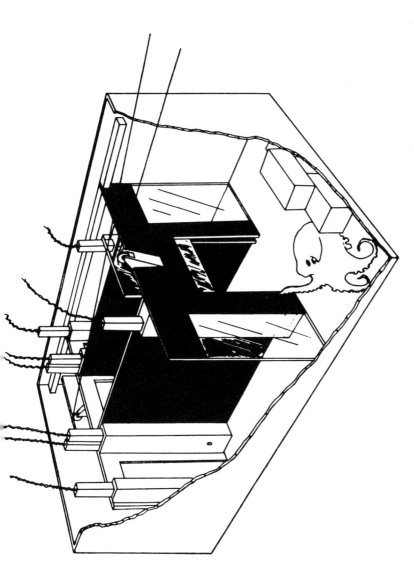

Fig. 14.8   A detour apparatus used in experiments with *Octopus vulgaris*. The animal was able to see a crab through one or other of two transparent windows. In order to reach the crab, it had to detour through an opaque corridor and make an appropriate turn at the far end of the corridor. The progress of the animal was monitored by photocells. Remotely controlled shutters were used to delay the animal in the corridor in some experiments (From Wells, 1970).

receptors in *Octopus* (Chapter 13)) leaves animals that are a trifle unsteady but still detour to the correct side as reliably as controls (Wells 1964c).

Annelids

Data about the performance of annelids in mazes is somewhat more plentiful than that available for molluscs. Yerkes (1912) and later Heck (1920) trained earthworms in a T maze, and there have been a number of more recent experiments (see Arbit, 1965; Dyal, 1973). In the present context, the older experiments are the more interesting; subsequent work has been concerned with phenomena such as latent learning, diurnal changes in learning ability, or the effect of drugs on learning and adds no information about the nature of the cues that the animals are using. Yerkes and Heck used the same sort of T maze; a 'wrong' turning was signalled by a sandpaper strip on the floor, and followed by an electric shock (Fig. 14.9). Trained worms continued to perform and naïve worms could be taught after removal of their cerebral ganglia, which suggests dependence on sense organs distributed along the body. Worms that had been trained to run to one side, persisted in turning to this side when the sandpaper and the electrodes had been moved to the previously 'correct' arm of the T. This shows that the animals were not detecting the electrodes at a distance, and implies learning to make a response.

In a series of experiments with polychaetes, Evans (1963) and Flint (1965) have shown that *Nereis virens, N. diversicolor* and *Perinereis cultrifera* will run in T mazes, provided they are rewarded for correct performance as well as punished for incorrect responses. They must be allowed to rest in a darkened chamber between trials, or they cease to run. Removal of the cerebral ganglia prevents learning and learned performance (Evans 1963). Removal of the antennae, palps, eyes and tentacular cirri stops correct performance, but naïve animals operated upon in this way before training will learn, and in fact do so in fewer trials than intact controls; 9 such worms averaged 57 trials to reach a criterion of 9 correct responses in 10 successive runs, compared with the 84 trials averaged by 33 controls (Fig. 14.10). Sensory deprivation was not complete in these experiments, since the nuchal organs, which are believed to be chemosensory, were left intact. Once again one is left with an incomplete analysis. Thus it looks as if the animals can learn on a

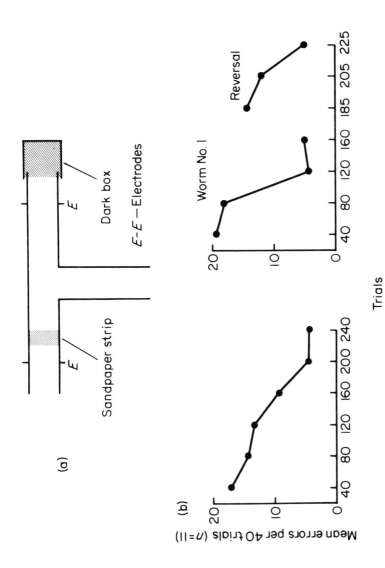

Fig. 14.9   T maze performance of earthworms. A repeat of Yerkes original experiment (Yerkes 1912) by Heck (1920). Lighting and other external cues were believed to be uniform. ((a) after Heck, 1920; (b) data from Heck, 1920.)

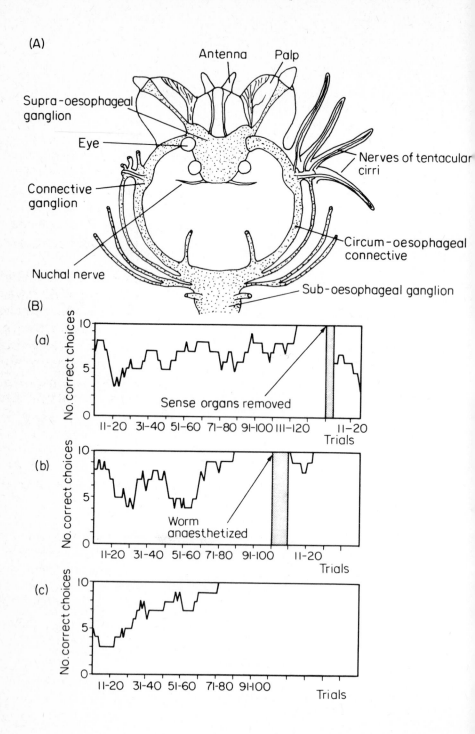

(A)

Antenna   Palp

Supra-oesophageal ganglion

Eye

Connective ganglion

Nerves of tentacular cirri

Nuchal nerve

Circum-oesophageal connective

Sub-oesophageal ganglion

(B)

(a)

No. correct choices

Sense organs removed

11-20 31-40 51-60 71-80 91-100 111-120   11-20
Trials

(b)

No. correct choices

Worm anaesthetized

11-20 31-40 51-60 71-80 91-100   11-20
Trials

(c)

No. correct choices

11-20 31-40 51-60 71-80 91-100
Trials

basis of proprioceptive information if they have to but, like mammals, use other sources of sensory information when they can. Both earthworms and polychaetes leave slime trails, and without careful cleaning at every trial (and in most cases—Flint's experiments included—there was no indication that this was done) one cannot be sure that the worms were not solving the problem in the gastropod way, by following their own tracks.

Earthworms have sometimes been found to show spontaneous alternation in their choices in the T maze situation and Jacobson (1963) reviews the few reports of this. Among vertebrates, spontaneous alternation has sometimes been held to indicate proprioceptive memory. There is, however, considerable uncertainty about the cues used, even by rats, where the phenomenon has been much investigated (e.g. Douglas, 1966) and its value as evidence of proprioceptive learning in worms is best noted as slight, in the absence of much more detailed analysis.

Planarians
Over the last fifteen years or so, interest in the chemistry of memory has generated a considerable literature on learning by planarians; a series of reviews is included in Corning and Ratner (1967) and a more recent account in Corning and Kelly (1973). These experiments have included a number of tests with T and Y mazes, and some rather variable results. Some authors reported learning, others found none (Bennett and Calvin, 1964). The key to the difference is said to be slime; planarians fail to learn to run mazes (or at best learn exceedingly slowly) if these are cleaned out at every trial (McConnell, 1967). An obvious implication is that the animals are learning to follow trails, as gastropods do. In fact there is no need to suppose that they follow any particular track; once they had made a few correct responses they would end up in the correct arm of the maze if they merely preferred to locomote on recently slimed areas, and there seems to be little doubt that this is so. McConnell (1967) reviews maze and other training experiments and concludes that the

---

Fig. 14.10   (A) The relevant parts of the central nervous system of *Nereis diversicolor*. (B) T maze performance of *N. diversicolor*. (a) Individual before and after removal of all anterior sense organs; (b) before and after sham operation; (c) after removal of all anterior sense organs. Results show the number of correct choices in 10 consecutive trials, running score. (From Flint, 1965.)

behaviour of the animals is so erratic in containers that have not been 'conditioned' by allowing planarians to crawl all over them, that 'we are forced to conclude that all results obtained from animals trained in clean apparatus may well have yielded spurious results'.

In this situation the planarian experiments cannot usefully be cited as evidence for proprioceptive learning since although the animals undoubtedly come to perform correctly in mazes (Corning and Kelly (1973) cite a large number of successful experiments), it would seem that there have rarely, if ever, been sufficiently stringent controls to eliminate a slime-following explanation.

14.3.3   Shape and size discrimination

Octopuses that have been blinded by section of their optic nerves can readily be trained to distinguish between objects by touch. The difficulty of any particular discrimination can be assessed in a number of different ways. A convenient measure is to assess the proportion of errors made at a stage in training when further improvement has become negligible or very slow, since performance is by then probably limited by inability to distinguish between the stimuli rather than capacity for attaching appropriate responses. When this sort of assessment is applied to the performance of octopuses required to distinguish between the objects shown in Fig. 14.11a several interesting facts emerge.

To begin with, it is clear that distinctions between objects of markedly different texture are rather easy for *Octopus*: Fig. 14.11b illustrates the progress of a typical training experiment. The textural discrimination between P1 (a cylinder with vertical grooves cut into it) and P4 (a similar, smooth cylinder) was soon learned, a score of 85 per cent or more correct responses being made after 40 trials. In contrast, the shape discrimination between a sphere (PS) and a cube (PC) proved difficult for the same octopuses. The animals treated the cube like the rough and the sphere like the smooth cylinder but were apparently unable to separate the two reliably, so that there was no improvement in performance over a further 70 trials.

Consideration of the proportion of errors made in a larger range of textural experiments (Fig. 14.12a) shows that the capacity to separate cylinders similar to those shown in Fig. 14.11a depends upon their difference in 'roughness'; here determined by the frequency of slots cut into their surfaces. Other characteristics of the

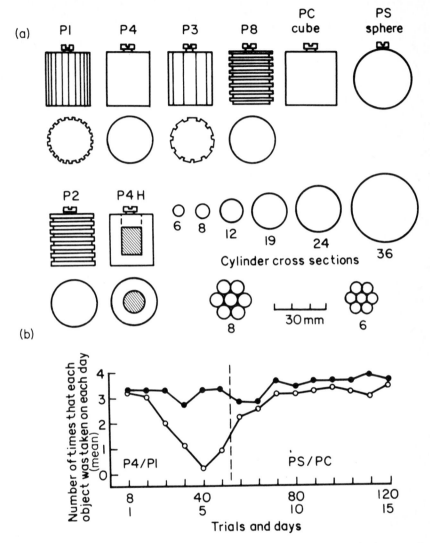

Fig. 14.11   (a)   Perspex   cylinders   and   other   shapes   used   in   tactile discrimination experiments with octopuses. P4H is a smooth cylinder, bored out and filled with lead; it was not distinguished from P4 which weighs one ninth as much in seawater. The compound cylinders 8* and 6* are bundles of rods of the stated unit diameter. All cross sectional measurements are given in mm. (b) a summary of the performance of 9 octopuses trained to distinguish by touch between P1 and P4 and subsequently between the sphere (PS) and the cube (PC). (●) shows the number of times the positive and (○) the number of times the negative object was taken. (From Wells, 1964a, 1966.)

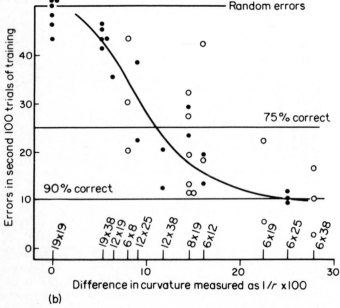

Fig. 14.12   The results of tactile training experiments with octopuses. a) each point shows the proportion of errors made in the last 40 trials of 12 days training at 8 trials per day. The objects used were a series of Perspex cylinders,

cylinders, such as the orientation of the slots, seemed to be irrelevant. In a further series of experiments with cylinders of different sizes (Figs. 14.11a, 14.12b) it was found that errors varied as the difference in radius of curvature of the objects; compound cylinders made up of narrower rods being treated as if of the same size as their components.

Taken together these results are most economically explained by supposing that the octopus's system for distinguishing between objects is based on the degree of distortion imposed upon the suckers used to grasp the objects. There is no need to suppose that the bend of the arm(s) as a whole, or the distribution of the sense organs in the suckers are taken into account (Wells 1964b).

A discriminatory system based on distortion alone should be incapable of distinguishing shape by touch. This seems to be true of *Octopus*. Cubes and spheres, as the experiment summarized in Fig. 14.11 shows, are difficult to distinguish. The small proponderance of 'correct' over 'incorrect' responses is explicable on a 'textural' basis; suckers in contact with the corners of the cube will be distorted, while all those in contact with a sphere can lie flat. It follows that a sphere should be equivalent to a flat surface in transfer experiments and that an ideal 'cube' (so far as the octopus is concerned) should have no flat surfaces to confound it with spheres. The latter possibility has been investigated in transfer experiments. Octopuses trained to take a cube and reject a sphere will readily accept a narrow rod and indeed make far fewer mistakes than before when this replaces the cube as the object to be taken in training experiments. Octopuses trained to take the sphere reject the rod (Wells, 1964a).

The evidence from tactile training experiments with *Octopus* suggests that there is no proprioceptive component to the cues by which the animal learns to recognize objects by touch. The animals examine objects that they touch and the arms move over them as they do so. One can contrast the operation with a hermit crab's examination of its potential home (Section 14.2.3). The hermit crab

---

some of which are shown in Fig. 14.11, b) animals were trained to distinguish between cylinders of differing diameter—the series shown in Fig. 14.11a. Results marked with open circles were from experiments including compound cylinders (also shown in Fig. 14.11a) and considered here as being the diameter of their component rods. ((a) from Wells and Wells, 1957; (b) from Wells, 1964b.)

discovers something about the shape and size of the aperture that it is examining, but the octopus seems to learn nothing of the shape or size of the objects that it manipulates.

It should be noted that *Octopus* undoubtedly has receptors sensitive to distortions of the arm musculature. If an arm is passively extended, the animal pulls back. Multipolar cells resembling the stretch receptors of crustaceans have been described from the muscles and are assumed to be the receptors controlling the reflex (Alexandrowicz, 1960; Graziadei, 1964). However, the behaviour of octopuses in weight-discrimination experiments shows quite clearly that information from these receptors is not available for learning. The animals compensate by increasing muscle tension to support heavy objects. But they never learn to use this as an indication of whether they should accept or reject the object concerned and they continue to make errors at randon for as long as training is continued (Wells, 1961).

The only observations that may be inconsistent with this picture of cephalopod proprioceptive learning abilities arise from studies of the shell-drilling habit of *O. vulgaris* in some parts of the world. Off Japan, and in the Bahamas, *Octopus* bores holes in the shells of other molluscs. The animals manipulate the shells and drill with the radula in an appropriate place on the spire. This itself is not surprising; they must orient crabs appropriately when they dismember them. What is more interesting is the existence of individual differences that may have arisen as a result of individual experience. At Bimini, each octopus drills at a characteristic point in the spire of the strombids on which they feed (Arnold and Arnold 1969; Wodinsky 1969). If it could be shown that this is truly a result of learning, rather than a reflection of individual structural variation, it would become necessary to suppose that the animal learns the shape of the conches that it is accustomed to feed upon. The region around the mouth is the least flexible part of the arms and the one region of the body where some sort of internal map of the spatial distribution of the tactile sense organs would remain stable as the animal moves about. Rowell (1966) has studied the activity of the peripheral nerves and interneurons in the arm nerve cords of *Octopus*. He found abundant evidence of mechanoreceptors responding to touch, but no indication of proprioceptors.

14.4 An attempt at synthesis

14.4.1 Proprioception and perception

In man, limb position sense depends upon receptors in and around the joints, and to a lesser extent upon receptors in the skin. Information from a second system of proprioceptors in the tendons and muscle spindles does not normally seem to penetrate to conscious levels (Merton, 1964; but see also Goodwin, McCloskey and Matthews, 1972).

Pringle (1963) has suggested that a similar state of affairs may be expected in Arthropods. Here too proprioceptors can be classified into those that would indicate the relative positions of parts and those monitoring muscle tension. The former include hair-plates and other organs measuring the positions of joints between sections of cuticle; the latter strain gauges indicating loads applied to the skeleton, as well as those registering muscle tension (see Table 17.1. p. 668).

In order to learn to repeat a movement, an animal must have feedback to tell it what it has done. This re-afferent input can be in any modality and may include proprioceptive information (Hinde 1970). In these instances, the relevent proprioceptive input would be that indicating bodily position achieved, rather than the load-variable input from stretch receptors.

Soft-bodied animals, like annelids and molluscs, have stretch receptors in their muscles (Chapter 11). But they seem to have no equivalent of the joint senses of arthropods and vertebrates. Indeed, it is difficult to imagine how any wholly soft-bodied organism could have a sensory system capable of monitoring the relative positions of parts of its own body independently of its strain detectors. Even if it were able to do so the number of points at which movement would have to be monitored might prove prohibitive. In arthropods and vertebrates considerable constraints are imposed by jointed skeletons. The relative positions of parts of the body can alter in a very limited number of ways. This opens up the possibility of fully centralized control, while the soft bodied animals are obliged to depend upon heirarchic systems to regulate the finer points of their movements.

14.4.2   Learning to make movements without proprioception

Insects will learn to adopt postures that avoid shocks given to their feet. This can be demonstrated in preparations consisting of one leg and the appropriate ganglion, and it would be reasonable to assume that it depends upon proprioceptive feedback. The motor output to the levator muscles will, however, change in the same way as a result of training when the apodemes are cut and the muscles attached directly to transducers, eliminating the proprioceptors that might have indicated leg position. The matter has been reviewed in Horridge (1965) and by Alloway (1973).

These experiments are relevant here as they provide yet another reason why one should be wary of attributing a change in behaviour to proprioceptive inputs where other obvious feedbacks are lacking. If the trained leg is learning at all (for a criticism of the method as an indicator of learning, see Bateson, 1974) it must be doing so on a basis of efference copy, associating motor output with its consequences, without reference to the position achieved as a result. This case arises in a group of animals known to use proprioceptive inputs in learning, and it makes one wonder to what extent similar states of affairs may be present in non-arthropod invertebrates. It highlights the possibility that learning to turn to one side in a T maze is at least sometimes based on efference copy rather than proprioceptive feedback.

14.4.3   Automation and the conscious control of movement

In respect of the possibility that very simple maze learning may be based on efference copy it is worth noting that while arthropods and vertebrates can be taught to run mazes of considerable complexity, no soft-bodied invertebrate has been convincingly shown to master a maze demanding more than a single correct choice at the first and only junction. A possible explanation arises from an examination of our own experience in learning to make skilled movements. A learned movement sequence is at first controlled consciously on a basis of information received through exteroceptors. With practice, elements of the sequence become automated, in the sense that one is no longer aware of controlling them in detail. It seems likely that cerebral control is short-circuited by cerebellar control (Marr, 1969). Something of the kind must happen as a rat or a cockroach learns a maze. The effect of automating elements of the problem is to clear the ground for cerebral action on further parts of the sequence, and for the abstraction of more general aspects of the situation—cognitive maps and so on. It will be noted that automation implies feedback, so that the system can chain movements, each movement setting up the context that triggers the next in the absence of cerebral orders to the contrary (Marr, 1969). But this feedback must be from receptors monitoring movement achieved, not simply movement intended. Feedback from stretch receptors could not supply this; being dependent upon load (which might vary, for example, with whether the animal had recently fed) it is too unreliable. The soft animals, in short, may fail to learn complex sequences of movement because they cannot automate. And they cannot automate because they lack proprioceptors that would indicate the relative positions of parts of their own bodies.

14.5   Conclusions

Among the invertebrates it is possible to answer a question that is strictly speaking unanswerable in vertebrates. Do animals use proprioceptive information in learning? The answer is 'yes' for at least some arthropods in some situations, and the indications are that arthropods in general regularly take this class of sensory input into account. By implication vertebrates do too, since they also have a double proprioceptive system, with receptors capable of signalling muscle tension and joint position independently of one another. For the rest of the invertebrates the indications are otherwise. Annelids, molluscs and the other animals with hydrostatic skeletons seem unable to use proprioceptive inputs in learning. They have stretch receptors but, apart from statocysts, seem to lack receptors capable of registering bodily position independently of muscle tension. Little is known about the use of information from the statocysts in learning by soft-bodied invertebrates but, in *Octopus*, which learns visual and tactile discriminations more readily than most, it seems that it is not involved.

References

Alexandrowicz, J. S. (1960) A muscle receptor organ in *Eledone cirrhosa*. *Journal of The Marine Biological Association of the United Kingdom,* **39,** 419–431.

Alloway, T. M. (1973) Learning in Insects except Apoidea. Ch. 8 pp. 131–171. In: *Invertebrate Learning.* Vol. 2. Corning, W. C. Dyal, J. A. Willows A. O. D. (eds), Plenum Press, New York.

Arbit, J. (1965) Learning in annelids and attempts at the chemical modification of this behaviour. *Animal Behaviour, Supplement* 1, 83–87.

Arnold, J. M. and Arnold, K. O. (1969) Some aspects of hole-boring predation by *Octopus vulgaris. American Zoologist,* 9, 991–996.

Barnwell, F. M. (1965) An angle sense in the orientation of a millipede. *Biological Bulletin. Marine Biological Laboratory, Woods Hole. Mass,* 128, 33–50.

Barnwell, F. H. and Brown, F. A. jr. (1964) Responses of planarians and snails. pp. 268–278. In *Biological effects of magnetic fields,* Barnothy M. F. (ed) Plenum Press, New York.

Bateson, P. P. G. (1974) Neurochemical correlates of environmental change and learning. *Biochemical Society Transactions,* 2, 189–193.

Bennett, E. and Calvin, M. (1964) Failure to train planarians reliably. *Neuroscience Research Progress Bulletin,* 2, 3–24.

Bock, A. (1942) Uber das Lernvermögen bei Asseln. *Zeitschrift für vergleichende Physiologie,* 29, 595–637.

Boycott, B. B. (1954) Learning in *Octopus vulgaris* and other Cephalopods. *Publiciazione della Stazione Zoologica di Napoli,* 25, 67–93.

Boycott, B. B. and Young, J. Z. (1950) The comparative study of learning *Symposia of the Society for Experimental Biology,* 4, 432–453.

Burger, M. L. and Mittelstaedt, H. (1972) Course control by stored proprioceptive information in millipedes. *Proceedings of the 3rd international symposium of Biocybernetics.* Drischel, H. and Dettmar, P. (eds), Fischer, Jena.

Cook, A., Barnford, O. S., Freeman, J. D. B. and Teideman, D. J. (1969) A study of the homing habit of the limpet. *Animal Behaviour,* 17, 330–339.

Cook, S. B. (1969) Experiments on homing in the limpet *Siphonaria normalis. Animal Behaviour,* 17, 679–682.

Corning, W. C. and Kelly, S. (1973). Platyhelminthes: The turbellarians. Ch. 4 pp. 171–224 In: *Invertebrate learning.* Vol. 1 Corning W. C., Dyal J. A. and Willows A. O. D. (eds), Plenum Press, New York.

Corning, W. C. and Ratner, S. C. (eds) *Chemistry of learning.* Plenum Press, New York.

Crisp, M. (1969) Studies on the behaviour of *Nassarius obsoletus* (Say) (*Mollusca.* gastropoda). *Biological Bulletin. Marine Biological Laboratory, Woods Hole, Mass,* 136, 355–373.

Dembrowska, W. S. (1926) Study on habits of the crab *Dromia vulgaris* M.E. *Biological Bulletin. Marine Biological Laboratory, Woods Hole, Mass.,* 50, 163–178.

Drzewina, A. (1910) Création d'associations sensorielles chez les crustacés. *Compte rendu hebdomadaire des séances de la Sociéte de biologie,* 68, 573–575.

Douglas, R. J. (1966) Cues for spontaneous alternation. *Journal of Comparative and Physiological Psychology*, 62, 171—183.

Dyal, J. A. (1973) Behaviour modification in Annelids. Ch. 5 pp. 225—290 In: *Invertebrate learning*. Vol. 1. Corning, W. C. Dyal, J. A. and Willows, A. O. D. (eds) Plenum Press, New York.

Evans, S. M. (1963) Behaviour of the polychaete *Nereis* in T mazes. *Animal Behaviour*, 11, 172—178.

Fischel, W. (1931) Dressurversuche an Schnecken. *Zeitschrifte für physiologie*, 15, 50—70.

Flint, P. (1965) The effect of sensory deprivation on the behaviour of the polychaete *Nereis* in T mazes. *Animal Behaviour*, 13, 187—193.

Frisch, K. von. (1967) *The Dance Language and Orientation of Bees*. Harvard University Press, Cambridge, Massachusettes.

Garth, T. R. and Mitchell, M. P. (1926) The learning curve of a land snail. *Journal of Comparative and Physiological Psychology*, 6, 103—113.

Gilhousen, H. C. (1927) The use of vision and of the antennae in the learning of crayfish. *University of California Publications in Physiology*, 7, 73—89.

Goodwin, G. M., McCloskey, D. I. and Matthews, P. B. C. 1972. A Systematic distortion of position sense produced by muscle vibration. *Journal of Physiology*, 221, 8P.

Graziadei, P. (1964) Muscle receptors in cephalopods. *Proceedings of the Royal Society, Series B*, 161, 392—402.

Le Guelte, L. (1969) Learning in spiders. *American Zoologist*, 9, 145—152.

Hazlett, B. A. (1971) Influence of rearing conditions on initial shell entering behaviour of a hermit crab (Decapoda Paguridea). *Crustaceana*, 20(2), 167—170.

Heck, L. (1920) Über die Bilding einer Assoziation beim Regenwurm auf Grund von Dressurversuchen. *Lotos*, 68, 168—189.

Henton, W. W. and Crawford, F. T. (1966) The discrimination of polarised light by the tarantula. *Zeitchrift für vergleichende Physiologie*, 52, 26—32.

Heyde, A. van der (1920) Über die Lernfähigheit der Strandkrabbe *Carcinus maenas*. *Biologisches Zentnalblatt*, 40, 503—514.

Hinde, R. A. (1970) *Animal Behaviour*. McGraw Hill, New York.

Horridge, A. (1965) The electrophysiological approach to learning in isolatable ganglia. *Animal Behaviour*, Supplement 1, 163—182.

Hullo, S. (1948) Role des tendances motrices et des donnés sensorielles dans l'apprentissage du labyrinthe par les blattes (*Blatella germanica*). *Behaviour*, 1, 297—310.

Jacobsen, A. L. (1963) Learning in flatworms and annelids. *Psychological Bulletin*, 8, 95—114.

Kinosita, H. and Okajima, A. (1968) Analysis of shell-searching behaviour of the land Hermit Crab, *Coenobita rugosus* H. Milne-Edwards. *Journal of the Faculty of Science, Tokyo University,* **2**, 293—358.

Krasne, F. B. (1973). Learning in crustacea. Ch. 7. pp. 49—130 In: *Invertebrate learning,* Vol. 2. Corning, W. C., Dyal, J. A. and Willows, A. O. D. (eds), Plenum Press, New York.

Lahue, R. (1973) The chelicerates. Ch. 6. pp. 1—48 In: *Invertebrate learning* Vol. 2. Corning, W. C., Dyal, J. A. and Willows, A. O. D. (eds), Plenum Press, New York.

Lindauer, M. and Martin, H. (1972) Magnetic effect on dancing bees, pp. 559—568. In: *Animal orientation and navigation Symposium.* Galler, S. R. *et al.* (eds) NASA, Washington.

McConnell, J. V. (1967) In: *Chemistry of learning.* Specific factors influencing planarina behaviour. Ch. 14, pp. 217—233 Corning, W. C. and Ratner, S. C. (eds) Plenum Press, New York.

Markl, H. (1964) Geomenotaktische fehlorientierung bei *Formica polyctena* Forster. *Zeitchrift für vergleichende Physiologie,* **48**, 552—586.

Markl, H. (1966a). Schwerkraftdressuren an honigbienen. I. Die geomeno-taktische fehloreientierung. *Zeitchrift für vergleichende Physiologie,* **53**, 328—352.

Markl, H. (1966b) Schwerkraftdressuren an honigbienen. II Die rolle der schwererezeptorcischen borstenfelder vergleichenderer gelenke für die schwerekompassozientierung. *Zeitchrift für vergleichende Physiologie,* **53**, 353—371.

Markl, H. and Lindauer, M. (1965) In: *The Physiology of insecta.* Physiology of insect behaviour, Ch. 1 Vol. 2 Rockstein, M. (ed). Academic Press, New York.

Marr, D. (1969) A Theory of cerebellar cortex. *Journal of Physiology,* **202**, 437—470.

Merton, P. A. (1964) Human position sense and sense of effort. *Symposia of the Society for Experimental Biology,* **18**, 387—400.

Mittelstaedt-Burger, M. L. (1972) Idiothetic course control and visual orientation. In: *Information processing in the visual system of* arthropods. Werner R. (ed), Springer, Berlin.

Moody, M. F. (1962) Evidence for the intraocular discrimination of vertically and horizontally polarized light by *Octopus. Journal of Experimental Biology,* **39**, 21—30.

Munn, N. L. (1950) *Handbook of Psychological research on the rat.* The Riverside press, Cambridge, Mass.

Peters, H. (1931) Die Fanghandlung der Kreuzspinne (*Epeira diademata* Cl.) *Zeitchrift für vergleichende Physiologie,* **15**, 693—747.

Pringle, J. W. S. (1963) The prioprioceptive background to mechanisms of orientation. *Ergbnisse der Biologie,* **26**, 1—11.

Reese, E. S. (1963) The behavioural mechanisms underlying shell selection by hermit crabs. *Behaviour*, **21**, 78–126.

Rowell, C. H. F. (1966) Activity of interneurons in the arm of *Octopus* in response to tactile stimulation. *Journal of Experimental Biology*, **44**, 589–605.

Sanders, G. D. (1970) The retention of visual and tactile discriminations by *Octopus vulgaris*. *Ph.D. Thesis,* University of London, England.

Schneirla, T. C. (1929) Learning and orientation in Ants. *Comparative Psychology Monographs*, **6**.

Schneirla, T. C. (1933) Some important features of ant learning. *Zeitschrift für vergleichende Physiologie*, **19**, 439–452.

Schneirla, T. C. (1941) Studies on the nature of ant learning. I. The characteristics of a distinctive initial period of learning. *Journal of Comparative Psychology*, **32**, 41–82.

Schneirla, T. C. (1943) The nature of ant learning. II. The intermediate stage of segmental maze adjustment. *Journal of Comparative Psychology*, **34**, 149–176.

Schöne, H. (1961a) Complex behaviour. pp. 465–520 In: *The Physiology of crustacea.* Waterman, T. H. (ed), Academic, New York.

Schöne, H. (1961b) Learning in the spiny lobster *Panulirus argus*. *Biological Bulletin. Marine Biological Laboratory. Woods Hole, Mass,* **121**, 354–365.

Schöne, H. (1965) Release and orientation of behaviour and the role of learning as demonstrated in crustacea. *Animal Behaviour*, Supplement 1, 135–144.

Sokolov, V. A. (1959) A conditional reflex in the gatropod mollusc *Physa acuta* (in Russian). *Vestnik*, Leningrad University. **8**, 82–86.

Sutherland, N. S. (1957) Visual discrimination of orientation and shape by the octopus. *Nature*, **179**, 11–13.

Sutherland, N. S. (1960) Theories of shape discrimination in *Octopus*. *Nature*, **186**, 848–860.

Thompson, E. L. (1917) An analysis of the learning process in the snail *Physa gyrina* Say. *Behaviour Monographs,* **3**.

Townsend, C. R. (1974) Mucous trail following of the snail, *Biomphalaria glabrata* (Say) *Animal Behaviour*, **22**, 170–177.

Turner, C. H. (1913) Behaviour of the common roach (*Periplaneta orientalis L*) on an open maze. *Biological Bulletin, Marine Biological Laboratory, Woods Hole, Mass,* **25**, 348–365.

Walker, J. J., Longo, N. and Bitterman, M. E. (1970) The Octopus in the laboratory; handling maintenance and training. *Behaviour Research Methods and Instrumentation*, **2**, 15–18.

Wells M. J. (1960) Proprioception and visual discrimination of orientation in *Octopus*. *Journal of Experimental Biology*, **37**, 489–499.

Wells, M. J. (1961) Weight discrimination by *Octopus*. *Journal of Experimental Biology*, **38**, 127–133.

Wells, M. J. (1964a) Tactile discrimination of shape by *Octopus*. *Quarterly Journal of Psychology*, **16**, 156−162.

Wells, M. J. (1964b) Tactile discrimination of surface curvature and shape by the octopus. *Journal of Experimental Biology*, **41**, 433−445.

Wells, M. J. (1964c) Detour experiments with octopuses. *Journal of Experimental Biology*, **41**, 621−642.

Wells, M. J. (1966) Learning in the octopus. *Symposia of the Society for Experimental Biology*, **20**, 477−507.

Wells, M. J. (1970) Detour experiments with split-brain octopuses. *Journal of Experimental Biology*, **53**, 375−389.

Wells, M. J. and Buckley, S. K. L. (1972). Snails and trails. *Animal Behaviour*, **20**, 345−355.

Wells, M. J. and Rowell, C. H. F. (1961) Retinal orientation and the discrimination of polarized light by octopuses. *Journal of Experimental Biology*, **38**, 827−831.

Wells, M. J. and Wells, J. (1957) The function of the brain of *Octopus* in tactile discrimination. *Journal of Experimental Biology*, **34**, 131−142.

Wells P. H. (1973) Honey-bees. Ch. 9 pp. 173−185 In: *Invertebrate Learning*. Corning W. C., Dyal J. A. and Willows A. O. D. (eds), Plenum Press, New York.

Wells, P. H. and Wenner, A. M. (1971) The Influence of food scent on behaviour of foraging honey-bees. *Physiological Zoology*, **44**, 191−209.

Willows, A. O. D. (1973) Learning in gastropod molluscs. Ch. 10, pp. 187−274. In: *Invertebrate learning*. Vol. 2. Corning, W. C., Dyal, J. A. and Willows, A. O. D. Plenum Press, New York.

Wodinsky, J. (1969) Penetration of the shell and feeding on gastropods by *Octopus*. *American Zoologist*, **9**, 997−1010.

Yerkes, R. M. (1912) The intelligence of earthworms, *Journal of Animal Behaviour*, **2**, 332−352.

Young, J. Z. (1964) *A model of the brain*, Oxford University Press.

15 Analysis of proprioceptive information

P. J. MILL and
R. N. PRICE

15.1 Introduction

15.1.1 Aims of modelling and types of model

Recently, considerable effort has been directed towards producing models of biological systems, and not least in the field of neurobiology. A model should be predictive and so provide a basis for further experimentation and, as far as possible, it should explain the properties of the system. It is generally easier to understand a system and, therefore, to model it accurately by breaking it down into its components. Mathematical functions can then be produced which describe the relationship between the input and output of each stage and these can then be assembled to produce an accurate model of the complete system.

Models which describe input-output relationships without regard for how the transformation is achieved, are termed non-parametric. This approach leads normally to the identification of the system under study with a known physical system and thus to an identification of those structural components of the system which account for the parameters of the model. Most of the quantitative work on mechanoreceptors has followed this 'black box' approach and, if the objective is to describe the information that a receptor supplies to the rest of the nervous system, for any stimulus, this non-parametric type of model is adequate.

A parametric model, on the other hand, involves a theoretical analysis of the system and leads to a description of system behaviour in terms of the basic physical equations for mass balance, energy balance etc. In other words, a model is produced with parameters which are functions of the physical processes involved in the system. This approach has only limited use in the analysis of proprioceptors at the present time because the parameters involved in the transducer and encoding processes cannot be identified easily. Parametric models will not be further considered in this chapter.

15.1.2   Linearity and system decomposition

Most of the mathematical techniques available for systems analysis can only strictly be applied to linear systems. For a system to be linear it must satisfy two conditions. Firstly, it must be *homogeneous*; that is, the output must be directly proportional to the input. Secondly, it must satisfy the *superposition* criterion; that is, if more than one input is applied to the system simultaneously, the resultant output should be the same as the sum of the outputs obtained when each input is applied separately. This latter criterion is not fulfilled by any system which has a threshold. These two conditions can be summarized by

input $k[u'(t) + u''(t)]$ produces output $k[y'(t) + y''(t)]$.

A corollary of the above is that a linear system also displays the property of *frequency preservation*. Thus, if a sinusoidal input of frequency ω is applied, the corresponding output should also be a sinusoid of frequency ω and differ from the input only in its magnitude (gain) and phase.

The input to a system can be considered as the sum of two components, the initial-state component $x(t_0)$ and the forced-input component $f(t)$. In a linear system there should be corresponding output components $y(t_0)$ and $y_f(t)$, which sum together to produce the total output. Consequently, if the initial-state conditions are zero, the total response of the system is the zero-state response $y_f(t)$. Conversely, if the forced input component is zero, the total response is the zero-input response $y(t)$. For a system to be linear the homogeneity and superposition criteria must apply to both components (Lathi, 1974).

Over most of their range tonic proprioceptors have an initial-state input which produces an identifiable zero-input response but phasic receptors have no output (at least in terms of action potentials) for all initial-state input conditions and, therefore, are never linear systems. Some analysis of non-linear systems is possible but there are no general methods available and each system must be considered separately. In practice no system is perfectly linear but only operates linearly over a limited range of inputs. However, linear systems analysis can usefully be applied to systems which only approximate to linearity or which behave linearly over only part of their range.

As far as proprioceptors are concerned, tonic units most closely approach linearity. The output from the slowly-adapting crustacean abdominal MRO for instance shows a reasonably linear relationship with mechanical input. However, while Krnjević and van Gelder (1961) and Wendler and Burkhardt (1961) observed a linear relationship between receptor-muscle tension and impulse frequency, Terzuolo and Washizu (1962) and Brown and Stein (1966) observed a linear relationship between receptor-muscle length and impulse frequency; yet all of these workers agreed that length and tension are not linearly related. Linearity of impulse frequency with length seems the more likely since Brown and Stein (1966) took the precaution of always applying stretches from the same initial length, thereby avoiding the problem of 'overstretch' (Wiersma, Furshpan and Florey, 1953) and the cumulative effects of adaptation. This is supported by a recent study (unpublished observations; Fig. 15.1).

As far as is known no truly phasic proprioceptors have been modelled mathematically. However, a rather slowly adapting, mechanoreceptive spine on the femur of the cockroach has been studied (Pringle and Wilson, 1952; Chapman and Smith, 1963; Crowe, 1967) and provides some insight into the problem of analysing phasic units. Figure 15.2 shows how input and output are related in this receptor. While it satisfies the homogeneity criterion, the graphs do not pass through the origin and thus the superposition criterion is not satisfied.

An alternative way to that discussed at the start of this section, in which a system response can be decomposed, is to separate the transient (natural) response from the steady-state response. The transient response results from the initial conditions of the system

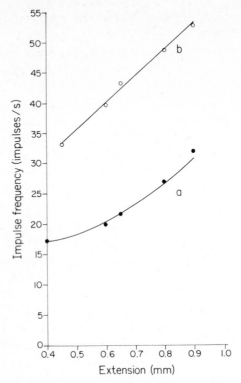

Fig. 15.1   The instantaneous impulse frequency (measured 12.5 s after the application of an extension) plotted against the level of sustained extension, for the slowly adapting MRO of the lobster *Homarus gammarus*. (a) Extension applied at an initial receptor length of 5.0 mm; (b) Extension applied at an initial receptor length of 6.0 mm. The same receptor was used for (a) and (b) and did not include the whole of the receptor muscle.

and decays with time while the steady-state response results from the applied input (forcing function). They represent, respectively, the complementary and particular solutions to the characteristic equation of the system.

15.1.3   Presentation of results

A proprioceptor can be represented by a block diagram;

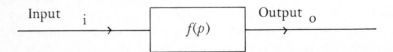

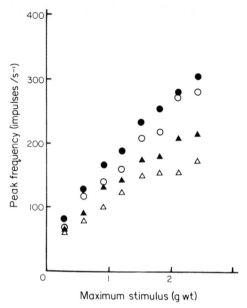

Fig. 15.2   Records obtained by subjecting the cockroach femoral spine to stimuli of constant frequency but with varying amplitude. (a) Ramps at 0.8 Hz; (b) Ramps at 0.5 Hz; (c) Ramps at 0.3 Hz; (d) sinusoidal stimuli at 1.0 Hz. (After Crowe, 1967).

This is an open-loop system and the ratio $o/i = f(p)$ is its transfer relationship some form of which, for a linear system, provides the most convenient model. The situation is normally complicated by feedback of the output (position) on the input (receptor length) so that a closed-loop situation exists.

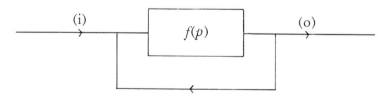

However, it is far simpler to study the receptor with the feedback loop opened and most workers have followed this approach.

    A certain amount of information about a system can be inferred from plots of the response (impulse frequency) against time, following the application of a stimulus. Each particular stimulus requires a separate graph and this limits the usefulness of this approach.

The dynamic behaviour of an nth order, linear, time-invariant system subjected to an input $i(t)$ and producing output $o(t)$ can be described by an nth order linear differential equation of the form

$$a_n \cdot \frac{d^n o(t)}{dt^n} + a_{n-1} \cdot \frac{d^{n-1} o(t)}{dt^{n-1}} + \ldots a_1 \cdot \frac{do(t)}{dt} + a_0 \cdot o(t) = i(t)$$

$a_n, a_{n-1}, \ldots a_0$, and a_1 are constants.

For most systems the right hand side of the equation will include derivatives of $i(t)$. Such an equation completely describes the system since, for any given input and initial conditions, the output can be determined but it is an unwieldy model and is seldom used. It does, however, demonstrate the strategy of modelling. That is, of finding suitable mathematical descriptions of the input and output which make calculating the system transfer relationship as simple as possible. The input and output signals can be waveforms which vary in time and a complete time-domain description of such a signal should specify its value at every instant of time. While this is possible for a simple waveform, of the square-wave or ramp type, complex waveforms can only be approximated by a polynomial expansion

$$f(t) = a_0 + a_1 t + a_2 t^2 + a_3 t^3 + \ldots + a_n t^n$$

This will fit the actual waveform at $n + 1$ points, if the constants are chosen appropriately, and the fit will improve as n is increased.

To simplify the data handling, it is usually necessary to *transform* the descriptions of input and output. Two of the most useful transforms are the Fourier and Laplace, both of which describe signals in the frequency-domain.

It is possible to consider a waveform as the sum of a series of sinusoids of appropriate frequency, amplitude and relative phase. Taking the Fourier transform of a signal is equivalent to representing a signal in this way and gives the signal spectum. Taking the Laplace transform of a signal is to represent it not only by sine and cosine functions but also by growing and decaying sinusoidal functions and exponential functions. Since the Laplace transform analyses a signal into oscillatory and non-oscillatory functions it is of more general applicability than the Fourier transform and is particularly useful for non-periodic signals. The Laplace transform $F(s)$ of a time function $f(t)$ is given by

$$F(s) = \mathscr{L}[f(t)] = \int_0^\infty f(t) e^{-st} \, dt$$

and is equal to $\sigma + j\omega$ (σ is the real part of s; ω is the imaginary part of s and is the frequency in radians s^{-1}; $j = \sqrt{(-1)}$. (For details of complex numbers see, for example, Lathé 1972). The inverse Laplace transform is given by

$$f(t) = \mathscr{L}^{-1}[F(s)] = \frac{1}{2\pi j} \int_{\sigma - j\infty}^{\sigma + j\infty} F(s)e^{st}\,ds$$

Fortunately, tables of Laplace transforms of standard functions are available. The Laplace transform has particular significance because the ratio of the Laplace transform of the output to the Laplace transform of the input is the transfer function $G(s)$ of the system.

$$G(s) = \frac{o(s)}{i(s)}.$$

The transfer function provides a particularly useful model since it gives information on both the gain and phase characterisitics and can be used to predict the system response to any input for which the Laplace transform is known. The transfer function assumes that the initial conditions (state and input) are zero (Section 1.2) but these can be dealt with separately and added to the predicted response. The overall response of a chain of black-boxes, for which only the transfer functions of the individual black-boxes are known, is simple to predict because of the additive property of transfer functions. It is important to appreciate that no such simple relationship as the transfer function exists in the time-domain and it is not possible to predict the output by multiplying the input waveform by some time-function. Multiplication in the frequency-domain is not equivalent to multiplication in the time-domain.

Since the Laplace transform of a signal represents that signal by an infinite set of terms of the form e^{st}, it is no surprise to find that the transfer function defines the effect of the system on each input of that form. If the Laplace transform of a signal is known, its' sinusoidal frequency spectrum can be derived by substitution of $j\omega$ for the complex variable s. This is equivalent to finding the Fourier transform of the signal. A similar procedure, when applied to the transfer function, yields the frequency response $G(j\omega)$ of the system. For any frequency ω, $G(j\omega)$ is the product of a gain term and a phase angle, and these are found from the modulus and argument, respectively, of $G(j\omega)$.

$$G(j\omega) = \frac{o(j\omega)}{i(j\omega)} = \frac{\text{output spectrum}}{\text{input spectrum}}.$$

The frequency response is another convenient model of a system and is often presented graphically, in the form of a Bode plot. A Bode plot consists of two graphs; one of the magnitude of the response, in decibels, plotted against the logarithm of the stimulus frequency and the other of the phase lag or lead, in degrees, plotted against the logarithm of the stimulus frequency (Fig. 15.3). From this it is possible to predict the steady-state response of the system to any sinusoidal input and the response to any input for which the spectrum is known. The latter is achieved by multiplying together the gain characteristics and adding the phase characteristics of the component frequencies of the input spectrum. This additive property of Bode plots and the fact that plots of many standard functions are available, often allows the transfer function to be estimated by drawing its' Bode plot.

An alternative graphical presentation of the frequency response is the Nyquist diagram in which the frequency response is plotted as a series of vectors on polar coordinates, with the gain at each frequency given by the distance from the origin and the phase angle by the angle with the abscissa (lag is plotted clockwise, lead anticlockwise; Fig. 15.4). The Nyquist diagram has an advantage in that it provides information on the stability of the system; that is, whether the transient responses of the system die away and, if so, how rapidly. Most open-loop systems are stable but it is possible to predict if the corresponding closed-loop system is also stable from the Nyquist diagram. If the open-loop transfer function is stable then the closed-loop system will be stable provided that the path of the Nyquist plot does not cross the abscissa outside the critical value of -1. An intersect beyond -1 represents a greater than unity gain at a phase lag of $180°$. In the closed-loop version of such a system a sine wave input would be reinforced by the returning signal and, since the returning signal is larger than the input, the amplitude would build up continuously and, even if the input were removed, the system would continue to oscillate.

15.2   Methods for testing and analysis

Three types of testing have been usefully applied to proprioceptors. In each case time-domain or frequency-domain descriptions are possible, but the latter are generally used.

(A)

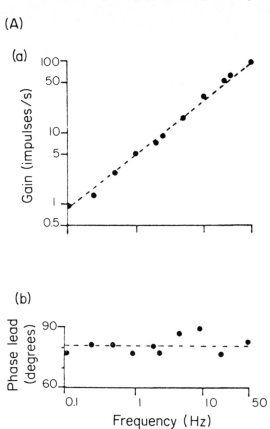

(a)

(b)

Fig. 15.3   Bode plots ((a) gain; (b) phase). (A) Femoral spine of cockroach. Sinusoidal forcing functions were used with a stochastic, band-limited auxillary function. The dashed lines indicate the responses predicted from the transfer function based on the step response; (B) Crayfish slowly-adapting MRO. Plots constructed from its' response to sinusoidal stretches. The amplitude is the peak response minus the resting response. (In (b) the phase lead is given by the vertical bars, the vertical lines indicating the uncertainty of these estimates. The horizontal, dashed line indicates the phase lead predicted by the transfer function derived from the step response. (C–E) Lobster slowly-adapting MRO. (C) Plots constructed from the response to sinusoidal stretches of small amplitude; (D) Plots constructed from its' response to sinusoidal stretches. Each symbol is a different preparation; (E) Plots constructed using a pseudo-random Gaussian sequence as a forcing function and analysis of input and output spectra. ((A) from French, Holden and Stein, 1972; (B) from Brown and Stein, 1966; (C) from Borsellino, Poppele and Terzuolo, 1965; (D) from Terzuolo and Knox, 1971; (E) unpublished observations.)

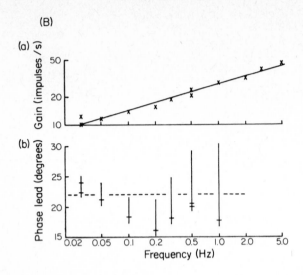

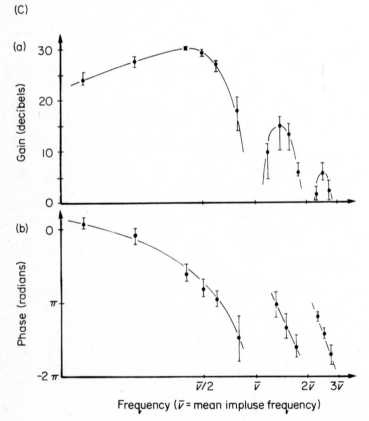

Figure Continued

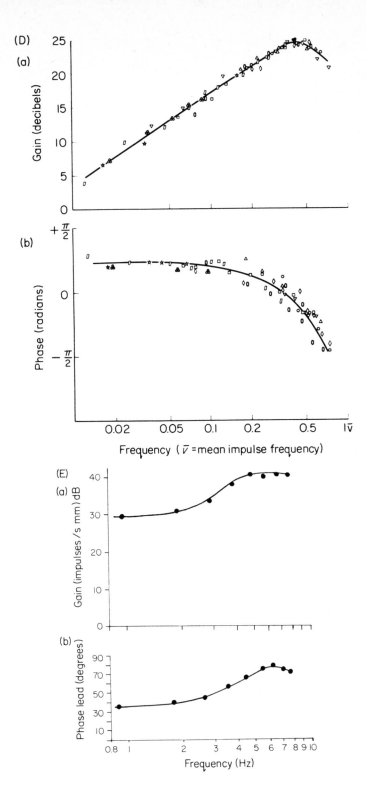

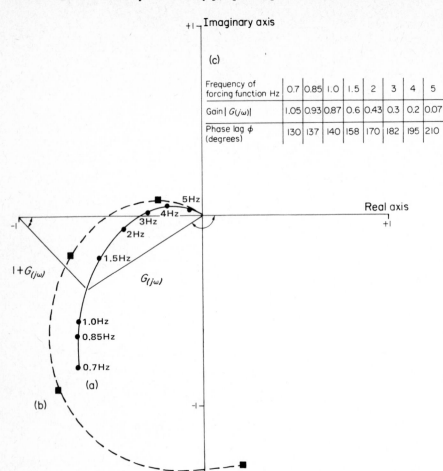

Frequency of forcing function Hz	0.7	0.85	1.0	1.5	2	3	4	5
Gain $\mid G_{(j\omega)} \mid$	1.05	0.93	0.87	0.6	0.43	0.3	0.2	0.07
Phase lag ϕ (degrees)	130	137	140	158	170	182	195	210

Fig. 15.4   Nyquist diagrams. (a) The open-loop Nyquist diagram for the system with the characteristics given in the table (c). The system is stable since the Nyquist plot does not enclose the $(-1,0)$ point. (b) The closed-loop Nyquist diagram for the same system as (a). This is found from $\overline{G(j\omega)} = [G(j\omega)] / [1 + G(j\omega)]$ (where $\overline{G(j\omega)}$ and $G(j\omega)$ are the closed and open-loop frequency responses respectively). e.g. at 1 Hz, $\overline{G(j\omega)} = (0.87 \cdot \angle{-140°}/0.65 \cdot \angle{-59°}) = 1.34 \cdot \angle{-81°}$.

15.2.1   Transient response methods

These methods use deterministic inputs, of the step, impulse and ramp type, which are readily described mathematically and are reasonably easy to apply to a proprioceptor. Brown and Stein (1966) used step changes in receptor length to determine the transfer function of the crayfish slowly-adapting MRO. The steps were all applied from the same initial receptor length and in each case showed an initial high impulse frequency in the response, followed by adaptation. The impulse frequency at any particular time after the application of the step was approximately proportional to the amplitude of the step and the time course of the response frequency y was best fitted by the equation

$$y = a x t^{-k}$$

(a is a constant for any one receptor and provides a measure of receptor sensitivity to unit length change; k is a constant, determining the rate of decay of the response; x is length change). The transfer function derived from this step response is given by

$$G(s) = \frac{\mathscr{L} \text{ output}}{\mathscr{L} \text{ input}} = \frac{\int_0^{\infty} e^{-st} a x t^{-k} \, dt}{\int_0^{\infty} e^{-st} x \, dt}$$

This is more conveniently expressed as a gamma function which is tabulated and can be looked-up in standard mathematical tables.

$$G(s) = \alpha \Gamma(1 - k) s^k .$$

This can be used to predict the system response to any input. Thus the Laplace transform of a ramp stretch of amplitude x, lasting t_1 seconds, with a constant velocity $v = x/t_1$ and with the stretch maintained at x for $t > t_1$, is given by

$$i(s) = \left(\frac{v}{s^2}\right) (1 - e^{-st_1}).$$

Multiplying this by the transfer function and finding the inverse Laplace transform, Brown and Stein predicted that the response frequency time course y would be given by

$$y = \frac{a v t^{1-k}}{1 - k} \qquad \text{for} \qquad t \leqslant t_1$$

and

$$y = a v \frac{(t^{1-k} - [t - t_1]^{1-k})}{1 - k} \qquad \text{for} \qquad t > t_1$$

which predicts that the impulse frequency during ramp application should increase according to the $(1 - k)$th power of the elapsed time and at peak stretch should decline as the difference between two power functions. Also, the peak frequency should increase as the kth power of the velocity. This is discussed in Section 15.4.1. When account was taken of the second order terms in x for the step response equation the ramp response predicted from the transfer function fitted the experimental curve very well (Fig. 15.5).

Pringle and Wilson (1952) applied step stretches to the femoral spine of the cockroach and found that the response frequency was best fitted by a series of exponential terms;

$$y = 24 + 165e^{-12t} + 120e^{-1.1t} + 42e^{0.175t}$$

The transfer function derived from this response is given by

$$G(s) = \frac{351s^3 + 789s^2 + 486s + 57}{s^3 + 13.3s^2 + 15.5s + 2.3}$$

From this, the response to a sinusoidal input was predicted and approximates closely to the experimental data. Chapman and Smith (1967) suggested that a transfer function of this type predicts in-phase modulation of the output for low frequency, sinusoidal forcing functions; with increasing gain and phase lead. This was contrary to the responses they observed which, over a wide range of frequencies, showed a constant phase lead. Chapman and Smith described the time course of the step response by a power function and produced a transfer function of the same form as that of Brown and Stein (1966) for the MRO. Crowe (1967) compared the predicted responses to ramp stretches from both models (Fig. 15.6) and found that his observed responses were most like those predicted by Pringle and Wilson's model.

However, French, Holden and Stein (1972), also using the step response of the receptor, supported Chapman and Smith, producing the same transfer function. Expressed as a gamma function it is given by $G(s) = B \ \Gamma \ (1 - K) \ s^{K}$, which is identical with that found by Brown and Stein (1966) for the slowly-adapting MRO.

A transient of particular interest is the impulse. This is a pulse of, theoretically, unit area and infinitely short duration and has the Laplace transform 1. Thus the Laplace transform of the impulse response is the transfer function of the system. In practice impulses are difficult to generate and apply to proprioceptors and the impulse response is, therefore, difficult to find directly. The superposition criterion allows any input signal to be represented by a series of impulses and the corresponding output as the sum of the individual impulse reponses. Provided the input can be resolved into a series of impulses the impulse response is a powerful model. However, integrating the individual impulse responses, in order to synthesise the output, involves finding the convolution integral of the impulse response. This is a difficult mathematical procedure but is made much simpler by considering the frequency-domain equivalent of the impulse response.

15.2.2   Frequency response methods

These analyse the steady-state response of a system when subjected to a continuous sinusoidal forcing function. To obtain a steady-state response the system must be stable and all the initial transients responses must die away. The response is expressed as a gain term and a phase shift relative to the input and in order to cover a useful range of inputs a large number of experiments is required. The 'sinusoidal transfer function' or frequency response, $G(j\omega)$ (Section 1.3) takes no account of the transient response of the system. It is related to the steady-state response $o'(t)$ of a system subjected to a sine wave input by

$$o'(t) = |G(j\omega)| \sin(\omega t + \arg G(j\omega))$$

(where $|G(j\omega)|$ is the modulus (gain term) and $\arg G(j\omega)$ is the argument (phase angle term) of $G(j\omega)$ (Section 1.3).

Borsellino, Poppele and Terzuolo (1965) and Terzuolo, Purple, Bayly and Handelman (1968) have applied this method of analysis to the crustacean slowly-adapting MRO (Fig. 15.3). Since the Bode plots are the same for inputs of different amplitudes, the length-impulse frequency relationship is linear for this receptor. The transfer function of the receptor is given by

$$G(s) = \frac{s^{\frac{1}{2}}}{s^2 + 1.4p\,s + p^2}$$

($p = 2\pi\bar{v}$ where $\bar{v}$ is the mean impulse frequency). Since the gain decreases as the input frequency approaches the mean impulse

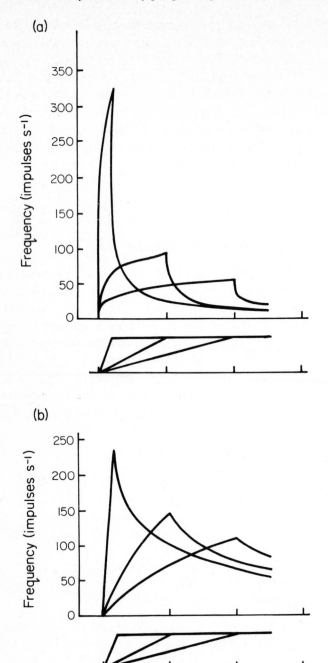

frequency the receptor is behaving, to some extent, as a holding device, which typically has a transfer function of the type

$$G(s) = \frac{1 - e^{-\frac{1}{v}s}}{s}$$

The Bode plot of a holding device has zero gain at the carrier frequency and its' harmonics and a phase lag at all frequencies. The phase lag at the carrier frequency and its harmonics is 90°.

The frequency response has special significance when one considers that many functions can be considered as the sum of a series of sinusoidal functions, of various periods, magnitudes and phases. If the input is considered in this way, the response is the sum of the individual frequency responses. This allows the output of a system to be predicted if its' frequency response is known and the spectrum of the input signal, the Fourier transform, can be found. The frequency response is only applicable to a stable system and is usually obtained for an open-loop system. However, for a unity-feedback system with open-loop frequency response $G(j\omega)$, the closed-loop frequency response $G(j\omega)$ is given by

$$\overline{G(j\omega)} = \frac{G(j\omega)}{1 + G(j\omega)}$$

15.2.3 Statistical response methods

This approach employs inputs which are non-deterministic, that is, they are random functions and can only be described in terms of probability statements and averages. The ideal forcing function is 'white noise', which is a random series of impulses and is characterized by having infinite power at all frequencies in its' spectrum. It cannot be realised in practice but signals can be generated which

---

Fig. 15.5 Predicted responses to ramp inputs for the cockroach femoral spine based on those obtained from step inputs. (a) responses calculated from the transfer function based on the power-function description of the step response; (b) responses calculated from the transfer function based on the description of the step response as the sum of a series of exponential terms. Lower traces in (a) and (b) indicate slopes of ramps used for the predictions shown above. (After Crowe, 1967).

approximate to it in their statistical properties. One such set of signals are the pseudo-random binary sequences (PRBSs). These can assume two possible states $\pm a$ and change or do not change state with equal probability, at discreet intervals of time Δt. (See Fig. 15.6) The sequence is periodic, with period $T = N\Delta t$. N is an integer given by $N = 2^n - 1$ (where n can be any integer and N is 15, 31, 63, 127, etc.). Since abrupt changes are difficult to apply to proprioceptors it is preferable to use a 'smoother' statistical input of the pseudo-random Gaussian type (Fig. 15.6).

Since the input is a random time function the particular input used is considered as only one of an infinite number of functions which might have occured and is known as the sample function. All possible functions are referred to as the sample space. It is assumed that all sample functions within the sample space exhibit the same

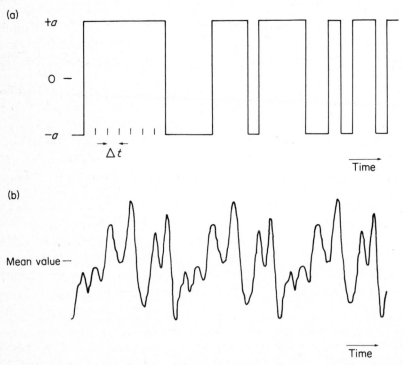

Fig. 15.6   (a) Part of a 63-bit pseudo-random binary sequence (PRBS). $+a$ and $-a$ are the two possible states of the signal; Δt is the bit interval. (b) Three cycles of a 127-bit pseudo-random Gaussian signal, derived by severe filtering of the equivalent 127-bit PRBS. The period of one cycle is 127 Δt.

statistical properties and that these properties are independant of time and show no trends. Such signals can be described in a variety of ways of which the probability density function is the most complete. However, a partial statistical description, such as a correlation function, will often suffice. The correlation of two random variables is the 'expected value' of their product and demonstrates whether one variable depends on the other in any way. It is often more convenient to find the correlation with their mean values subtracted and this is called the covarience. The covariance when normalised to the range -1 to $+1$ is the correlation coefficient.

The correlation of two random variables which are samples, taken at different times, of the same sample function is the *autocorrelation function* (ACF). This is equivalent to finding the time average of the products of the values of the time function τ seconds apart, as τ is varied from zero to some large value. If the value of the time function τ seconds from now is greatly dependent on the present value then the ACF will be large. If the random variables are taken from different random processes the equivalent function is the *cross-correlation function* (CCF), which provides a measure of the dependence of one signal (the receptor output) on another (the stimulus). If the two signals are completely independent of each other (uncorrelated) the CCF is zero.

An input signal can be represented by a series of impulses with the output equal to the sum of the individual impulse responses (Section 2.1) and it is possible to determine the transfer function using this form of input. However, the deconvolution is difficult and it is better to consider the frequency-domain charateristics of the correlation functions; an approach known as *spectral analysis* (e.g. Jenkins and Watts, 1968). Spectral analysis is concerned with the *power spectrum* (power density spectrum) $\Phi(\omega)$ of a signal which illustrates how the power in a signal is distributed between the various frequency components that make it up. The mean power of a sine wave at each frequency is considered but this spectrum does not give any information concerning the phase relationships of the frequency components. Consequently, the power spectrum can not be used to reconstruct the waveform. However, the Fourier transform provides the spectrum of a time function (Section 1.3) and since the ACF of a statistical signal is a time function it has a Fourier transform. Furthermore, the Fourier transform of the ACF of a statistical signal is its' power spectrum.

The power spectrum of the output from a linear system can only depend on the gain of the system and the power spectrum of the input signal and it can be shown that if the input power spectrum is $\Phi_{xx}(\omega)$ the output power spectrum $\Phi_{yy}(\omega)$ is given by

$$\Phi_{yy}(\omega) = \Phi_{xx}(\omega) \cdot |G(j\omega)|^2$$

In other words, each input component is multiplied by the square of the system gain at the relevant frequency. Given the input power spectrum and the output power spectrum, both of which can be found from the input and output signals, it is then possible to define the gain part of the frequency response of the system.

In practice the signals employed in spectral analysis are finite. This is equivalent to viewing the data through a rectangular window and has the effect of convolving the spectral components with a (sine $(x))/x$ function. This and the necessity to use sampled-data versions of the input and output signals result in the need to use the discrete Fourier transform which produces a line spectrum. That is, a series of spectral estimates, made only at descrete frequencies. Such spectra suffer from the problem of leakage of spectral components into adjacent frequencies. This can be compensated for by the use of a Tukey lag window (Blackman and Tukey, 1959) which has the effect of including the spectral estimates of a band of frequencies in the estimate at a nominal centre frequency. This produces what is known as a smoothed spectrum. The bandwidth of the window must be appropriate to the detail of the spectrum it is necessary to resolve.

Although the ACF and its' spectrum provide no information on phase, the CCF does retain phase information and it also has a spectrum, the cross power density spectrum $\Phi_{xy}(\omega)$. Thus

$$\Phi_{xy}(\omega) = C_{xy} - jQ_{xy}(\omega)$$

(where both $C_{xy}(\omega)$ and $jQ_{xy}(\omega)$ can be found from the CCF). It can be shown that the frequency response of a system is given by

$$G(j\omega) = \frac{\Phi_{xy}(\omega)}{\Phi_{xx}(\omega)} = \frac{C_{xy}(\omega) - jQ_{xy}(\omega)}{\Phi_{xx}(\omega)}$$

$G(j\omega)$ can also be expressed in the form

$$G(j\omega) = A(\omega) \cdot e^{j \cdot P(\omega)}$$

$A(\omega)$ is the modulus of $G(j\omega)$ and the gain of the system and

$$A(\omega) = \frac{\sqrt{[C_{xy}{}^2(\omega) + Q_{xy}{}^2(\omega)]}}{\Phi_{xx}(\omega)}$$

while $P(\omega)$ is the argument of $G(j\omega)$ and the phase of the system and

$$P(\omega) = \arctan -\frac{Q_{xy}(\omega)}{C_{xy}(\omega)}$$

By following this route it is possible to find the frequency response of a receptor and its' Bode diagram, using a statistical signal as the input. Fig. 15.7 shows the strategy adopted in applying spectral analysis to the slowly-adapting MRO of the lobster; Fig. 15.8 examples of the results (unpublished observations).

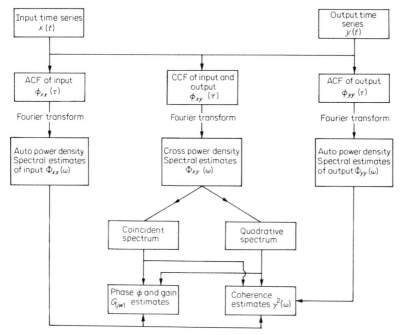

Fig. 15.7 Strategy for applying spectral analysis to the input and output of a system. (See Figs. 15.3E, 15.8 for an example of the implementation of this scheme.)

(a)

Output

Input

1mm

1s

(b) ϕ_{xx}

0.1

0.5

0

-0.2

-0.4

$\Delta\tau$

10

20

Lag number

(c) ϕ_{yy}

20

10

Lag number

(d) ϕ_{xy}

1.0

0.5

0

-0.2

-0.4

10

20

Lag number

(e) $\Phi_{xx}\left(\dfrac{mm^2}{Hz}\right)$

10^0

10^{-1}

10^{-2}

0

10

5

Frequency (Hz)

Spectral line placings

20

(f) $\Phi_{yy}\left(\dfrac{Impulses\ s^{-1}}{Hz}\right)$

75

25

50

10

10

5

Frequency (Hz)

Spectral line placings

20

(g) Coherence

1.0

0.8

0.6

0.4

0.2

0

10

5

Frequency (Hz)

Spectral line placings

20

The *coherence* (or squared coherence spectrum) $\gamma^2(\omega)$ provides a measure of the correlation between the two time series as a function of frequency. If the input and output are poorly correlated the coherence is low, while a value of 1 indicates maximum correlation. Thus it is a measure of the output-signal to noise ratio and a low coherence indicates that the phase and gain estimates have been calculated from noisy data and reduces confidence in these estimates.

Spectral analysis has been applied to a neuron analogue (French and Holden, 1971c), the femoral chordotonal organ of the locust (Holden and French, 1971) and the femoral spine of the cockroach (French *et al.*, 1972). This group of workers have employed the direct method of spectral estimation, by which the Fourier transforms of the input and output time series are found directly, to produce the input and output power spectra, without the intermediate stages of calculating covariance functions. The direct approach is facilitated by use of the fast Fourier transform (Cooley and Tukey, 1965) which considerably reduces the number of computations necessary and allowed French *et al.* to apply spectral analysis in an on-line context (e.g. French and Holden, 1971b).

15.3   Sampling techniques

15.3.1   General considerations

So far the input and output have been considered as continuous signals but, if digital computers are to be employed to do the necessary data analysis, sampled representations of these signals are required. In order to conserve and reproduce the harmonic content of

---

Fig. 15.8   (a) Oscilloscope trace of typical output from the lobster slowly-adapting MRO when subjected to five cycles of pseudo-random Gaussian input; (b) Autocorrelogram (ACF) of the input; (c) Autocorrelogram (ACF) of the output; (d) Crosscorrelogram (CCF) with the output delayed relative to the input. The shift in peak correlation reflects a phase shift by the receptor. (B−D) are from the same experiment and represent an average of four cycles. The lag Δt is 0.04 seconds in all cases. (e) Input Power Density Spectrum based on estimates made from (b); (f) Output Power Density Spectrum based on estimates made from (c). For (e) and (f) a Tukey window of 2.3 Hz was employed. The phase and gain estimates derived from the Power Density estimates in (e) and (f) are given in Fig. 15.3E. (g) Cohence Spectrum.

a continuous signal, it is preferable to sample at some fixed time interval, which is chosen after consideration of the signals' bandwidth and the maximum frequency, which it is necessary to detect. The sampling procedure is liable to distort the signal spectrum and invalidate any information derived from it. This can be minimized by allowing the continuous signal to modulate the amplitude of a train of equispaced narrow pulses, Dirac pulses. This is equivalent to point sampling the signal at regular intervals. The sampled-signal spectrum is then the convolution of the spectrum of the Dirac pulse train and the continuous-signal spectrum. It can be shown that this sampled-signal spectrum is of the same form as the continuous-signal spectrum but is repeated indefinitely along the frequency axis at intervals of $2\pi/T$ radians per second. (T is the sample interval) The total energy in the two spectra is the same. Consequently the peak power of each repeated sub-spectrum in the sampled-signal spectrum must be considerably less than in the continuous-signal spectrum. If compensation is made for this, sampling with a Dirac comb does not distort the signal spectrum. If the upper or lower frequency components of the continuous-signal spectrum is greater than π/T, adjacent repetitions of the sub-spectra in the sampled-signal spectrum will overlap and exaggerate the power in the high frequency components. This distortion is known as aliasing. It can be avoided by filtering the continuous signal to eliminate high frequency components before the sampling is done or by using a sufficiently high sampling rate. If the highest frequency component of a continuous signal is $\hat{\omega}$ radians per second or $\hat{f}$ Hz (i.e. $\hat{f} = \hat{\omega}/2\pi$) aliasing cannot occur if $\hat{\omega}$ is less than π/T. That is,

$$\hat{f} = \frac{\hat{\omega}}{2\pi} < \frac{1}{2T} \ .$$

This specifies the minimum sampling rate $2\hat{f}$ which is required to produce an adequate representation of a continuous signal. $\hat{f}$ is usually known as the Nyquist frequency and given the symbol F_n.

15.3.2   Input representation

The input to a proprioceptor will usually take the form of length or tension changes. A description of the continuous signal will usually be available for deterministic inputs but for non-deterministic inputs a sampling procedure is often necessary. Although sampling can often be avoided with stochastic signals, of the PRBS type, by using correlation techniques, inputs of the pseudo-random Gaussian type (Fig. 15.7) do require sampling. (Section 15.3.1).

15.3.3 Output representation

Where a train of action potentials is considered as the output of the proprioceptor the problem arises of how to produce a series of equispaced samples of this spike train with the same information content as the spike train itself. Bayly (1968) has shown that a spike train contains high frequency components with considerable power and if the spike train is considered as a continuous function a very high sample rate is necessary to avoid aliasing. The sample interval would need to be less than the duration of an action potential and this would present difficulties in data handling and implementation. Thus it is profitable to consider methods which may allow lower sample rates to be used without loss or change of the encoded information.

It is possible to filter the spike train with a low pass filter prior to sampling, and so remove the high frequency components, but the spectral estimates of the spike train then depend on the shape of the spikes. This is unlikely to be biologically significant and is very dependant on the recording methods. Alternatively the spike train can be considered as a point process, that is as a time series of identical events of very short duration, rather than as a continuous signal. This allows various parameters of these events to be considered. The most commonly used is instantaneous frequency (the reciprocal of the interspike interval) and this can be plotted as a function of time in a number of ways (Fig. 15.9) and then sampled at regular intervals.

The track and hold method introduces a delay into the signal which will affect the cross spectral estimates and introduce spurious troughs into the frequency response function of a system at multiples of the mean carrier frequency (Terzuolo and Bayly, 1969; McKean, Poppele, Rosenthall and Terzuolo, 1969). Borsellino *et al.* (1965) treated the spike train from a crustacean MRO with a first order filter and then a track and hold device. Consequently, the transfer function they derived for the receptor included that of the spike train treatment. This had to be subtracted to produce the true transfer function of the receptor.

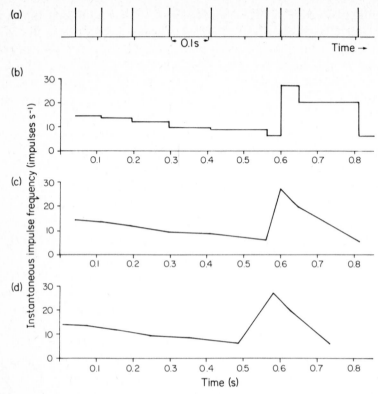

Fig. 15.9   Various methods of producing instantaneous frequency plots. (a) the impulse train from which (b–d) were produced. (b) 'Track and hold'. The instantaneous frequency is plotted at the impulse occurence time and assumed to maintain that value until re-set by the next impulse. (c) 'Track and linear interpolation'. Plotted as B, but the instantaneous frequency is assumed to change linearly between impulse occurences. This is probably the most widely used type of instantaneous frequency plot. (d) as (c), but the instantaneous frequency of an impulse is plotted at a time mid-way between its occurence time and that of the previous impulse.

French and Holden list in greater detail the objections to the sampling procedures so far discussed and have supplied a more suitable alias-free method for sampling spike trains (French and Holden, 1971a). Thus Nelsen (1965) has shown that the spectrum of a spike train can be approximated by the spectrum of a series of Dirac pulses and this allows each spike to be considered as a Dirac function occuring at the same time. Each Dirac pulse is then convolved with the sampling function. This function is equivalent to the impulse response of a perfect low pass filter of the same bandwidth as the Nyquist frequency F_n associated with the sample interval to be used, and in this way aliasing is avoided no matter what the sample interval. (Fig. 15.10) but since the function extends from $-\infty$ to $+\infty$ it is non-realizable. However, it can be made finite without altering its spectral characteristics but is still symmetrical about the spike occurance time and, therefore, the convolution must be worked out for past and future time. This requires off-line (or semi-on-line) computation. The resulting continuous signal for the complete spike train can then be sampled with a Dirac comb. This method has been used in spectral analysis of a receptor by French *et al.*, and Stein (1972) and by ourselves (unpublished observations; Fig. 15.8.)

A comparison of the same spike train sampled by a number of different methods is illustrated in Fig. 15.11.

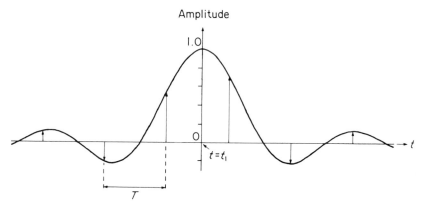

Fig. 15.10   An impulse occuring at time t_1 is convolved with the function [sin $(2\pi\ Fn\ t)$] / [$2\pi\ Fn\ t$] (where Fn is the Nyquist frequency associated with the sample interval T). The resultant time function shown can be sampled at regular intervals. The amplitude is zero at $t = t_1 \pm nT$ (where n is any integer). (After French and Holden, 1971b.) Fig. 15.11f illustrates a spike train sampled using this method.

15.4   Evaluation of models

15.4.1   Model accuracy

All the modelling techniques so far considered have only limited applicability; for example they only apply to linear systems which do

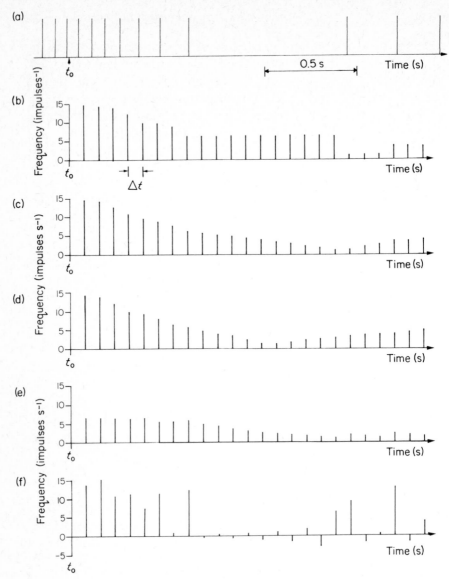

Fig. 15.11   (a) Portion of a spike train from the lobster tonic MRO. (b) Sampled version of (a) produced from a 'track and hold' plot of instantaneous impulse frequency (Fig. 15.9b). (c) Sampled version of (a) produced from a 'track and linear interpolation' plot of instantaneous impulse frequency (Fig. 15.9c). (d) As (c), but in this case the instantaneous impulse frequency associated with each spike is plotted at the preceding mid-interspike interval (Fig. 15.9d). (e) Sampled version of (a) after first order filtering with a time constant of 0.5 seconds. (The value of the time constant was chosen arbitrarly and is not necessarily the most suitable). (f) Sampled version of (a) after perfect low pass filtering. (Fig. 15.10). The sample interval Δt is 0.08 seconds in each case.

not vary with time. Receptors are unlikely to conform to these specifications exactly and models of receptors must be tested in order to assess their accuracy. The various models discussed are inter-related and one must ask which models provide the most accurate description of the systems behaviour and which approach provides this model most conveniently. This evaluation can be achieved by comparing the receptor response predicted by the model with the experimentally obtained response, for a number of different inputs. Comparing models of the same receptor, which have been derived by different methods, is also useful.

As described in section 15.2.1 Pringle and Wilson (1952); Chapman and Smith (1967); Crowe (1967) and French *et al.* (1972) have all produced transfer functions for the femoral spine of the cockroach based on transient response methods.

The frequency response derived from the transfer function of French *et al.* (see equation, p. 618) is given by

$$G(j\omega) = B\Gamma(1-k)f^k e^{jk \cdot \pi/2}$$

(where f is the frequency of the forcing function). The Bode plot produced from this is in close agreement with that obtained by sine wave testing the spine (Fig. 15.3A). French *et al.* also used spectral analysis to produce a Bode plot and this was also in close agreement with the results of sine wave tests.

Thus for the cockroach femoral spine transient response methods, frequency response methods and statistical response methods can all be used to produce similar and accurate models. Spectral analysis is the most convenient method to use since it can be made to provide information on a wide range of frequencies in one test and the period required for testing is short. It also has better spectral resolution than is usually possible when each frequency of sine wave has to be tested separately and spectral analysis does not require the input to be rigorously controlled but merely to be accurately monitored. Disagreement only occurs with regard to the best description of the step response (Section 2.1). Over a short period of time a power function seems to provide the best fit. Sensory adaptation is known to result from at least three separate decay processes in mechanoreceptors; the application of a constant stretch is accompanied by decaying tension, the application of a constant tension is accompanied by a decaying receptor potential and a constant receptor potential is accompanied by a declining impulse frequency (Nakajima and Onodera, 1969a, b). This probably explains why, over a longer time period, the description as a series of exponential decays, with different time constants, provides a better fit.

Brown and Stein (1966) compared their model of the slowly-adapting MRO of the crayfish with experimentally observed data by determining the value of the decay constant k using step (in fact a fast ramp with a rise time of 40 ms), ramp and sinusoidal stretches and the values they obtained ranged from 0.210 to 0.346.

The frequency response of the receptor derived from the step response predicts a constant phase lead of $90k$ degrees. This fits well with the observed results (Fig. 15.3B). In this case the step response would seem to provide the most accurate way to derive the model and the deviations in k produced by other routes probably reflect the effect of non-linearities in the receptor. These probably arise largely as a result of the visco-elastic nature of the receptor muscle. The forces developed in viscous materials are very dependant on the rate of application of stretch as well as the extent of the stretch. If this is a major cause of the deviations from linearity in this receptor then the deviations should be most marked in dynamic situations. Indeed, the fact that the phase lead measured at half the peak value is much greater than when measured at 0.9 peak value supports this idea. It may well be that the linear relationship between length change and impulse frequency in this receptor has no parametric basis. The direct stimulus to the dendrites of this receptor is probably not length change but tension in the receptor muscle and the length-tension relationship for a visco-elastic material is non linear. The impulse frequency is a logarithmic function of the receptor potential and, if the receptor potential is linearly related to the tension, the non-linearities between the length change and impulse frequency will tend to cancel each other out. This would create the impression of linearity between length change and impulse frequency. Borsellino *et al.* (1965), Terzuolo *et al.* (1968) and Terzuolo and Knox (1971) have applied sine wave testing to the slowly adapting MRO of the lobster (Section 15.2.2) (Fig. 15.3C, D). These workers all used the same methods, employing only very small displacements, of the order of a few microns, and using a first order filter and a track and hold device to handle the spike train. The most striking disparity between the Bode diagrams occurs in the phase plots (Fig. 15.3). Terzuolo *et al.* (1968) shows a phase lag at all frequencies above 0.3 $\bar{v}$ approaching 180 degrees at $\bar{v}$. $\bar{v}$ seems to be of the order of 8–15 Hz. This is considerably broader than the band of frequencies tested by Brown and Stein and it may be that their results correspond to those up to a frequency of only 0.25 $\bar{v}$. This could also account for the failure of the gain to level off in Brown and Steins' model. There are structural differences between the stretch receptors of the crayfish and the lobster. Notably, the receptor muscle of the lobster in the region of the sensory dendrites is replaced completely by a mass of 'connective tissue'. (Alexandrowicz, 1951, 1952; Florey and Florey, 1955; Bodian and Bergman, 1962). However, Terzuolo *et al.* (1968) used both crayfish and lobsters and do not report any differences in the behaviour of these receptors and so presumably this can not account for the disparity between the models. Statistical inputs have been applied to this same receptor in the lobster and spectral analysis used to obtain gain and phase estimates for it (unpublished observations) (Fig. 15.3E). The results match neither of the others very closely. At all frequencies examined there is a considerable phase lead which never exceeds 90 degrees. Spectral analysis has only provided estimates up to 10 Hz but sine wave testing at higher frequencies has never revealed a phase lag, although the observed phase lead is usually less than that indicated by the phase estimates from spectral analysis. Using a complex curve-fitting technique (Levy, 1959) a transfer function has been produced from this Bode plot. Preliminary studies suggest that it is of the form

$$g(s) = k(1 + \tau s)$$

k, the static gain is of the order of 0.5 and the constant, τ, associated with the dynamic response is of the order 0.15. This transfer function has no denominator, implying that the phase and gain of this system increase indefinitely with increase in frequency. This is not realizable but emphasizes that this transfer function is only a non-parametric model of the receptor over the range tested. If the range covered were extended, the transfer function would be a more complete description of the receptor.

15.4.2   Applications of models

Assuming that a model of a proprioceptor can be produced which fits the requirements of being a concise and accurate description of the receptors performance characteristics, are there any uses to which the model can be put?

Does the model, albeit non-parametric, provide any insight into the mechanisms underlying the receptors' behaviour or draw attention to features of the receptor that require further investigation? Terzuolo *et al.* (1968) and Terzuolo and Knox (1971), using intracellular electrodes for stimulation and recording, were able to look separately at the transducer and encoder transfer relationships and how they contribute to the overall characteristics of the crustacean MRO. The Bode plot of the encoder was discovered by applying sinusoidally modulated currents to the receptor neuron to mimick the receptor potential. From this it is apparent that the holding device character of the receptor is due to the encoder and it is this that introduces the phase lag into their model of the receptor. The gain is maintained over a wider frequency range than it would be if the encoder process was merely a zero-order hold device, such as a relaxation oscillator. By subtraction of the encoder characteristics from those of the complete receptor the characteristics of the transducer were obtained and could be checked against direct measurements of the length change-receptor potential relationship, when tetrodotoxin was applied to supress spike generation. The Bode plot of the transducer process indicates that transduction introduces a considerable phase lead into the system and this directs attention back to the length-tension relationship for the receptor muscle, and hence to its' mechanical properties; and to the relationship between tension and receptor potential. The dynamic characteristics of crustacean muscle, and this receptor muscle in particular, require thorough investigation.

Brown (1967) has shown that the behaviour of the stretch receptor in the crayfish is different when the receptor muscle is contracting rather than passively stretched. The differences are more marked during the dynamic phases of stimulation. It would be interesting to see what affect receptor muscle contraction has on the transfer relationship of this receptor.

Terzuolo *et al.* (1968) investigated the effect of postsynaptic inhibition on the frequency response of the stretch receptor, and shown that it affects both the gain and phase characteristics. It reduces the system gain and increases the phase, producing a phase lead of 90 degrees at the lower and of the frequecy range.

Another consideration is how models of sense organs can be linked with other models to synthesize networks which will improve our understanding of how in an animals' behaviour is organized.

Modelling of a reflex would seem a good starting point, but despite the many reflexes known to be associated with invertebrate propioceptors no such models have yet been produced. The reflexes associated with the slowly-adapting MRO of the crayfish abdomen are well defined and hypotheses about the role of this receptor, as an error detector for abdominal position, are at a stage where a model of the reflexes would be useful (Fields and Kennedy, 1965; Fields, 1966; Fields, Evoy and Kennedy, 1967) (Sections 2.8, 2.9). Before such a model can be produced the input-output relationship of crustacean muscle requires more thorough investigation. In vertebrates the modelling of reflexes has progressed much further (e.g. Poppele and Terzuolo, 1968; Partridge and Kim, 1969).

Acknowledgements

The previously unpublished work was supported in part by an S.R.C. grant (B/SR/4219) to P.J.M.

References

Alexandrowicz, J. S. (1951) Muscle receptor organs in the abdomen of Homarus vulgaris and Palinuras vulgaris. *Quarterly Journal of Microscopical Science*, 92, 163–199.

Alexandrowicz, J. S. (1952) Receptor elements in the thoracic muscles of Homarus vulgaris and Palinuras vulgaris. *Quarterly Journal of Microscopical Science*, 93, 315–346.

Bayly, E. J. (1968) Spectral analysis of pulse frequency modulation in the nervous systems. I.E.E.E. *Transactions on Bio-medical Engineering*, BME–15, 257–265.

Blackman, R. B. and Tukey, J. W. (1959) Measurement of Power Spectra from the Point of Communication Engineering. New York: Dover.

Bodian, D. and Bergman, R. A. (1962) Muscle receptor organs of crayfish: Functional-anatomical correlations. *Johns Hopkin Hospital Bulletin*, 110, 78–106.

Borsellino, A., Poppele, R. E. and Terzuolo, C. A. (1965) Transfer functions of the slowly adapting stretch receptor organ of Crustacea. *Cold Spring Harbour Symposium of Quantitative Biology*, 30, 581–586.

Brown, M. C. (1967) Some effects of receptor muscle contraction on the responses of slowly adapting abdominal stretch receptors of the crayfish. *Journal of Experimental Biology*, 46, 445–458.

Brown, M. C. and Stein, R. B. (1966) Quantitative studies on the slowly adapting stretch receptor of the crayfish. *Kybernetik*, 3, 175–185.

Chapman, K. M. and Smith, R. S. (1963) A linear transfer function underlying impulse frequency modulation in a cockroach mechanoreceptor. *Nature*, **197**, 699–700.

Cooley, J. W. and Tukey, J. W. (1965) An algorithm for the machine calculation of complex Fourier series. *Mathematics of Computation*, **19**, 297–301.

Crowe, A. (1967) Studies on the transfer function of a cockroach mechanoreceptor. *Comparative Biochemistry and Physiology*, **20**, 13–25.

Fields, H. L. (1966) Proprioceptive control of posture in the crayfish abdomen. *Journal of Experimental Biology*, **44**, 455–468.

Fields, H. L., Evoy, W. H. and Kennedy, D. (1967) The reflex role played by the efferent control of an invertebrate stretch receptor. *Journal of Neurophysiology*, **30**,

Fields, H. L. and Kennedy, D. (1965) The functional role of muscle receptor organs in crayfish. *Nature*, **206**, 1235–1237.

Florey, E. and Florey, E. (1955) Microanatomy of the abdominal stretch receptors of the crayfish (*Astacus fluviatilis* L.). *Journal of General Physiology*, **39**, 63–85.

French, A. S. and Holden, A. V. (1971a) Alias-free sampling of neuronal spike trains. *Kybernetik*, **8**, 165–171.

French, A. S. and Holden, A. V. (1971b) Semi- on-line implementation of an alias-free sampling system for neuronal signals. *Computer Programs in Biomedicine*, **2**, 1–7.

French, A. S. and Holden, A. V. (1971c) Spectral analysis of information transfer through single units. *Fourth Annual Winter Conference on Brain Research*, Aspen, Colorado.

French, A. S., Holden, A. V. and Stein, R. B. (1972) The estimation of the frequency response function of a mechanoreceptor. *Kybernetik*, **11**, 15–23.

Holden, A. V. and French, A. S. (1971) Spectral analysis of receptor dynamics. *Third Winter Meeting of the Canadian Physiological Society*, Banff, Alberta.

Jenkins, G. M. and Watts, D. G. (1968) *Spectral Analysis and its applications.* Holden-Day, San Francisco.

Krnjević, K. and van Gelder, N. M. (1961) Tension changes in crayfish stretch receptors. *Journal of Physiology*, **159**, 310–325.

Lathi, B. P. (1974) Signals, Systems and Controls. Intext: New York.

Levy, E. C. (1959) Complex curve fitting. *Institute of Radio Engineers Transactions*, AC–4, 37–43.

Mercer, R. (1973) Some problems encountered in the development of techniques for modelling mechanoreceptors. *Ph.D. Thesis*, University of Leeds.

Nakajima, S. and Onodera, K. (1969a) Membrane properties of the stretch receptor neurons of the crayfish with particular reference to the mechanisms of sensory adaptation. *Journal of Physiology*, **200**, 161–185.

Nakajima, S. and Onodera, K. (1969b) Adaptation in the generator potential in the crayfish stretch receptors under constant length and constant tension. *Journal of Physiology*, **200**, 187–204.

Nelsen, D. E. (1964) Calculation of power density spectra for a class of randomly jittered waveforms. *Research Laboratory of Electronics, M.I.T.*, QPR–74, 168–179.

Partridge, L. D. and Kim, J. H. (1969) Dynamic characteristics of the response in the vestibulomotor reflex. *Journal of Neurophysiology*, **32**, 485–494.

Poppele, R. E. and Terzuolo, C. A. (1968) Myotatic reflex: its input-output relation. *Science*, **159**, 743–745.

Pringle, J. W. S. and Wilson, V. J. (1952) The response of a sense organ to a harmonic stimulus. *Journal of Experimental Biology*, **29**, 220–234.

Terzuolo, C. A. and Knox, C. K. (1971) Static and dynamic behaviour of the stretch receptor organ of Crustacea. In: *Handbook of Sensory Physiology.* Volume 1, Loewenstein, W. R. (ed), Ch. 16, pp. 500–522. Springer-Verlag.

Terzuolo, C. A., Purple, R. L., Bayly, E. and Handelman, E. (1968) Postsynapyic inhibition – its action upon the transducer and encoder systems of neurons. In: *Structure and Function of Inhibitory Neuronal Mechanisms*, von Euler, C. (ed), pp. 261–275. Oxford: Pergamon.

Terzuolo, C. A. and Washizu, Y. (1962) Relation between stimulus strength, generator potential and impulse frequency in the stretch receptors of Crustacea. *Journal of Neurophysiology*, **25**, 56–66.

Wendler, L. and Burkhardt, D. (1961) Zeitlich abklingende Vorgänge in der Wirkungskette zwischen Reiz und Erregung (versuche ein abdominalen Streckreceptoren dekapoder Krebse). *Zeitschrift für Naturforschung*, **166**, 464–469.

Wiersma, C. A. G., Furshpan, E. and Florey, E. (1953) Physiological and pharmacological observations on muscle receptor organs of the crayfish, Cambarus clarkii Girard. *Journal of Experimental Biology*, **30**, 136–150.

16 Central processing of proprioceptive information

C. A. G. WIERSMA

16.1 Introduction

The information received by the central nervous system from proprioceptors is handled in two ways. One is by more or less direct connections to motor elements, the other by way of interneurons. The most studied and best understood processing of proprioceptive input is that for localized reflexes, specifically the myotatic one, in which interneurons do not play a part. In this case the motoneurons involved are monosynaptically connected with sense organs, usually located either in the muscles themselves, or in the tendons, or in special organs spanning the joints; but in some cases exteroreceptors, such as hairs bent by the movement, may be involved. Note that even in a simple monosynaptic reflex excitation of one set of muscles is normally accompanied by inhibition of the antagonistic set, and thus even at this level questions about the neuronal pathway followed arise. Another complication is that the primary sensory fibres entering the CNS often branch, so that they also make connections with elements other than the motor fibres of the body segment in which they are located. (Note that the motoneurons, though usually present in this ganglion, are known to be in some instances located in a neighbouring ganglion of the abdominal cord). This gives rise to a number of queries about the significance of such a distribution and, as we will see, these are as yet only partially answered.

The interneurons excited by proprioceptive input often have synaptic connections in more than one segmental ganglion. One of the results of this type of connectivity is that there is a great variety of interneuronal units with regard to the number and the distribution of the sensory fields of segments represented by each. However, these features are quite constant and are the main means by which interneurons can be identified in different specimens.

The same types of interneuronal representation are found in the integration of other sensory inputs, such as touch of hairs, and also in multimodal interneurons, in which proprioceptive input is accompanied by visual and/or tactile input. For a general review of the subject see Wiersma and Roach (1976).

16.2   The distribution of primary proprioceptive fibres

Originally it was generally accepted that primary sensory fibres in invertebrates ended in the ganglion they entered. This view has now been disproved in so many instances that it should never be taken for granted unless it is based on direct anatomical and physiological evidence. The information about the types of sensory neurons which show a spread to other ganglia is more complete for the crayfish, especially in regard to proprioceptive input, than for other animals. Proprioceptive signals from swimmerets have been found to run posteriorly for one segment in the abdominal cord and anteriorly for at least two (Wiersma and Hughes, 1961). No information is available for the more anterior appendages, except that proprioceptive fibres from the first swimmeret can be found in the cord between the 5th thoracic and 1st abdominal segment (Wiersma and Bush, 1963). This finding makes it likely that similar bundles of primary proprioceptor fibres from the pereiopod joints are present in the thoracic cord.

The pathway followed by the axons of the abdominal stretch receptors is better known. They may well be the sensory axons with the greatest spread in any central nervous system, as they run both to the last ganglion of the abdominal chain and the 'midbrain,' though they do not appear in the optic nerves. This distribution is like that of interneurons, of which many spread throughout the CNS, with several including the optic nerves in their pathways. As a matter of fact, other properties of the abdominal stretch receptors, as illustrated specifically by the known functional connections of the slow receptors, are also more like those of interneurons than of ordinary primary fibres. They receive inhibitory input from 2 or 3 centrifugal units by way of synapses on their dendrites in the receptor muscles, and are elements of at least two feedback loops, since their output inhibits not only specific extensor motor fibres but also the neighbouring stretch receptors, especially those in the next anterior ganglion. Though the distribution of hair fibres also involves, in many cases, pathways through several ganglia, none are known to have as wide a distribution as the stretch receptors.

In other arthropods spread of sensory fibres to other ganglia is also present. For instance in the isopod *Ligia* sensory fibres of the pereiopods have been shown to run in the ipsilateral connectives both forward and backward to the next ganglion, but not further (Alexander, 1970). In scorpions a sensory hair fibre has been shown to make synaptic connections in several ganglia with interneurons (Sanjeeva-Reddy and Rao, 1970). By degeneration experiments sensory fibres from the antenna of the bee have been shown to spread as far as the contralateral suboesophageal ganglion (Pareto, 1972). For the tympanic organ of the cricket it was observed that the 4 peripheral bundles entering the CNS end diversely in different neuropiles of the meso- and meta-thoracic ganglia (Kalmring, Rheinlaender and Rehbein, 1972). But here, as in most previously described cases, it was not shown that any single unit divides and in this instance that seems less likely than in others, such as the swimmeret proprioceptor axons.

In molluscs and annelids the amount of spread shown by sense cells is less clear. This is partially because in both phyla there are doubts about the primary nature of the neurons, especially those with extensive sensory fields, such as mechanoreceptors in the leech; and partially because local peripheral ganglia occur in molluscs as well as in certain annelids. Whether one wants to define the sense cells involved as cutaneous exteroreceptors or as proprioceptors is a rather moot point. In the arms of the octopus it seems that the proprioceptors are only involved in strictly localized reflex contractions, indicating that their pathways are short.

16.3   Problems arising from the distribution of primary fibres

The fact that the primary fibres spread to other ganglia produces a considerable complication in determining what effects are caused by their activity. The question arises whether the connections in the other ganglia involve motor fibres and, if so, whether they are excitatory or inhibitory. As yet, there is insufficient knowledge about this aspect to permit any general statement and it appears likely to vary in different instances. With regard to connections with interneurons, a somewhat less vague picture is present. It has been found, at least for hair fibres of the crayfish abdomen, that connections with the same interneuron are made in several ganglia, and that these are additive (Kennedy and Mellon, 1964). This has the consequence that when the incoming sensory information into one ganglion is too weak to give rise to a threshold discharge, similar and simultaneous stimulation of the hair fibres entering the next posterior ganglion can be sufficient to cause spiking of the interneuron at the anterior locus. Such a mechanism provides an excellent means of preventing the triggering of the interneuron by weak stimuli from one segment, but causing it when stimuli of the same weak intensity are present in neighbouring segments. How the interneuron will react when the stimuli are so strong that they cause spike initiation in several ganglia will be discussed later.

The spread of primary sense cells to neighbouring ganglia is likely to be useful in the integration of myotatic reflexes of the stimulated part with those of the neighbouring segments, as shown for the stretch receptors. However, when primary fibres run to stations far removed from the stimulus location, as is also the case for the crustacean stretch receptors, yet another function is indicated. At present it is still uncertain what the functions of their connections in the brain are. At one time it was speculated that they might influence the reflex excitability of the medial giant fibres which, when triggered, cause a tail flip. But this view has lost much in attractiveness after it was shown that, once swimming has started, the giant fibres are no longer involved in triggering the separate tail bends (Schrameck, 1970). Nevertheless, it may well be that their information does influence the excitability of the medial giant fibre or that of other swimming command fibres originating, as shown by Bowerman and Larimer (1974), in the brain. One could also hypothesize that they are the means by which the 'midbrain' stays aware of the movement and position of each separate segment, and might thus 'know' which abdominal segments are flexed and which extended. However, the 5 N-cells of the thorax (Wiersma and Pilgrim, 1961), which are homologous to the slow abdominal stretch receptors, also send similar branches forward and backward, but they are not involved in signalling tail bending. Their tasks appear to be more of a vegetative nature, such as signalling pressure in the thoracic cavity. Therefore, the latter hypothesis is also not very attractive. These considerations may serve to demonstrate that in many cases the information received by the CNS cannot be judged as to its functional significance without extensive knowledge of where it goes and what it does to other neuronal, or for that matter neuroendocrine, elements.

16.4   Feature selection by peripheral sense cells and interneurons

As will be evident from previous chapters, the proprioceptors of arthropods are very specific in the type of joint movement to which they respond. The two types of stretch receptors of the abdominal segments are not a very good example, for though their differences in response to current injected in their cell bodies illustrates an inherent property by which they respond differently to an equivalent input, specificity of their attachments is certainly another feature by which they will be selectively stimulated under natural conditions. The differences between the movement and position fibres of the joint organs illustrate that there is a fundamental difference in the type of change perceived, movement or position, without obviously different attachments, which are all on the elastic strands. Furthermore, within each group, the thresholds of spike generation vary. In this case, there is no indication that additional feature selection is performed by interneurons; instead there are indications that in many cases different types of primary sensory fibre make contact with the same interneuron, which would negate at least part of the peripheral selectivity. As we will see, there are other instances in which interneurons do provide for additional feature extraction, such as in insect acoustic interneurons, in which the directionality of the sound is signalled by excitatory influence from one side and inhibitory effects from the other. In hearing there is also evidence for extraction of features not directly provided by differences between primary sense cells, but performed by the neural synapses and this is even more marked in the visual system. In vision fast movement, for instance, is signalled by certain interneurons which respond only shortly to fast moving objects, presumably because their inputs from the photosensitive cells respond only to the first incoming disturbance.

Loss of feature information is well illustrated by an interneuron reacting to activation of the abdominal slow stretch receptors, of which the input connections appear to be restricted to the neuropile of the 6th abdominal ganglion (Hughes and Wiersma, 1960). This interneuron can be activated by any of the stretch receptors but its firing rate is low for any one input, and becomes only slightly higher when several, or even all, are firing maximally by extreme flexion of the abdomen. Furthermore, extreme extension of the tail, which is signalled by certain axons most likely belonging to primary sense cells with endings in the perineural sheath of the nerve cord itself, can also cause this interneuron to fire. However, the anterior stretch receptors of the thorax, including the N-cells, are not connected with this interneuron, so one may hypothetically conclude that the function of this interneuron is to provide for corrective measures when extreme extensions or flexions of one or more of the abdominal segments occur.

In keeping with the interneuronal representation of hair territories on different parts of the body, interneurons reacting to proprioceptive input are mostly influenced by several joint organs, often of more than one appendage. In Table 16.1 the interneurons of the type responding to movement of joints on the ipsilateral side of the body are enumerated. As can be seen from the table, interneurons specific for one joint of an appendage are present. Such interneurons are known for the antenna and for the cheliped; in the case of the antenna, especially for movement of the long flagellum on its base (for which both a phasic and a tonic interneuron are specific) and for outward movement of its squame. Flexion of the propodite of the cheliped is signalled by a specific interneuron, thus indicating claw closure. These two movements may be of special functional significance: the one monitors the position in space of the flagellum, which plays a significant role in investigating the environment; whereas closure of the claw is of importance in prey catching and fighting. Both these appendages also have interneurons which are limited to them but respond to motion at more than one joint. For example, for the antenna an interneuron responds to movement at all basal joints, including that of the squame; whereas one interneuron for the claw integrates the information from several

Table 16.1 Table of ipsilateral proprioceptive interneurons in the circumoesophageal connectives of the crayfish

Single joint fibres

    Eyestalk: Basal joint, phasic (C109).

    Antenna: Joint of antenna flagellum on base of antenna in any direction, phasic (C107 and 121, distinguished by their locations in the cross-section). Ditto backward movement only, tonic (C44). Joint of squame, ipsilateral, tonic response to outward and upward movements (C23).

    Cheliped: Pro-dactylopodite joint, flexion only. Phasic on passive movements, but continues when claw grasps an object actively (C62).

    Coxo-basipodite joint, flexion only, phasic (C69).

    Same joint extension only, phasic (C42).

    Uropod: Basal joint, phasic (C54).

Multiple joint fibres of single appendages

    Antenna: Several joints of ipsilateral antenna, tonic (pro-, carpo-, meropodite and squame) (C88).

    3rd Maxilliped: Responds to 'vibration' especially well for mero-carpopodite joint. Might therefore be exclusive for this joint or perhaps the myochordotonal organ and would then belong to single joint group (C1).

    Cheliped: Flexion of joints, especially mero- carpopodite but also carpo-propodite, phasic (C61). Extension of most joints, including coxo-basipodite, mero-carpopodite, carpo-propodite and pro-dactylopodite. Tonic, stops only when all are flexed (C50).

Interneurons from single, homologous joints of successive appendages

    Eye, antennule, antenna: Basal joint, phasic (C32).

    Antennule and antenna: Flagella joint. Outward movement, phasic, (C39).

    Ditto inward movement, phasic (C47).

    Somewhat tonic for slow inward movements (C96).

    3rd maxilliped, 1st to 4th pereiopods: Basal joints flexion, phasic (C30).

    3rd maxilliped to uropod: Basal joints, extension, tonic (C7).

    All pereiopods (perhaps also 3rd maxilliped): Mero-carpopodite joint, flexion, phasic (C79).

    Pereiopods: Basal joint, flexion, phasic (C27).

    2nd to 4th Pereiopods: Mero-carpopodite joint, flexion, tonic (C58).

Interneurons from more than one joint of several appendages

    Antennules-antenna: Joints of all basal segments, phasic (C141).

    Same but tonic (C116).

    Eye, antennule, antenna: All joints, phasic (C126).

    Same but tonic (C127).

    3rd maxilliped and all five pereiopods: Coxo-basipodite joint flexion mainly but likely not exclusively, tonic (C82).

    1st–3rd pereiopods, maybe 3rd maxilliped: Any joint, flexion, phasic (C26).

(From Wiersma, 1958; Wiersma and Mill, 1965)

joint receptors signalling extension at all joints, and keeps firing as long as one of them is extended. As the table shows, no neurons specific for single joints of most other appendages were found. However, as in the case of interneuronal representation of touch areas, segmental input to proprioceptive interneurons can be selective for single joints of successive appendages, involving from two to all appendages on one or both sides of the body. In the case of joints, it is especially the basal joints which have this type of representation, whereas often interneurons responsive to peripheral joints have input from several of each appendage. However, there are exceptions. Thus, outward movement of the antennule flagellum is represented by a unit that, in addition, responds to outward movement of the antennal flagellum, the latter joint therefore being represented twice specifically. Fig. 16.1 shows the reaction of an interneuron with multisegmental ipsilateral representation to separate and combined stimulation of the pereiopods to which it reacts. The frequency of firing goes up with an increase in the number of legs furnishing input. However, it may be noticed that because this type of unit collects input in successive ganglia, collisions of the impulses can occur, so that the input in the ganglion nearest to the lead-off electrode, in this case the cheliped, will have an advantage over other inputs. There is thus a built-in non-proportionality between input and output relationships in these types of integrations.

(a)

(b)

(c)

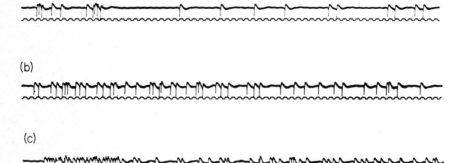

Fig. 16.1   Records of tonic discharges of fibre C82, responding to flexing of the joint between coxopodite and basipodite, showing increase in frequency with increase in number of legs bent. (a) 5th leg alone; (b) 3rd to 5th legs; (c) 2nd to 5th legs. Number of impulses (a) 20; (b) 44; (c) 64. Time 60 Hz. (From Wiersma, 1958.)

16.5   More complex interneurons

In addition to the identified ipsilateral interneurons shown in the table, a smaller number of similarly reacting interneurons with contralateral proprioceptive input has been found. They behave quite like the ipsilateral ones, and are thus their mirror images. In addition a number react to bilateral inputs. In some cases the input from the contralateral side is restricted to a small part of the body and such units have then an asymmetric sensory field. In this class units with a contralateral response from the uropod-telson area are most numerous. However, most interneurons with bilateral input have a symmetrical input from one to many body segments. They again follow the same rules as the ipsilateral ones in that they may react to one or to several appendage joints located in various numbers of consecutive appendages. It is likely that this class is not homogeneous, for though in several instances stimulations on either side show equal efficiency, in others the contralateral side, especially after previous stimulation, is less effective than the ipsilateral one.

Other complications are interneurons which react to both touch and proprioceptive inputs. Such multimodal units have been found for single swimmerets in the abdominal cord, and for the 4th pereiopod in the circumoesophageal connectives.

A unit with a reaction that allows it to be assigned a more specific behavioural significance is the bilateral proprioceptive fibre (C13) called Sherrington's fibre. This unit is excited by ipsilateral extension of the coxo-basal joint of any or all of the five pereiopods; whereas flexion inhibits or lowers the discharge rate. The contralateral legs, on the other hand, increase this neuron's firing rate when they are flexed and inhibit it when extended. The two interneurons, one on each side, will thus inform the brain of the body's position in space; one increasing, the other decreasing, its activity when the body is tilted on the legs, as for instance by gravity. Of course, similar information is provided by the statocysts, and it is not known whether these furnish inputs to Sherrington's fibre.

16.6   Pathways followed by the incoming information: cascading principle

We have already mentioned the strong indication that single, primary sensory fibres make connections with the same interneuron in different segments, which lead to complications in its response pattern. Short mention has also been made of bilaterally reacting interneurons, whose ipsilateral input is more reactive and less fatiguable than its contralateral one. Though not directly proven, the cause of the latter may be that the unit receives its ipsilateral input from primary sensory fibres, whereas the input from contralateral sensory fields is provided by a 'heterolateral' interneuron. This raises the possibility that the ipsilateral input of such bilateral interneurons consists of two sources, namely primary fibres and the homolateral partner of the interneuron which provides the heterolateral input. That this mixed type of input to interneurons does indeed exist, has been proven by experiments from Kennedy's laboratory, especially by Zucker (1972a, 1972b). He investigated, using intracellular recordings, the pathways by which the lateral giant fibres are stimulated from touch cells on the abdominal surface and found that sensory fibres end directly on the giant fibres, giving rise to an early excitatory post-synaptic potential (EPSP). In addition several tactile interneurons, which are also stimulated by the same primary input, give rise to a later EPSP. To further complicate the picture the tactile interneurons showed additional mutual interconnections. Fig. 16.2 illustrates this for interneurons A (A6), which reacts to unisegmental ipsilateral hairs, B (A63) which responds to trisegmental ipsilateral hairs, and C (A64), a bilateral interneuron, which responds to stimulation of the entire abdomen. It was found, using separate intracellular stimulation, that A stimulates C and that B and C

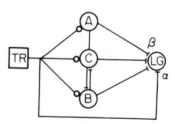

Fig. 16.2   Schematic circuit of elements and connections concerned with generating a rapid tail flexion to phasic mechanical abdominal disturbances. Large circles represent single neurons; the square encompasses a population of tactile receptors. Small open circles are chemical synapses. Electrical junctions are shown as bars. The separate pathways generating the α- and β-components of the lateral giant response are indicated. TR—tactile receptors; A, B, C, identified tactile interneurons; LG—lateral giant. (After Zucker, 1972a.)

stimulate each other, and that all converge on the giant fibre. All synapses involved are electrical with the exception of the ones made by the primary fibres onto the interneurons, which are chemical. The parallel computing, and cascading aspects of such connectivities, added to the differences in facilitation and defacilitation in the synaptic transfers, make such networks very versatile but also extremely difficult to study. For instance it is known that in the described system more interneurons than the three shown in the figure participate, but not precisely how these are connected. Whether the giant fibre will be brought to firing level, which does not seem to happen when activated by any single pathway, becomes very dependent on features which do not necessarily depend on the strength of the sensory hair stimulation as such.

Another difficulty in exploring the fate of incoming proprioceptive information is that, besides interneurons responsive solely to this information, there are a number which combine it with responsiveness to touch. Since the latter is usually the more readily evoked reaction, one has to be certain for this type of unit that joint bending suffices. This can be done by accommodating the surrounding area to touch by repetitively stroking all surface areas and only then bend or extend the joint. Even so some doubt may remain that it is really the joint organs and not other similar muscular or exoskeleton deformation receptors which are involved.

Most multimodal fibres are units with very large receptive areas, which indicates that they will be mainly involved in regulating the overall reactivity of the nervous system. However, for some swimmerets a multimodal interneuron restricted to single segmental input has been found.

An interesting example to demonstrate the difficulty in deciding whether input is solely from touch or if this is combined with proprioception is a mechanoreceptive unit in the optic nerve of the crayfish. Here, what turned out to be a single unit was described under two separate labels. Under code-number 014 it was thought to respond solely to joint manipulations of the pereiopods, whereas as 017 it was thought to respond especially to touch of their hairs. In many preparations only one or the other unit was found; in a few first 017 appeared, whereas later 014 was found. From recordings in which other already localized optic units were present, 014 and 017 had to be in close proximity. Since both units were among those producing often very large signals, one would expect that they should be found very frequently together in a single penetration of the nerve. However, this was never the case; whereas large fibres which are further apart were frequently registered simultaneously. Therefore, the conclusion that 014 and 017 represent two different aspects of the same unit is certain. The strong depression of the hair input, shown by the '014 form', indicates the presence of an inhibitory mechanism and, therefore, again considerable complexity in the pathway involved.

16.7   Statocysts, excited state, and branching patterns

Very little is yet known about interneurons reacting to statocyst stimulation. The only ones reported so far react very phasically and may have, in addition, other input from the antennules and other head parts. Since statocyst output has a profound influence on body position adjustment, and thus on leg extension and flexion, tonic interneurons are strongly indicated. That they have not so far been encountered is quite possibly because of experimental difficulties in changing the position of the preparation with regard to gravity. In the optic nerve, however, where needles can be used for monitoring axon traffic, fibres are encountered which show that tonic statocyst discharges are part of their inputs. These units are distinguished by the fact that their visual fields are space constant. Their normal fields are invariably pointed upwards as long as the statocysts are present, but become enlarged downwards when the statocysts are removed. Whereas the statocysts' inhibitory influence might be transmitted by interneurons, all indications are that the primary axons of the statocysts, which are equivalent to fibres from hairs, are responsible. For on the one hand, other hair fibres of the head region are present in the optic nerve, and on the other, the shift in sensory field with variation in body position is extremely accurate.

Some of the interneurons with statocyst input have been more closely investigated. Taylor (1970) implanted an electrode in a circumoesophageal connective and found a large fibre, presumably identical with C4, which had been shown to react phasically to contralateral statocyst stimulation and to touch of some hairs on the first segment of the antennule. Under implanted conditions it was quite sensitive to vibratory stimuli when the animal was at rest. An outstanding feature shown by this fibre is that during an excited state of the animal it shows inhibition, whereas most interneurons become more reactive in this condition. For instance, Aréchiga and Wiersma (1969) showed, with a similarly implanted electrode, that a multimodal visual and 'vibratory' sensitive interneuron's reactivity to both of its inputs waxed and waned with the alertness of the animal (Fig. 16.3).

Staining interneurons with the aid of intracellular dye injections is a promising approach to further identification of them by their course and the location of their neuropilic fields. For proprioceptive interneurons this method has been used by Wilkens and Larimer (1973) on the just mentioned interneuron (C4) and on a neighbouring large fibre, which has not received a code number, but is presumably that described by Wiersma and Mill (1965) as a bilateral fibre for movement of the flagella of the antennules and antennae (*loc. cit.*, p. 76). As Fig. 16.4 shows, both their cell bodies are on the contralateral side, but the neuropile of C4 (LHG) is restricted to that side, whereas the bilateral fibre (MHG) has a large one on the ipsilateral and a smaller one on the contralateral side. Whether the authors' suggestion that the first one is primary and the second one secondary in nature is correct would need further evidence about the nature of the postsynaptic potentials, especially those of the ipsilateral input. They found, as has been generally true

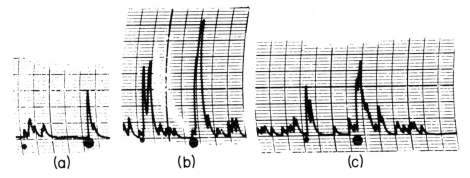

(a)   (b)   (c)

Fig. 16.3   The effect of 'awakening' on a multimodal interneuron in an oesophageal connective (C-120). The small dots indicate mechanical stimulation by a drop of water near the tail; the large dots a visual stimulus (moving object). Note that both responses went up roughly proportionally with awakening (b) and decreased again when the animal became more drowsy (c). Level of awareness is shown by changes in background activity. Time interval between A and B a few seconds; between B and C a few minutes. Vertical lines, 20 s; horizontal lines provide logarithmic scale of number of impulses. Integrated polygraph record. (From Aréchiga and Wiersma, 1969.)

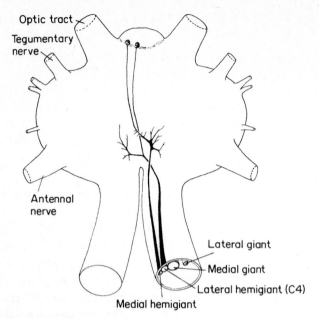

Fig. 16.4   A reconstruction of two large interneurons descending in the circumoesophageal connectives from their branching structures in the brain of the crayfish as shown by Procion yellow injections. (From Wilkens and Larimer, 1973.)

for other interneurons studied by dye injections, that the branching patterns of the neuropiles of the same interneurons from different animals are as much alike as are the positions of their axons in the cross-sections of the CNS.

16.8   Hearing

Integration by interneurons responding to airborne sound has been studied rather intensively, especially in insects. The primary sound receptors are sense cells which, in many instances, either can be directly correlated with tactile cells of the skin or with proprioceptive cells of the joints; whereas in others neither homology is satisfactory. The subject is further complicated by the fact that there is no sharp division between vibration and sound receptors and vibration sensitivity can also be transmitted by either externally or internally located sense cells.

Whereas some interneurons in decapod Crustacea may be quite

sensitive to vibration, true hearing appears restricted to land living forms, of which the land crab *Ocypode* is the best investigated instance (Horch, 1971). In this crab the cuticle overlying the myochordotonal organs, which are located near the ischio-meropodite joint (Section 6.2.2; 6.8), is very thin in the walking legs, though thick in the chelipeds. When electrodes were introduced into the circumoesophageal connectives spikes were found in certain interneurons when sound (e.g. 1 KHz) was presented. The input is bilateral and by blocking all eight organs spiking is prevented. Of special interest is the observation that only extended legs 'hear'; this is because the muscle which is connected with the myochordotonal organs is relaxed when the leg is flexed.

A similar change in sensitivity due to the position of a part has been found in moths with a palp-pilifer organ, which functions as a sound receiver only when the median wall of the palp makes contact with the pilifer (Roeder, Treat and Vande Berg, 1970). The location of this hearing organ on the antenna is only one of the many places in which sound sensitive cells occur in insects (Section 12.3). Even in rather closely related groups of Orthoptera the tympanic organs can be either located in the last thoracic segment or in the tibia of the front legs. In both cases subgenual organs, which are directly related to joint receptors, have an additional influence.

This heterogeneous nature of the 'ears' brings up an interesting problem for interneuronal connectivity. From the way the interneuronal connections of similar receptors in different segments are made, it is logical to assume that, as is the case in *Ocypode*, they are connected to the same interneuronal units, whether the ear is at the base of the forewing, at the base of the hindwing, or in both places. However, that a thoracic tympanic organ would be connected with an interneuron homologous to one connected with the organ in the legs, or an organ in the head is much less likely. Nonetheless, as we will see, there are some arguments to be made in favour of this possibility. Perhaps future developmental studies may be able to show what the true situation is. Insects with the tympanum in the thorax are the locust and the cricket. The tympanal nerve enters the third abdominal ganglion, but then runs to four different neuropiles, two in the metathoracic and two in the mesothoracic ganglion. No crossing occurs. Since there are four types of sensory fibres, it is possible that each type runs to a different central destination, but this question has not yet been solved.

16.8.1 Orthoptera

A considerable variety of interneurons have been described by Kalmring and coworkers (Kalmring, 1971; Kalmring, Rheinlaender and Römer, 1972; Kalmring, Rheinlaender and Rehbein, 1972) in the thoracic ganglia and the suboesophageal connectives of the locust. As can be seen from Table 16.2, they differ from each other in various aspects, such as firing threshold, frequency range, influence of contralateral input and firing type. The list is presumably not complete, for with intracellular recording by other workers a type of interneuron was found which was hyperpolarized both by touch of the abdomen and by sound (Hoyle and Burrows, 1973). It is not well known how well the reported differences represent those under normal conditions when other input sources are present. In the investigation quoted most of the other possible inputs for sound stimuli were removed. It was found by other workers (Yanagisawa, Hashimoto and Katsuki, 1967) that the reaction spectrum of an

Table 16.2 Characterization of the acoustic type of neurons in the ventral nerve cord of *Locusta migratoria*

Type	Frequency range KHz*	Type of response	Characteristics	Information content
A	3—9	Phasic	Maximum discharges at 60—80 db; somewhat influenced by contralateral tympanic organ.	Stimulus duration for medium intensities; frequency-dependent; directional sensitivity.
	10—40	Tonic		
B	4—40	Tonic or phasic	Tonic discharges near threshold only, for higher intensities phasic; short stimuli cause bursts; no contralateral influence.	Soft sounds; short stimuli; directionality only by tonic discharge pattern
C	3—40	Tonic	Firing rate increases with stimulus duration and/or intensity; nonspecifically influenced by contralateral side.	Stimulus duration and intensity; independent of frequency; directional sensitivity.

Continued

Table 16.2 (*cont.*)

Type	Frequency range KHz*	Type of response	Characteristics	Information content
C_1	3–40	Tonic	As C but threshold 10–20 db higher; high spontaneous discharge rate.	Duration and intensity of stimulus; independent of stimulus direction and frequency.
D	3–30	Phasic	On discharge to sound stimuli of any duration, intensity, or frequency.	Stimulation rate; some directional sensitivity present.
E	7–30	Phasic to tonic	High threshold (60 db); tonic for high intensities; discharges independent of frequency; almost no contralateral influence.	High intensities; directional sensitivity present.
F	10–40	Very tonic	Tonic discharges, increasing with the duration of the stimulus, low threshold; long latency, at high intensities; after-discharges; strong contralateral influence.	High frequencies; stimulus duration; directional sensitivity very pronounced.
G	3–30	Tonic or phasic	Broad band phasic responses; however, limited tonic discharge area which increases in threshold with frequency; mutual bilateral influence.	Signal band specific for frequency-intensity relationship; directionality present only for medium frequencies and intensities.
H	0.3–12,	Tonic	Tonic discharges; high threshold, increasing with frequency; not from tympanal organ.	Low frequencies, high intensity; duration and intensity of stimulus.

* Note that 40 KHz was the highest frequency tested.
Data from Kalmring, Rheinlaender and Römer (1972).

interneuron in an intact animal widened considerably to lower frequencies when other inputs were made nonfunctional, which indicates that these sources, among which are those from legs and abdominal receptors, have an inhibitory influence on the 'tympanic' interneuron. Obviously, considerable difficulty results from this type of information transfer when comparing results of different investigations. It may well be necessary for definitive comparison between interneurons to be made to study their discharge patterns in non-tethered animals, using a wide variety of sound stimuli, especially normal song patterns and other behaviourally important sounds.

In the brain, interneurons with characteristics similar to those in the connectives have been found, as well as some even more complexly reacting ones. It is likely that the latter represent units in which additional inputs, especially from visual stimuli, are integrated with those from sound stimuli. However, the interneurons in the cord have often not been critically studied with regard to such influences, and it should also be noted that the methods used do not usually reveal whether their signals are truly ascending and not descending. In the Tettigoniidae and the Gryllidae the 'ears' are located in the front legs and thus the 'tympanic' nerve enters the prothoracic ganglion. In both families, communication by sound between the singing male and the attracted female is well established. As in the locust other pathways, especially from abdominal receptors, that influence the responses to tympanally stimulated elements are indicated. Thus, it was found (Jones and Dambach, 1973) that removal of the tympanal organs, antennae and cerci in the cricket did not abolish the inhibitory effect on chirping by that of another male and that units in the neck connectives kept responding to lower (5 KHz) sounds. With the more anterior location of the main hearing organ one might expect that sound-sensitive interneurons would show descending branches to the more posterior thoracic ganglia, and this is indeed the case. In several species a large fibre, responsive to high frequency sounds, has been described and called the T-fibre because it gives off an anterior branch going to the brain and a posterior one to the posterior thoracic ganglia. This branch stops short of the abdominal chain. Abdominal interneurons responsive to low frequencies (less than 2KHz) have been reported. These fibres make connections with forward going interneurons but themselves stop in the metathoracic ganglion. The T-fibres are likely to be involved in sound source location, since in many species they show mutual inhibition. These units thus resemble the G-fibres of the locusts rather closely. In the cricket there appears to be no mutual inhibition between them, and correspondingly, the location of sound sources is reported to be much poorer. Records from the cervical connectives of female crickets during male singing show that some interneurons respond to the pulses and others to the chirps (Stout and Huber, 1972). The 'pulse coding' units respond to each pulse in a chirp with a short burst, whereas the 'chirp coding' units continue their activity for the duration of each chirp (Fig. 16.5), a type of feature extraction also seen in moths (Section 16.8.2). Whether the preference shown by hybrid females to the hybrid male song above either parental song (Hoy and Paul, 1973) is also mirrored at this interneuronal level is as yet unknown.

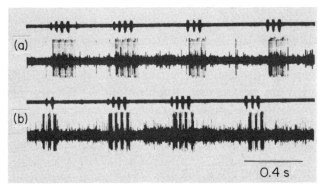

Fig. 16.5   A comparison of the 'chirp coding' neuron (a) with the 'pulse coding' neuron (b) in a female cricket. Note that 'a' fires between pulses, whereas 'b' is silent at those times, but answers each pulse with a higher frequency discharge than 'a'. (From Stout and Huber, 1972.)

16.8.2   Lepidoptera

Whereas hearing in Orthoptera appears to be mainly, but not exclusively, used for conspecific attraction; in Lepidoptera, especially in moths, the main function is for escape from predators. As in the Orthoptera, the hearing organs have different locations. In the hawk moths, (Sphingidae) the members of one subfamily have palp-pilifer organs on the head, but the members of other subfamilies show similar evasive behavioral reactions, for which the appropriate sense organ(s) is not yet known. As mentioned above the pilifer organ changes in sensitivity with the degree of contact between the palp and pilifer. In the noctuid moths, on the other hand, the 'ear' is located on the posterior wall of the metathorax. In both instances, very few sensory cells are involved in sound reception. In the noctuids there are at most three cells, of which the 2 called A cells are stimulated by the high frequency sounds of bat cries. (The 3rd or B cell appears to be inhibited, perhaps directly, by the A cells.) The two A cells differ in threshold and in the broadness of their spectrum, but have the same 'best frequency.' In certain butterflies sense organs are present at the wing bases but either base may have the main organ, depending on the species. In all these species, interneurons responding to sounds have been reported, with those of the noctuids being the most extensively studied. Here Roeder and his coworkers (see Roeder, 1966) have distinguished between three types in the thoracic cord which have been called the 'pulse-marker,' the 'train-marker' and the 'repeater.' The pulse-marker interneuron has a low threshold and fires soon after the onset of a high frequency sound burst, but with only a single spike for each burst. However, for repetitive bursts it can follow a repetition rate of up to 40 bursts per second. This is an ipsilateral interneuron and its latency depends on the sound intensity. The train-marker gives a more or less continuous discharge as long as trains of sound pulses of around 20 per second are present. The frequency of its discharge is not dependent on the finer structure of the sound. At low repetition rates of the pulses it remains silent. The 'repeater' mirrors the discharge pattern of the A-cells rather faithfully. It discharges repetitively during each burst and is silent between bursts.

In the brain, more complexly reacting interneurons were encountered. Some units show lapses in their reactivity, which are sometimes correlated with visual stimuli. A large interneuron was found to be inhibited when the sound level became high, whereas other units kept increasing in firing rate with sound intensity.

In sphingids, descending auditory responses were found in the prothoracic ganglia. It will be some time before the total circuitry of the interneurons and the sound evoked behavioural reflexes can be correlated. In this respect it is important to note that headless moths can show changes in flight pattern when high frequency sound is presented (Treat, 1955). Such reactions could come about by thoracic connections of the same interneurons which go to the brain, or even by direct ones with motor fibres, and according to the cascading principle both might well be present.

References

Alexander, C. G. (1970) Studies on the nervous system of an isopod crustacean, *Ligia oceanica. Comparative Biochemistry and Physiology*, **33**, 323—332.

Aréchiga, H. and Wiersma, C. A. G. (1969) The effect of motor activity on the reactivity of single visual units in the crayfish. *Journal of Neurobiology*, **1**, 53—69.

Bowerman, R. F. and Larimer, J. L. (1974) Command fibres in the circumoesophageal connectives of crayfish. II. Phasic responses. *Journal Experimental Biology*, **60**, 119—134.

Horch, K. (1971) An organ for hearing and vibration sense in the ghost crab *Ocypode. Zeitschrift für vergleichende Physiologie*, **73**, 1—21.

Hoy, R. R. and Paul, R. C. (1973) Genetic control of song specificity in crickets. *Science*, **180**, 82—83.

Hoyle, G. and Burrows, M. (1973) Neural mechanisms underlying behavior in the locust *Schistocerca gregaria*. II. Integrative activity in metathoracic neurons. *Journal of Neurobiology*, **4**, 43—67.

Hughes, G. M. and Wiersma, C. A. G. (1960) Neuronal pathways and synaptic connexions in the abdominal cord of the crayfish. *Journal of Experimental Biology*, **37**, 291—307.

Jones, M. D. R. and Dambach, M. (1973) Response to sound in crickets without tympanal organs (*Gryllus campestris* L.). *Journal of Comparative Physiology*, **87**, 89—98.

Kalmring, K. (1971) Akustische Neuronen im Unterschlundganglion der Wanderheuschrecke *Locusta migratoria. Zeitschrift für vergleichende Physiologie*, **72**, 95—110.

Kalmring, K., Rheinlaender, J. and Römer, H. (1972) Akustische Neuronen im Bauchmark von *Locusta migratoria. Journal of Comparative Physiology*, **80**, 325—352.

Kalmring, K., Rheinlaender, J. and Rehbein, H. (1972) Akustische Neuronen im Bauchmark der Wanderheuschrecke *Locusta migratoria. Zeitschrift für vergleichende Physiologie*, **76**, 314—332.

Kennedy, D. and Mellon, D. Jr. (1964) Receptive-field organization and response patterns in neurons with spatially distributed input. In *Neural Theory and Modeling*. Reiss, R. F. (ed.), Ch. 20, pp. 400—413. Stanford University Press, Stanford.

Pareto, A. (1972) Die zentrale Verteilung der Fühlerafferenz bei Arbeiterinnen der Honigbiene, *Apis mellifera* L. *Zeitschrift für Zellforschung und mikroskopiche Anatomie*, **131**, 109—140.

Roeder, K. D. (1966) Auditory system of noctuid moths. *Science*, **154**, 1515—1521.

Roeder, K. D., Treat, A. E. and Vande Berg, J. S. (1970) Distal lobe of the pilifer: An ultrasonic receptor in Choerocampine hawkmoths. *Science*, **170**, 1098—1099.

Sanjeeva-Reddy, P. and Rao, K. P. (1970) The central course of the hair afferents and the pattern of contralateral activation in the central nervous system of the scorpion, *Heterometrus fulvipes*. *Journal of Experimental Biology*, **53**, 165—169.

Schrameck, J. E. (1970) Crayfish swimming: Alternating motor output and giant fiber activity. *Science*, **169**, 698—700.

Stout, J. F. and Huber, F. (1972) Responses of central auditory neurons of female crickets (*Gryllus campestris* L.) to the calling song of the male. *Zeitschrift für vergleichende Physiologie*, **76**, 302—313.

Taylor, R. C. (1970) Environmental factors which control the sensitivity of a single crayfish interneuron. *Comparative Biochemistry and Physiology*, **33**, 911—921.

Treat, A. E. (1955) The response to sound in certain Lepidoptera. *Annals Entomological Society of America*, **48**, 272—284.

Wiersma, C. A. G. (1958) On the functional connections of single units in the central nervous system of the crayfish, *Procambarus clarkii* Girard. *Journal of Comparative Neurology*, **110**, 421—471.

Wiersma, C. A. G. and Bush, B. M. H. (1963) Functional neuronal connections between the thoracic and abdominal cords of the crayfish, *Procambarus clarkii* (Girard). *Journal of Comparative Neurology*, **121**, 207—235.

Wiersma, C. A. G. and Hughes, G. M. (1961) On the functional anatomy of neuronal units in the abdominal cord of the crayfish, *Procambarus clarkii* (Girard). *Journal of Comparative Neurology*, **116**, 209—228.

Wiersma, C. A. G. and Mill, P. J. (1965) 'Descending' neuronal units in the commissure of the crayfish central nervous system; and their integration of visual, tactile and proprioceptive stimuli. *Journal of Comparative Neurology*, **125**, 67—94.

Wiersma, C. A. G. and Pilgrim, R. L. C. (1961) Thoracic stretch receptors in crayfish and rocklobster. *Comparative Biochemistry and Physiology*, **2**, 51—64.

Wiersma, C. A. G. and Roach, J. L. M. (1976) Principles in the organization of invertebrate sensory systems. In *Handbook of Physiology, The Nervous System*, Brookhart, J. M. and Mountcastle, V. B. (eds.), American Physiological Society. In press.

Wilkens, L. A. and Larimer, J. L. (1973) Sensory interneurons: Some observations concerning the physiology and related structural significance of two cells in the crayfish brain. *Tissue & Cell*, 5, 393—400.

Yanagisawa, K., Hashimoto, T. and Katsuki, Y. (1967) Frequency discrimination in the central nerve cords of locusts. *Journal of Insect Physiology*, 13, 635—643.

Zucker, R. S. (1972a) Crayfish escape behavior and central synapses. I. Neural circuit exciting lateral giant fiber. *Journal of Neurophysiology*, 35 599—620.

Zucker, R. S. (1972b) Crayfish escape behavior and central synapses. II. Physiological mechanisms underlying behavioral habituation. *Journal of Neurophysiology*, 35, 621—637.

17 *Summary*

P. J. MILL

In the Preface it was stated that, for the purposes of this volume, Lissmann's (1950) definition of proprioception would be followed in its widest context. By now the reader will be more than aware of the immense complexities of form and function exhibited by this group of mechanoreceptors and an attempt will be made to highlight some of the problems which have emerged.

Classification of proprioceptors

It is possible to divide proprioceptors into various categories such as chordotonal organs, stretch receptors, etc. and to some extent this has been done in the lay-out of the preceding chapters. Although considerably more work is required on their ontogeny, it is useful to attempt to produce some order out of the mass of structural and functional information available. The terms 'Type I' and 'Type II' were introduced by Zawarzin (1912) to differentiate bipolar (uniterminal) from multipolar (multiterminal) receptor cells and Pringle (1961, 1963) and Finlayson (1968) have both used these types as a basis for further subdivisions. In his earlier paper, Pringle (1961) suggested a hypodermal origin for the type I receptors, which are all derived from a ciliary structure; the type II receptors being possibly derived from cells of central nervous origin.

665

However, as Pringle pointed out, there is no correlation between derivation and function, with receptors of both types serving to detect position, movement and force. Primitively each receptor was probably capable of monitoring all aspects, but separation of function occurs in many instances (e.g. the phasic and tonic crustacean MROs–Chapter 2) (Pringle, 1963). Furthermore, the whole system of sense organs and associated reflexes has evolved to meet the requirements of the animal and any conceptual distinction which an investigator can make between the detection of position, movement and force is not necessarily that made in the animal's physiological machinery. 'The evolution of its (the animal's) organization has produced responsiveness to the environment so as to ensure survival and, of this process, the gathering of information is only one part' (Pringle, 1963).

There are a few apparent exceptions to the multiterminal arrangement of the type II receptors. The plecopteran stretch receptors and the longitudinal stretch receptor of dragonfly larvae each have only one dendrite (Section 4.2). Although an ultrastructural study may reveal the presence of small side branches, their presence has certainly not been indicated by methylene blue staining. Finlayson (1968) has suggested that such receptors have subsequently lost their branching. An alternative explanation is that, since they are found in fairly primitive insect orders, the uniterminal arrangement may represent the primitive form of type II receptor.

Pringle (1963) has suggested that there is another category of proprioceptors (Type III), which includes hair plates (Section 1.4), which has been superimposed on the system of type I and type II receptors where very precise knowledge of relative movement and position is required. Thus the type I and type II receptors have evolved as part of the mechanism of movement coordination and hence are functionally part of the *effector* rather than the receptor system. In contrast, the type III receptors are primarily associated with orientation and provide the animal with an integrated 'picture' of the relative position of parts of its body.

Those hair-plates which monitor attitude between different body regions (Sections 12.3.1, 12.4.1) and the statocysts and canal systems, which are partially enclosed hair-plates which have become associated with aggregations of dense material (Chapters 12, 13) certainly fulfil the requirements for type III, in that they are all concerned with the detection of linear and or angular accelerations as well as being derived from structures which, presumably, primitively had a different (tactile) function. However, this type of information is also supplied by some type I receptors. It may be in the form of general proprioceptive information from the limbs (Section 12.3.1) or be produced by specialized receptors such as the crustacean myochordotonal organs (Section 6.8), Johnston's organ (Section 12.3.2) or the campaniform sensilla at the base of the halteres (Section 12.4.2). Thus the orientation sense has been provided for in different instances by the adaptation of different receptor types. Nevertheless the most sophisticated orientation receptors, the statocysts and canal systems, which are only found in aquatic animals (Chapters 12, 13) all belong to type III. The converse situation also occurs in that the hair plates at the bases of the legs in insects play a role in the coordination of locomotion as well as of posture (Section 8.4.2). Thus, where necessary, receptors which are primitively tactile have subsumed a role more usually filled by type I receptors. It would thus seem useful, to subdivide Pringle's type I and type III receptors on the basis of their primary function (Table 17.1).

Functional role

It appears that the main role of proprioceptors is to provide a detector system which will signal deviation from some centrally programmed 'position' and initiate the appropriate correction sequences. The 'position' may be a dynamic (changing) one, as determined for example by a locomotory programme (coordination of movement), or a static, postural one (control of orientation). Pringle (1961) has distinguished between length detector systems in which the regulatory muscle is in parallel with the receptor, and hence corresponds with the extrafusal fibres of the vertebrate muscle spindle system, and force detector systems, in which the receptor is in series with the appropriate muscle. He also raised the possibility that the hair-plates represent a third detector system which has a more widespread effect, involving various muscle groups.

We probably know most about the functioning of length-detecting systems. In all cases the reflex elicited produces excitation of the muscle in parallel with the receptor, which thus restores the 'desired' or programmed position. This is often referred to as a 'resistance reflex' since it opposes movements away from the desired position.

Table 17.1   Classification of proprioceptors

	Type I		Type II	Type III	
Origin	Differentiated from hypodermis		Migrated from c.n.s.?	Differentiated from hypodermis	
Structure	Uniterminal		Normally multiterminal	Uniterminal	
Association	Scolopale and Cuticle		Internal tissues	Cuticle	
	(Type Ia)	(Type Ib)		(Type IIIa)	(Type IIIb)
Function	Primarily co-ordination of movement	Primarily co-ordination of orientation	Co-ordination of movement	Orientation and control of movement	Orientation
Examples	Joint chordotonal organs	Myochordotonal organs	Stretch receptors	Coxal hair plates	Body hair plates
	Campaniform sensilla	Campaniform sensilla at	Muscle receptor organs		Statocysts
	Slit sensilla	base of halteres	Tendon receptors?		Canal systems
	Lyriform organs	Johnstons organ			

After Pringle (1963.).

In many systems there is, in addition, inhibition of the antagonistic muscle. This occurs for example in reflexes elicited by the scorpion post-abdominal MROs (Section 4.6.2) and by the crustacean and insect joint chordotonal organs (Sections 6.8, 8.4.2), but not apparently in the crustacean MRO systems (Sections 2.8.3, 3.6).

Furthermore, in those systems in which the receptor includes a muscular element (corresponding to the intrafusal fibres of the vertebrate muscle spindle) there is generally feed-back to this muscle. Thus, in the crustacean thoracico-coxal MRO system, stretch of the receptor evokes an excitatory reflex to the receptor muscle, possibly to maintain the gain of the sensory system when contraction of the extrafusal muscle is acting so as to unload the system (Section 3.4.1). Alternatively, in the crustacean and insect abdominal MRO systems this feedback is inhibitory (Sections 2.8.3, 4.6.1) and, in the former at least, this produces a decrease in reflex strength as the desired position is approached. Another apparent alternative occurs in the myochordotonal system of brachyuran limbs, where feedback to the receptor muscle is positive when movement is away from the 'set' position and non-existent or negative when movement is towards this position. However, this receptor muscle is shared by two receptors which are primarily responsive to opposing directions of movement. In astacurans, where there is only one receptor involved, there is self-excitation when the receptor muscle is stretched.

In some cases the reflexes operate on receptor stretch only (e.g. crustacean MRO, chordotonal and myochordotonal systems); in others they operate on receptor relaxation also, with consequent reversal of sign (e.g. scorpion post-abdominal MRO system). In the crustacean legs, where movement needs to be monitored in both directions during locomotion, many joints have two receptors which are stretched by opposite directions of movement.

In the caterpillar and scorpion MRO systems, contralateral reflexes reinforce ipsilateral ones and stretch of the caterpillar MRO also elicits a reflex in the anterior ipsilateral segment (Section 4.6). The crustacean abdominal tonic MRO inhibits the receptor muscle in the anterior ipsilateral segment (Section 2.9.1).

It is important that the reflexes described above should not interfere with the central programme if the peripheral events are as, or are proceeding as, proscribed; yet they must produce an immediate effect once a non-programmed deviation occurs. It is thus pertinent to ask how this can be achieved and it should be obvious that there are two possible mechanisms: either the sensory input to the central nervous system must be prevented or else its effect neutralized, except when the refiex is required to operate. Both methods have been utilized.

The crustacean abdominal tonic MRO system effectively controls itself. During tail flexion the self-inhibitory reflex has the effect of inhibiting the excitatory reflex to the extrafusal (extensor) muscles. Furthermore, since the flexion occurs in a posterior-anterior direction and since the anterior receptor muscle inhibitory reflex is even stronger than the self-inhibitory one, this effect is reinforced. Since this MRO system is only stretch responsive there is no sensory discharge during extension, unless of course a programmed extension is prevented (Section 2.9.1).

In contrast the crustacean joint chordotonal organs cannot control their own output since they are devoid of a muscular component. The reflexes which they produce are only observed when the limb is moved passively or when active movement is impeded; and yet the sensory input to the central nervous system appears to be the same whether movement is active or passive. Thus the effect of the sensory input must be controlled centrally, presumably by comparison of the 'actual' with the 'expected' input. This could involve retention of motor information as suggested by von Holst (1954) who termed this retained command an 'efference copy' (Section 6.9). In the case of a locomotory system the information must be retained for a significant period of time and since in such a rhythmic system the motor output is continually changing so must also be the efference copy. If the control system is associated with posture the problem is somewhat more simple since the motor output will still be the same when the input from the receptors is received for comparison, but even here as soon as a correction sequence is initiated some retention of information will be needed. An alternative explanation is that, during rhythmical activity such as locomotion, the motor programme is so prescribed that the sensory information can be directly compared with the outgoing motor score, thereby negating the need to retain information centrally.

It is difficult to see whether a chordotonal organ system or a stretch receptor system has any particular advantages over the other for the control of locomotion and comparative work is urgently needed on groups which rely on non-chordotonal organs in this situation; the arachnids for example. Possibly the large number of sensory cells associated with limb chordotonal organs has some connection with their inability to control their own outputs. Thus different sensory cells could operate over different parts of the range. However, such range fractionation is apparently limited and anyway there would still appear to be far more cells than are required for this. Alternatively perhaps they are associated with providing postural control at these limb joints. The myochordotonal organs have evolved at the most important peripheral joint, as far as posture is concerned, where particularly fine control is required.

Comparatively little detailed knowledge is available of the reflexes involving force detecting receptors such as campaniform and slit sensilla, although the former at least are known to elicit compensatory reactions when, for example, an insect is walking up a slope (Section 8.4.2). Tendon receptors have only recently been discovered in arthropods and their role in crustaceans is apparently to dampen the resistance reflexes produced by the joint chordotonal organs and they may be of particular importance when the system is loaded (Section 10.6). One of the crustacean limb cuticular stress detectors is certainly involved in the autotomy reflex (Section 7.4).

The receptors primarily concerned with orientation tend to have more widespread effects, although this is not always the case (e.g. the myochordotonal organs of crustacean legs and the dipteran halteres), and their removal results in the loss of orientation responses (Chapters 12, 13). It is almost certainly detector systems of this type which are involved in learning responses (Chapter 14).

Transduction and the optimal stimulus

It is important to differentiate between the stimulus applied to the receptor and the effect this has on the dendritic terminations, which latter it is generally agreed are the site of the transduction process. As outlined in Section 1.4.3 the two principle contenders for the effective stimulus are dendritic stretch and dendritic compression and both apparently are effective in different receptors.

As far as the nerve cord sheath receptors (Section 1.4.3) and the crustacean muscle receptor organs (Sections 2.5.3; 3.3.3; 3.4.1) are concerned the optimal stimulus is receptor stretch producing dendritic stretch. Also, in the crustacean joint chordotonal organs it seems most likely that receptor stretch and receptor relaxation both produce dendritic stretch of specifically sensitive sensory cells (Section 6.8). Dendritic compression on the other hand is almost certainly the optimal stimulus for the campaniform sensilla and for one at least of the dendrites innervating the slit sensilla. For hair plate sensilla both dendritic compression and dendritic stretch have been advocated as the optimal stimulus resulting from bending of the hair (Section 1.4.3).

The available evidence appears to indicate dendritic stretch for length detecting systems and dendritic compression for force detecting systems. However, this seems somewhat unlikely since it is difficult to envisage how dendritic compression could be the optimal stimulus for, say, tendon receptors. Possibly compression is only the optimal stimulus for those receptors which have a dendritic tip modified into a tubular body.

Site of spike initiation

In those proprioceptors with central cell bodies action potentials do not occur. Instead information is transmitted along the axons as graded, decremental potentials (Chapter 3). As far as those with peripheral cell bodies are concerned the site of spike initiation has been studied in the tonic abdominal MRO of the crayfish (Section 2.5.4) in insect hair sensilla (Section 1.4.3) and in the crustacean chordotonal organ (Section 6.7.4). In the tonic MRO the spikes arise in the axon fairly close to the soma, while in the other two cases the site is dendritic. In the chordotonal organs it probably lies in the scolopale region.

In the preceding chapters many problems have been raised which face the neurobiologist working on proprioceptors and a few of these have been briefly discussed in this summary. What should by now be obvious is that, in spite of the amount of effort which is being applied to this rapidly growing field of study, there is still much to be done and learnt, and work over the next few years should produce considerably deeper insight into the nature of proprioception.

References

Finlayson, L. H. (1968) Proprioceptors in the invertebrates. *Symposia of the Zoological Society of London*, **23**, 217–249.

von Holst, E. (1954) Relations between the central nervous system and the peripheral organs. *British Journal of Animal Behaviour*, **2**, 89–94.

Lissmann, H. W. (1950) Proprioceptors. *Symposia of the Society for Experimental Biology*, **4**, 34–59.

Pringle, J. W. S. (1961) Proprioception in arthropods. In *The Cell and the Organism*. (eds.) Ramsay, J. A. and Wigglesworth, V. B., pp. 256–282, Cambridge University Press.

Pringle, J. W. S. (1963) The proprioceptive background to mechanisms of orientation. *Ergbnisse der Biologie*, **26**, 1–11.

Index

675